T0269419

THE POLITICAL ECOLOGY OF OIL AND GAS ACTIVITIES IN THE NIGERIAN AQUATIC ECOSYSTEM

THE POLITICAL ECOLOGY OF OIL AND GAS ACTIVITIES IN THE NIGERIAN AQUATIC ECOSYSTEM

Edited by

PRINCE E. NDIMELE
Lagos State University, Lagos, Nigeria

ACADEMIC PRESS
An imprint of Elsevier

Academic Press is an imprint of Elsevier
125 London Wall, London EC2Y 5AS, United Kingdom
525 B Street, Suite 1800, San Diego, CA 92101-4495, United States
50 Hampshire Street, 5th Floor, Cambridge, MA 02139, United States
The Boulevard, Langford Lane, Kidlington, Oxford OX5 1GB, United Kingdom

Notices
Knowledge and best practice in this field are constantly changing. As new research and experience broaden our
understanding, changes in research methods, professional practices, or medical treatment may become necessary.

Practitioners and researchers must always rely on their own experience and knowledge in evaluating and using
any information, methods, compounds, or experiments described herein. In using such information or methods
they should be mindful of their own safety and the safety of others, including parties for whom they have a
professional responsibility.

To the fullest extent of the law, neither the Publisher nor the authors, contributors, or editors, assume any liability
for any injury and/or damage to persons or property as a matter of products liability, negligence or otherwise, or
from any use or operation of any methods, products, instructions, or ideas contained in the material herein.

British Library Cataloguing-in-Publication Data
A catalogue record for this book is available from the British Library

Library of Congress Cataloging-in-Publication Data
A catalog record for this book is available from the Library of Congress

ISBN: 978-0-12-809399-3

For Information on all Academic Press publications
visit our website at https://www.elsevier.com/books-and-journals

Working together
to grow libraries in
developing countries

www.elsevier.com • www.bookaid.org

Publisher: Candice Janco
Acquisition Editor: Anneka Hess
Editorial Project Manager: Emily Thompson
Production Project Manager: Surya Narayanan Jayachandran
Cover Designer: Mark Rogers

Typeset by MPS Limited, Chennai, India

Contents

II

THE EFFECTS OF CRUDE OIL EXPLORATION ON THE SOCIO-CULTURAL AND ECO-ECONOMICS OF NIGERIAN ENVIRONMENT

16. Trade-off Analyses of Ecosystem Services in Nigerian Waters

PRINCE EMEKA NDIMELE, ADENIRAN AKANNI, JAMIU
ADEBAYO SHITTU, LOIS OYINDAMOLA EWENLA AND
OLUWANIFEMI ESTHER IGE

17. Land Use/Land Cover Change in Petroleum-Producing Regions of Nigeria

SAHEED MATEMILOLA, OLUDARE HAKEEM ADEDEJI AND
EVIDENCE CHINEDU ENOGUANBHOR

18. Petroleum Industry Activities and Climate Change: Global to National Perspective

MICHAEL ADETUNJI AHOVE AND
SEWANU ISAAC BANKOLE

III

PETROLEUM INDUSTRY CHALLENGES AND THEIR SOLUTIONS

19. Politics of State/Oil Multinational Alliance and Security Response

FIDELIS ALLEN

20. Reactions to Petroleum Exploration From Oil-Bearing Communities: What Have We Learned?

EBINIMI J. ANSA AND OJO A. AKINROTIMI

21. The Political Economy of the Amnesty
Program in the Niger Delta Region
of Nigeria and Implications for
Durable Peace

BABATUNDE ABDUL-WASI MOSHOOD, TARILAYEFA
EBIMO DADIOWEI AND BAMIDELE FOLABI SETEOLU

22. Sustainable Exploration of Crude
Oil in Nigeria

ADEKUNLE ADEDOYIN IDOWU AND
TAIWO MARY LAMBO

23. Dealing with Oil Spill Scenarios
in the Niger Delta: Lessons from the Past

CHIBUIKE ALLISON, GODWIN ORIABURE, PRINCE EMEKA
NDIMELE AND JAMIU ADEBAYO SHITTU

24. Remediation of Crude Oil Spillage

PRINCE E. NDIMELE, ABDULWAKIL O. SABA, DEBORAH O.
OJO, CHINATU C. NDIMELE, MARTINS A. ANETEKHAI
AND EBERE S. ERONDU

IV

CASE STUDIES

About the Editor

Dr. Prince E. Ndimele has a PhD in Fisheries Management, with a focus on Hydrobiology, Environmental Toxicology and Ecological Restoration. He lectures in the Department of Fisheries, Faculty of Science, Lagos State University, Ojo, Lagos State, Nigeria. He is a fellow of the Europe-Africa Marine Earth Observation Network (EAMNet), winner of the 2016 edition of the Ecologists in Africa grant by the British Ecological society, recipient of 2017 ASLO Global Outreach Initiative, a member of several learned societies such as: Fisheries Society of Nigeria (FISON), Association for the Sciences of Limnology and Oceanography (ASLO), Society of Wetland Scientists (SWS), American Fisheries Society (AFS), British Ecological society (BES), Global Lake Ecological Observatory Network (GLEON) and Society for Ecological Restoration (SER). His research interests are Environmental Toxicology, Ecosystem Modelling and Ecological (Wetland) Restoration.

List of Contributors

Oludare Hakeem Adedeji Federal University of Agriculture, Abeokuta, Ogun State, Nigeria

Michael Adetunji Ahove Lagos State University, Lagos, Nigeria

Adeniran Akanni Lagos State Ministry of Environment, Ikeja, Lagos, Nigeria

Ojo A. Akinrotimi African Regional Aquaculture Center of the Nigerian Institute for Oceanography and Marine Research (ARAC/NIOMR), Port Harcourt, Rivers State, Nigeria

C.G. Alimba University of Ibadan, Ibadan, Nigeria

Fidelis Allen University of Port Harcourt, Port Harcourt, Nigeria

Chibuike Allison Appleseed Consulting Limited, Lagos, Nigeria

John Olurotimi Amigun Federal University of Technology, Akure, Nigeria

Martins A. Anetekhai Lagos State University, Ojo, Lagos State, Nigeria

Ebinimi J. Ansa African Regional Aquaculture Center of the Nigerian Institute for Oceanography and Marine Research (ARAC/NIOMR), Port Harcourt, Rivers State, Nigeria

G.A. Ataguba University of Agriculture, Makurdi, Nigeria

Folalu O. Awe Lagos State University, Lagos, Nigeria

Bolaji Benard Babatunde University of Port Harcourt, Port Harcourt, Rivers State, Nigeria

Sewanu Isaac Bankole Ogun State Institute of Technology, Ogun, Nigeria

B. Chidinma Briggs Pre-natal Section of the Maternal and Childcare Center, Amuwo-Odofin, Lagos, Nigeria

I. Lucky Briggs ESPAM-Formation University, Cotonou, Republic of Benin

K.S. Chukwuka University of Ibadan, Ibadan, Nigeria

Tarilayefa Ebimo Dadiowei College of Education, Sagbama, Nigeria

Isa Olalekan Elegbede Brandenburg University of Technology, Cottbus-Senftenberg, Germany

Evidence Chinedu Enoguanbhor Brandenburg University of Technology Cottbus – Senftenberg, Cottbus, Germany

Ebere S. Erondu University of Port Harcourt, Rivers State, Nigeria

Lois Oyindamola Ewenla Lagos State University, Ojo, Lagos State, Nigeria

Oluwaseun Omolaja Fadeyi University of Trier, Trier, Germany

A.O. Giwa-Ajeniya Lagos State University, Lagos, Nigeria; Universiti Malaysia Terengganu, Kuala Terengganu, Malaysia

Monday Ilegimokuma Godwin-Egein University of Port Harcourt, Port Harcourt, Nigeria

Bidemi M. Green University of Lagos, Lagos, Nigeria

Adekunle Adedoyin Idowu Federal University of Agriculture, Abeokuta, Nigeria

Oluwanifemi Esther Ige Lagos State University, Ojo, Lagos State, Nigeria

Bibobra Ikporo Niger Delta University, Amasoma, Nigeria

W.A. Jimoh Federal College of Animal Health and Production Technology, Ibadan, Nigeria

Anthony Kola-Olusanya Osun State University, Osogbo, Nigeria

Taiwo Mary Lambo Lagos State University, Ojo, Nigeria

Saheed Matemilola Brandenburg University of Technology Cottbus – Senftenberg, Cottbus, Germany

Ikechukwu C. Mbachu Brandenburg University of Technology, Cottbus-Senftenberg, Brandenburg, Germany

Gabriel Olarinde Mekuleyi Lagos State University, Lagos, Nigeria

Oluwaseyi Y. Mogaji National Institute for Freshwater Fisheries Research, New-Bussa, Niger State, Nigeria

Babatunde Abdul-Wasi Moshood Lagos State University, Ojo, Nigeria

Richard Mulwa University of Nairobi, Nairobi, Kenya

Chinatu C. Ndimele University of Ibadan, Oyo State, Nigeria

Prince Emeka Ndimele Lagos State University, Ojo, Lagos State, Nigeria

Ike Nwachukwu Michael Okpara University of Agriculture, Umudike, Abia State, Nigeria

Napoleon Ogbon Ogbarode Federal University of Petroleum Resources Effurun, Delta State, Nigeria

Deborah O. Ojo University of Ibadan, Oyo State, Nigeria

Atei Mark Okorobia University of Port Harcourt, Port Harcourt, Nigeria

Sylvester Okotie Federal University of Petroleum Resources Effurun, Delta State, Nigeria; Federal University of Petroleum Resources Effurun, Warri, Nigeria

Stephen Temegha Olali Niger Delta University, Wilberforce Island, Bayelsa State, Nigeria

Oluwatosin M. Olarinmoye Lagos State University, Lagos, Nigeria

Sunday Omovbude University of Port Harcourt, Port Harcourt, Nigeria

Mgbeodichinma Eucharia Onuoha Technical University Bergakademie Freiberg Saxony, Freiberg, Germany

John Onwuteaka Rivers State University of Science and Technology, Port Harcourt, Nigeria

Godwin Oriabure Appleseed Consulting Limited, Lagos, Nigeria

Abdulwakil O. Saba Lagos State University, Ojo, Lagos State, Nigeria

Bamidele Folabi Seteolu Lagos State University, Ojo, Nigeria

Jamiu Adebayo Shittu Lagos State University, Ojo, Lagos State, Nigeria

I.O. Shotonwa Lagos State University, Lagos, Nigeria

Akeem O. Sotolu Nasarawa State University, Keffi, Nigeria

Udensi Ekea Udensi University of Port Harcourt, Port Harcourt, Nigeria

Obih A. Ugwumba University of Ibadan, Ibadan, Nigeria

Michael Uwagbae Wetlands International-Nigeria, Port Harcourt, Nigeria

Ijeoma Favour Vincent-Akpu University of Port Harcourt, Port Harcourt, Rivers State, Nigeria

Peace C. Wilfred-Ekprikpo Nigerian Institute for Oceanography and Marine Research, Lagos, Nigeria

Nenibarini Zabbey University of Port Harcourt, Port Harcourt, Rivers State, Nigeria

Foreword

The discovery of crude oil in Nigeria is a watershed event that signified a remarkable change in the political and economic history of the country. Prior to this period, the Nigerian economy was sustained by agriculture and each region specialized in the production of certain crops for which Nigeria was a major global producer. The eastern region was renowned for oil palm cultivation, while the northern and western regions were popular in growing cocoa and groundnut, respectively. Other cash and food crops grown in different parts of the country, which contributed to the national GDP, included rubber, cassava, yam, etc. Fisheries was also a major source of livelihood, especially for the riverine population residing predominantly in the coastal Niger Delta and Lagos.

Crude oil exploration in Nigeria brought enormous economic prosperity, while at the same time severe environmental degradation. The Nigerian economy grew geometrically during the oil boom era but not much attention was paid to the negative impacts of petroleum exploration on the physical, social, economic, and cultural well-being of the inhabitants of the oil-producing region and their environment. With the passage of time, environmental degradation became severe and affected fisheries, which is the major source of livelihood for the people living in the Niger Delta. This economic stagnation resulted in youth restiveness which has engulfed the region in the last two decades. In a bid to solve the problem, the Nigerian government under the leadership of President Umaru Yar'Adua implemented an amnesty program that successfully disarmed the militant youths, trained them in vocational jobs, and gradually integrated them back into the larger society.

This book looks at the interrelationship between the politics of crude oil exploration and its impacts on the aquatic and human ecology of the Niger Delta region by chronicling the history of petroleum exploration in Nigeria, examining the impacts of petroleum industry activities on the lives of the indigenous population, as well as identifying the challenges facing the Nigerian oil industry and proffering possible solutions.

This is a book that will certainly be of interest to decisionmakers, policymakers, researchers, regulators, nongovernmental organizations, practitioners in the oil industry, and other professionals nationally and internationally. I congratulate the authors and editor as I recommend the book to all.

Prof. Olanrewaju A. Fagbohun, PhD
Vice-Chancellor, Lagos State University, Lagos, Nigeria

Introduction

Pre-independence Nigeria was an agriculture-driven economy that was sustained by farm produce from different regions of the country. The North majored in the production of groundnut, the West was known for cocoa, and the East specialized in palm oil production. The other regions of the country were not left out; the middle belt produced different types of crops like yam and cassava. This is the reason it is described as the "food basket of Nigeria."

The Niger Delta region was not just a haven for fish and other aquatic organisms, due to the abundance of natural waters in the region; there were also farmers cultivating food and cash crops, like rubber and timber. These different regions of the country were self-sustaining, engaged close to 70% of the population, and produced the bulk of the raw materials required by the industrial sector. All these changed with the discovery of crude oil in commercial quantity in present-day Bayelsa State, in the Niger Delta region of the country, on the eve of independence.

The political emancipation of Nigeria, heralded by the country's independence on October 1, 1960, also came with economic freedom. Just as the country was breathing the air of freedom from decades of slavery from colonial masters, crude oil, which was very valuable, and the most-sought natural resource in the world, was struck in commercial quantity in 1958, after 20 years of search by the Shell Petroleum Development Company (SPDC), then known as the Shell d'Arcy, an affiliate of the Shell-BP. Earlier attempts by a German international oil company, the Nigerian Bitumen Corporation, had failed.

The political and economic landscapes of Nigeria changed with the discovery of crude oil. A country which was predominantly agro-centric metamorphosed into a petro-centric economy, with the attendant benefits and consequences. The government was so wealthy from the sale of crude oil that the period is described in the politico-economic history of Nigeria as the "oil boom era."

There was much infrastructural development, and Nigeria became an important figure in global politics, especially in Africa, which earned her the name "Giant of Africa." Indeed, she merits the appellation, because she played, and is still playing, a leading and pivotal role in African politics. Petro-dollar contributed significantly to shaping the political idiosyncrasy of the country, as the political class saw politics as a lucrative venture for financial aggrandizement.

Crude oil exploration also caused massive destruction of the ecosystem in the petroleum-producing Niger Delta region. Incessant gas flaring and oil spill incidences, as a result of crude oil exploration, became commonplace, and have rendered the aquatic ecosystems of the region uninhabitable. This has adversely affected the socio-economic capacities of the inhabitants who depend on fisheries and other aquatic ecosystem services for sustenance. It has also raised serious climate change questions, as well as affects the unique but fragile biodiversity of the region. The implication of this is widespread poverty and severe food insecurity, as a majority of the population depends on fish as a major source of animal protein. Biotic and abiotic factors have further exacerbated the ecological and environmental damage, and as a result have led to changes in the aquatic ecosystems and their services to human well-being. This situation has caused hunger, aggressiveness, desperation, and low life expectancy among the people.

From the fore-going, it becomes imperative to evolve a holistic process or mechanism that will address the many ramifications (political, social, economic, and environmental) of the Niger Delta problem. The Nigerian government has implemented an amnesty program that successfully disarmed the militant youths, trained them in vocational jobs, and gradually integrated them back into the larger society. However, in order to have a lasting peace in the region, the core cause of the problem, environmental degradation, must be tackled. This book traces the genesis of the problem, highlights the impacts on different segments of the society, and proffers solution.

The book is divided into four sections. The first section, Background Information on Petroleum Industry Activities and the Nigerian Environment, is intended to familiarize readers with the Nigerian aquatic, economic, social and political environments by tracing the historical trajectory of crude oil discovery and exploration in the country. The second section examines the effects of crude oil exploration on different segments of the Nigerian environment, while the third highlights the challenges of the Nigerian oil and gas industry in its six decades of existence, as well as proffers solutions to those problems. The forth section discusses case studies that address the massive devastation of the Niger Delta by anthropogenic activities, and evaluates the ecosystem services provided by this biodiversity hotspot.

Prince Emeka Ndimele, PhD
Lagos State University, Lagos, Nigeria

BACKGROUND INFORMATION ON PETROLEUM INDUSTRY ACTIVITIES AND THE NIGERIAN ENVIRONMENT

1

The Nigerian Environment

Oluwatosin M. Olarinmoye[1], Obih A. Ugwumba[2] and Folalu O. Awe[1]

[1]Lagos State University, Lagos, Nigeria [2]University of Ibadan, Ibadan, Nigeria

1.1 NIGERIA: GEOGRAPHICAL SITUATION, DEMOGRAPHICS, POLITICAL HISTORY, AND ECONOMIC SIGNIFICANCE

The federal republic of Nigeria (Fig. 1.1) is a country situated on the west coast of Africa, abutting the Gulf of Guinea, lying between 9.0820°N, 8.6753°E. Nigeria has a land area of approximately 924,000 km² and shares land borders with Benin to the west, Niger to the north, Chad to the northeast and east, and Cameroon to the east and southeast. The country also has sovereign control over a marine-exclusive economic zone of 217,313 km² (Figs. 1.1 and 1.2). With a population estimated at 173 million people (World Bank, 2015), Nigeria is the most populous country on the continent, accounting also for the larger part of the population (approximately 47%) of the West African subregion. The implications of this large population for land use, animal and plant exploitation, and biodiversity will be discussed later in this chapter. The totality of Nigeria's extensive terrestrial and aquatic area contains several ecological and climactic zones, econiches, and correspondingly abundant and diverse faunal and floral species adapted to their habitats. In addition to these biotic resources, abiotic/nonliving resources and physical conditions, including water, solar radiation, winds, soil, and exploitable minerals, directly or indirectly affect the abundance, spatial and temporal distribution, and diversity in terms of species richness and spatial distribution of endemic plants and animals.

Nigeria's nationhood commenced with its becoming a British protectorate in 1901.This state of affairs remained until 1914 when the protectorate transformed to a bi-status entity; the colony and protectorate of Nigeria, comprising northern and southern provinces, and the Lagos colony. From this point on, Nigeria rapidly progressed to partial and total independence as the federation of Nigeria in 1954, and the federal republic of Nigeria in 1960. Further evolution of the country into an ever-increasing number of states has continued since independence, effectively balkanizing even largely similar population groupings into smaller, largely economically unviable entities along language and ethnic lines, of which Nigeria abounds at the present time.

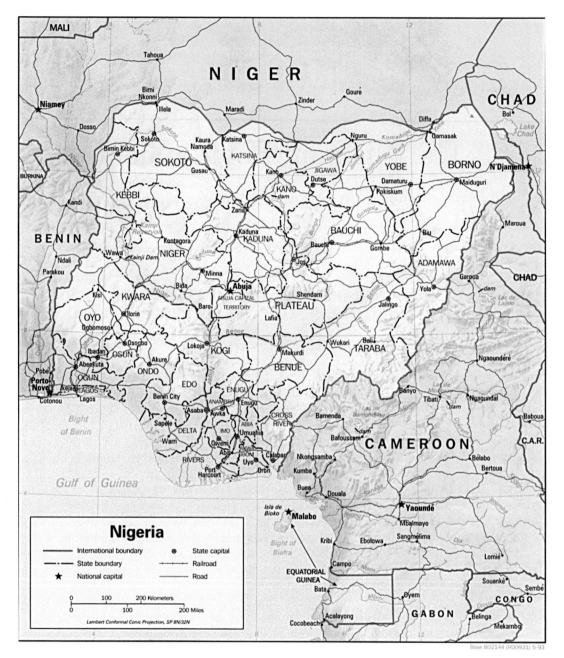

FIGURE 1.1   Map of Nigeria showing federating states and border delineations with neighboring countries.
Source: University of Pennsylvania African studies center. http://www.africa.upenn.edu/CIA_Maps/Nigeria_19877.gif

I. BACKGROUND INFORMATION ON PETROLEUM INDUSTRY ACTIVITIES AND THE NIGERIAN
ENVIRONMENT

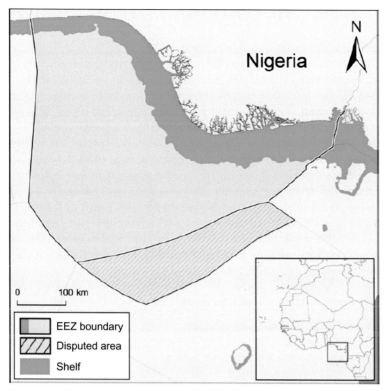

FIGURE 1.2   Nigeria's Exclusive Economic Zone (EEZ, 217,000 km²) and shelf area (to 200 m depth). *Source: Etim, L., Belhabib, D., & Pauly, D. (2015). An overview of the Nigerian marine fisheries and a re-evaluation of its catch data for the years 1950-2010. In: Fisheries catch reconstructions: West Africa, Part II, Belhabib, D. and Pauly, D. (eds). Fisheries Centre Research Reports 23(3), pp. 66-76 (Etim, Belhabib, & Pauly, 2015).*

Economically at the subregional and continental levels, the country's large population is also a driver for consumer demand largely met by imports from other countries as local production hardly suffices to meet demands of the populace due to a decline in production capacities, coupled with massive population growth over the years; loss of the strength of the local currency, the Naira, against other international currencies of trade and exchange, especially the dollar; decline in international oil prices; internal security problems; and other factors affecting local production of virtually all products and services.

Nigeria remains a continental and international powerhouse, having the "dubious" credentials of being the largest producer of oil and gas on the continent, with the second-largest proven oil reserves after Libya. In addition, Nigeria is the largest producer of crude oil in the world, a position seriously undermined by the current fluxes in international oil values, and the influence of the refractoriness of the nonaligned countries to OPEC-determined oil production quotas, coupled with other factors, including international brigandage, illegal oil transfers, and sales on international waters.

I. BACKGROUND INFORMATION ON PETROLEUM INDUSTRY ACTIVITIES AND THE NIGERIAN ENVIRONMENT

1.2 CLIMATIC, VEGETATION, AND ECOLOGICAL ZONES

Climate is defined as the annual weather conditions of an area averaged over a series of years. In general, the climactic factors of an area primarily determine the available floral and faunal species and ecological interactions between and within available life forms. For an extensive area as encompassed within the land and aquatic boundaries of Nigeria, with several climatic zones, and corresponding regional vegetation types, it is easily conceived that there would be a correspondingly diverse amount of animal and plant types based on habitual proclivities, primarily determined by climate and species adaptations to the various climactic zones and regions. Nigeria, as a consequence of its tropical situation, generally has a warm climate that is more humid in the south due to higher precipitation and that gets progressively drier northwards. Two seasons are generally accepted, the wet/rainy season lasting from mid- March to November in the south and from May to October in the north, and the dry/harmattan season, which spans the remainder of the year (Oyenuga, 1967). Since temperature fluctuations are minimal between the north and south (32–38°C), climatic indices such as relative humidity and rainfall form the valid basis of seasonal differentiations. Vegetation types mirror this trend closely with the country being divided into seven ecological zones on the basis of characteristic vegetation types (Oyenuga, 1967). These are, from south to north (Fig. 1.3):

1. The mangrove forest and coastal vegetation
2. The tropical high forest zone
3. The derived Guinea savannah with relict forest
4. The Southern Guinea savannah zone
5. Montane zone
6. The Sudan savannah
7. The Sahel savannah

The ecological concepts, such as abiotic restrictions, interspecies interactions (e.g., food webs), human activities, and latitudinal diversity that factor in being responsible for species diversity/richness (Rosenzweig, 1995) is true for Nigeria and as postulated, the vegetation types and zones in Nigeria are closely related to precipitation and temperature, transiting from mangrove swamps in the wettest coastal parts of the country through to the grass plains and the near-desert transitory Sahel savannah in the north of Nigeria, a fact mediated by human populations and activities also closely linked to the habitability (Mendez, 2014) of the various zones.

1.3 FLORAL AND FAUNAL DIVERSITY

Nigeria is home to a large and varied number of endemic animal and plant species, found exclusively in Nigeria. These resources are used by the population as food, building, and trade items, mostly in an unrestricted and indiscriminate manner. With an exponentially increasing population and a concomitant and proportionate increase in life requirements, the effects of the latter activities are becoming increasingly obvious in the

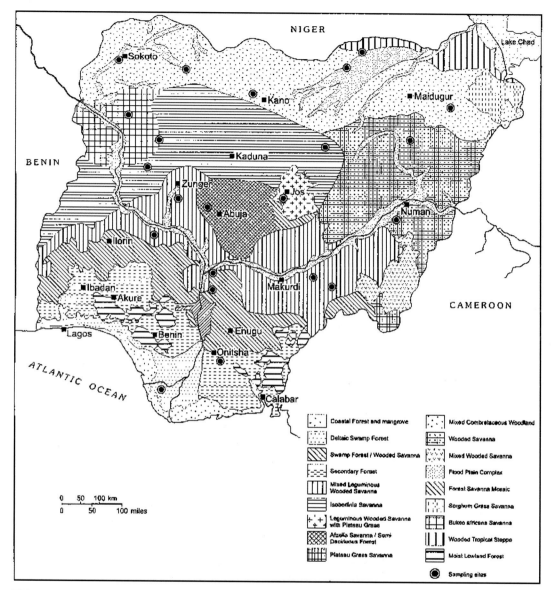

FIGURE 1.3 Map of Nigeria showing ecological zones. *Source: Owa, S.O., Dedeke, G.A., Morafa, S.O.A., & Yeye, J.A. (2003). Abundance of earthworms in Nigerian ecological zones: Implications for sustaining fertilizer-free soil fertility. African Zoology. 38(2):235–244 (Owa, Dedeke, Morafa, & Yeye, 2003).*

form of species depletion in terms of numbers and variants. Habitat encroachment due to the expansion of existing communities and the siting of new ones is complementary to the progress of diversity loss. Borokini (2014) documented 91 plant species from 44 families as endemic to Nigeria, whereas the fifth national biodiversity report (2015), referencing several sources, reported four mammalian, four avian, and 24 fish species as endemic to Nigeria. The current status of these species as it concerns abundance, and the threat/extinction status are sketchy at present. The IUCN (2016) red list defines 73 animal species as endangered.

1.4 THE DELTA

The Niger River delta, bounded by the Atlantic coast, is in southwestern Nigeria (Fig. 1.4). The delta, formed by several tributaries of the River Niger at the points of its emptying into the Atlantic Ocean, is reputed to be the second-largest such geographical feature in the world, with a coastline of 450 km (Awosika, 1995). With a surface area of 112,110 km^2, the Niger delta constitutes approximately 12% of Nigeria's total surface area and is home to an estimated 26 million inhabitants (15% of the total population) making it

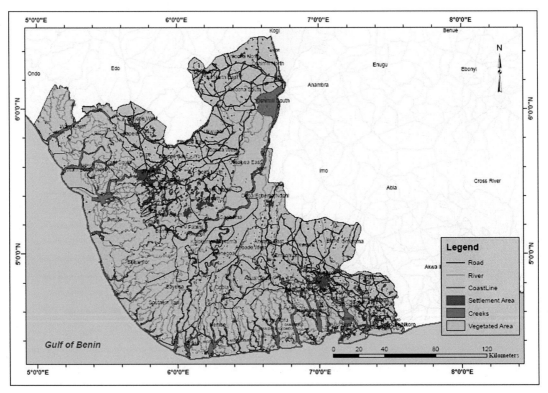

FIGURE 1.4   Map of the Niger Delta showing networks of drainage systems, road networks, and vegetation and settlements. *Source: Amangabara, G.A., Obenade, M. (2015). Flood vulnerability assessment of Niger delta states relative to 2012 flood disaster in Nigeria. American Journal of Environmental Protection, 3(3): 76-83 (Amangabara and Obenade, 2015).*

one of the most densely populated tracts of land worldwide (Fig. 1.3). The delta has large deposits of naturally occurring materials, including crude oil and bitumen, among others.

An understanding of the political ecology of the Niger Delta situation becomes necessary in order to conceptualize the problem and offer solutions to ameliorate for the rejuvenation and sustained development of the delta. For the latter purpose, a framework for the elucidation of conflict situations suggested by the Africa Peace Forum will be used here to enhance a basic understanding as concerns the relationships between the four cardinal points for conflict analysis as suggested (i.e., conflict profile, actors, causes, and dynamics of conflict).

1.5 NATURAL RESOURCES OF NIGERIA

There are approximately fourteen major wetland belts in Nigeria. World Bank (1995) identified four different ecological zones: fresh water swamps, lowland rain forests, mangroves, and barrier island forests. Hutchful (1985) had earlier classified Nigeria into two ecological zones: (1) the coastal area of the mangrove vegetation, transversed by many freshwater ecosystems such as rivers, creeks, and tributaries in the south and (2) the tropical rainforest in the northern reaches of the delta. Hutchful (1985) further subdivided these two ecological zones into: (a) a salt water riverine area that adjourns the coast where the Niger and its tributaries flow into the sea; and (b) a freshwater riverine area, which is further inland. Nigeria has the largest mangrove forests in Africa and the third largest in the world, and mangrove vegetations are found in the Niger Delta (Ebeku, 2005). It is important to note that the mangrove swamps occupy a central position in a sensitive and complex ecosystem providing valuable services to the inhabitants. The artisanal fishing industry depends on this ecosystem because the mangrove is the breeding ground for many fish species found in the Niger Delta, and the fishing industry in turn is a major source of livelihood for rural dwellers. The Niger Delta is blessed with natural resources, including crude oil, and it is also rich in biodiversity, making it an ecological hotspot as there are faunal and floral species endemic to the area that are not found in any other place in the world.

1.6 FOREST RESOURCES OF THE NIGER DELTA REGION OF NIGERIA

The Niger Delta has a large forest reserve that is rich in economic trees used as timber. Some of these economic trees are: mahogany (*Khaya* sp.), red mangrove (*Rhizopora* sp.), abura (*Hallea ledermanmi*), iroko (*Milicia excelsa*), and cotton tree (*Ceiba pentandra*). Other common species in the Niger Delta are *Pycnanthus angolensis*, *Ricinodendron heudelotii*, *Hallea ledermannii*, *Uapaca* spp., *Treculia africana*, *Albizia adianthi-folia*, *Lophira alata*, *Irvingia gabonensis*, *Sacoglottis gabonensis*, *Ficus vogeliana*, and *Klainedoxa gabonensis* (McGinley & Duffy, 2007). These trees are commonly used for fuel wood, building poles, saw logs, and transmission poles.

1.7 FISHERIES RESOURCES OF THE NIGER DELTA REGION OF NIGERIA

The Niger Delta is considered to be one of the richest ecosystems in the world in terms of biodiversity and requires protection because it is home to 36 families and nearly 250 species of fish, of which 20 are endemic. Some of the common fish species found in the delta are the catfish, tilapia, shellfish, barracuda, denticle, finfish, herring, and croakers (Ekeke, Davies, & Alfred-Ockiya, 2008). Fisheries resources of the Niger Delta can be placed into three groups: freshwater, marine, and aquaculture resources. Ajayi and Talabi (1984) reported that more than 70% of the fish stocks targeted by the industrial fishery are caught in coastal zones of the Niger Delta region. The Niger Delta has abundant aquatic ecosystems consisting of fresh, brackish, and marine waters, which are home to numerous fin-fish as well as nonfish fauna that support artisanal fisheries, and contribute approximately half of the total domestic fish supply in Nigeria (Akankali & Jamabo, 2012). The marine fisheries are dominated by small pelagic species that account for more than 50% of total fish catch (FAO, 1997). Previous studies reported approximately 199 species in 78 families of fin-fish and shellfish in the marine and brackish waters (Sikoki, Hart, & Abowei, 1998; Tobor, 1990). The study by Arimoro and Ikomi (2009) reported 57 taxa of aquatic insects, some of which are edible and act as bioindicators of water quality. Others faunas encountered are edible aquatic organisms such as barnacles, crabs, and periwinkles (World Bank, 1995).

The brackish water systems (creeks, estuaries, and lagoons) in the Niger Delta occupy a total area of approximately $4800 \, km^2$, with approximately $2267 \, km^2$ of estuaries and $937 \, km^2$ of coastal lagoons (Lowenberg & Kunzel, 1991). Some of the popular species exploited by artisanal fishermen are *Ilisha africana* (West African shad), *Ethmalosa fimbriata* (bonga), *Sardinella maderensis* (flat sardine), and some Carangids. The most abundant and the most widely exploited by artisanal fishermen is the bonga fish (*E. fimbriata*). Clupeids are also heavily exploited. Shellfish/crustaceans are commonly found near the mouths of the lagoons and rivers, where they feed on rich organic sediment matters. Species normally seen are the white shrimp (*Nematopalaecom hastaus*), brackish water prawn (*Macrobrachium vollenhonenii*), and the pink shrimp (*Panaeus notialis*) (Dublin-Green & Tobor, 1992; Ajayi & Talabi, 1984). An estuary species, white shrimp account for approximately 81% of shrimp landings in the Niger Delta (Enin, Lowenberg, & Kunze, 1996). Shrimps are the most valued shellfish/crustacean resources abundant in the region.

1.8 CONFLICT PROFILE

Several documents and publications already document in detail the history of the exploitation of the oil resources of the Niger Delta and the agitations and conflicts by the various nationalities of the area against the latter, and no attempt is made here to delve into this area of contemporary national history. Instead, our focus will be on the effects of the activities of extractive industries on the environment, especially the aquatic, the extents of deterioration as an aspect of spills and other incidental causal phenomena leading to further degradation of the delta as largely uncontrolled oil mining continues. The discovery and extraction of oil in the Niger Delta have been something of a mixed blessing, if not literally a curse to the teeming peoples of the area. Oil spills, water pollution (especially of the creeks, rivers, and groundwater), the highest gas flaring rates in the world, and the destruction of livelihoods and artisanal native industries tied to inland water harvesting for fish and other natural foods, coupled with the social disruptions and violent agitations alluded to earlier, have made human existence in the region extremely tenuous and below international well-being standards. Oils spills are a continuing phenomenon (Fig. 1.5), with no sign of abatement as the regulatory bodies, including the NNPC, DPR, and other ancillaries, remain ineffective in the regulation of oil drilling companies. It is estimated that 9–13 million tonnes of crude oil has been discharged into the environment of the Niger Delta in the last 50 years (Kadafa, 2012). Six thousand eight hundred and seventeen (6817) spills were recorded, equating to 3 million barrels between 1976 and 2001 (15 years), according to the UNEP (2006), equating to approximately 115,000 barrels per year for the period. The Shell Petroleum Development Company (SPDC), for a similar time span of 17 years (1990–2007), reported 284,000 barrels, equivalent to 28,000 barrels per year, as spilled (Zabbey, 2009) during routine extraction, and as inadvertent and malicious activities. The latter figures from the SPDC smack of deliberate misrepresentation and under-reporting. The latter is not exclusive to the SPDC, as the majority, if not all, of the oil companies deliberately under-report spill figures routinely. Spill information from joint venture partners (i.e., the oil companies) and from government are therefore unreliable, as motives such as seeking to limit their legal liabilities in the case of compensation claims by affected communities are drivers of such unsavory actions.In the worst-case scenarios, spills are never reported or arbitrarily branded as minor with hardly any efforts taken to contain or limit the effects of such spills.

Natural gas generated routinely during oil extraction is a very important natural resource which, if properly harnessed and channeled to end user individuals and companies, is a clean and relatively green energy source. This, however, is not the case in Nigeria, where venting/flaring reduction initiatives have been ineffective, mainly because of a lack of enforcement, inadequate demand, insecurity, infrastructural deficiencies, and so on. Gas flaring and venting are, respectively, the burning or release into the air without burning of natural gas associated with crude oil extraction. The World Bank (2017), on the basis of remote sensing data, estimates that the annual gas flaring volumes in Nigeria stand at approximately 8 billion cubic meters of gas, making Nigeria the seventh-largest gas flarer in the world. The damage described as "chronic and cumulative" due to the combination of extensive oil spillage, and unmitigated gas flaring have had profound ill effects on the inhabitants and environment of the Niger Delta, resulting in poor health status due to respiratory and other disease syndromes attributable to flaring, increased carbon and greenhouse emissions, and the destruction of fishing grounds and arable agricultural land, which has turned the delta into a veritable wasteland counted as one of the five most petroleum-polluted environments in the world. Although the petroleum resources of the Niger Delta have proven to be a readily milkable cash cow for the Nigerian state, they have, paradoxically, become a source of misery to the inhabitants of the area in focus. It is clear that the current state of things as they exist are simply unsustainable, but despite this glaring fact, several factors basically hinged on the insufficiency

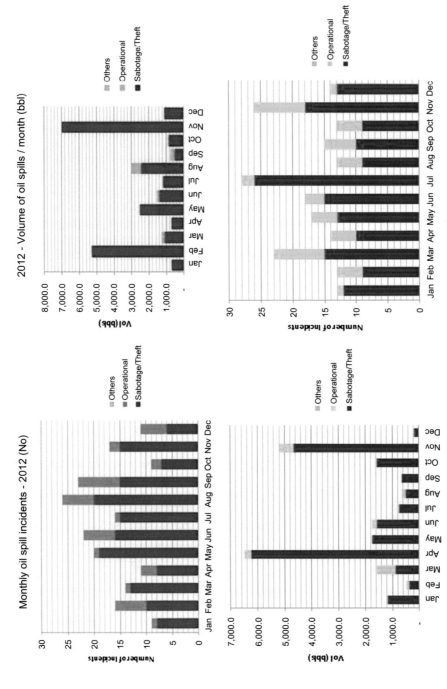

FIGURE 1.5   Oil spill statistics for 2012 and 2014. *Source: Shell Petroleum Development Company of Nigeria Limited (SPDC).*

of government regulation of the oil companies, failures of adherence to international conventions on green energy, and exploitation impact reduction ensure the persistence of the current state of things. In keeping with the statement that: "crises are not accidental phenomena but are the ultimate result of a long gestation period" (Roux-Dufort and Metais, in Sessay, 2009). The gestative emergence of armed groups, sporadic and enduring banditry, and armed conflict between these militant groups and the government forces sent in to pacify them are all direct offshoots of the situation now commonly referred to as the Niger Delta problem. It is a situation that has taken approximately 50 + years to develop into one of the most serious national security problems in Nigeria since independence (Okoli, 2013).

1.9 ACTORS, AND CONFLICT DYNAMICS

The actors in the current analysis include as a matter of course the people/inhabitants of the area, the oil companies, and the government of Nigeria. The activities of these stakeholders and their motivations are central to understanding real-time scenarios and are seemingly fixed over time. Nonhuman factors such as flora and fauna, oil extractive equipment, infrastructure, and wastes could also be considered actors, as they mediate the activities of the former "purely human" group. All actors have their objectives and motivations for their activities, and these are not always in the best interests of other actors, which is a root cause of resource conflict. For instance, the oil companies driven by the maximization of profits are interested in extracting as much oil as possible in the shortest time without having to bear responsibility for ancillary damage to the environment and people accruable to their activities. This is at cross purposes to the inhabitant actors, whose livelihoods depend on the degraded environment, and whose existence is then threatened by the actions of the oil companies. In addition, expectations of a profit throwback (i.e., to compensate for the depredations to life and existence), when threatened, act as a catalyst for conflict emergence. The actions and motivations of the latter two actor groups are basically antithetical to each other and could only be mediated by the actions of the third group in this case, the statutory bodies of government. The action or inaction of this group mediated by such factors as institutional refractoriness, inaction based on a lack of facts, or political will, play major roles in escalation/deescalation/persistence/resolution cycles. The Ogoni crisis of the Niger Delta, which culminated in the killing of the poet environmental activist, Ken Saro-Wiwa, is illustrative of the complex interplay between the aforementioned three "human" actor groups. The larger world (i.e., countries, the UN, NGOs) play complementary mediatory and sometimes inflammatory roles outside of the national ambit, roles that are often more immediately beneficial in dissenting communities and groups but often of limited scope and duration.

1.10 CONCLUDING NOTES

Nigeria remains a country encompassing a wide variety of natural biotic and abiotic resources, the exploitation of which could adequately support and sustain national growth and individual prosperity. However, the converse remains the case. The ecology of the unregulated utilization of common resources drives the contemporary extinction of floral and faunal resources. Other global factors, such as climate change and the ever-increasing international demand for fossil fuel products, further compound the doomsday scenario constantly harped on by first-tier sufferers of the ill effects of exploitation (communities and populations) and the concerned international commons. Species diversity and richness in the delta are under increasing pressure, and corresponding depreciations in both of these indices have been recorded for animal and plant species and varieties endemic to the region. The presence of large oil deposits in the delta, rather than being a source of developmental currency, has become instead a resource "albatross" as a consequence of largely unregulated exploitation of oil deposits. The impacts have been inimical to the interests of the people in the main, and although several in-depth analyses of the situation have been done in recent years, there is an urgent need for an evaluation of the extents of environmental damage with the aim of informing remediation and amelioration efforts to halt and roll back such degradation for the long-term sustainability of animal and plant stocks in the interest of dependent human populations and economies.

References

Ajayi, T. O., & Talabi, S. O. (1984). The potential and strategies for optimum utilization of the fisheries resources of Nigeria. *Nigerian Institute for Oceanography and Marine Research (NIOMR) Technical Paper, No. 18*, 25.

Akankali, J. A., & Jamabo, N. A. (2012). Effects of flooding and erosion on fisheries resources in Niger Delta, Nigeria. *European Journal of Scientific Research, 90*(1), 14–25.

Amangabara, G. A., & Obenade, M. (2015). Flood vulnerability assessment of Niger delta states relative to 2012 flood disaster in Nigeria. *American Journal of Environmental Protection, 3*(3), 76–83.

Arimoro, F. O., & Ikomi, R. B. (2009). Ecological integrity of upper Warri River, Niger Delta using aquatic insects as bioindicators. *Ecological Indicators, 9*, 455–461.

Awosika, L. F. (1995). Impacts of global climate change and sea level rise on coastal resources and energy development in Nigeria. In J. C. Umolu (Ed.), *Global Climate Change: Impact on Energy Development*. Nigeria: DAM TECH Nigeria Limited.

Borokini, T. A. (2014). A systematic compilation of IUCN red-listed threatened plant species in Nigeria. *International Journal of Environmental Sciences, 3*(3), 104–133.

Dublin-Green, C.O., & Tobor, J.G. (1992). Marine resourcesand activities in Nigeria. *Nigerian Institute for Oceanography and Marine Research (NIOMR)* Technical paper. No 84, 24pp.

Ebeku, K. S. A. (2005). Biodiversity conservation in Nigeria: an appraisal of the legal regime in relation to the Niger Delta area of the country. *Journal of Environmental Law, 16*, 361–375.

Ekeke, B. A., Davies, O. A., & Alfred-Ockiya, J. F. (2008). Sand dredging impact on fish catch in Bonny River Estuary, Nigeria. *Environmental Research Journal, 2*, 299–305.

Enin, U. I., Lowenberg, U., & Kunze, T. (1996). Population dynamics of the estuarine prawn (*Nematopalaemon hastatus* Aurivillius 1898) of the Southeast coastof Nigeria. *Fisher Research, 26*, 17–35.

Etim, L., Belhabib, D., and Pauly, D. (2015). An overview of the Nigerian marine fisheries and a re-evaluation of its catch data for the years 1950-2010. In: Fisheries catch reconstructions: West Africa, Part II, Belhabib, D. and Pauly, D. (eds). Fisheries Centre Research Reports 23(3), pp. 66-76.

Fifth National Biodiversity Report. (2015). Available online from: https://www.cbd.int/doc/world/ng/ng-nr-05-en.pdf.

Food and Agriculture Organisation of the United Nations (FAO). (1997). Review of the State of the World Fishery Resource: Marine Resource. *FAO Fisheries Circular* No. 920 Firm/C920, Rome, 173pp.

Hutchful, E. (1985). Oil companies and environmental pollution in Nigeria. In C. Ake (Ed.), *Political Economy of Nigeria* (pp. 113–140). London and Lagos: Longman.

International Union for Conservation of Nature (IUCN) Red List of Threatened Species. Version 2016-3. Available online from: www.iucnredlist.org. Downloaded on 03 July 2017.

Kadafa, A. A. (2012). Environmental impacts of oil exploration and exploitation in the Niger Delta of Nigeria. *Global Journal of Science Frontier Research Environment & Earth Sciences., 12*(3), 19–28.

Lowernbeg, U., & Kunzel, T. (1991). Investigation on the trawl fishery of the Cross River Estuary, Nigeria. *Journal of Applied Ichthyology, 7*, 44–53.

McGinley, M., & Duffy, J. E. (2007). Species diversity. In J. Cutler (Ed.), *In Encyclopedia of Earth.* Cleveland. Washington, DC: Environmental Information Coalition, National Council for Science and the Environment.

Mendez, A. (2014). What is habitability and how is it measured? <http://phl.upr.edu/library/notes/what_is_habitability>. Accessed 04.06.16.

Owa, S. O., Dedeke, G. A., Morafa, S. O. A., & Yeye, J. A. (2003). Abundance of earthworms in Nigerian ecological zones: Implications for sustaining fertilizer-free soil fertility. *African Zoology., 38*(2), 235–244.

Okoli, A. (2013). The political ecology of the Niger delta crisis and the Prospects of Lasting Peace in the Post-Amnesty Period. *Global Journal of Human Social Science, 13*(3), 37–46.

Oyenuga, V. A. (1967). *Agriculture in Nigeria.* Rome, Italy: Food and Agriculture Organization of the United Nations.

Rosenzweig, M. L. (1995). *Species Diversity in Space and Time.* Cambridge: Cambridge University Press.

Sessay, A. (2009). Theory and Practice of Crisis Management. Lecture Delivered To The Participants of Course 17 of the National Defence College, Abuja On June 4th, 2009. Spectrum Books Limited.

Sikoki, F. D., Hart, A. I., & Abowei, J. F. (1998). Gillnet selectivity and fish abundance in the lowerNun River, Bayelsa State, Nigeria. *Journal of Applied Sciences and Environmental Management, 1*(1), 13–19.

Tobor, J.G. (1990). The fishing industry in Nigeria: status and potential for self sufficiency in fish production. *NIOMR Tech. Paper* No. 54.

United Nations Development Programme (UNEP). (2006). Africa environment outlook 2: our environment, our wealth. <http://wedocs.unep.org/handle/20.500.11822/9626>.

World Bank (1995). *Defining an environmental strategy for the Niger Delta.* Nigeria: World Bank Industry and Energy Operations Division, West Central Africa Department.

World Bank. (2015). "Nigeria: overview." Countries. Retrieved 8/22/2016, 2016, from http://www.worldbank.org/en/country/nigeria/overview.

World Bank. (2017). Nigeria's Flaring Reduction Target: 2020. (n.d.). Retrieved April 07, 2017, From http://www.worldbank.org/en/news/feature/2017/03/10/nigerias-flaring-reduction-target-2020.

Zabbey, N. (2009). Impacts of oil pollution on livelihoods in Nigeria. Paper presented at the conference on "Petroleum and pollution- how does that affect human rights?". Orgarnisers: Amnesty international, forum Syd, friends of the earth. Kulturhuset, Stockholm, Sweden.

Further Reading

Ubom, R. M. (2010). Ethnobotany and biodiversity conservation in the Niger Delta, Nigeria. *International Journal of Botany, 6*, 310–322.

I. BACKGROUND INFORMATION ON PETROLEUM INDUSTRY ACTIVITIES AND THE NIGERIAN ENVIRONMENT

2

The Historical Trajectory of Crude Oil Exploration and Production in Nigeria, 1930–2015

Atei Mark Okorobia[1] and Stephen Temegha Olali[2]

[1]University of Port Harcourt, Port Harcourt, Nigeria

[2]Niger Delta University, Wilberforce Island, Bayelsa State, Nigeria

2.1 BACKGROUND

Historically, crude oil exploration and production in Nigeria began when the nation was still under British colonialism. This leads one to wonder why the British showed the level of interest they exhibited, bearing in mind that the industrial sector was one aspect of the Nigerian economy that they were most reluctant to develop. To the British, then, the industrialization of Nigeria would defeat one of their principal objectives of colonization. That is, it was better for them to preserve the colony simply as a source of raw materials and ready market for their industrial outputs. To encourage local industrialization was to create unnecessary rivalry against their economic interest.

It was against this backdrop that the British undertook to formulate and execute various industrial policies and programs not only to destroy or discourage local industrialization, but also to encourage the consumption of British manufactured goods. A popular example was the British attitude towards the indigenous distilleries and breweries. It is said that in the freshwater swamp area of the Niger Delta, because of the abundance of raphia-palms, the Ijo and migrants from elsewhere in Nigeria had perfected a technique for distilling alcohol from the palm-wine prior to the advent of colonialism. Under colonialism, the local gin remained popular and continued to be patronized by the people in spite of the existence of imported substitutes such as J.J.W. Peters, Hasekamps, and Scheidem Schnapps, among others. To create a ready market for their imports, the colonial authorities decided to brand the local gin as "illicit." It was treated as a dangerous drug or beverage. Here, the Christian missionaries were also very cooperative with the colonial political authorities. They gave a religious dimension or explanation to the "ills" associated with the consumption of the local gin, particularly among their converts and some of the Western-educated elite.

As with the local distilleries, the salt industry also suffered under colonial rule. In spite of the fact that the coastal area of Nigeria had well-developed salt industries capable of meeting all the salt needs of the nation and beyond, the British never encouraged it. Rather, they tried to undermine, and in some instances, eventually destroyed the industry. From the available records, it would appear that this policy even predated the formal hoisting of the British flag. In the 18th century, e.g., each English vessels doing business with the Ibani-Ijo people (now in Rivers State) had up to 50 tons of salt. The dumping of English salt in the area was such that the people almost completely abandoned the local industry in favor of the foreign salt (Alagoa, 1977: 358).

While this remained the major policy thrust, there were few instances where, driven either by enlightened self-interest or other very compelling situations, the British had to encourage some form of industrial activities. It was for this reason that a historian, Lawal (1987: 115), sees "colonial industrialization as a tokenism." For example, during the World Wars (1914–18 and 1939–45), it became inevitable for Britain to encourage some form of rudimentary industrialization. Some sawmills were, for instance, erected to exploit and process the timber resources of the region for the production of furniture and other items, instead of continuing with the peace-time policy of importing such items. After the wars, however, some of these industrial projects were withdrawn, overhauled, or allowed to die. This is understandable. The British were ill-prepared to introduce ideal industrialization because that would be in conflict with their imperialist goals. They believed that they were only destined to guide the natives along the "path of progress" while contemplating a time when the people would, out of their own volition and efforts, manufacture for themselves what they considered necessary (Lawal, 1987: 116).

Despite the foregoing revelations, it has to be conceded that, at least for the Niger Delta, there was an industrial activity that the British initiated that has not only survived to the present but has, indeed, become the fulcrum on which many other sectors of the economy, polity, and society revolve. This is the petroleum industry, responsible for the exploration and production of crude oil and other associated by-products. The purpose of this paper is to examine, diachronically, the origin and evolution of this industry in Nigeria.

2.2 THE DAWN OF CRUDE OIL EXPLORATION AND PRODUCTION IN NIGERIA, 1930s–2015s

Although there had been earlier interest by an international oil company from Germany (the Nigerian Bitumen Corporation) in the prospect of Nigeria as a potential oil-producing country, it was the Shell Petroleum Development Company (SPDC) (then known as the Shell d'Arcy, an affiliate of the Shell-BP) which repioneered the search for crude oil from 1937. From its base at Owerri, the company was then investigating a completely unknown territory, as no crude oil had been discovered anywhere in West Africa. The search area initially covered Nigeria in its entirety. In the course of time, however, this was narrowed down to 103,600 km^2 around the Niger Delta basin. The search was interrupted temporarily because of World War II, but from 1946, exploratory activities resumed (SPDC, 1982).

The company drilled its first exploration well at Ihuo near Owerri (now in Imo State) in 1951. However, no oil was found there. They then moved over to Akaata-1 2 years later; some oil was found there. Unfortunately, the quantity was not commercially viable. Additional searches were all fruitless until January 1956, when the company struck its first commercially viable oil well at Itokopiri, a land jointly owned by the Otabagi and Otuogidi communities in the Oloibiri clan of the present Ogbia Local Government Area of Bayelsa State. Towards the end of the same year, a second discovery was made at Afam in the Eleme clan, sometimes classified as part of the larger Ogoni ethnic nationality in the present Rivers State of Nigeria.

To convey the crude oil to Port Harcourt where oil tankers could load it for export overseas, pipeline connections between Oloibiri and Port Harcourt were constructed. The first cargo of crude oil was shipped in February 1958. At that time, production was approximately 4000 barrels daily, and Shell had invested more than forty million naira in the industry (SPDC, 1982) (A recent Nigerian National Petroleum Corporation (NNPC) source, "History of the Nigerian Petroleum Industry," says that Nigeria joined the ranks of oil producers in 1958 when its first oil field came on-stream, producing 5100 bpd, a figure slightly higher than the SPDC estimate.).

The discovery of oil in the Niger Delta promoted Shell to move its base from Owerri in the Igbo heartland to Port Harcourt. This was because, unlike the other cities in the delta proper, Port Harcourt, though young, already had all the infrastructure necessary for the expansion and success of the company's activities, including adequate harbor and port facilities to handle the freighting of materials and heavy equipment. The city also had an airport and telecommunication links with the world. To provide more facilities for handling the increasing production, however, they decided to build a terminal at Bonny in the Eastern Niger Delta. The first phase of the terminal was completed in 1961 when four tanks were constructed with a combined total capacity for 300,000 barrels of crude oil. The terminal was connected to Bomu and Afam by pipelines (SPDC, 1982). It is noteworthy that, unlike other industries established by the British to exploit Nigeria's natural resources, especially precious minerals, whose production saw marked decline after their departure, the petroleum industry continued to develop, attracting greater and greater attention, particularly from the early 1970s, to become the major source of Nigeria's external revenues.

2.3 THE AGE OF "OIL RUSH"

The SPDC success in striking oil in commercial quantity in the late 1950s led to the dawn of the Age of "Oil Rush" in the Niger Delta. Thus, especially from 1965, other major oil exploration and exploiting companies such as the Gulf Oil Company, Mobil Unlimited, Texaco, Nigerian Agip Oil Company (NAOC), Safrap and Sunray, Ashland, Pan Ocean, Teneco and Philips, among others, flocked to the region. The SPDC ceased to be the sole explorer and exploiter of Nigeria's crude oil. They were now to compete with these companies and many more in the course of history.

Historically, then, a number of events stand out as landmarks in the oil and gas sector. First, in 1908, the Nigerian Bitumen Co. & British Colonial Petroleum commenced operations around Okitipupa; in 1938, Shell D'Arcy was granted exploration license to prospect for oil throughout Nigeria; in 1955, Mobil Oil Corporation started operations in Nigeria; in 1956, the first successful well was drilled at Oloibiri by Shell D'Arcy; and in the same year, the company changed her name to Shell-BP and several years later, to SPDC Nigeria Limited. Other equally significant developments in the history of crude oil exploration and production include the first shipment of oil from Nigeria in 1958, the 1961 commissioning of Shell's Bonny Terminal, the dawn of Texaco Overseas operations in Nigeria in 1961, the 1962 commencement of Safrap (Elf) and the NAOC operations in Nigeria, and the 1963 Elf discovery of oil in the Obagi field. The Gulf Oil Company also started its first production about this time. In 1965, the NAOC found its first oil at Ebocha, while the Phillips Oil Company started operations in what was then known as the Bendel State, (now Delta and Edo States) in 1966. That year also saw Elf commence production in Rivers State with 12,000 bpd. The NNPC also recorded that, in 1967, Phillips drilled its first well at Osari-I. This turned out to be dry, but a continuous effort led to her first oil discovery at Gilli-Gilli-I in the same year. The year 1968 is remembered as the year Mobil Producing Nigeria Limited was formed, and as the year the Gulf's Terminal at Escravos was commissioned. Although no major event could be recalled for 1969, the following year, 1970, was when Mobil started production from four wells at the Idoho Field.

Other significant events were that the NAOC started production, as did the regulatory body in the sector, the Department of Petroleum Resources Inspectorate. The year 1971 is remembered as the year the Shell's Forcados Terminal as well as that of Mobil at Qua Iboe were commissioned; while in 1973, the first Participation Agreement was signed to give the Federal Government 35% shares in the oil companies. Also in 1973, Ashland secured a working understanding with the then-NNOC (NNPC), while Pan Ocean Corporation drilled its first discovery well at Ogharefe-I. In 1974, the Second Participation Agreement was signed to raise the Federal Government equity share to 55%, while Ashland's first oil discovery was made at Ossu-I. The year 1975 saw the first oil lifting from Brass Terminal by the NAOC and the upgrading of the DPR to Ministry of Petroleum Resources. In 1976, Pan Ocean commenced production via Shell-BP's pipeline at the rate of 10,800 bpd, while in 1977, the federal government established the NNPC by Decree 33 to absorb the Nigerian National Oil Company and the Ministry of Petroleum Resources. The year 1979 was a very memorable one. The Third Participation Agreement was signed to raise the FGN equity participation to 60%, and within the same year, the Fourth Participation Agreement was signed to nationalize British Petroleum's shareholding, leaving NNPC with 80% equity and Shell with 20% in the joint venture and changed the company's name to SPDC of Nigeria.

In 1984, another understanding was reached to consolidate the NNPC/Shell Joint Venture. In 1989, 10 years after the third and fourth agreements were signed, the Fifth Participation Agreement was signed, giving the Federal Government represented by the NNPC 60%, the SPDC 30%, Elf 5%, and the NAOC 5%. The year 1993 witnessed the Sixth Participation Agreement giving the NNPC 55%, SPDC 30%, Elf 10%, and the NAOC 5%; as well as the coming on-stream of Elf's Odudu blend, offshore OML 100. The oil and gas sector saw a major way forward in 1995. Shell nigeria exploration and production company (SNEPCO) started the drilling of their first exploration well, while the Nigeria Liquefied Natural Gas's final investment decision was also made. This laid the foundation for the first shipment of gas out of the Bonny terminal in 1999 by the nigeria liquefied natural gas limited (NLNG). In 2001, production started in the Okono offshore field, and in 2002, several fundamental policy decisions were made. The downstream sector of the petroleum industry was liberalized, and the NNPC commenced a retail outlet scheme. Several upstream, midstream, and downstream ventures and subsidiaries were also introduced.

Going by the records of NNPC (2012), between the late 1960s and early 1970s, Nigeria attained a production level of over 2 million barrels of crude oil per day. Although production figures dropped in the 1980s due to the economic slump, the year 2004 saw a total rejuvenation of oil production to a record level of 2.5 million barrels per day. The active participation of these companies raised the hydrocarbon sector to become the dominant sector of Nigeria's economy by contributing over 75% of the Gross Domestic Product of the country since after the Nigerian Civil War (1967−70).

Again, the immediate postcivil war era also coincided with the rise in the world oil price, and Nigeria was able to reap easier wealth from its oil and gas resources. To further maximize her gains from her hydrocarbon resources, Nigeria joined the Organization of Petroleum Exporting Countries (OPEC) in 1971 and established the NNPC in 1977, a state-owned and -controlled company that is a major player in both the upstream and downstream sectors. Table 2.1 shows the percentage share of crude oil production by some of the oil companies in the oil rush between 1958 and 1981, the period for which there was proper record-keeping.

2.4 THE DYNAMICS OF OIL EXPLORATION AND PRODUCTION IN NIGERIA

Having seen in a nutshell the origin and contributions of the petroleum sector to the revenues of Nigeria, it is reasonable to explain in some detail the forces that had propelled the direction of the industry generally, and the exploratory and production aspects in particular. What specific factors have influenced the volume of crude oil explored and produced in Nigeria? How has Nigeria fared as a crude oil−producing nation? Our investigations reveal that a number of local and global events from the 1930s onward, influenced the exploratory and production activities of the industry.

Key among these were the Nigerian Civil War, especially from 1967 to 1969; the Middle East Crisis, especially from late 1973 when the world witnessed the Arab oil producers' decision to withdraw crude oil shipment to all nations they saw as their strategic enemies on account of such nations' support for the State of Israel. At about this same time, the OPEC, which Nigeria had joined as a newly independent country, decided to cause a change in the crude oil pricing mechanism. In later years, the militant agitations of the aggrieved people of the crude oil−bearing communities of the Niger Delta would also exert enormous negative impact on the quantity of oil produced. During the civil war, for instance, there was marked decline in the exploratory and production activities, partly due to the fact that many of the oil-bearing communities of the Niger Delta were under the control of the secessionist forces of the defunct Republic of Biafra, leaving only a few oilfields under the control of the federal government of Nigeria.

Later, in the early 1970s, the interplay of the Middle East Crisis as well as the OPEC's decision to cause a change in the crude oil pricing mechanism propelled the price of crude oil to about USD15 per barrel (Imomoh, 2006: 4). In Nigeria, according to Ikein (1991: 21),

TABLE 2.1   Percentage Share of Crude Oil Production by Companies, 1958–81

Coys	Percentage of total production										
Year	Shell	Gulf	Mobil	Elf	Texaco	Agip	Ashland	Pan Ocean	Tenneco	Philips	Total
1958	100	–	–	–	–	–	–	–	–	–	100
1959	100	–	–	–	–	–	–	–	–	–	100
1960	100	–	–	–	–	–	–	–	–	–	100
1961	100	–	–	–	–	–	–	–	–	–	100
1962	100	–	–	–	–	–	–	–	–	–	100
1963	100	–	–	–	–	–	–	–	–	–	100
1964	100	–	–	–	–	–	–	–	–	–	100
1965	90.27	9.73	–	–	–	–	–	–	–	–	100
1966	84.97	12.21	–	2.82	–	–	–	–	–	–	100
1967	76.56	17.15	–	6.29	–	–	–	–	–	–	100
1968	30.66	69.34	–	–	–	–	–	–	–	–	100
1969	65.59	34.41	–	–	–	–	–	–	–	–	100
1970	72.98	21.39	5.00		0.22	0.41	–	–	–	–	100
1971	72.36	18.10	4.73	1.64	0.68	2.49	–	–	–	–	100
1972	66.48	17.92	9.16	3.04	0.56	2.84	–	–	–	–	100
1973	63.01	17.76	10.83	3.13	0.41	4.85	–	–	–	–	100
1974	62.00	16.34	10.97	3.71	0.09	6.88	–	–	–	–	100
1975	62.04	13.47	10.90	4.05	0.41	8.78	0.35	–	–	–	100
1976	59.53	14.14	11.14	3.67	1.67	8.92	0.47	0.46	–	–	100
1977	58.25	13.85	10.64	3.79	2.53	10.14	0.33	0.47	–	–	100
1978	56.85	13.71	10.54	4.05	2.26	11.59	0.46	0.51	0.03	–	100
1979	57.13	16.43	10.76	3.38	2.35	9.09	0.35	0.33	0.17	0.04	100
1980	56.61	16.69	10.38	4.17	2.10	8.91	0.41	0.39	0.27	0.07	100
1981	51.39	19.55	11.18	5.06	2.39	8.79	0.66	0.49	0.41	0.08	100

Source: NNPC (2012). Petroleum Statistics Handbook.

oil revenues sky-rocketed from N176 million in 1970 to N1.4 billion in 1973 and to N12.86 billion by 1980. The petroleum sector has not only grown tremendously, it has also become the dominant foreign income earner in the Nigerian economy. Indeed, by 1980, it had come to account for over 80% of the total government revenue and 96% of the total export earnings. Table 2.2 shows the yearly oil exports and proceeds from 1958 to 2007.

TABLE 2.2   Nigeria's Crude Oil Export 1958—2007 (Thousand Barrels)

Year	Export	Export as % of production	Oil (Nm) revenue	% of crude oil in total exports 1960—97	Remark (source)
1958	1820.3	97.1	1,784,000	NA	NNPC
1959	3957.4	96.6	5,270,000	NA	NNPC
1960	6243.5	98.1	8,414,000	2.7	CBN
1961	16,506.0	98.2	22,664,000	NA	NNPC
1962	24,679.8	99.9	34,412,000	NA	NNPC
1963	27,701.3	99.3	40,352,000	NA	NNPC
1964	43,431.6	98.7	64,352,000	NA	NNPC
1965	96,985.0	97.6	136,194,000	NA	NNPC
1966	139,550.0	91.3	183,884,000	32.4	NNPC
1967	109,274.9	93.8	142,100,000	NA	NNPC
1968	52,129.9	98.5	77,695,000	17.5	NNPC
1969	197,245.6	100.0	301,365,390	NA	NNPC
1970	283,455.4	96.9	509.6	57.6	CBN
1971	542,545.1	97.0	1053.0	73.7	CBN
1972	650,979.7	99.7	1176.2	82.0	CBN
1973	723,313.8	96.5	1893.5	83.1	CBN
1974	795,710.0	96.6	5365.7	92.6	CBN
1975	627,639.0	96.3	4555.1	93.6	CBN
1976	736,823.0	97.0	6321.7	93.6	CBN
1977	744,413.4	92.2	7072.8	93.4	CBN
1978	667,387.1	96.3	5461.6	89.1	CBN
1979	818,726.9	96.6	10,166.8	93.8	CBN
1980	698,163.5	92.6	13,523.0	96.1	CBN
1981	488,715.4	85.4	10,453.2	96.9	CBN
1982	360,410.2	76.7	9207.9	97.5	CBN
1983	341,360.9	68.1	7507.2	94.3	CBN
1984	401,155.8	79.0	8209.7	97.3	CBN
1985	454,800	97.2	10,915.1	97.1	CBN
1986	453,800	97.1	8244.5	93.8	CBN
1987	389,456	97.2	19,027.0	92.9	CBN

(Continued)

I. BACKGROUND INFORMATION ON PETROLEUM INDUSTRY ACTIVITIES AND THE NIGERIAN ENVIRONMENT

TABLE 2.2   (Continued)

Year	Export	Export as % of production	Oil (Nm) revenue	% of crude oil in total exports 1960–97	Remark (source)
1988	435,797	82.3	20,933.8	86.7	CBN
1989	507,332	80.9	41,334.4	92.2	CBN
1990	548,249	82.9	54,713.2	97.6	CBN
1991	689.9	NA	60,316	96.6	TELL
1992	711.3	NA	115,392	97.9	TELL
1993	695.4	NA	106,192	97.7	TELL
1994	696.2	NA	160,192	97.4	TELL
1995	715.4	NA	324,548	97.3	TELL
1996	681.9	NA	369,190	98.2	TELL
1997	855	NA	416,811	97.6	TELL
1998	806.4	NA	289,532	NA	TELL
1999	774.7	NA	500,000	NA	TELL
2000	828.3	NA	1,340,000	NA	TELL
2001	859.6	NA	1,707,600	NA	TELL
2002	725.9	NA	1,230,900	NA	TELL
2003	844.1	NA	2,074,300	NA	TELL
2004	900.0	NA	3,354,800	NA	TELL
2005	923.5	NA	4,762,400	NA	TELL
2006	814.0	NA	6,109,000	NA	TELL
2007	880	NA	6,700,000	NA	TELL

NNPC Petroleum Statistics Hand Book, *1982; Central Bank of Nigeria* Annual Report and Statement of Accounts for the years ended 31st December 1970–90 *(6 vols) and* Tell Magazine, *February 18, 2008.*

2.5 SOME CONSEQUENCES OF OIL EXPLORATION AND PRODUCTION IN NIGERIA

The sharp increase in the volume of crude oil production along with the corresponding rise in crude oil prices brought a sudden boost to Nigeria's foreign income base, giving her leaders the false impression that the nation had arrived, that she had acquired the economic power to finance her developmental plans, and that she could now transport herself from being a poor to an advanced, or at least, a medium-power state. Indeed, the easy petrodollars Nigeria earned gave her such a false image of affluence and power abroad that she began to assume leading roles in international affairs, particularly at the United Nations, the Commonwealth, the Organization of African Unity (now the African Union), the Economic Community of West African States, the Nonaligned Movement, and the OPEC. Ironically, back at home, many of her citizens, particularly those in the rural areas and the oil-producing Niger Delta region in particular, continued to reel in utter indigence and backwardness (Okorobia, 2012: 25–37). It was at this time that a former Nigerian military Head of State, General Yakubu Gowon, is reported to have boasted that the challenge before Nigeria was no longer the scarcity of funds, but how to manage the abundance that was available. Unfortunately, rather than investing this unparalleled windfall in laying a sustainable foundation for the development of Nigeria, the various administrations of the oil boom era and beyond, namely, General Yakubu Gowon (July 1966–July 1975), General Murtala Mohammed (July 1975–February 1976), General Olusegun Obasanjo (February 1976–September 1979), Alhaji Shehu Aliyu Shagari (October 1979–December 1983), General Muhammadu Buhari (December 1983–August 1985), General Ibrahim Badamasi Babangida (August 1985–August 1993), Chief Ernest Shonekan (August 1993–November 1993), Sani Abacha (November 1993–January 1998), General Abdusalami Abubakar (January 1998–May 1999), Chief Olusegun Obasanjo (May 1999–May 2007), Alhaji Umaru Musa Yar'Adua (May 2007–5th May 2010), and President Goodluck Ebele Jonathan (6th May 2010–29th May 2015), resorted to neglecting the agricultural sector while promoting a tragic culture of reckless expenditure and import-driven consumption, transforming the blessing of oil into a curse (Okorobia, 2013: 59–73). The neglect of agriculture was not total, as some of these administrations embarked on various agricultural programs, but these projects did not live beyond the administrations that established them. In addition, they were conduit pipes for embezzlement of funds, which further enriched the upper or ruling class. Some of these agricultural programs are "Operation Feed the Nation" (OFN) by General Olusegun Obasanjo and "Green Revolution" by Alhaji Shehu Aliyu Shagari, and President Goodluck Ebele Jonathan is known to have initiated the most successful 7fertilizer distribution program to farmers in postindependence Nigeria.

The growth in petrodollars led to other consequences. In one respect, it created an overdependence on the oil and gas sector and a rapid decline of agriculture and other nonoil exports. As indicated on Table 2.3 below, from 1965 to 1966, agriculture and other nonoil exports contributed as much as 54.9% of Nigeria's revenue, whereas petroleum contributed only 4.82%. By 1990, the situation had been so transformed negatively, that the revenue from the oil/gas sector rose to 81.8% while agriculture and nonoil exports fell drastically to only 18.2%.

Similarly, Table 2.4 illustrates the decline in earnings from agricultural exports from 1960 to 1974 as a consequence of the growing neglect of that sector.

The most telling effect of the neglect of agriculture was decline in local food production, be it food crops, livestock, or fish. It also led to the decline in all other sectors relying on agriculture. During the first decade of military rule, 1966–77, three remarkable trends occurred: (1) a steady rise, up to about 1965/66 (the terminal phase of the First Republic), followed by a significant decline thereafter (the beginning of military rule), (2) a significant drop in the output of food crops and rubber, as well as palm oil, and (3) wide fluctuations in total food crop production, and variability even of individual crops (Nzimiro, 1985: 28).

Again, according to Nzimiro (1985:28), in the 1977/78 fiscal year, the military spent only 0.3% of its total recurrent, and about 4% of its capital expenditure on agriculture, whereas defense attracted 10.7% of the total expenditure, even though it was a time of

TABLE 2.3   Contributions by Crude Oil and Agriculture/Nonoil Exports to the Revenue of the Federal Republic of Nigeria, 1965–90 (In Percentages)

Year	Agriculture/nonoil exports (%)	Crude oil (%)
1965–66	54.9	4.82
1970	36.0	33.19
1975	21.51	51.61
1980	9.54	77.76
1982	4.92	88.20
1984	26.3	73.
1985	25.3	74.7
1986	34.1	65.9
1987	24.2	75.8
1988	23.3	76.7
1989	17.3	82.2
1990	18.2	81.8

Source: Central Bank of Nigeria Annual Report and Statement of Accounts for the years ended 31st December 1970-90 (6 volumes).

TABLE 2.4   The Decline in Earnings From Agricultural Exports, 1960–74

Year	Value (₦ million)	% of GDP	% of total agriculture
1960	282.5	12.6	19.8
1961	283.0	11.9	19.3
1962	260.0	9.9	16.2
1963	285.9	10.2	17.1
1964	303.9	10.4	18.1
1965	327.3	10.6	19.3
1966	292.5	9.1	16.4
1967	264.5	8.7	15.4
1968	263.7	8.4	15.3
1969	278.7	8.5	16.0
1970	280.5	8.0	15.7
1971	265.2	2.8	7.8
1972	190.1	1.7	5.3
1973	278.4	2.3	8.3
1974	297.8	2.3	9.2

Source: Kayode, M. O., & Usman, Y. B. (1985). The economic and social development of Nigeria. In: Proceedings of the national conference on Nigeria since independence, Zaria, March, 1983, Vol. II. Kayode, M. O. and Usman Y. B. (Eds.). Panel on Nigeria Since Independence History Project, Zaria.

peace. This development raises a number of questions as to the desirability of investing such colossal sums of money on defense to the neglect of other critical sectors such as agriculture, education, and socioeconomic infrastructure. Whereas the petroleum sector was greatly expanded and intensively exploited under military rule, the agricultural sector was allowed to dwindle. Thus, in the 1970s, Nigeria became a mono-product economy dominated by crude oil.

The obvious message here is that the military leadership utilized the nation's oil wealth for self-aggrandizement. Going by the assessment of Nzimiro (1985: 21–25), their expenditure on armament exceeded the combined expenditure on health, education, housing, urban development, and environmental sanitation, agriculture, and rural development. A former Permanent Secretary, Ministry of Defense, Mr. F.I. Adesonoye, is reported to have explained that the military governments' emphasis on building a proto-pentagon was the only way they could hold the Nigerian populace in political subjugation, appropriating the national wealth for development of military bureaucracy and power, having depoliticized the masses with no opposition against heavy expenditure on arms (Nzimiro, 1985: 24).

The full implications of the military misadventure in governance are many and complex. In one respect, we find that the oil boom transformed only some sections of the urban communities while afflicting severe pains on the rural masses. This in turn triggered rural-urban drift and its attendant consequences. Foreign exchange trafficking, smuggling, and other forms of scandalous and corruptive habits became rife, leading to a depletion of the foreign reserves (Obasi, 2000: 209–210). Again, the over-reliance on crude oil has also increased the vulnerability of the economy for external shocks in the oil market, leading to the accumulation of foreign debts to cover shortfalls in expected revenue. This has been encouraged by banks that favor lending to crude oil exporters because their loans are backed by regular oil supply. Furthermore, it has led to dereliction in the collection and administration of other forms of taxation. This has not only undermined the links between the citizenry and the government, but has also exacerbated alienation, exclusion, and non-accountability on the part of the state authorities. It became obvious that Nigeria had lost her bearings.

The time-tested wisdom that a people cannot continue to eat their cake and have it, or that whatever a people sow, more of the same will they reap, stood staring at the faces of Nigerians. Food scarcity mounted and, whether they liked it or not, the military woke up to the consciousness that the nation was starving, as the urban demographic challenges amplified. The rural condition equally worsened as they began to experience scarcity of appropriate labor, rising cost of hired agricultural labor, and shrinkage in land area cultivated due to mounting cost of farming without government assistance. There was a clear cry for a change in the nation's agricultural policy. There were really very few options open to the government of the day. It was either they ignored the reality before them and face the consequences or confront it head-on to give the nation some hope. The Obasanjo administration decided in 1976 to revive agriculture through the OFN program. We know from hindsight that, as lofty as this program sounded, it was designed to fail, as the bureaucrats controlled and directed it for their own benefit. We know, e.g., that some of the government functionaries, including the then military Head of State, General Obasanjo himself, who initiated the project, retired and became consultants to the very same government, paying to themselves huge consultancy fees for surveys and analyses, based on capitalist methods of agriculture fit for industrial and not Nigerian peasant societies. Some of them formed companies with overseas firms manufacturing fertilizers and turned the entire exercise into a huge and scandalous import spree. Thus, some individuals and multinational firms made huge profits from the venture, to the detriment of the masses that were the initial target.

Consequently, the prices of foodstuffs and the national food import bills continued to soar. The resultant inflation, as expected, took a greater toll on the poor than on the rich, while rural-urban migration, inflation, deindustrialization, jobs losses, and crime were accentuated (Okorobia, 2013: 59–73).

As Nigeria became more and more dependent on the petroleum sector for national survival, there also grew a new culture of environmentally destructive wastes, especially in the Niger Delta where all the key oil exploratory and exploitative facilities are situated. In the gas subsector, e.g., Nigeria has come up with the worst record of gas flaring. Recent statistics indicate that Nigeria alone accounts for approximately 19% of the total amount of gas flared in the world. The implications of this are many. The huge revenue that would have been realized from the proper utilization of the flared gas is lost forever, and experts have said that gas, if well developed, is even more valuable than oil (Okorobia, 2000).

For the Niger Delta, especially the rural and coastal communities, the environmental and economic implications of the oil/gas industry on the region are generally negative because much of even the so-called "benefits" of the hydrocarbon business in the area have been concentrated mainly in towns at the delta periphery—e.g., Port Harcourt, Uyo, and Warri—not in the wetlands (Udofia, 2001: 3–4). For the generality of the oil/gas—bearing communities, the only benefit they see occasionally is the opportunity to be recruited as casual and unskilled workers in the servicing firms working for the major multinational companies in the industry. Perhaps, the few other areas of positive impacts are that during exploratory activities, itinerant traders specializing in general consumer goods enjoy a period of transient boom. We are not sure whether the generally unfair and inadequate monetary compensations paid on surface rights over acquired land (which itself is usually gotten only after serious clashes and sometimes legal battles) is to be counted as part of the economic benefits (Okorobia, 2013). Therefore, it is clear that the direct economic benefits from the oil/gas activities to the rural oil-bearing communities are very marginal. Unfortunately, in terms of the economic costs, these people bear much of the brunt, including the loss of biodiversity, loss of arable land, increase in land and water traffic mishaps, improper disposal of dredge spoils and increase in localized flooding, the creation of unfulfilled expectations in the communities, and the generation of oil-induced inflation. Others include frequent oil pollution and the intensification and internationalization of the Niger Delta struggle for economic and environmental justice. Also to be included in the list of their pains are the disorientation of the youths and the collapse of many traditional leadership structures and the increase in social vices, especially prostitution, theft, illegal oil-bunkering, kidnaping-for-ransom, piracy, increase in oil-induced health problems such as dermatitis, tetanus, gastroenteritis, sexually transmitted diseases, alcoholism, drug abuse, and respiratory infections. The over-crowding of existing health and transport infrastructures and residential accommodation, as well as the damage to cultural and historical properties of the region, are the other consequences of

I. BACKGROUND INFORMATION ON PETROLEUM INDUSTRY ACTIVITIES AND THE NIGERIAN ENVIRONMENT

the activities of the petroleum industry on the land and people of the Niger Delta (Okorobia, 2013).

On the whole, Nigeria, like other petro-states in Africa, has come to be characterized by a state of functional failure and social disorder, as even the minimal social and economic infrastructure which, ideally, should have been taken for granted by now, such as the utilities, transport, and communication facilities are virtually absent or comatose. Unemployment and oil-induced inflation, environmental degradation, collapse of the traditional social and political structures leading to a state of near anarchy, youth unrest, and general sense of insecurity, especially in the oil-bearing Niger Delta region, have compelled an agitated people to strive to achieve control of their oil and gas resources. The agitation itself began as a struggle for an autonomous political status within the Nigerian federation but has ultimately ended up as an arms struggle for resource control. This twist is principally because of the many failed attempts by the region to receive a fair share of the benefits accruing from the oil and gas resources in their land (Okorobia, 2013).

Another dominant, and perhaps unique, pathology of the Nigerian petro-state is the tendency among the elites of the three largest ethnic nationalities, Hausa/Fulani, Yoruba, and Igbo to collaborate among themselves, in spite of their occasional squabbles, to resist the principles of derivation, a tool they had created when the agro-products from their regions, cocoa (from the West), groundnut and cotton (from the North), and oil-palm produce (from the East) were the principal sources of federally collected revenue. Thus, through their resistance to the derivative principle, the crude oil–bearing Niger Delta region, occupied mainly by ethnic minorities such as the Ijo, Itsekiri, Urhobo, Andoni, Ogoni, Abua, Odual, Ogbia, Engenni, Ekpeye, Ikwerre, and Etche as well as the Efik, Ibibio, and some Igbo elements has been denied much of the revenue they deserve to address their peculiar developmental challenges. Consequently, federal revenue allocated on the basis of the derivative principle has been drastically reduced from the original 100% in 1953 to 1.5% in 1980, and later increased slightly to 13% since 1999 (Okorobia, 2000: 243).

Among the best-known fallout of the crude oil exploration and exploitation process in Nigeria is the increase in corruption and the rabid struggle for the public treasury, both under the military and civilian elites. In the process, governance at the state and local levels has been grossly denuded of the appropriate funds. The oil revenue was squandered on unviable and ostentatious projects that were either very poorly executed or abandoned, and their cost grossly inflated. A culture of waste through importation was introduced until the ports became congested and the country had to pay a fortune in demurrage (Okorobia, 2013). Even the NNPC and the Ministry of Petroleum Resources, which cater to the interest of the federal government in the industry, have not fared any better. The Crude Oil Sales Tribunal of 1980 set up by the Second Republican government under Alhaji Shagari and headed by Justice Ayo Irikefe on the missing N2.8 billion from the account of the NNPC pointed out that:

> It is very difficult if not impossible to determine the volume of crude oil being taken away from this country (because) the companies make the NNPC oil inspectors happy ... They give them enough of imported beer to drink. And while drinking beer, oil is being pumped and lifted out of our country (*Newswatch*, May 9, 1988).

2.6 CONCLUSION

In summary, we note that crude oil exploration and production in Nigeria began under British colonialism. This was in spite of the fact that generally, the industrial sector was one aspect of the Nigerian economy the British were most reluctant to develop. To them, the industrialization of Nigeria would defeat one of their principal objectives of making the colonies dependent on the metropole. This notwithstanding, they initiated the petroleum industry, which has not only survived to the present, but has, indeed, become the leading sector in terms of its ability to attract foreign exchange. It is the petroleum industry that is responsible for the exploration and production of crude oil and other associated products. Relying largely on secondary sources and personal observations, this paper examines, diachronically, the origin and evolution of the petroleum sector in Nigeria from 1930 to 2015. On the whole, the paper discovers that crude oil exploration and production in Nigeria from 1930 to 2015 witnessed many vicissitudes. From being a marginal player at the dawn of Nigeria's nationhood, the hydrocarbon industry rose to become the dominant sector from the immediate postcivil war era, contributing more than 75% of the nation's foreign income for the longest period of Nigeria's history as an independent nation. Unfortunately, this has also created the negative and unintended consequence of turning Nigeria from being an essentially agro-state into a petro-state, with the attendant consequences of neglect of agriculture and the nonoil sector.

References

Alagoa, E. J. (1977). *The teaching of history in African universities*. Accra: AAU Publications.

Ikein, A. A. (1991). *The impact of oil on a developing country: The case of Nigeria*. Ibadan: Evans Brothers Nigeria Ltd.

Imomoh, E.U. (2006). The Nigerian oil and gas industry: The last 40 years and the next 40. In *Paper presented at the Inaugural lecture in Honour of Late Prof. Chi Ikoku, on April 20th 2006*. University of Port Harcourt.

Kayode, M.O., & Usman, Y.B. (1985). The economic and social development of Nigeria. In M.O. Kayode & Y.B. Usman (Eds.), *Proceedings of the national conference on Nigeria since Independence. Zaria March, 1983 Vol. II*. Zaria: Panel on Nigeria since Independence History Project.

Lawal, A. A. (1987). Industrialisation as tokenism. In T. Falola (Ed.), *Britain and Nigeria: Exploitation or development?* (pp. 115–116). London: Zed Books.

Nigerian National Petroleum Corporation (NNPC). (2012). History of the Nigerian Petroleum Industry, Abuja.

Nzimiro, I. (1985). *The green revolution in Nigeria or modernization of hunger*. Nigeria: Zim Pan African Publishers.

Obasi, N. K. (2000). Overview of the political economy 1960–1999. In H. I. Ajaegbu, B. J. St Matthew-Daniel, & O. E. Uya (Eds.), *Nigeria: A people united, a future assured. vol. 1, a compendium*. Abuja, 209-210: Federal Ministry of Information.

Okorobia, A. M. (2000). *The Niger delta: A developmental history*. Port Harcourt: Niger Delta Heritage Centre Publication.

Okorobia, A. M. (2012). The Abuja factor in the Niger Delta crises, 1990-2010. *Journal of Education and Sociology, (Baku, Azerbijan)*, 3(2), 25–37.

Okorobia, A. M. (2013). The oil curse: A history of the fall of the non-oil sector in Nigerian economy, 1956-2006. *Reiko International Journal of Business and Finance (Special Edition)*, 4(6), 59–73.

Shell Petroleum Development Company (SPDC). (1982). *Shell in Nigeria's briefing note*, Lagos.

Udofia, U. (2001, July 16–20). Impact of oil pollution on women, children and communities. In *Paper presented at the world conference of mayors on the theme: oil industry environment pollution: managing women children and environmental issues* (pp. 3–4). Eket: Royalty Hotel.

Further Reading

African Peer Review Mechanism. (2008). *Country review report federal republic of Nigeria*. APRM Country Review Report No. 8.

Central Bank of Nigeria. *Annual report and statement of accounts for the years ended 31st December 1970-90.*

News Watch Nigeria's Weekly Newsmagazine. 2008, March 10.

Tell Nigeria's Independent Weekly, Special Edition. 2008, February 18.

I. BACKGROUND INFORMATION ON PETROLEUM INDUSTRY ACTIVITIES AND THE NIGERIAN ENVIRONMENT

3

The Physical and Chemical Components of Nigerian Crude Oil

I.O. Shotonwa[1], A.O. Giwa-Ajeniya[1,2] and Gabriel Olarinde Mekuleyi[1]

[1]Lagos State University, Lagos, Nigeria [2]Universiti Malaysia Terengganu, Kuala Terengganu, Malaysia

3.1 INTRODUCTION

Oil, from a Nigerian and a global viewpoint, plays a pivotal role in impacting the economic and political components of any nation. Nigeria's oil industry, which came into the limelight in the mid-1950s, only began to flourish her economically after the civil war in 1970. However, it equally created many conflicts in the economic, educational, environmental, social, ethnic, regional, and religious sectors. One major cause of the high degree of instabilities is the turmoil and hardship encountered by communities that are located close to where oil wells are exploited (Odularu & Okonkwo, 2009). Most of these communities suffer greatly from environmental contamination and an upset of the ecological balance. Moreover, Nigeria's unwholesome dependence on crude oil totally made her neglect her vibrant agriculture and manufacturing industries (Nøstbakken & Jadhav, 2014; Yusuf & Samuel, 2013). It is our opinion that the oil boom came with so many opportunities and prospects that it was very easy to be blinded by them. Therefore, these prospects are the determining factors and impetus that guide the development of this chapter. This chapter is divided into three sections: types of crude oil in Nigeria, composition and properties of crude oil in Nigeria, and refinery products and by-products of crude oil in Nigeria.

3.2 TYPES OF CRUDE OIL IN NIGERIA

Nigeria received a flurry of universal recognition from the Organization of Petroleum Exporting Countries (OPEC) as Africa's largest producer of crude oil. This is a sequel to the discovery of crude oil by the Consortium of Shell and British Petroleum at Oloibiri in the Niger Delta region in 1956. This happened after close to 50 years of exhaustive exploration by licensed companies (Yusuf & Samuel, 2013). These events were followed by the setting up of oil fields and wells in alignment with the policies of the Petroleum Resources Department. In the year 1958, the first oil field started production of crude oil (Nøstbakken & Jadhav, 2014).

It is interesting to note that Nigeria's crude oil, which is classified as light and sweet, has improved her economy tremendously (*Moscow 21st World Petroleum Congress*, 2014). This is attributed to its paraffinic and low sulfur content, all of which are embraced by consumer refineries in the United States and Europe (Dickson & Udoessien, 2012). The major classes of crude oil are so named in accordance with their export terminals. They are Bonny light (whose terminal is located in the city of Bonny in Rivers State, South-south Nigeria), Qua Iboe (Qua Iboe terminal is situated on the eastern side of the Qua Iboe River Estuary), Brass Blend (it is produced from a refinery located on the Brass River, which is a part of River Niger in the Niger Delta region of Nigeria), Escravos (which is located close to the Escravos site in Warri South Local Government Area of Delta State), and Forcados (whose terminal is located in a small town in the Burutu Local Government Area of Delta State) (Dickson & Udoessien, 2012). The less important or minor crude oil types in Nigeria include Antan Blend, Bonny medium, Odudu Blend, Pennington light, Ukpokiti, Bonga, Yoho Blend, Agbami, Abo, Oyo, Okono Blend, Amenam Blend, Atam Blend, Okwori, Okoro, Ima, Obe, Okwuibome, Ebok, and Asaratoru. Bonny light has the highest demand of all the classes, and this is not unconnected to its highly desired grade, which is a function of its low sulfur content, low corrosive impact on infrastructural designs for refineries, and the vehemently low impact of its refinery by-products on the environment (Odularu & Okonkwo, 2009; Wilberforce, 2016). Therefore, it has received accolades as a major source of income for Nigeria as a country (Badmus, Oyewola, & Fagbenle, 2012; *Moscow 21st World Petroleum Congress*, 2014).

Table 3.1 below shows a summarized compilation of the properties of the major categories of crude oil in Nigeria:

3.3 COMPOSITION AND PROPERTIES OF CRUDE OIL IN NIGERIA

There are four main types of crude oil based on densities and toxicity levels as determined by their volatilities (Karras, 2010).

Light Distillates: These crude oils possess very high volatility and thus are capable of evaporating within a very short time (window of a few days). They diffuse at a very fast rate, thus decreasing toxicity levels. They include petroleum naphtha and ether, heavy and light virgin naphtha, kerosene, gasoline, and jet fuel.
Middle Distillates: These crude oils exude moderate volatility and are thus less evaporative and toxic. They are from a petroleum industry perspective referred to as grade 1 and grade 2 fuel and diesel fuel oils. Other examples include light crude marine gas oils and virtually all domestic fuels.
Medium Oils: These fall into the category of crude oils sold on local market floors nowadays. They are low-volatile oils that require very stringent cleanups, thus resulting in increased level of toxicity.

TABLE 3.1   Properties of the Major Classes of Crude Oil in Nigeria

	Bonny light	Qua Iboe	Brass Blend	Escravos	Forcados
API gravity (°)	35.4	36.0	Varies but 36.3@stable conditions	33.0–33.5	30.0
Sulfur content (% bbl/mt)[a]	0.20	0.10	0.13	0.15–0.18	0.20
Conversion factor	7.53	7.45	7.46	7.50	7.22
Pour point	−18°C	60°F	−36°C	7°C	−27°C
Total Acid Number—TAN (mg KOH/g)[b]	0.27	0.32	0.29	0.53	0.34
Nickel (wppm)[c]	3.6	4.1	2.0	4.8	3.9
Vanadium (wppm)[c]	0.4	0.3	< 2.0	0.5	1.0
Viscosity (cSt)[d]	@ 50°C, 2.90	@ 20°C, 5.71	@ 40°C, 5.20	@ 40°C, 4.75	@ 50°C, 4.60
Specific gravity	0.85	≈0.85	≈0.85	N/A	N/A
Naphtha	N/A	N/A	N + 2A >70	N/A	N/A

[a]Barrel per metric ton.
[b]Milligram per gram.
[c]Weight parts per million.
[d]Centistoke.

Heavy Fuel Oils: In terms of volatility and toxicity, heavy fuel oils are worse than medium oils. Examples include intermediate and heavy marine oils, grade 3, 4, 5, and 6 fuel oil (a strong equivalence of Bunker B and C) (Karras, 2010; Santos, Loh, Bannwart, & Trevisan, 2014).

Oils can also be categorized by virtue of sources and quality. One of them, called the *OPEC Basket* oil, is a combination of crude oils variants from seven countries (Nigeria, Venezuela, the Mexican Isthmus, Saudi Arabia, Algeria, Indonesia, and Dubai). The OPEC is a global organization created in 1960 to pass legal-binding policies that control and implement the importation and exportation of oil within its jurisdiction. Other oil types in this category that are somewhat foreign to Nigeria are the West Texas Intermediate and Brent blend (Bina & Vo, 2007).

3.3.1 Physical and Chemical Properties of Crude Oil in Nigeria

The physico-chemical properties of Nigeria's crude oil samples vary from one oil field to another. This is attributed to the fluctuating quantity of hydrocarbons (alkanes, alkenes, alkynes, cyclo compounds, and aromatics) and their derivatized forms (the presence of heteroatoms such as nitrogen, sulfur, and oxygen (Onyenekenwa, 2011) as well as organic compounds with carboxylic (−COOH) and alcohol (−OH) functionalities) (Onyema & Manilla, 2010). They also contain varying composition of heavy metals that have been confirmed as major pollutants in oil-producing regions (Isah, Alhassan, & Garba, 2017; Madu, Njoku, & Iwuoha, 2011). The variation in properties inevitably leads to carrying out a thorough analysis of the physico-chemical properties of crude oil variants. The following crude oil parameters are vital in the classification and specification of crude oil blends: pour point and kinematic viscosity as functions of temperature, density, metal contents, API gravity, water and salt contents (%), nitrogen and sulfur contents (%), and asphaltene (%) (Dickson & Udoessien, 2012; Riazi, 2005; Wilberforce, 2016).

Nigeria currently boasts more than fourteen commercially available crude oil blends. These commercially available blends come in major and minor categories. The major blends are Bonny light, Qua Iboe, Brass Blend, Escravos light, and Focados blends. Other minor blends include Bonny medium light, Pennington light, Amenam Blend, Yoho light, Erha blend, Bonga blends, and Agbami light. Others are Antan Blend, Odudu Blend, Ukpokiti, Bonga, Abo, Oyo, Okono Blend, Ima, Obe, Okwuibome, Ebok, and Asaratoru (Badmus et al., 2012; Dickson & Udoessien, 2012; *Moscow 21st World Petroleum Congress*, 2014).

3.3.1.1 *Specific Gravity, Api Gravity, and Sulfur Content*

The classification of crude oil as heavy or light is determined by a standard scale called the American Petroleum Institute (API) gravity. It uses the index that is based on the relative density of oil as one of the criteria for oil classification. Depending on the nature of the oil, API gravity greater than 10 will float on water (immiscible liquids with oil being the upper organic layer and water the lower aqueous layer) whereas oil with API gravity less than 10 will form the lower aqueous layer. On the other hand, there have been reports about API gravity being used to classify crude oil as light (>31), medium ($22-31$) and heavy (≤20). Specific gravity of crude oil is simply described as the ratio of the density or mass of a specific crude oil blend to the density or mass of a reference substance, which in most cases is water. It is also the ratio of the weight of a volume of the crude oil blend to the weight of an equal volume of water. Literature reports have affirmed that most of the crude oil blends obtained from Nigeria are light crude oils. Light crude oil samples are in high demand and are of higher market value in Nigeria than their Heavy counterparts. On the contrary, heavy crude oils are characterized by low H/C ratios and very high levels of specific gravity, viscosity, asphaltene, sulfur, nitrogen, and heavy metals. This corroborates reports that API gravity of crude oil often increases as the specific gravity decreases (API, 2011; American Society for Testing and Materials (ASTM), 2011). Sulfur content (expressed in percentage) in crude oil determines its crude sweetness or sourness. Sweet crude oil has Sulfur content less than 0.5% whereas those with more than 0.5% are considered sour. However, Nigeria's crude oil is sweet, and one advantage this offers is the drastic reduction in its corrosion/pollution potentials, which leads to a reasonable cost of production. Moreover, this makes it more suited for the production of most valuable refined products. Reports have also shown that API gravity has an inverse relationship with sulfur contents of crude oil blends (Al-Salem, 2015; Dickson & Udoessien, 2012).

3.3.1.2 *Pour Point, Viscosity, Water, Salt, Nitrogen, and Asphaltene Contents*

Viscosity of crude oil is a measure of its ability to flow from one point to another. The majority of crude oil samples in Nigeria are light and have relatively low viscosity. This indicates that they are easily transported through pipes that connect oil wells to refineries. However, the implication of this property is that they have the intrinsic ability to flow rapidly during spillage, resulting in massive environmental pollution, which is often high temperature–dependent (Odilinye, 2012).

An absolute grip of nitrogen and water contents of any Nigerian crude oil sample is critical in the refining, procurement, and sales of crude oil. These parameters are also connected to the level of corrosion encountered in Nigeria's refineries. Nigeria's crude oils have appreciably low water and nitrogen contents, which expose refineries in Nigeria to mitigated risks associated with corrosion. Pour point is a measure of the low-temperature flow (viscosity) of crude oil blends (Salam, Alade, Arinkoola, & Opawale, 2013). Pour points of heavy, viscous crude oil blends are above 5°C whereas those for light, less viscous crude oil blends can be as low as −15°C. These pour point values are indicative of crude oil blends' facile utilization in low-temperate operations (Stratiev, Petkov, & Stanulov, 2010). Asphaltene is one of the high molecular weight, high boiling point and C/H ratio involatile refinery products that is converted into a number of useful secondary products. It has been reported that its high concentration in crude oil blends results in heavy oil blends with high viscosity and pour points (Kukurina & Rosanova, 2013).

Salt content is an important index for refining operations. This is because salt content of crude oil blends is mainly sodium chloride dissolved in the aqueous phase of the oil or as a suspension in the oil phase. Thus it is ideal to desalt crude oil blends before distillation to prevent salt particles from adsorbing on heat transfer surfaces. The adsorbed particles are capable of reducing the thermal efficiencies of the distillation procedure via buildup of deposits that will block refinery equipment. Therefore, high values of any of these parameters indicate high corrosion tendency of crude oil. However, Nigeria's crude oil samples have a very low content of salt, which make them favorable targets for both local and international marketers (Cani et al., 2016).

3.3.1.3 *Heavy Metals*

Levels of most of the trace metals found in crude oil in Nigeria are generally low except for nickel, iron, and vanadium. These validate reports that light crude oil samples in Nigeria usually contain relatively low trace metal contents compared to the heavy counterparts. The inference drawn from these reports is that crude oil samples from Nigeria, especially those with high concentrations of nickel and vanadium, exude very high tendencies to contaminate the environment. This contamination is on the threshold with nickel and vanadium such that a high vanadium-nickel heavy metal concentration ratio in soil and water bodies is strongly suggestive of the presence of crude oil contamination (Wilberforce, 2016).

The crude oil blends in Nigeria can be categorized as light-sweet crude oil blends. They flow and spread out rapidly as well as possess low levels of water, salt, pour point, and trace metals. These topographical and developmental properties of Nigeria's crude oil blends account for their precedence in indigenous and international oil markets as well as in refinery-based operations.

3.4 REFINERY PRODUCTS OF NIGERIA'S CRUDE OIL

Petroleum, which is popularly referred to as "crude oil," is a naturally occurring, unrefined oil believed to have evolved from rocks or earth. It is a viscous, awful-smelling liquid (whose color alternates from colorless to green, yellow, and black) that is extracted from natural underground reservoirs. It contains a complex mixture of gaseous, liquid, and solid hydrocarbons. Some of these hydrocarbons are derivatized by the presence of heteroatoms such as nitrogen, sulfur, and oxygen (Onyenekenwa, 2011). Some are also in the form of organic compounds with diverse functionalities such as carboxylic (−COOH) and hydroxyl (−OH) groups (Onyema & Manilla, 2010). Reports have it that it contains certain heavy metals such as nickel and vanadium, which unfortunately are major sources of pollution to crude oil−dominated regions (Madu et al., 2011). Crude oil is "necessarily" refined into many very useful primary and secondary products that have positively impacted the fields of medicine, food, textile, cosmetics, polymers, transportation, and energy. The word *necessarily* is herein emphasized because the refining process is pivotal to bringing these products into existence. Therefore, it is important to use this review to (1) consider the procedure for proper understanding of the nature of the products separated out at each stage of the process and (2) evaluate the intermediate and final products from the refining process (Badmus et al., 2012; Onyenekenwa, 2011).

Crude oil has very low end-use value, and this necessitates the need to convert it into a considerable range of well-refined finished products. Refining crude oil to finished products is primarily to maximize the value it adds to the economy. These refining processes are carried out in a refinery, and penultimate to refining is the extraction and drilling of crude oil from the ground and then transferring it through pipelines or by ships and barge to a refinery where the crude oil constituents are separated into purer, useable fractions (Odilinye, 2012). The modern practice of refining crude oil commenced in the 19th century between 1847 and 1859 and extended into the early 20th century. Most of the discoverers were astute entrepreneurs who saw the need and made effort to turn the foul-smelling crude oil into useful, valuable products (Matveichuk, 2004; McKain & Bernard, 1994; Russell, 2003).

Crude oil refineries are robust, capital-demanding manufacturing structures with intensely complicated processing stages. They possess very peculiar physical configuration, economical benefits, and operating principles (Saepudin, Sukarno, Soewono, Sidarto, & Gunawan, 2010). All these peculiarities are immensely connected to the location of the refinery as well as the drilling points, magnitude of investments, demands, and supply for crude oil and its refined products by local and/or export marketers, environmental laws/standards, and marketplace facilities. Local production of crude oil is stirred by the federal government of Nigeria through four subsidiary refineries under the auspices of the Nigerian National Petroleum Corporation (NNPC). These refineries are Warri Refinery and Petrochemical Company (WRPC), Old Port Harcourt Refinery, New Port Harcourt Refinery and Petrochemical Company (PHRC), and Kaduna Refinery and Petrochemical Company (KRPC). A considerable number of different crude oils usually discovered from different geographical origins are processed in varying volumes (lesser or greater) in Nigeria's and other continents' refineries (Badmus et al., 2012). The stages of transforming crude oil into finished, purer, and reusable fractions are divided into two steps:

1. Crude oil samples are separated into primary fractions (based on range of boiling points and variation in the number of carbon atoms present in the fraction) via industrialized, large-scale fractional distillation.
2. Innumerable sequential physical and chemical techniques are used to effect the transformation of primary fractions into finished, useable secondary products or by-products. This is triggered by the increasing demand for secondary products with better efficiency than the primary products obtained via the distillation of crude oil (Kukurina & Rosanova, 2013; Onyenekenwa, 2011).

The first stage of involves the separation of sand and other solid particles from the crude oil. The crude oil is then channeled to a reservoir where it is separated into two immiscible fractions; the upper organic layer (crude oil) and the lower aqueous layer (water). The former is pumped through strategically laid pipelines to the refinery where it is subjected to consistent fractional distillation (Seneviratne, 2007). Fractional distillation aids the separation of the crude oil into different fractions of varying boiling point ranges, a physical property that is a close function of the number of carbon atoms in the primary fractions and also to the states (solid, liquid, and gaseous) of the primary fractions (Kukurina & Rosanova, 2013). It is established that the proportions of primary fractions separated out from crude oil change with the class of crude oil subjected to the distillation process. The boiling points of primary fractions obtained from the fractional distillation of petroleum fall in the range 20°C to >400°C (Kukurina & Rosanova, 2013; Wang, Wang, Leng, & Chen, 2016).

The earlier statements "Innumerable sequential physical and chemical techniques are used to effect the transformation of primary fractions into finished, useable secondary products or by-products. This is triggered by the increasing demand for secondary products with better efficiency than the primary products obtained via the distillation of crude oil" create a suspicious cloud over the efficiency of the fractional distillation process! Why is there the need for further transformation of primary refined fractions into by-products or secondary products? This question is better answered from the viewpoint of the public/consumption demand for secondary products with better efficiency than the primary products. For instance, fractional distillation of crude oil in Nigerian refineries yields approximately 10% of petrol, which is regrettably small compared to the demand from the entire populace for energy production as well as for exportation (nature in action and nothing more!). Nature has made it even more pathetic and disappointing to affirm that the higher–boiling point fractions (majorly solids and semisolids) are produced in yields that far outweigh their demand by the entire populace and government (Stanley, 2009). Therefore, it becomes inevitably reasonable to generate secondary products from the excessive primary products to cover up for the lapses created by the likes of kerosene and petrol, which come in generally low yields after crude oil is subjected to fractional distillation. In fact, it has become standard practice globally to produce quite a number of refined products such as petrol in multiple grades to meet the varying demands in terms of standards and specifications such as sulfur and nitrogen contents, octane ratings, and others. Most countries have embarked on attaining this feat by configuring their refineries to maximize the production of petrol (a lower–boiling point distillate) at the expense of other higher—boiling point refined distillate such as gas, heavy oils, and diesel oils whose demands by the public is on the very low side. All thanks to the very fast demand petrol is enjoying in Nigeria as a country and on a global scale (Kukurina & Rosanova, 2013). Below is Table 3.2, which shows a compilation of primary fractions obtained from crude oil via fractional distillation, their properties (boiling point ranges, carbon distribution), and their uses:

TABLE 3.2   Compilation of Primary Fractions Obtained From the Distillation of Crude Oil and Their Properties and Uses

Primary petroleum fractions	Boiling point range ($^{\circ}$C)	Number of carbon atoms per molecule	Major uses
Natural gas	0–30	C_1–C_4	Fuel and the synthesis of raw materials for chemical industries
Petroleum ether	20–90	C_5–C_7	Solvent used majorly in pharmaceutical industries and as inhalant drug
Crude naphtha	30–150	C_5–C_{14}	Solvents, cleaning fluids, paints and varnishes
Gasoline of petrol	70–90	C_6–C_{18}	Fuel for internal combustion engines in cars and bikes; Solvent for dry-cleaning
Petrol	70–200	C_6–C_{10}	Fuel for internal combustion engines in motorbikes and boats
Ligroin	90–120	C_7–C_8	Solvent in dry-cleaning and as a laboratory solvent
Fuel	100–200	C_5–C_{10}	Fuel for automobiles and power-generating plants
Benzene	120–160	C_5–C_{10}	Solvents for dry-cleaning and also utilized in the paint and oil industries
Kerosene	150–300	C_{10}–$C_{18}C_{10}$–C_{38}	Fuel for Jet engines, central heating systems, lamps and stoves, manufacturing oil gas
Gas oil	>275	C_{12}–C_{20}	Fuel for diesel engines
Heavy oil	>300	C_{18}–C_{38}	Fuel for marine engines
Diesel oil	300–400	C_{15}–C_{25}	Fuel for locomotives
Lubricating oil	300–400	C_{20}–C_{24}	Lubricant and manufacture of shoe polishes and candles
Paraffin wax	>400	C_{21}–C_{30}	Electrical insulators and lubricants
Lubricating oil, waxes	Nonvolatile oil	>C_{20}	Lubricant candles
Asphalt and bitumen	Solid residue	>C_{40}	Road surfaces and roofing
Pitch			Toilet goods and ointments
Coke			As fuel and as paints and varnishes

C_1–C_4 are gases, C_5–C_{10} are mostly liquids, while the rest are solids and viscous liquids, which can be classified as semisolids from a physical state viewpoint.

3.5 BY-PRODUCTS OF CRUDE OIL IN NIGERIA

The physical and chemical properties of the various classes of hydrocarbons depend on the number of carbon atoms present in the molecule as well as the nature of the chemical bonds between them. Carbon atoms readily form bonds (single, double, and triple) with one another via catenation as well as with hydrogen and heteroatoms. One major characteristic used to describe fractions of crude oil is the carbon-to-hydrogen (C/H) ratios. For example, as one goes down Table 3.2 above, it is noticed that the C/H ratios increase, thus indicating the change in trend from light (less dense) fractions to heavy (denser) fractions. With reference to the chemistry of crude oil refining, the lower the C/H ratio of crude oil fraction, the more reduced the cost of running the refinery for efficient crude oil processing. The distribution of carbon atoms and the number of heteroatoms in a sample of crude oil are also determinants of the yield and quality of the refined fraction as well as the economic value of the crude oil sample (Riazi, 2005; Santos et al., 2014).

Nigeria's refineries, just like every other refinery in North America, Asia, the Middle East, and South America, are experiencing rapid growth in the demand for lighter petroleum fractions. In terms of configuration, Nigeria's refinery complexity revolves around (1) complicated stages in the refining processes and (2) a numerical grade (assigned a complexity of 1) that describes the capacity and capital strength of the refining processes downstream (component of the petroleum industry that monitors retailers, consumers, oil tankers, etc.) of the crude distillation unit. It is established that the higher the complexity of a refinery's operation, the greater its investment capacity. Moreover, a refinery's higher complexity determines its ability to add value to the crude oil by converting heavy crude oil fractions into light congeners with better quality specifications (*Oil and Gas for Beginners: A guide to the oil and gas industry*, 2013).

The configuration of a refinery to align with a target complexity is attained by two major schemes: conversion (which is popularly referred to as *cracking*) and deep conversion (popularly termed *coking*). The former exists either as catalytic cracking or hydrocracking. These processes convert heavy crude oil fractions (such as gas or heavy oil, which naturally come in yields higher than their demands and possess low-octane–rated naphthas) into lighter crude oil feedstocks (which naturally come in yields lower compared to demands and possess high-octane blends) such as petrol, jet fuels, diesel oil, and petrochemicals. Cracking thus improves the natural yields of lighter products to meet the demands of local and international markets. It occurs in the presence of certain conditions such as high pressure in bars and fairly high temperatures of 400–800°C in the presence of hydrogen and specific catalysts. Its only disadvantage is that it still inevitably produces heavy, low-value products such as asphalt. Deep conversion, on the other hand, involves not only catalytic and hydro-cracking but goes a step further to convert (destroy) very heavy and least valuable residual oil into lighter, straight-run petrol refinery streams via the use of coking units. The term *coking* comes from the total reduction of hydrogen during the deep conversion process, resulting in a form of carbon called "coke." The coke obtained comes in two or three forms, depending on reaction mechanism, time, temperature, and the nature of the crude feedstocks. Coking produces additional lighter crude streams to other lighter feeds produced by catalytic cracking and reforming processes (Auta, Ahmad, & Akande, 2012; Isah et al., 2017; Olori, 2015; Onyenekenwa, 2011).

Petrochemicals, as the name implies, are chemical compounds derived from petroleum fractions. The two major categories of petrochemicals are olefins and aromatics. As stated earlier, refineries transform primary petroleum fractions into petrochemicals via cracking and reforming processes. Specifically, olefins (generally known as alkenes) are produced by cracking natural gases such as ethane and propane in the presence of steam, whereas aromatics such as benzene, xylene, and toluene are produced by catalytically reforming naphtha. These categories of petrochemicals have become the building blocks for quite a wide range of materials useful for man's survival and in organic chemistry. Below is a compilation of petrochemicals and their products and uses in today's modern society (Table 3.3) (*Oil and Gas for Beginners: A guide to the oil and gas industry*, 2013).

On a conclusive note, efforts made to reduce the sulfur contents in refinery streams or finished products such as diesel or residual fuels have caused a hike in the expended

TABLE 3.3   A Compilation of Petrochemicals and Their Products and Uses

Petrochemicals	Products	Uses
Terephthalic acid	Terylene polymeric fibers	Clothing, sheets, sails and plastic bottles
Methylbenzene	Benzene, caprolactam	Benzene: used as industrial chemicals for pesticides and drugs, detergents, dyes, resins, rubber, and lubricants
		Caprolactam: precursor to nylon 6 fibers and resins via ring-opening polymerization
1,4-dimethylbenzene	Benzene-1,4-dicarboxylic acid	Commodity chemical used as precursor for polyethylene terephthalate used to make clothes and plastic bottles
Phenylamine	Aniline-based dye pigments	Rubber processing chemicals, herbicides, dyes, pigments, and antioxidants
Benzene	Styrene, cyclobenzene, phenylamine, Phenol, xylene-based compounds	Styrene: building block for plastic materials
		Cyclobenzene: lubricating oil composition
		Phenylamine: manufacture of precursor to polyurethane
		Phenol: antiseptic, synthesis of dyes, aspirin, and used as explosives
		Xylene-based compounds: solvents used in rubber and leather industries
Phenol	Bakelite-based resins	Possess nonelectrical conductivity and heat resistance properties utilized in insulators and telephone casings. Used as jewelry, kitchenware, and firearms

(*Continued*)

TABLE 3.3 (Continued)

Petrochemicals	Products	Uses
Ethene	Poly(ethene), ethylbenzene, epoxyethane, ethanol, 1,2-dichloroethane	Poly(ethene): packaging in the form of plastic bags, films, and geomembranes
		Etylbenzene: solvent for coating and building block for plastic and manufacture of styrene
		Epoxyethane: production of ethane-1,2-diol
		Ethanol: fuel and a major constituent of alcoholic beverages
		1,2-dichloroethane: to produce vinyl chloride, the main precursor for polyvinyl chloride (PVC) production
Methane	Hydrogen cyanide, ethyne, synthesis gas, haloalkanes, and carbon black	Hydrogen cyanide: fumigation, mining, electroplating, and as a chemical warfare agent
		Ethyne: fuel for oxyacetylene flame used in welding metals, prepare ethanol and vinyl chloride monomer used as precursor for PVC
		Synthesis gas: fuel gas
		Haloalkanes: solvent, methylating agent in organic synthesis and as refrigerant
		Carbon black: color pigment, reinforcing fillers in rubber products and used in water pipes and electric cable installation
Chloroethene	Polymers (e.g., poly(chloroethene) and poly(chloroethane)	Packaging and labeling, textile and in laboratory equipment
Ethylbenzene	Phenylethene	Effective building block for plastic products
Propene	Polypropene, polyacrylonitrile, propan-2-ol, Propane-1,2,3-triol, methyl-buta-1,3-diene, (1-methylethyl) benzene	Polypropene: plastic materials used in packaging
		Polyacrylonitrile: copolymer in fabrics and used as component of shock-proof materials
		Propan-2-ol: sterilizer and disinfectant in hospitals and veterinary institutions, used as fungicide and bactericide
		Propan-1,2,3-triol: solvent, antifreeze, plasticizer, and sweeteners
		Methy-buta-1,3-diene: a major component in rubber used for car tires and in the manufacture of copolymers such as polyamides
		(1-methylethyl) benzene: used as intermediate in the synthesis of industrially relevant chemicals such as phenols
Propane-1,2,3-triol	Alkyd-based resins	Used in paints and in molds for casting and as binders in commercial oil-based coatings
Propan-2-ol	Propanone	

TABLE 3.3   (Continued)

Petrochemicals	Products	Uses
		Used as building block in organic synthesis and an active ingredient in nail polish remover and as paint thinner
Propanone	Perspex	Thermoplastic used in sheet form as shatter-resistant replacement for glass
Ethanol	Ethanal	Precursor in the synthesis of acetic acid and ethyl acetate
Epoxyethane	Ethane-1,2-diol	Antifreeze in car engines and in poly(ethylene terephthalate) synthesis
1,2-dichloroethane	Chloroethene	Precursor to PVC
2-methylpropene	2-methylpropan-2-ol	Solvent, paint remover, and petrol octane booster
1-(methylethyl) benzene	Phenol and phenolic resins	Coatings, composites, and adhesives
Buta-1,3-diene	Butadiene-based polymer materials	As binder in pigmented coatings
Phenylethene	Poly(phenylethene)	Used in containers, refrigerator parts, as packaging and insulator

refinery energy as well as a tremendous increase in the emission of CO_2 gas as by-product from refineries. The increased refinery energy coupled with CO_2 emission stems from the total combustion of additional natural and still gases (Al-Salem, 2015; Karras, 2010). From an environmental viewpoint, CO_2 is a greenhouse gas (Kwasniewski, Blieszner, & Nelson, 2016) that is strongly connected to global warming, and the alarming surge for global warming by CO_2 gas with its resulting energy disasters have called for novel technologies for electrochemically reducing it to fuels and other important petrochemicals (Weng et al., 2016; Fenwick, Gregore & Luca, 2014).

References

Al-Salem, S. M. (2015). Carbon dioxide (CO_2) estimation from Kuwait's petroleum refineries. *Process, Safety and Environment Protection, 95*, 38–50.

American Petroleum Institute (API). (2011). *Specification for Materials and Testing for Petroleum Products. API Production Dept. API 14A, Eleventh edition. Dallas: 20-21. AOAC (1984) Official Methods Analytical Chemistry 10th ed.*

American Society for Testing and Materials (ASTM). (2011). Standard test method for substances in crude oil. *Annual Book of Standards, 2*, 287.

Auta, M., Ahmad, A. S., & Akande, H. F. (2012). Comparative studies of traditional (non-energy integration) and energy integration of catalytic reforming unit using pinch analysis. *Nigerian Journal of Technological Development, 9*(1), 1–10.

Badmus, I., Oyewola, M. O., & Fagbenle, R. O. (2012). A review of performance appraisals of Nigerian federal government-owned refineries. *Energy and Power Engrgy, 4*(January), 47–52.

Bina, C., & Vo, M. (2007). OPEC in the epoch of globalization: An event study of global oil prices. *Global Economy Journal, 7*(1), 1−49.

Cani, X., Malollari, I., Beqiraj, I., Manaj, H., Premti, D., & Liçi, L. (2016). Characterization of crude oil from various oilfields in Albania through the instrumental analysis. *Journal of International Environmental Application & Science, 11*(2), 223−228.

Dickson, U. J., & Udoessien, E. I. (2012). Physicochemical studies of Nigeria' s crude oil blends. *Petroleum and Coal, 54*(3), 243−251.

Fenwick, A. Q., Gregore, J. M., & Luca, O. R. (2014). Electrocatalytic reduction of nitrogen and carbon dioxide to chemical fuels: Challenges and opportunities for a solar fuel device. *Journal of Photochemistry and Photobiology B: Biology, 152*(Pt. A), 47−57.

Isah, A. G., Alhassan, M., & Garba, M. U. (2006). Feed quality and its effect on the performance of the fluid catalytic cracking unit (a case study of Nigerian based oil company). *Leonardo Electronic Journal of Practices and Technologies1, 9*, 113−120 , ISSN: 1583-1078

Karras, G. (2010). Combustion emissions from refining lower quality oil: What is the global warming potential? *Environmental Science & Technology, 44*(24), 9584−9589.

Kukurina, O.S., & Rosanova, Y.V. (2013). *English for Specific Purposes. Oil Refining.*

Kwasniewski, V., Blieszner, J., & Nelson, R. (2016). Petroleum refinery greenhouse gas emission variations related to higher ethanol blends at different gasoline octane rating and pool volume levels. *Modeling and Analysis, 10*, 36−46.

Madu, A. N., Njoku, P. C., & Iwuoha, G. A. (2011). Extent of heavy metals in oil samples in escravous, abiteye and malu platforms in delta state Nigeria. *Public Journals of Agriculture & Environmental Studies, 2*(2), 41−44.

Matveichuk, A. A. (2004). *Intersection of oil parallels: Historical essays. Moscow: Russian Oil and Gas Institute.*

McKain, D. L., & Bernard, L. A. (1994). *Where it all began: The story of the people and places where the oil industry began—West Virginia and South- eastern Ohio.* Parkersburg: W.Va.

21st World Petroleum Congress, Moscow (2014). Theme: Responsibly energising a growing world.

Nøstbakken, L., & Jadhav, S. (2014). Petroleum resource management: The role of institutional frameworks and fiscal regimes in value creation. A comparative analysis for Norway and Nigeria.

Odilinye, I.O.H. (2012). *Pollution levels of some heavy metals and total petroleum hydrocarbons (TPH) in soil samples from umuorie oil spill site, Ukwa West Local Government Area of Abia State.* M.Sc. Thesis, University of Nigeria, Nsukka, 87 pages.

Odularu, G. O., & Okonkwo, C. (2009). Does energy consumption contribute to economic performance? Empirical evidence from Nigeria. *Journal of Economics Internship and Finance, 1*(2), 44−58.

Oil & Gas for Beginners; A guide to the oil & gas industry. (2013). Industry Update: Deutsche Bank Markets Research.

Olori, T. O. (2015). *Studies of regeneration of catalytic reforming catalyst.* Zaria, Nigeria: Department of Chemical Engineering, Ahmadu Bello University.

Onyema, M. O., & Manilla, P. N. (2010). Light Hydrocarbon correlation of niger delta crude oils. *Journal of American Science, 6*(6), 82−91.

Onyenekenwa, C. E. (2011). A review on petroleum: Source, uses, processing, products and the environment. *Journal of Applied Science, 11*, 2084−2091.

Riazi, M. R. (2005). *Characterization and properties of petroleum fractions.* West Conshohocken, PA: ASTM International.

Russell, L. S. (2003). *A heritage of light: Lamps and lighting in the early canadian home.* Toronto: University of Toronto Press.

Saepudin, D., Sukarno, P., Soewono, E., Sidarto, K. A., & Gunawan, A. Y. (2010). Oil production optimization in a cluster of gas lift wells system. *Journal of Applied Science, 10*, 1705−1713.

Salam, K. K., Alade, A. O., Arinkoola, A. O., & Opawale, A. (2013). Improving the demulsification process of heavy crude oil emulsion through blending with diluent. *Journal of Petroleum Engineering, 2013*, 1−6.

Santos, R. G., Loh, W., Bannwart, A. C., & Trevisan, O. V. (2014). An overview of heavy oil properties and its recovery and transportation methods. *Brazilian Journal of Chemical Engineering, 31*(03), 571−590.

Seneviratne, M. (2007). *A practical approach to water conservation for commercial and industrial facilities.* Oxford: Elsevier, Technology and Engineering.

I. BACKGROUND INFORMATION ON PETROLEUM INDUSTRY ACTIVITIES AND THE NIGERIAN ENVIRONMENT

Stanley, I. O. (2009). Gas-to-liquid technology: Prospect for natural gas utilization in Nigeria. *Journal of Natural Gas Science & Engineering, 1,* 190−194.

Stratiev, D., Petkov, K., & Stanulov, K. (2010). Evaluation of crude oil quality. *Petroleum and Coal, 52*(1), 35−43.

Wang, S., Wang, Y., Leng, F., & Chen, J. (2016). Stepwise enrichment of sugars from the heavy fraction of bio-oil stepwise enrichment of sugars from the heavy fraction of bio-oil. *Energy and Fuels, 30*(3), 2233−2239.

Weng, Z., Jiang, J., Wu, Y., Wu, Z., Guo, X., Materna, K. L., . . . Wang, H. (2016). Electrochemical CO2 reduction to hydrocarbons on a heterogeneous molecular cu catalyst in aqueous solution. *JACS, 138,* 8076−8079.

Wilberforce, J. O. (2016). Levels of heavy metal in bonny light crude oil. *IOSR Journal of Applied Chemistry, 9*(7), 86−88.

Yusuf, A., & Samuel, T. (2013). Shell D′ arcy exploration & the discovery of oil as important foreign exchange earnings in Ijawland of Niger Delta, C. 1940s-1970. *Arabian Journal of Business and Management Review, 2*(11), 22−33.

The Oil and Gas Industry and the Nigerian Environment

Sylvester Okotie[1], Napoleon Ogbon Ogbarode[1] and Bibobra Ikporo[2]

[1]Federal University of Petroleum Resources Effurun, Delta State, Nigeria [2]Niger Delta University, Amasoma, Nigeria

4.1 INTRODUCTION

The oil and gas industry is one of the sectors that has contributed to the world positively in numerous ways. According to the international energy agency, in 2012, 40.7% of oil and 15.2% of natural gas were contributed to global energy. They provide fuel for our cars, airplanes, and ships for mobility as well as for operation of other industrial engines. They are used as heating source in our homes, cooking of our foods, generation of electricity that powers our day-to-day activities, and other human endeavors that make life comfortable. In addition, the by-products from crude oil serve as raw materials for many industries such as pharmaceutical, petrochemical, solvents, pesticides, fertilizers, and plastics, among others. In Nigeria, crude oil was first discovered in commercial quantity in 1956 by Shell British Petroleum at a village in present day Bayelsa State called Olobiri, and actual commercial production began in 1958. The desire to meet the global energy need has prompted the intensification of the search for the black gold. It comes with many benefits as well as challenges. The gains include economic development, but the environment also suffers degradation and loss in economic and social values.

Nigeria is one of the global players in oil and gas, and she has the largest natural gas reserve in Africa. She also has the second largest oil reserve in Africa, with four oil refineries and an estimated total refining capacity of 445,000 barrels per day (Onuoha, 2008). However, the discovery and exploration of crude oil in Nigeria has left a bitter tale on the inhabitants of the Niger Delta region where most of Nigeria's crude oil reserves are concentrated. The environment has been degraded with loss of fauna and flora, some of which are endemic to the region.

The modern petroleum industry began in 1859, when the American oil pioneer E.L. Drake drilled a producing well on Oil Creek in Pennsylvania at a place that later became Titusville. Many wells were drilled in the region. Kerosene was the chief finished product, and kerosene lamps soon replaced whale oil lamps and candles in general use. Little use other than as lamp fuel was made of petroleum until the development of the gasoline engine and its application to automobiles, trucks, tractors, and airplanes. Today the world is heavily dependent on petroleum for power, lubrication, fuel, dyes, drugs, and many synthetics. The widespread use of petroleum has created serious environmental problems.

4.2 OIL AND GAS RESERVES

The global development of oil and gas fields today depends solely on the amount of recoverable (reserves) fluid discovered in subsurface formation (reservoir). Li (2011) stated that there are more than 65,000 oil and gas fields of all sizes in the world. Ivanhoe and Leckie (1993) reported that about 1500 of these fields, which account for about 94% of global crude oil reserve, come from the giant and major fields concentrated in the Near and Middle East, Commonwealth of Independent States, and Venezuela. Table 4.1 shows the world's largest oil reserves by country. Hydrocarbon reserves comprise both conventional and nonconventional oil deposits. The nonconventional reserves are mostly found in America, and they are difficult to exploit, but with recent technology advancement in the United States, Canada, and other areas, it is now possible to crack the shale oil (nonconventional) reserves.

Reserves are seen as the heart of the oil and gas business in the countries where they are found, and it grows with time as new discoveries are added to the existing ones. Reserves can be defined as the estimated quantities of hydrocarbon such as crude oil, condensate, natural gas (associated or nonassociated gases) that are anticipated to be commercially recoverable with the use of established technology to known hydrocarbon accumulations from a given date forward under existing economic conditions, established operating conditions, and current government regulations with a legal right to produce and install facilities to deliver the products to the market. Reserves are further classified according to the degree of certainty or probabilities associated with the estimates, and whenever a new field is discovered, the bid to acquire such an oil block is based on the classification of the reserves. The reserves classifications are: proved and unproved (probable and possible reserves).

Proved reserves are those quantities of hydrocarbon reserves based on analysis of geological and engineering data that can be estimated with a reasonable high degree of certainty (at least 90%) to be commercially recoverable from a given date forward from known reservoirs and under current economic conditions, operating methods, and government regulations. In general, reserves are considered proved if the commercial producibility of the reservoir is supported by actual production or formation tests. Probable reserves are attributed to known accumulation having a 50% level of confidence of recoverable oil whereas possible reserves have a 10% confidence level of oil recovery.

TABLE 4.1 Global Distribution of Oil Reserves by Country

Rank	Country	Reserves (billion barrels)
1	Venezuela	298.4
2	Saudi Arabia	268.3
3	Canada	171.0
4	Iran	157.8
5	Iraq	144.2
6	Kuwait	104.0
7	Russia	103.2
8	United Arab Emirates	97.8
9	Libya	48.36
10	Nigeria	37.07
11	United States	36.52
12	Kazakhstan	30.0
13	Qatar	25.24
14	China	24.65
15	Brazil	15.31
16	Algeria	12.2
17	Mexico	9.812
18	Angola	9.011
19	Ecuador	8.832
20	Azerbaijan	7.0

From Jessica Dillinger. < http://www.worldatlas.com/articles/the-world-s-largest-oil-reserves-by-country.html > .

4.3 METHODS OF RESERVES ESTIMATION

The oil and gas business globally is a high risk and challenging venture because not all exploration culminates into successful oil and gas discovery (reserves). Sometimes, when oil companies do hydrocarbon prospect, they end up drilling dry holes. There are different techniques available today for estimating oil and gas reserves. These are: analogy, volumetric (geology method), material balance, decline curve analysis, and reservoir simulation (Petrobjects, 2003).

The estimation of these reserves is usually associated with some level of uncertainties (uncertainty in seismic predictions, gross rock volume of a trap, rock properties (net-to-gross, fluid saturation, and porosity), fluid properties, recovery factor, etc.), and when these uncertainties are not properly factored into the hydrocarbon prospect evaluation, it could result in an incorrect estimation of the reserve. This means that the value of reserve estimation is a key driver for exploration and production companies to decide whether to develop or abandon the prospect based on their set criteria. Therefore, in estimating oil and gas reserves, we rely on the integrity, skill, and judgment of the evaluator, who makes a decision based on the amount of data available, the complexity of the formation geology, and the degree of depletion of the reservoir(s).

The selection of the appropriate method to estimate reserves and resources and the accuracy of the estimates depend largely on the following factors: the type, quantity, and quality of geoscience, engineering, and economic data available for technical and commercial analyses; geologic complexity; the recovery mechanism; stage of development; and the maturity or degree of depletion.

4.4 REASONS FOR ESTIMATING RESERVES

- For exploration, development, and production
- To negotiate property sales and acquisitions
- To determine market value
- To design facilities
- To obtain financing
- Evaluation of profit/interest
- Government regulation and taxation
- Planning and development of national energy policies
- Investment in oil/gas sector
- Reconciling dispute or arbitration involving reserves

4.5 ACTIVITIES OF THE GLOBAL OIL AND GAS INDUSTRY

There are two schools of thought about the activities of the global oil and gas industry. One school believes that the activities are upstream and downstream. The other school of thought split the activities into upstream, midstream, and downstream. The upstream sector involves the exploration, drilling, and production of oil and gas from the reservoir to the surface production facilities. The activities and processes that constitute the midstream include: oil and gas transportation (transmission) processes and modes, pipeline operations, oil and gas storage facilities, wholesale marketing of crude oil, natural gas, and liquefied natural gas (LNG), and value creation in the midstream activities. The downstream sector involves the refining of crude oil and/or raw natural gas derived from the upstream sector, transportation to retail facilities, as well as the marketing and distribution of the products obtained from the upstream sector. The downstream facilities include petrochemical plants, oil refineries, natural gas distribution companies, retail outlets (i.e., gas stations), etc. The downstream sector reaches customers via products such as: gasoline or petrol, kerosene, jet fuel, diesel oil, heating oil, fuel oil, lubricants, waxes, asphalt, fertilizers, plastics, liquefied petroleum gas, natural gas, as well as hundreds of petrochemicals.

Furthermore, the activities of the oil and gas industry are usually integrated, requiring several disciplines. Exploration geology/seismology, well or drilling engineering, which can also incorporate the mud and cementing operations to its program, completions engineering, petrophysics, reservoir engineering, production engineering, process engineering, economics, and environmental, to mention only a few, work together to get the oil to the surface.

Exploration involves the search for oil and gas in the subsurface formations beneath the earth crust by petroleum geologists and geophysicists. The determination of the actual location of petroleum reservoirs is a very difficult task and is probably the most challenging aspect of the petroleum industry job. Finding or discovering a petroleum reservoir beneath the earth crust involves three major activities:

- Geologic surveying
- Geophysical surveying
  - Gravity survey
  - Magnetic survey
  - Seismic survey
  - Remote sensing
- Exploratory drilling activities (wildcatting)

Discovery of the oil and gas beneath the earth crust does not make sense to investors; their concern is how to get access to this resource. Hence, drilling technology is employed. It begins with spudding of the well to initiate a hole to the target zone through which the oil and gas can be accessed. Reservoir engineering estimates the amounts of oil and gas deposit in the surface formation. They continuously monitor the reservoir, collect relevant data, and interpret these data to be able to determine the present conditions of the reservoir, estimate future conditions, and control the flow of fluids through the reservoir with an aim to increase recovery factor and accelerate oil recovery (Okotie & Onyekonwu, 2015). The reservoir engineer is saddled with responsibility similar to that of a medical practioner to make sure that the reservoir does not go below expected performance (fall sick) and even if it falls sick, he looks for a way to bring it back to full performance throughout the entire life of the reservoir or project.

4.6 OPERATIONAL ENVIRONMENTS OF THE GLOBAL OIL AND GAS INDUSTRY

There are basically two environments where oil and gas activities take place globally: the onshore and offshore environments. The same processes are required in both environments to explore and exploit oil and gas from the reservoir to the surface production facilities, but the architectural design of offshore environment is far more complex than the onshore environment. Apart from cost, the offshore environment requires the installation of subsea facilities such as the riser system, loading buoy, mooring lines, and subsea trees, among others.

The global exploration and production of oil and gas has gradually moved from onshore to the offshore environment (North Sea, Gulf of Mexico, Gulf of Guinea, etc.), requiring the development of new technology in order to exploit the huge hydrocarbon deposits in this region that are inaccessible with the existing onshore or shallow water technologies. Due to the depletion of onshore and offshore shallow water reserves, exploration and production operations are advancing into frontier deep water basins and are currently pushing into a more hostile and remote environment (ultra-deep water), which has become a challenge to the oil and gas industry. Deep water oil drilling is attractive to the oil companies despite the hostile nature of the environment because of the little income/share that accrues to government from oil revenue from this area and the inability of government to regulate their activities. It has fewer disturbances by communities on personnel and facilities as well as a reduction in oil theft.

There are different structures employed today in the drilling and production of hydrocarbons in the offshore environment. Over the years, as oil and gas exploration advanced into deep and ultra-deep waters, fixed structures were no longer relevant in drilling and production. Hence, there are several floating structures available today in ultra-deep offshore operations. Below are the drilling and production platforms:

4.6.1 Offshore Production Platforms

- Floating Production and Storage Systems (FPSs)
- Floating Production, Storage, and Offloading Systems (FPSOs)
- Fixed Platform

4.6.2 Offshore Drilling Structures

- Stationary Platforms
  - Compact Platform Rigs
  - Tender Platform Rigs
- Bottom-Supported Mobile Platform
  - Swamp Barges
  - Jack-ups
  - Submersible
  - Spar
  - Semisubmersible
- Floating Drilling Platform
  - Semisubmersible
  - Drillship

These structures serve several functions such as exploration and development drilling. They are employed in the processing of hydrocarbon reservoir fluid to meet customer specification. They can be used as a wellhead or riser platform, used as chemical, gas, and water injection platforms, and they also serve other functions such as utilities, living quarters, flare and storage platform, among others. They are unique in design because of their limit to the area (water depth) of operation. It therefore implies that some factors must be considered prior to the selection of offshore platforms. These factors are:

- The depth of the water or ocean
- Environmental criteria (wind, wave, icebergs, storms, and current movement)

- Type and density of sea water
- The depth from the seabed to the target zone (drilling depth)
- Operational limitations
- Type of well
- The weight, size, et cetera, of the topside equipment
- Construction constraint (space availability)
- The movement of drilling rig from one location to another (rig move)
- Economic analysis (cost)

4.7 GLOBAL OIL AND GAS TRADE (INDUSTRY AND MARKET PARTICIPANTS)

Generally, crude oil, natural gas, and petroleum products are global commodities and, as such, their prices are always in flux because they are determined by supply and demand factors on a worldwide basis. The market is largely oligopolistic in nature. This is an economic condition in which there are few suppliers of a product such that one supplier's actions can have a significant impact on prices and on its competitors. The excess supply is usually restricted by a cartel.

The main participants in the oil and gas industry and in the market of oil and gas vary in both functions and interest. The list includes:

- Resource owners
- Those in dire need of the resource
- Producers
- Middle men who act as the product courier
- Those who offer their expert services to those who regulate the activities

Nevertheless, the major drivers of the oil and gas industry will fall into one of the categories listed below:

- The Governments
  - They provide the policies and regulations for the operations of the industry
  - They collect royalties
  - In some cases they own the oil and gas
  - They also participate and operate as majors in some countries
- OPEC
  - This is a cartel of oil exporting countries
- The Major Oil and Gas Companies
  - Operations of these companies include both upstream and downstream activities
  - They own the major assets of the oil and gas industry
  - They own the majority of the oil and gas fields
  - They own the majority of the production and processing facilities in the upstream and the downstream
  - They provide the fund and, in conjunction with the service companies, drive the ever-changing innovations and technology of the industry
  - They control a major portion of the world total reserves and production

- Downstream Companies
    Operations of these companies include:
  - Refining of crude into petrol, diesel, kerosene, etc.
  - Production of gas
  - Bottling of gas
  - LNG
  - Raw materials for the petrochemical industry
  - Petrochemicals
- Marginal Field Operators
  - Indigenous participation and building indigenous capacity in the upstream sector
  - Reduce the rate of abandonment of depleting oil fields by international oil companies (IOCs).
- Service Companies
    Operations of these companies include:
  - The experts and owners of the industry technology
  - Owns and controls NO assets in the industry
  - Suppliers of all materials needed in the industry operation

  Examples: Schlumberger, Halliburton, Baker Hughes, WOG, Hexagon, etc.

4.8 NIGERIAN ECOSYSTEM

This chapter will not be complete without looking at the Nigerian environment. The Nigerian vegetation is made up of two main and broad vegetation belts. The heavily forested south with humid tropical conditions graduates into savannah with hot and dry conditions in the north. Some isolated mountainous vegetations are found in the high plateau in the central and far eastern parts of the country. However, the discussion on the global oil and gas industry and the Nigerian environment will be limited to the Niger Delta region of Nigeria where exploration and production of oil and gas take place. Francis, Ikemefuna, and Ekwoaba (2012) reported that the Niger Delta is a region that is economically strategic to Nigeria as a nation because oil contributes more than 90% of aggregate foreign exchange earnings for Nigeria and 80% of the federal government's revenue. They further added that the Niger Delta region is the country's proverbial "goose that lays the golden egg."

Exploration of crude oil has had negative impacts on the Niger Delta ecosystem since oil prospecting began in that part of Nigeria. Oil and gas activities have the potential for varieties of impacts on the ecosystem. Environmental challenges vary. The nature and degree of impact depends upon:

1. The stage, situation, size, and complexity of the activities involved in process or project; execution
2. The nature and sensitivity of the environment
3. The techniques used for the control, contamination/pollution prevention, and mitigation
4. Above all, the effectiveness of the planning put into the project

FIGURE 4.1   The Nigerian business survival characteristic features. From Jessica Dillinger. <http://www.worldatlas.com/articles/the-world-s-largest-oil-reserves-by-country.html>.

From the business perspective, the Nigerian environment has all the component of a host community that is beyond the control of corporations but which will affect their functioning and survival as business enterprises. This business survival characteristic feature is depicted in Fig. 4.1.

4.9 THE NIGER DELTA ECOSYSTEM

The oil and gas exploration and production activities are taking place in the Niger Delta region of Nigeria. This region has contributed significantly with other oil-producing regions of the world to meet the global energy demand and as such, have played an important role in the global economy. The Niger Delta region consist of nine states (Rivers, Akwa Ibom, Delta, Bayelsa, Cross River, Edo, Ondo, Imo, and Abia) in the south-south region of Nigeria (Fig. 4.2). According to Congressional Research Service (CRS) (2008), approximately 31 million people of more than 40 ethnic groups, including the Annangs, Bini, Efik, Esan, Ibibio, Ekpeyes, Esan, Etches, Ibeno, Igbo, Okpe, Ikwerre, Agbo, Okwale, Ahoada, Ukwuani, Bonny, Okrika, Andoni, Annang, Oron, Ijaw, Itsekiri, Ndoni, Obolo, Yoruba, Isoko, Urhobo, Ukwuani, Kalabari, Eleme, Auchi, Benin, and Ogoni to mention a few, are among the inhabitants of the Niger Delta, with about 250 different dialects.

The Niger Delta region accounts for more than 23% of Nigeria's total population in 2005 (Twumasi & Merem, 2006; Uyigue & Agho, 2007). According to Awosika (1995), the Niger Delta is located in the Atlantic coast of Southern Nigeria, and it is the world's second largest delta, with a coastline of approximately 450 km, which ends at the Imo River entrance, which constitutes the eastern boundary. Anifowose (2008) stated that the region is approximately 20,000 km², and it is the largest wetland in Africa and the third largest in the world. Approximately 2370 km² of the Niger Delta area consists of rivers, creeks, and estuaries. Stagnant swamps cover approximately 8600 km²; the mangrove swamp spans approximately 1900 km², and it is the largest mangrove swamp in Africa (Awosika, 1995).

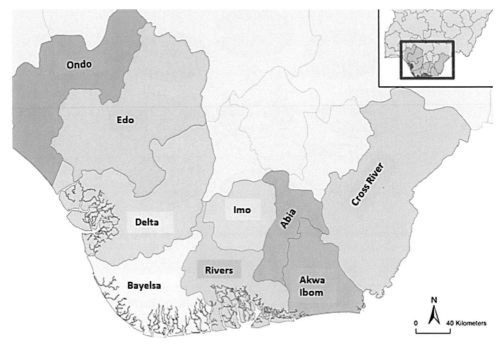

FIGURE 4.2   Map of the Niger Delta states of Nigeria.

The Niger Delta ecosystem has been reported to be one of the 10 most important wetland and coastal marine ecosystems of the world (Uluocha & Okeke, 2004) and one of the major biodiversity hotspots on the planet (Ebeku, 2004; Iyayi, 2004). The region is rich with diverse ecosystems comprising four ecological zones, including coastal ridge barrier islands, mangrove swamps, freshwater swamps, and lowland rainforests (Ogbe, 2005). Each of these ecologies provides habitation for different species of plants, fish, reptiles, mammals, and minerals.

4.10 ENVIRONMENTAL ISSUES OF THE NIGER DELTA

A tabular synopsis of the human, fauna, and flora of the region is given in Table 4.2. The Niger River alone is home to 36 families and nearly 250 species of fish, of which 20 are endemic.

4.11 IMPACTS OF THE OIL AND GAS INDUSTRY ON THE NIGER DELTA ECOSYSTEM

Prior to the discovery of crude oil in the Niger Delta, the inhabitants lived peacefully while carrying out their daily activities. The major occupations of the Niger Delta people

TABLE 4.2 Tabular Synopsis of the Human, Fauna, and Flora of the Niger Delta Region of Nigeria

Item	Number	Resident	
Ethnic Nationalities	≥140	Rivers, Bayelsa, Delta, Abia, Akwa Ibom, Cross River, Edo, Imo and Ondo States	Annangs, Bini, Efik, Esan, Ibibio, Ekpeyes, Esan, Etches, Ibeno, Igbo, Okpe, Ikwerre, Agbo, Okwale, Ahoada, Ukwuani, Bonny, Okrika, Andoni, Annang, Oron, Ijaw, Itsekiri, Ndoni, Obolo, Yoruba, Isoko, Urhobo, Ukwuani, Kalabari, Eleme, Auchi, Benin, and Ogoni
Fauna	≥96 Mammals 25 Birds 27 Reptiles ≥200 Fishes	Resident in the four main ecologies	Reptiles, chimpanzees, red colobus monkeys, hippopotamus, black-fronted duiker, Sclater's guenon
Flora	339 of forest and homestead garden plants species	In all the four ecologies	Mahogany (*Khaīya* sp.), red mangrove (*Rhizopara* sp.), abura (*Hallea ledermanmi*) iroko (*Milicia excelsa*), etc.
Rivers	•24		Benin, Brass, Escravos, Forcados, Ramos, Dodo, Pennington, Digatoro, Bengatorn, Kulama, fishtown, Nun, Okpare, Sangara, Andoni, Imo, Bonny, Orashi, etc.

FIGURE 4.3 Farm produce from the Niger Delta.

before crude oil discovery were farming and fishing (Fig. 4.3). The operations of the oil companies have greatly affected the people living in the host and nearby communities. Oil spills and gas flaring as a result of the activities of the oil companies contaminate water and food and cause ill-health, environmental degradation, loss of livelihood, and even displacement from their ancestral homes. Adelana, Adeosun, Adesina, and Ojuroye (2011) opined that the extent to which a family is affected when the income from farming or fishing is reduced due to the activities of oil companies is dependent on their livelihood structure. A family that is solely dependent on farming and fishing as a means of livelihood is likely to suffer severe socioeconomic shock and would be left with no other option than to seek alternative sources of livelihood to meet their needs.

FIGURE 4.4   Geophone.

These main resources of the Niger Delta that have suffered degradation include land, renewable resources, and humans. The major sources of degradation are discussed in the following subsections.

4.11.1 Vegetation Clearing

Clearing of vegetation is carried out to prepare seismic lines and oil and gas pipeline trenches. Clearing destroys vegetation such as rainforests and mangroves, which are rich in biodiversity. and also reduces the habitat area.

4.11.2 Seismic Operations

Seismic operations involve the use of sound waves to locate petroleum traps. The sound waves are created on the surface of the ocean or land by detonated explosive charge that penetrates subsurface rock layers. Geophysicists use the time taken for the explosive sound waves to be reflected back to determine the structure of the bedrock and whether they are the types of rock that could contain oil and gas. The reflected sounds are received at the surface by a geophone (onshore) (Fig. 4.4) or hydrophone (offshore). However, this massive dynamiting of geological structure produces a narcotic effect and mortality of fish and other fauna. On the other hand, the destabilization of sedimentary materials associated with dynamite shooting causes increment in turbidity, blockage of filter-feeding apparatuses in benthic fauna, reduction of photosynthetic activity due to reduced light penetration, et cetera.

4.11.3 Pipeline Construction

Pipeline construction (Fig. 4.5) involves the clearing of vegetation and excavation of soil up to 3−6 ft. This requires a right of way of about 20−30 ft. In the process, the vegetation

FIGURE 4.5   Pipeline construction.

and the habitat of animals are destroyed, and the animals are driven into new habitats. Arable lands for farming are also destroyed. Fig. 4.5 is an underground pipeline in Effurun, Delta State, running across a University environment, claiming vast land and thus infringing on the students' right of way.

4.11.4 Dredging

The process of dredging involves the excavation of large parcels of sand in the floor of rivers, lakes, swamp, sea, and land by lifting or sucking it up and dumping it on the adjacent land or water. The process is employed to reclaim swampland, get a portion of the sea turned into land, fill lakes, and solidify land for road construction (Fig. 4.6). It is also employed in the deepening of rivers so that large ships can access ports. In the process, sea grasses and marine animals living in the sea are totally eradicated. It has also been known to have caused death of large aquatic species such as turtle. Studies by Rim-Rukeh, Koyeme, Ikhifa, and Okokoyo (2007) and Ohimain, Imoobe, and Bawo (2008) reveal that dredging is responsible for the physicochemical changes in the water of the Niger Delta, particularly pH, total dissolved solids, and total suspended solids, conductivity, turbidity, sulfate, dissolved oxygen, and biological oxygen demand.

4.11.5 Industrial Emission

All industrial equipment emits gases that are flared to the environment (Fig. 4.7), contributing to the global warming and causing several health-related challenges.

Carbon dioxide: This gas causes the greenhouse effect and acid rain.

FIGURE 4.6   Dredging of oil-spilled land in a village in the Niger Delta.

FIGURE 4.7   Gas flaring in the Niger Delta.

Sources of Emission
1. *Direct emissions* are produced through:
    a. Burning fuels for power and heat
    b. Chemical reactions
    c. Leaks from industrial processes or equipment
2. *Indirect emissions* are produced by:
    a. Burning fossil fuels at a power plant to generate electricity for operations in oil and gas facilities

Both direct and indirect emissions make oil industry the second largest contributor of greenhouse gases to the environment.

4.11.6 Underground Water Contamination

The oil and gas activities introduce pollutants into groundwater through surface and subsurface sources such as produced water, fracture chemicals, sludge pit, and mud pit, among others. Whenever the contaminants seep into the formation, it alters the physical, thermal, chemical, and biological quality of groundwater. Humans and animals drink this water to satisfy their thirst and by so doing expose themselves to several diseases and sometimes death. This scenario is worsening in the oil-rich villages of the Niger Delta, where good drinking water sources are not readily available because of pollution. Inhabitants of these communities fetch water from the contaminated running streams for drinking. Fig. 4.8 shows contaminated water from an aquatic ecosystem in the Niger Delta. Table 4.3 shows the sources, sites, contaminants, and means of surface contamination into groundwater.

There are also subsurface entrance contaminants, which include:

1. **Injection Fluids**

   Injected fluids migrate through formation fractures, channeling, and casing leaks to introduce fracturing chemicals, water, natural gas, workover fluids, and enhance oil recovery (EOR) fluids into the groundwater layer.
2. **Oil and Gas Reservoir Fluids and Brine Formation Below Aquifer**

   The fluids from these two sources can migrate through formation fractures and channeling to introduce oil, natural gas, and brine into groundwater.
3. **Drilling Activities and Materials**

   Chemicals and heavy metals used in composing drilling mud such as water-based mud, oil-based mud, and synthetics drilling fluids have been known to pollute the groundwater. Furthermore, the mud filtrates containing additives such as ferrochrome lignosulfate (chromium pollution) and lead compounds (lead pollution) are of major concern. Drilling processes resulting in underground blow-out, fracture, circulation loss, filtrate loss, and excessive overbalance, which may lead to excessive pressurization of the annulus of the well, could cause filtrate migration. Oil mud contamination containing oxidized asphalt, organic acids, alkali, stabilizing agents and low toxic oils are major sources of groundwater contamination. Cuttings and blow-out discharges on the surface also percolate into groundwater. Table 4.4 illustrates the details of the subsurface contaminants.

FIGURE 4.8   Contaminated water body in a village in the Niger Delta.

TABLE 4.3   Means of Contaminating Surface Water in Oil-Producing Region

Source	Site	Means of entrance into groundwater	Contaminants
Produced water	Discharged on surface	Leaking well cover	Bacteria
		Defective well	Heavy metals
		Casing cement	Biological oxygen demand
		Infiltration	Chemical oxygen demand
Drilling chemicals	Discharged on surface in slush pits	Infiltration	Toxic materials
			Heavy metals
			Biological oxygen demand
			Chemical oxygen demand
Workover operations	Discharged on surface in slush pits	Infiltration	Toxic materials
			Heavy metals
			Biological oxygen demand
			Chemical oxygen demand
Cuttings and other solid waste surface disposal	Drilling site	Infiltration	Toxic materials
			Heavy metals
			Biological oxygen demand
			Chemical oxygen demand

4.12 EFFECTS OF OIL AND GAS ON THE PEOPLE AND THEIR HABITAT

Invariably, the ecosystem contamination and degradation get transferred to people living in the environment. The more than 140 ethnic nationalities living in the Niger Delta region have been known to suffer untold hardship due to loss of fishing areas, loss of thousands of coastal lands and farmlands, exposure to varying diseases, childbirth defects, and chronic illnesses. The major causes of these woes are attributed to extreme levels of pollution of arable land, drinking water, and water bodies by oil spills.

The effluents from oil and gas production have resulted in the ill health of the inhabitants of these operating areas, which have greatly affected their economic (farming and fishing) performance. This case is worsened by the absence of good health centers in the

TABLE 4.4 Subsurface Entrance Contaminants

Sources	Site	Means of entrance into groundwater	Contaminants
Well construction	Drilling fluids, chemicals, and equipment	Filtration Circulation loss Rust from equipment	Oxidized asphalt, organic acids, alkali, stabilizing agents, and low toxic oils. Bacteria, Heavy metals, Biological oxygen demand, Chemical oxygen demand, chromium and lead pollution
Drilling processes	Excessive overbalance	Formation fracture	
Well completions	Completion fluids	Casing leaks	Bacteria
		Defective casing shoe	Heavy metals
			Toxic materials
			Biological oxygen demand
		Channeling	Chemical oxygen demand
Workover operations	Workover fluids and chemicals	Formation fractures	
Injection wells	Injection fluids: water, natural gas, nitrogen, other EOR fluids	Casing leaks Defective casing shoe Channeling Formation fractures	Nitrogen and phosphorus cause tiny water algae to bloom, or grow rapidly. These require oxygen to decompose at death, resulting in water oxygen deficiency leading to many aquatic animals' death
Hydraulic fracturing	Injections Chemicals Diesel Solid Chemicals	Fractures High-pressure migration of water, sand, proppants, and chemicals during and after the fracturing	Toxic materials, cancer-causing agents Biological oxygen demand Chemical oxygen demand Degree of pollution depends on proximity to groundwater, formation permeability, and nature of chemical mix
Failed casing cementing and wellbore bottom seal	Drilling and producing wells	Channeling Infiltration Fractures	Toxic materials Biological oxygen demand Chemical oxygen demand
Oil spillage and blow-out emissions	Drilling, production, and transportation	Infiltration and runoff	Bacteria, biological oxygen demand, chemical oxygen demand Toxic materials

I. BACKGROUND INFORMATION ON PETROLEUM INDUSTRY ACTIVITIES AND THE NIGERIAN ENVIRONMENT

areas. To further aggravate issues, the existing centers are located far from many communities, and the roads leading to these health centers are in very poor state.

4.12.1 Damage to Farmlands

A report from an unpublished source shows that the majority of the farmland in the villages along the coastal region of Delta State has been ravaged by spilled oil due to the activities of the oil and gas industry. When such areas are cultivated, the output often has one form of deformity or the other. Personal discussions with locals reveal that if the rate of farmland destruction is not stopped, it is possible that in 10 years' time, there might not be farmland available for cultivation in these areas (Fig. 4.9).

4.12.2 Damage to Fishery and Wildlife

Fig. 4.10 shows the extent of damage done to aquatic ecosystems by oil spills in Niger Delta. The leftmost panel of the figure is a picture of two fishermen who set their traps (middle panel of the figure) in an oil-polluted river. The catches (rightmost panel of the figure) were coated with crude oil, and this might affect their market value because they constitute a health hazard to consumers.

FIGURE 4.9   Damaged farmlands in the Niger Delta region of Nigeria.

FIGURE 4.10   Damage to aquatic life by oil spillage in a village in Niger Delta.

FIGURE 4.11   Gas flaring.

4.12.3 Gas Flaring

Gas flaring is the process of venting unwanted natural gas and burning it (Fig. 4.11). The burning process produces carbon dioxide, carbon monoxide, and sometimes soot, depending on the efficiency of burning and the hydrocarbon composition of the natural gas. Aniefiok and Udo (2013) opined that poor efficiency in the flare systems often result in incomplete combustion, which produces a variety of volatile organic compounds, polycyclic aromatic hydrocarbons, and inorganic contaminants. Gas flaring causes thermal pollution; destruction of vegetation, soil, and crops; human illness; and discomfort due to fumes, among other catastrophies (Francis et al., 2012). Studies have revealed that gas flaring is a major source of CO_2 emission, which is one of the gases responsible for climate change.

In developed countries that have hydrocarbon in commercial quantities, gas flaring has been banned and offenders pay severe penalties, which has made flaring of associated and nonassociated natural gas a less attractive alternative for gas management. Thus, oil companies are frequently asked to devise a means to contain their excess gas production. Besides, the gas can be used as a secondary drive to enhance depleted oil reservoirs. So, the oil companies are forced to reinject the excess gas for production of their crude oil reserves as in the case of the Mobil Oso Gas condensate plant in Nigeria, which provides gas for reinjection into nearby oil reservoirs, supplies methane as fuel to power their gas turbines, and sends the remaining propane plus to their natural gas liquid plant at Bonny Island instead of flaring and wasting the gas.

4.12.4 Illegal Refinery Operation

The issue of oil theft and illegal refining of crude has become the order of the day in Niger Delta, especially amongst the youths. Due to a survival instinct triggered by deprivation and loss of livelihood, communities in this region have come to adopt unconventional technologies in refining stolen crude oil, which they sell. This activity is not safe, and it leads to such things as pollution, environmental degradation, and loss of revenue. Fig. 4.12 shows a fire outbreak from one of the local refineries in Niger Delta.

FIGURE 4.12   Fire outbreak from illegal refineries in Niger Delta.

FIGURE 4.13   Poor maintenance of wellhead.

4.12.5 Oil Spillages

One of the most devastating consequences of oil and gas activities on the Niger Delta ecosystem since the commencement of petroleum exploration and production in Nigeria has been the incidents of oil spill. Major sources of oil spillage in Niger Delta can be categorized into operational, sabotage, infrastructure decay, and human error.

4.12.5.1 Operational Causes

These causes include accidental discharges (tank accident), discharges from refineries, equipment failure, and poor maintenance of infrastructure (Fig. 4.13).

FIGURE 4.14 Sabotage of oil facility in the Niger Delta.

TABLE 4.5 Shell Petroleum Development Company Spill Statistics for January 2016 to June 2016

No.	Incident/facility	Date	Terrain	Causes	Spilled volume (barrel)
1	16″ Nun-River–Kolo Creek Pipeline at Aguobiri	04/Feb/2016	Swamp	Sabotage	3
2	28″ Bomu–Bonny Pipeline at K-Dere / Kpor	09/Jan/2016	Swamp	Sabotage	72
3	28″ Bomu–Bonny Pipeline at Opobo Riser Owokiri	08/Jan/2016	Swamp	Operational	0.1
4	16″ Nun-River–Kolo Creek Pipeline at Oporoma	07/Feb/2016	Swamp	Sabotage	2
5	Kolo Creek well 3 T Flowline at Otuasega	19/Jan/2016	Land	Sabotage	0.2
6	28″ Bomu–Bonny Pipeline at Bodo City	11/Feb/2016	Swamp	Sabotage	0.1
7	Ubie Well 5 S/L Flowline at Idu-Ekpeye	26/Jan/2016	Land	Sabotage	0.4
8	6″ Mininta–Ahia Bulk Line at Akpabu	09/Feb/2016	Land	Accidental 3rd Party Damage	1135
9	28″ Nkpoku–Bomu Pipeline at Kporghor / Gbam	18/Feb/2016	Land	Sabotage	24
10	6″ Seibou Bulkline-2 at Azagbene	23/Feb/2016	Swamp	Operational	50
11	28″ Bomu–Bonny Trans Niger Pipeline at Ayaminima-Oloma	16/Feb/2016	Swamp	Sabotage	35
12	20″ Opukushi–Brass Creek at Tamogbene	21/Apr/2016	Swamp	Sabotage	114
13	4″ Belema Well 1 L Flowline at Belema	24/Feb/2016	Swamp	Sabotage	9
14	Soku Well 21 L Wellhead at Soku	08/Mar/2016	Water	Sabotage	0.8
15	36″ Nkpoku–Bomu Trans Niger Pipeline at Rumuochiolu-Eneka	22/Mar/2016	Land	Sabotage	0.6
16	Etelebou Flow Station at Ogboloma	17/Mar/2016	Land	Operational	72

(Continued)

I. BACKGROUND INFORMATION ON PETROLEUM INDUSTRY ACTIVITIES AND THE NIGERIAN ENVIRONMENT

TABLE 4.5   (Continued)

No.	Incident/facility	Date	Terrain	Causes	Spilled volume (barrel)
17	Soku Well 21 S/L Wellhead Slot at Oluasiri	28/May/2014	Swamp	Sabotage	0.7
18	12″ Oguta–Egbema Pipeline at Eziorsu	26/Mar/2016	Land	Sabotage	83
19	Oguta Well 17 L Flowline at Umuorodogwum Oguta	31/Mar/2016	Land	Operational	0.6
20	16″ Nun-River–Kolo Creek Pipeline at Oporoma	10/Apr/2016	Swamp	Sabotage	280
21	20″ Kolo Creek–Rumuekpe Trans Niger Pipeline at Odau	06/Apr/2016	Land	Sabotage	348
22	Obigbo North Well 38 L Flowline at Umuebulu Obigbo	22/Apr/2016	Land	Operational	0.4
23	Ibaa Manifold 8″ Header at Ibaa	06/May/2016	Land	Operational	304
24	12″ Imo River–Ogale Pipeline at Owaza	17/May/2016	Land	Sabotage	330
25	Obigbo Well 28 at Imeh	18/May/2016	Land	Sabotage	220
26	16″ Egbema–Assa Pipeline at Ekpeagah	05/Jun/2016	Land	Sabotage	10
					3094.5

4.12.5.2 Sabotage

Sabotage involves the act of deliberate vandalization of pipelines for the purpose of oil theft for bunkering or getting the government's attention. The vandals tap the pipeline to extract the oil for sale. In the process, the pipeline is damaged or destroyed, and this can go unnoticed for days. If repair is not effected in good time, it could result in environmental degradation (Fig. 4.14).

The leftmost panel of the figure shows three lines transferring products that were vandalized by members of the communities, and the result is seen in the middle and rightmost panels of the figure. Table 4.5 is a list of notable spillages from Shell facilities between January 2016 and June 2016.

4.13 CONCLUSION

There is no doubt that the oil and gas industry has affected the Nigerian environment. The influence has been stories of woes and joy. The foregoing review points in one direction: an abundant natural resource that has been exploited at the expense of the people in the area where the resource is located. The issues that include utter disregard for the wellbeing of the people, corruption and squandering of the oil money, ecosystem degradation, and water and air pollution are glaring. Regrettably, until recently, the government of Nigeria has paid lip service to these issues. The result is militancy, kidnaping, and many other vices. However, government is working hard to reverse this ugly trend. An amnesty program was initiated to rehabilitate militants, and the outcome has been encouraging. Some agencies such as the Niger Delta Development Commission were set up to bring development to the oil-producing region. Oil companies are beginning to live up to their corporate social responsibilities, but a great deal still needs to be done to correct the errors of the past.

References

Adelana, S. O., Adeosun, T. A., Adesina, A. O., & Ojuroye, M. O. (2011). Environmental pollution and remediation: Challenges and management of oil Spillage in the Nigerian coastal areas. *American Journal of Scientific and Industrial Research*, 2(6), 834–845.

Aniefiok, E. I., & Udo, J. I. (2013). Gas flaring and venting associated with petroleum exploration and production in the Nigeria's Niger Delta. *American Journal of Environmental Protection*, 1(4), 70–77.

Anifowose, B. (2008). Assessing the impacts of oil and gas transport on Nigeria's environment. In *U21 Postgraduate Research Conference Proceedings 1*, University of Birmingham, UK.

Awosika, L. F. (1995). Impacts of global climate change and sea level rise on coastal resources and energy development in Nigeria. In J. C. Umolu (Ed.), Global climate change: Impact on energy development. Nigeria: DAMTECH Nigeria Limited.

Congressional Research Service (CRS) (2008): Report for Congress, Nigeria: Current Issues. Updated 30 January 2008.

Ebeku, K. S. A. (2004). Biodiversity conservation in Nigeria: An appraisal of the legal regime in relation to the Niger Delta area of the country. *Journal of Environmental Law*, 16, 361–375.

Francis, C. A., Ikemefuna, C. O., & Ekwoaba, J. O. (2012). Conflict and environmental challenges facing the oil companies in Nigeria. *International Journal of Business and Management Tomorrow*, 2(3), 2–8.

Ivanhoe, L. F., & Leckie, G. G. (1993). Global oil, gas fields, sizes tallied, analyzed. *Oil and Gas Journal*, 15, 87–91.

Iyayi, F. (2004). An integrated approach to development in the Niger Delta. In *A paper prepared for the Centre for Democracy and Development (CDD)*.

Li, G. (2011). *World atlas of oil and gas basins* (p. 20). Oxford: Wiley-Blackwell.

Ogbe, M.O. (2005). Biological resource conservation: A major tool for poverty reduction in Delta State. In *A key note address delivered at the Delta State maiden Council meeting on Environment with all stakeholders held at Nelrose Hotel, Delta State, Nigeria*, 28 and 29 September, 2005, 16 p.

Ohimain, E. I., Imoobe, T. O. T., & Bawo, D. D. S. (2008). Changes in water physico-chemical properties following the dredging of an oil well access canal in the Niger Delta. *World Journal of Agricultural Sciences*, 4, 752–758.

Okotie, S., & Onyekonwu, M.O. (2015). Software for reservoir performance prediction. In: *Nigerian Society of Petroleum Engineer Annual International Conference and Exhibition, 4-6 August, Lagos, Nigeria*. https://doi.org/10.2118/178288-MS.

Onuoha, F. C. (2008). Oil pipeline sabotage in Nigeria: Dimensions, actors and implications for national security L/C. *African Security Review*, 17(3), 99–115.

Petrobjects. (2003). *Petroleum reserve estimation methods*. www.petrobject.com.

Rim-Rukeh, A., Koyeme, J., Ikhifa, G. O., & Okokoyo, P. A. (2007). Chemistry of harvested rainwater in the refinery area of Warri, Nigeria between November, 2003 and October, 2005. *International Journal of Chemistry*, 16(2), 65–74.

Twumasi, Y., & Merem, E. (2006). GIS and remote sensing applications in the assessment of change within a coastal environment in the Niger Delta Region of Nigeria. *International Journal of Environmental Research & Public Health*, 3(1), 98–106.

Uluocha, N. O., & Okeke, I. C. (2004). Implications of wetlands degradation for water resources management: Lessons from Nigeria. *GeoJournal*, 61(2), 151–154.

Uyigue, E., & Agho, M. (2007). Coping with climate change and environmental degradation in the Niger Delta of Southern Nigeria. Community Research and Development Centre Nigeria (CREDC). http://priceofoil.org/content/uploads/2007/06/07.06.11%20-%20Climate_Niger_Delta.pdf

I. BACKGROUND INFORMATION ON PETROLEUM INDUSTRY ACTIVITIES AND THE NIGERIAN ENVIRONMENT

The Nigerian Economy Before the Discovery of Crude Oil

Sylvester Okotie

Federal University of Petroleum Resources Effurun, Delta State, Nigeria

5.1 INTRODUCTION

The socioeconomic development of a nation depends on the value of all goods and services produced within that nation. It is therefore the primary role of the government of every nation to maintain economic growth and stability via strong microeconomic policies and a stable political institution, among others. The government uses the gross domestic product (GDP) to evaluate the economic production or output, which implies that a nation's economy is sustained and raised by active and adequate utilization of its human and natural resources.

Over the years, a school of thought has reasoned that oil has been a blessing to Nigeria since its discovery in commercial quantity in 1956 by Shell D'Arcy in a village called Oloiribi in present day Bayelsa State. Another school of thought holds the view that crude oil discovery has been nothing but a curse to the country. As a result of the production of oil, which started far back in 1958, the economy of Nigeria has experienced remarkable and tremendous growth. Petroleum currently accounts for about 95% of Nigeria's foreign exchange earnings. This has been the view of the school of thought that says crude oil is a blessing to Nigeria, and they opine that the growth of agriculture in the 1900s was unsatisfactory. According to Rilwan (2010), the proceeds from the sale of oil have considerably sustained Nigeria's economy. He asserted that the advent of petroleum has improved the economy of Nigeria as a result of deployment of oil money into several projects across the nation. It has had and it is still having its negative effects on the environment and man. These negative effects are exacerbated by poor management and negligence of other sources of income such as agriculture, manufacturing, and others.

5.2 THE ECONOMY OF NIGERIA PRIOR TO THE OIL BOOM ERA (AGRARIAN ECONOMY)

Prior to oil discovery in 1956, agriculture was the mainstay of the Nigerian economy, which made her famous across the globe. Her major exports are cash crops such as rubber from Delta State in south-south region; groundnut, hide, and skin produced by the northern region; cocoa and coffee from the western region; and palm oil and kernels from the eastern region of the country. According to the National Bureau of Statistics (2010), agriculture provided employment for about 30% of the populace. Chigbu (2005) stated that there is a need to overhaul and service the agricultural sector, which is the powerhouse of the world economy. He also said that agriculture provided more than 80% of Nigerian export earnings in the 1960s, 65% of the total output of the GDP generated by employment, and approximately 50% of government revenue. The Central Bank of Nigeria brief (2010) reported that agriculture only accounts for 34.6% of the GDP presently compared to its leading contribution of about 50% of the government revenue in the past. Olomola (2007) stated that the agricultural sector has been transformed by commercialization at the small-, medium-, and large-scale enterprise levels. Besides, agriculture was the bedrock of Nigeria's economic development but today, development in that sector has been truncated, which led Arigbabu (2013) to state that Nigeria's agriculture has fizzled out to the level that the nation is now a net importer of food where it was once known for its large exportation of agricultural produces at the international market.

According to the United States Department of State (2005), Nigeria is no longer a major exporter of cocoa, groundnut, rubber, and palm products. Cocoa production, mostly from obsolete varieties and over-aged trees, is stagnant at around 150,000 tonnes annually. Twenty-five years ago, cocoa production in the country was about 300,000 tonnes. Sekumade (2009) also stated that there has been a similar decline in groundnut, palm oil, and the other major export crops. The share of agricultural products in total exports has plummeted from over 70% in 1960 to less than 2%. The decline was largely due to the phenomenal rise of oil shipments, but also reflected the fall in the output of products such as cocoa, palm oil, rubber, and groundnuts, of which Nigeria was once a leading world producer.

It is worthy to note that the oil boom has blinded the eyes of successive governments from agriculture from 1960 to date and as such, the agricultural sector has only been sustained at the peasant level because of the huge revenue derived from sale of crude oil. Nigeria had been a fertile ground for farming before the emergence of oil flow stations in the Niger Delta region. The people of this region solely depended on farming and fishing as their major sources of livelihood. However, because the environment is no longer conducive for agricultural activities, which has resulted in continuous importation of food items such as sugar, rice, and wheat, the nation's economy continued to nose-dive, causing severe hardship on the citizens.

Furthermore, Nigeria has attracted investments from different parts of the world due to the presence of agricultural produce found in different parts of the country. These produces are beans, cocoa, coffee, cassava, yam, cashew nuts, groundnuts, kolanut, melon,

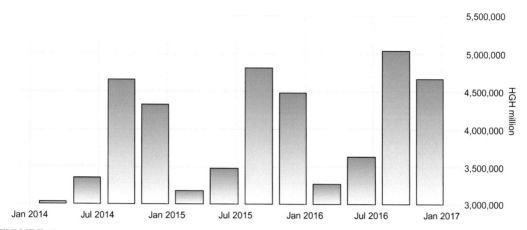

FIGURE 5.1   Nigerian gross domestic product from agriculture. *From National Bureau of Statistics. (2010). Labour force statistics. National Bureau of Statistics. Retrieved June 22, 2015 (National Bureau of Statistics, 2010).*

palm oil, rubber, and rice, among others. In addition, livestock production is also practiced in Nigeria. The data of the Nigerian GDP from agriculture reported by the National Bureau of Statistics showed that the GDP in the fourth quarter of 2015 was 4,481,257.62 million NGN and decreased to 3,274,725.01 million NGN in the first quarter of 2016. Fig. 5.1 shows the trend of the Nigerian GDP from agriculture for the period January 2014 to January 2017.

The report also included GDPs from other aspects of the Nigerian economy aside from the oil and gas sector. It showed that agriculture was the second largest, surpassed only by services. Furthermore, the value added to the GDP from agriculture from 1981 to 2015 according to the World Bank national accounts data and OECD (Organisation for Economic Co-operation and Development) national accounts data files are presented in Table 5.1. The pictorial representation showing the variation of the GDP over the years is shown in Fig. 5.2 below.

In a bid to stop total dependence on oil and gas and diversify the economy, several agricultural programs were established by various governments from the 1970s to date. In the late 1970s, the administration of General Olusegun Obasanjo floated the "Operation Feed the Nation", while in the early 1980s, the government of Alhaji Shehu Shagari established the "Green Revolution". In the regime of the immediate past president, Goodluck Jonathan, another program called "the Agriculture Transformation Agenda" was launched, which had significant impacts in the agricultural sector. The adminstration of the current president, Muhammadu Buhari, has also introduced several agricultural programs to help boost the economy and stop the over-reliance on crude oil. The current administration is making efforts to stop importation of food items into the country by attracting foreign investments into the Nigerian agricultural sector.

TABLE 5.1   Value-Addition From Agriculture (% of GDP)

Year	%GDP	Year	%GDP
1981	28.518	1999	35.306
1982	32.414	2000	26.034
1983	35.470	2001	33.754
1984	39.917	2002	48.566
1985	39.207	2003	42.707
1986	40.331	2004	34.210
1987	37.258	2005	32.755
1988	41.647	2006	31.999
1989	32.156	2007	32.714
1990	31.525	2008	32.850
1991	31.224	2009	37.050
1992	27.267	2010	23.894
1993	33.900	2011	22.289
1994	38.811	2012	22.054
1995	32.061	2013	20.996
1996	31.134	2014	20.236
1997	34.031	2015	20.858
1998	39.048		

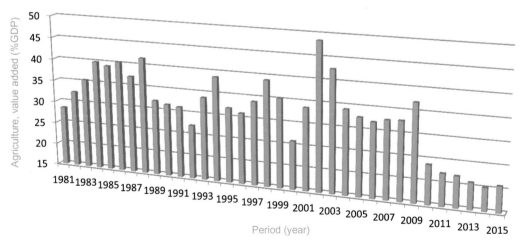

FIGURE 5.2   Distribution of value-addition by agriculture to GDP. *From World Bank and Organisation for Economic Co-operation and Development (OECD). (2016). Agriculture value added "% of GDP". In* World Bank national accounts data, and OECD National Accounts data files. *Available at: <data.worldbank.org> (World Bank and Organisation for Economic Co-operation and Development (OECD), 2016).*

5.3 THE ROLE OF AGRICULTURE IN THE NIGERIAN ECONOMY

The role of agriculture in the country cannot be overemphasized, especially now that the country is in recession due to a fall in the price of oil globally. Agriculture was the mainstay of the Nigerian economy, making her self-sufficient in food production prior to the discovery of petroleum. Agriculture was Nigeria's highest earner of foreign exchange, but the discovery of crude oil changed the situation and Nigeria now imports foods that would have been produced in various parts of the country. In a statement made by the current Governor of Kogi State, Yahaya Bello, through his deputy, Mr Simon Achuba, he opined that revenue generated from agriculture could replace revenue from oil and gas within a decade or fund the gap created by the dwindling fortunes of crude oil and its accompanying economic challenges only if the government and citizens of Nigeria take agriculture seriously. The crude oil industry is not only enticing to young graduates because of the high remuneration packages offered by these companies, but it has also done incalculable damage to the environment of the host communities. Some communities in the Niger Delta have been destroyed by oil and gas exploration and production activities to such an extent that the lands can no longer support crop farming. Therefore, it is believed that if Nigeria must diversify her economy and thus, reduce over-reliance on oil and gas, especially at this time of recession, the following roles of agriculture in Nigeria should be considered:

- Development of an agro-based industry in the country that will supply the raw material needs of these companies and by so doing boost the output of the existing firms and also attract direct foreign investments into the country.
- Other sectors of the economy will be improved as a result of the development of the agricultural sector, thereby reducing the level of poverty and stabilizing the economy of the country.
- Establishment of functional agricultural farms requires a workforce. This implies that the agricultural sector is capable of reducing the level of unemployment ravaging the country and prevents or limits rural-urban migration. Recently, it has been observed that young unemployed graduates venture into fish farming, creating employment for themselves and reducing the national unemployment rate. It is worthy to note that if people can be encouraged to go into fish farming and other forms of agriculture, the economy of Nigeria will improve immensely as jobs would be created.
- Agriculture is the second highest foreign exchange earner for the country after crude oil (Fig. 5.3), which implies that the agricultural sector is capable of attracting foreign investors and thus strengthens the Nigerian currency (Naira) against other major currencies in the world.
- There has been the development of a continuous monitoring and evaluation scheme for all established agricultural programs in the country.

Agricultural products of Nigeria can be divided into two main groups: food crops, produced for home consumption, and export products, which generate revenue for the farmer. Prior to the Nigerian civil war, the country was self-sufficient in food but importation of food increased substantially after 1973. Bread, made primarily from US wheat, replaced

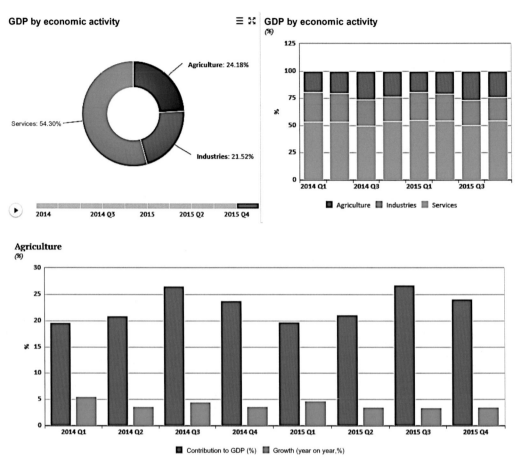

FIGURE 5.3   Contribution of agriculture to the national GDP. *From National Bureau of Statistics. (2010).* Labour force statistics. *National Bureau of Statistics. Retrieved June 22, 2015.*

domestic crops as the cheapest staple food for much of the urban population. The most important food crops are yams and manioc (cassava) in the south and sorghum (guinea corn) and millet in the north. In 1999, production of yams was 25.1 million tons (67% of world production); manioc, 33.1 million tons (highest in the world and 20% of global production); cocoyams (taro), 3.3 million tons, and sweet potatoes, 1.56 million tons. The 1999 production estimates for major crops were as follows (in thousands of tons): sorghum, 8443; millet, 5457; corn, 5777; rice, 3399; peanuts, 2783; palm oil, 842; sugar cane, 675; palm kernel, 565; soybeans, 405; and cotton lint, 57. Many fruits and vegetables are also grown by Nigerian farmers.

Although cocoa is the leading non-oil foreign exchange earner, growth in the subsector has been slow since the abolition of the Nigerian Cocoa Board. The dominance of small holders in the cocoa subsector and the lack of farm labor due to urbanization hold back production. Nigeria has the potential to produce over 300,000 tons of cocoa beans per year, but production only amounted to 145,000 tons in 1999. Rubber is the second largest non-oil foreign exchange earner. Despite favorable prices, production has fallen from 155,000 tons in 1991 to 90,000 tons in 1999. Low yield, aging trees, and lack of proper equipment have inhibited production.

Agricultural exports (including manufactured food and agricultural products) decreased in quantity after 1970, partly because of the discouraging effects of low world prices. In 1979, the importation of many foods was banned, including fresh milk, vegetables, roots and tubers, fruits, and poultry. The exportation of milk, sugar, flour, and hides and skins was also banned. During the years 1985–87, importation of wheat, corn, rice, and vegetable oil was banned as declining income from oil encouraged greater attention to the agricultural sector. In 1986, government marketing boards were closed down and a free market in all agricultural products was established. In 2001, agricultural exports totaled $323.5 million. Exports of cocoa beans that year totaled $210.4 million; cotton lint, $21 million.

5.4 CONTRIBUTION OF OIL TO THE NIGERIAN ECONOMY

As highlighted above, Nigeria's economy was sorely dependent on agriculture before the discovery by Shell D'Arcy of oil in commercial quantity at Oloibiri in 1956 in the Niger Delta region of the country. Two years later, crude oil production in the country started from the same oil field with about 5100 barrels per day, making Nigeria an oil-producing nation. Today, Nigeria is ranked 10th in the world in crude oil reserves. Thus, since the discovery of crude oil in commercial quantity, the economic structure of Nigeria has changed considerably. In the oil boom era (1970s), Nigeria enjoyed huge revenue from oil production due to the rise in world oil prices, making her the wealthiest African nation.

During the oil boom era, the revenues generated from oil sales were channeled into several capital projects leading to massive employment and manpower development in the country. During this period, the government and people of Nigeria focused solely on oil, which culminated in the neglect of the agriculture and manufacturing sectors of the economy. In 1976, there was a rapid fall in the prices of the crude oil, which was due to the oil price shock, leading to a reduction in oil revenue of all the oil-exporting countries, including Nigeria (Fig. 5.4). Prior to that time, embezzlement of project funds and reckless spending was the order of the day. These corrupt actions led to the abandonment of some projects embarked upon during the oil boom period, while those that were completed could not be maintained. Table 5.2 shows the revenue generated from crude oil since Nigeria started exporting petroleum to other countries in the world. Fig. 5.4 indicates that the revenues generated over the years from crude oil are not constant. Between 1980 and 1981, there was a drastic drop of revenue, from 82.30% to 31.22%.

The data in Table 5.2 shows that the socio-economic development of Nigeria has been on a steady increase. Presently, more than 90% of Nigeria's GDP comes from revenues accruing from crude oil sales. As stated by Francis, Ikemefuna, and Ekwoaba (2012), the Niger Delta is a region that is economically strategic to Nigeria as a nation. The reason is because oil contributes more than 90% of aggregate foreign exchange earnings for Nigeria and 80% of federal government revenue. Odularu (2008) also reported that in the year

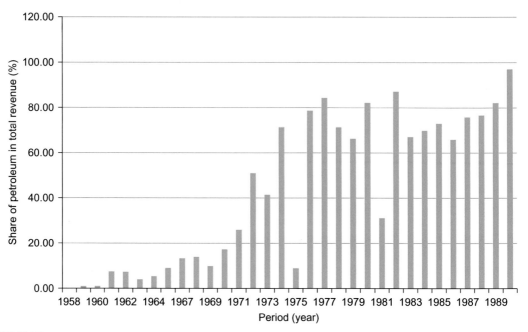

FIGURE 5.4 Contribution of the petroleum sector to the economy of Nigeria. *From Nigerian National Petroleum Corporation (NNPC). (1992). Contribution of federal ministry government and oil companies to oil producing areas: Vol. 1, Lagos (NNPC, 1992).*

TABLE 5.2   Contribution of Petroleum Sector to Nigerian Economy

Fiscal year	Fed. Govt. total revenue (US$)	Revenue from petroleum (US$)	Share of petroleum in total revenue (%)	Fiscal year	Fed. Govt. total revenue (US$)	Revenue from petroleum (US$)	Share of petroleum in total revenue (%)
1958	154,632	122	0.08	1975	5,177,370	463,816	8.96
1959	177,648	1776	1.00	1976	5,861,600	4,611,700	78.68
1960	223,700	2452	1.10	1977	7,070,400	5,965,500	84.37
1961	228,962	17,070	7.46	1978	8358400	5,965,500	71.37
1962	231,638	16936	7.31	1979	7252400	4,809,200	66.31
1963	249,152	10,060	4.04	1980	12,273,400	10,100,400	82.30
1964	299,132	16,084	5.38	1981	15,813,100	4,936,900	31.22
1966	321,870	29,175	9.06	1982	10,143,900	8,847,800	87.22
1967	339,196	44,976	13.26	1983	10,811,400	7,253,000	67.09
1968	300,176	41,884	13.95	1984	11,738,500	8,209,700	69.94
1969	299,986	29,582	9.86	1985	15,041,800	10,975,100	72.96
1970	435,908	75,444	17.31	1986	12,302,000	8,107,300	65.90
1971	755,605	196,390	25.99	1987	25,099,800	19,027,000	75.81
1972	1,410,811	720,185	51.05	1988	27,310,800	20,933,800	76.65
1973	1,389,911	576,151	41.45	1989	50,272,100	41,334,400	82.22
1974	2,171,370	1,549,383	71.36	1990	4,765,700	4,624,400	97.04

TABLE 5.3   Contributions of Agriculture, Oil, Service and Manufacturing to Nigeria's GDP

Year	GDP (N'm)	Agric (N'm)	Oil (N'm)	Service (N'm)	Manufacturing (N'm)
1960	2233	1417.6	7	303	108
1961	2361.2	1456.6	21.2	340.2	122.8
1962	2597.6	1605.8	29	365.6	146.4
1963	2755.8	1673.8	28.8	389.8	173
1964	2894.4	1676.4	42.2	453.6	173.6
1965	3110	1691.6	106.8	476.4	214.6
1966	2374.8	1855	129	508	233
1967	2752.2	1527.8	71.8	415.6	194.2
1968	2656.2	1415.2	43	497.2	198.6
1969	3549.3	1711.7	230.5	642.6	281.8
1970	5281.1	2576.4	489.6	851.9	378.4
1971	6650.9	3033.7	944.2	979.2	415.8
1972	7187.5	3092.7	1144	1031.3	511.1
1973	8630.5	3261.2	1899.2	1252.2	622.4
1974	18,823.1	4377.9	4108.7	2782.33	1589.02
1975	21,475.24	5872.92	4165.5	3619.77	1170.44
1976	26,655.78	6121.96	6105.91	4164.6	1464.3
1977	31,520.34	7041.64	7071.6	4755.61	1695.58
1978	34,540.1	8033.55	7539.39	5105.54	2915.82
1979	41,974.7	9213.14	10687.66	5478.28	3815.57
1980	49,632.32	10,011.46	14,137.35	6157.84	5162.21
1981	47,619.66	13,580.32	10219.8	9005.04	4699.95
1982	49,069.29	15,905.5	8512.94	9633.17	5047.61
1983	53,107.38	18,837.19	7388.73	10,109.16	5542.96
1984	59,622.53	23,799.43	9037.44	10,849.45	4847.51
1985	67,908.55	26,625.21	11375.15	12,338.3	6422.64
1986	69,146.99	27,887.19	9558.86	13,455.84	6591.12
1987	105,222.8	39,204.22	26,722.84	14,550.52	7468.45
1988	139,085.3	57,924.38	29,859.19	16,745.33	11,017.78
1989	216,797.5	69,713	76,530.31	21,656.53	12,475.51

(*Continued*)

I. BACKGROUND INFORMATION ON PETROLEUM INDUSTRY ACTIVITIES AND THE NIGERIAN
ENVIRONMENT

TABLE 5.3   (Continued)

Year	GDP (N'm)	Agric (N'm)	Oil (N'm)	Service (N'm)	Manufacturing (N'm)
1990	267,550	84,344.61	100,223.4	27,425.6	14,702.4
1991	312,139.7	97,464.06	116,525.8	31,355.45	19,356
1992	532,613.8	145,225.3	246,828	44,227.32	27,004.01
1993	683,869.8	231,832.7	242,109.7	60,863.26	38,987.14
1994	899,863.2	349,244.9	219,109.3	98,336.16	62,897.69
1995	1,933,212	619,806.8	766,518	151,822.9	105,289.6
1996	2,702,719	841,457.1	1,157,911	194,941.2	132,897.7
1997	2,801,973	953549.4	1,068,979	221,391.9	144,107
1998	2,708,431	1,057,584	736,795.3	299,450.1	141,496.4
1999	3,194,015	1,127,693	1,024,464	373,576.2	150,946.5
2000	4,582,127	1,192,910	2,186,682	471,814.6	168,037
2001	4,725,086	1,594,896	1,669,001	572,666.2	199,079.3
2002	6,912,381	3,357,063	1,798,823	692,179.5	236,825.5
2003	8,487,032	3,624,579	2,741,554	843,690.5	287,739.4
2004	11,411,067	3,903,759	4,247,716	124,672.7	349,316.3
2005	14,572,239	4,773,758	5,664,883	1,620,112	412,706.6
2006	18,564,995	5,940,237	6,982,935	2,143,487	478,524.1
2007	20,657,318	6,757,868	7,533,043	2,502,831	520,883
2008	24,296,329	7,981,397	9,097,751	2,785,655	585,573
2009	24,794,239	9,186,306	7,418,149	3,106,820	612,308.9
2010	29,205,783	10,273,652	9,747,355	3,430,112	647,822.8

Source: Central Bank of Nigeria Statistical Bulletin. (2010). Volume 21 (Central Bank of Nigeria Statistical Bulletin, 2010).

2000, crude oil and gas accounted for more than 98% of Nigeria's export and about 83% of federal government revenue. Nigeria's proven oil reserves are estimated to be 35 billion barrels, natural gas reserves are 1000 trillion feet and her crude oil production is around 2.2 million barrels per day. The mono-economic system of Nigeria has had significant impacts on all facets of her national life. This is even more pronounced in recent times due to a drastic drop in global oil prices. This shortfall in earnings has affected the government and populace in terms of balance of payment, rising external debts, and debt servicing burden, as well as the inability of the country to import crucial capital and intermediate goods to execute her development projects.

Table 5.3 shows the contribution of agriculture, oil, service, and manufacturing sectors to Nigeria GDP from 1960 to 2010 as extracted from the Central Bank of Nigeria (CBN) statistical bulletin. It indicates that the revenue generated from agriculture contributes significantly to the national GDP. It was the highest source of foreign exchange earnings before the oil boom era and has competed favorably over the years with the revenue from petroleum.

5.5 CONCLUSION

The growth and stability of any national economy greatly depends on the GDP of that nation. Nigeria is blessed with vast land and water for agricultural activities and also huge oil and gas reserves. Therefore, Nigeria should be one of the richest and most industrialized nations in the world. However, a combination of factors, including corruption, reliance on crude oil as the major source of revenue, and neglect of agriculture have left the majority of her populace in extreme poverty. There were initiatives by various governments after independence in 1960 to sustain and improve the agricultural sector; however, these initiatives were short-lived because of corruption and lack of political will to effectively executive the projects. It is possible to return to the pre-oil boom era where agriculture was the mainstay of the economy by embarking on diversification of the economy and encouraging the teaching of agriculture in schools at all levels. Also, farmers should be given incentives to make their profession attractive.

References

Arigbabu, S. (2013). *Oral history*. Port Harcourt, Nigeria.

Central Bank of Nigeria Statistical Bulletin. (2010). Volume 21.

Chigbu, U. E. (2005). The case of agriculture as the only saviour to the Nigeria's dying economy. Retrieved July 25, 2011 from <http://www.nigerianwillagesquare.com/articles/guest/2005/03/agric-as-onlv-saviour-to-html>.

Francis, C. A., Ikemefuna, C. O., & Ekwoaba, J. O. (2012). Conflict and environmental challenges facing the oil companies in Nigeria Niger-Delta region. *International Journal of Business and Management Tomorrow*, 2(3), 2−8.

National Bureau of Statistics. (2010). *Labour force statistics*. National Bureau of Statistics. Retrieved June 22, 2015 Abuja, Nigeria. Available at: www.nigerianstat.gov.ng/pages/download.

Nigeria National Petroleum Corporation (NNPC). (1992). *Contribution of federal ministry government and oil companies to oil producing areas: Vol. 1*, Lagos, Nigeria.

Odularu, G. O. (2008). Crude oil and Nigerian economic performance. *Oil and Gas Business Journal*, Ufa State Petroleum, Russia. Available at: <http://www.ogbus.ru/eng/authors/odularo/odularo_1.pdf>.

Olomola, A. S. (2007). Strategies for managing the opportunities and challenges of the current agricultural commodity booms in SSA. In *Seminar papers on managing commodity booms in sub-saharan Africa: A publication of the AERC senior policy seminar IX*. African Economic Research Consortium (AERC), Nairobi, Kenya.

Rilwan, L. (2010). *Oral history*. Abuja, Nigeria.

Sekumade, A. B. (2009). The effects of petroleum dependency on agricultural trade in Nigeria: An error correlation modeling (ECM) approach. *Scientific Research and Essay*, 4(11), 1385−1391.

United States Department of State. (2005). *The political economy of income distribution in Nigeria*. Available at: <www.dailyglobalwatch.com>.

World Bank and Organisation for Economic Co-operation and Development (OECD). (2016). Agriculture value added "% of GDP". In *World Bank national accounts data, and OECD National Accounts data files*. Available at: <data.worldbank.org>.

I. BACKGROUND INFORMATION ON PETROLEUM INDUSTRY ACTIVITIES AND THE NIGERIAN ENVIRONMENT

The Oil Boom Era: Socio-Political and Economic Consequences

Mgbeodichinma Eucharia Onuoha[1] and Isa Olalekan Elegbede[2]

[1]Technical University Bergakademie Freiberg Saxony, Freiberg, Germany [2]Brandenburg University of Technology, Cottbus-Senftenberg, Germany

6.1 INTRODUCTION

Commodity costs have for some time been remarkable for their unpredictability, making exceptional economic conditions in major producing nations. One regularly contemplated relationship in such nations is the impact of natural resource booms on non-resource economic activity (Brock, 2014). Nigeria, one of the largest oil-exporting countries in Africa, has a rapidly growing economy; the nation takes after a resource-based development technique driven by the production and exportation of oil. With the instability of worldwide oil prices and frequently unpredictable development of Nigeria's economy, this chapter is intended to inspect the harm of Nigeria's reliance on crude oil and its impacts on economic development. Confronted with higher, quick, and foreseen revenue after 1973, the Nigerian government had a scope of decisions: It could expend the present expansion at a later time, or spread the expanded utilization after some time. In the event that wealth was to be put aside for future utilization, this could be accomplished through expanded interest in foreign assets or local physical capital. The choices on these issues depend on communications between the social rate of time inclination, the relative rates of return amongst physical and money-related resources, and assumptions with respect to oil prices and, thus, future incomes (Pinto, 1987). There is no arguing the fact that Nigeria is the giant of Africa, the rallying point of African politics and economy, the most populous black country on the planet, and she is blessed with various natural resources such as gas, coal, gold, crude oil, etc. In reality, oil exploration and production in Nigeria have taken aggregate control of the economy and it is absolutely impossible to talk about the economy of the nation without offering a plan of action for crude oil. There is no chatting of the post-autonomy legislative issues of the nation without specific mention of crude oil. There is no discussing governmental issue of the nation without recognizing the part of oil. At the end of the day, oil is the typical equator or denominator of any talk in Nigeria and has in this way occupied an overwhelming part in both economic and legislative issues of the nation; to put it plainly, the political economy of Nigeria revolves around crude oil (Akinyetun, 2016).

For the past 50 years, oil has maintained a very crucial role in the Nigerian economy. The substantial oil production in Nigeria in the mid-1970s elevated the status of the nation to one of the prominent 10 oil exporters in the world. Since then, Nigeria has been fully involved in the boom cycles of the international oil market. However, several years of oil money have not emancipated the populace from penury nor liberated the economy from what appears to be continual redundancy in the nonoil economy. Is this record the inevitable result of the supposed resource reviles, or have misinformed strategies added to moderate development? The commensurate response has a direction on current policies: Nigeria has reformed its macroeconomic policies recently and has progressively witness quite appreciable economic performance. Traditional clarifications of poor performance in oil-rich nations prompted the supposed Dutch disease, named after Holland's poor record in dealing with its gaseous petrol riches in the 1960s. Spending out of oil riches increases interest for the non-tradable, thus attracting appreciable assets into that sector (Budina, Pang, & van Wijnbergen, 2007).

6.2 NIGERIA AND HER CRUDE OIL PERFORMANCE

Nigeria can be classified as a nation that is fundamentally rustic and that relies on essential item trades, particularly oil items. Since Nigeria's independence in 1960, she has encountered a series of religions and ethnic crises amplified by the huge inconsistencies in educational, economical, and ecological progress in the north and south. These phenomena could not be completely attributed to the actual discovery of oil in the nation, which influences and is influenced by social and economic components. The discovery of crude oil has affected the Nigerian economy both emphatically and unfavorably (Odularu, 2008). Dependence on oil and the charm it produced in relation to extraordinary riches through government contracts also generated economic imbroglio. The nation's high affinity to import indicates that approximately 80% of government consumptions are reinvested into foreign trade. Modest purchaser imports coming about because of a constantly hyperbolic Naira, coupled with extreme domestic production costs due largely to sporadic power and fuel supply, have declined modern limit usage to fewer than 30%. More plants in Nigeria would have come up from moderately low work costs (10%−15%). Domestic producers of materials and pharmaceuticals have lost their contending strength in conventional local markets. However, there are indicators that a few makers have begun to address their grievances. Following the discovery of crude oil by Shell D'Arcy Petroleum, initial production commenced in 1958 from the organization's oil field in Oloibiri in the Eastern Niger Delta. As at the late 1960s and mid-1970s, Nigeria had successfully recorded a daily production level of more than 2 million barrels of crude oil. In spite of the fact that production figures dropped in the 1980s as a result of economic droop, the oil production level sprouted up again in 2004 to a remarkable daily level of 2.5 million barrels (Odularu, 2008).

Advancement in procedures has augmented production to approximately 4 million barrels per day in 2010. The importance of oil export and production to the Nigerian economy cannot be jettisoned. This is because it accounts for approximately 90% of her total income. However, this overwhelming crude oil business has relegated agribusiness, a former backbone of the economy until the mid-1960s to the background. As much as the discovery of crude oil in the Niger Delta is a positive sign in the country, it often flagged a peril of grave outcome; oil incomes lubricated existing ethnic and political pressure and really "copied" the nation. This pressure attained its climax with the civil war that incapacitated the nation between 1967 and 1970. Oil production by joint venture (JV) partners represents approximately 95% of petroleum production in Nigeria. Shell drills approximately 50% of Nigeria's crude oil and operates the biggest JV partnership in Nigeria, where government holds 55% share through the Nigerian National Petroleum Corporation (NNPC). The other JV partnership where NNPC holds 60% share involves Chevron Texaco, Exxon Mobil, Total, and ENI/Agip.

Crude oil production in Nigeria is characterized by moderately little fields and many wells, with each well delivering in the vicinity of 50 and 50,000 barrels of oil per day of which roughly 65% of the oil delivered is light, sweet crude. In December 1992, the Mobil OSO field came on-stream with a production of more than one hundred thousand barrels per day, which actually was not counted toward Nigeria's OPEC quantity. Shell is in charge of the greater part of the oil produced in Nigeria (from its approximately 1000 different fields). Before the allocation of recent oil wells, the Shell JV had concessions covering 26,000 km^2 in onshore and swampy regions and 4700 km^2 in its offshore regions. The oil stock of this venture remains at approximately 11 billion barrels, while productivity capacity is more than 1.05 million barrels per day. The Chevron and Mobil JV takes second position, with the Mobil JV producing offshore from Eket in Akwa Ibom state, while Chevron Nigeria Limited is also offshore but with its operational base in Delta state.

Obviously, the same rent-seeker relentlessly attacks the reinvigoration of the local refineries, and thus compelled Nigeria to rely on importation of refined products to provide for its local need. Currently, Nigeria has four active refineries, namely: Kaduna refinery with 110,000 bpd capacity, Warri refinery with 125,000 bpd capacity, and the first and second Port Harcourt refinery with a daily production capacities of 60,000 and 150,000 barrels, respectively. These put together imply that Nigeria had the potential of refining oil in quantities as much as 445,000 barrels per day. In recent years, the oil business has immensely contributed to the Nigerian economy. Their contributions include energy supply to industry and business, increase in government profits and foreign exchange stock, improved gross domestic product (GDP), employment opportunities, and a provision of local consumption of goods and services. Despite the fact that the industry's value added is also contributing to boosting the nation's GDP, the latter is not really proportional to economic development. The rise in the oil business' value is basically a pointer to the critical increment in crude oil production since the end of the civil war and, more particularly, of the remarkable increment in oil prices since 1973, which are indifferent of the level of development of the local economy. The oil sector has been tormented by different issues that undermined its ideal development for decades. However, the petroleum segment of the Nigerian economy in the 1990s was confronted and is still confronted with the following issues: public control and administration, poor financing of investments, communal disturbances (i.e., smuggling and redirection of oil-based goods), fraudulent domestic marketing habits, and products debasement, among others (Odularu, 2008). It could without much of a stretch be seen that Nigeria for all intents and purposes has no control on the export trade. However, there seems to be some rays of hope considering the actions of the federal government in spite of the insecurity of the oil market. It is necessary on the part of government to look for alternative sources to supplement oil as the major source of income. Some areas have been proposed by experts as a solution that can help rescue the present economic circumstance. These are agriculture, mining, and manufacturing (Ijeh, 2010).

The Nigerian oil industry is principally overseen by the government whose operator is the NNPC. The NNPC (which replaced the Nigerian National Oil Corporation—the NNOC), is in charge of research, oil exploration and production, transportation and refining, and marketing of the oil-related commodities. The NNOC earlier set up in April 1971 aimed at taking control of the oil division as stipulated in the 1970—74 development plans. In her ambition and vision to become a key player in the oil business globally, Nigeria registered as a member of the Organization of Petroleum Exporting Countries (OPEC) in July 1971. This was a period when OPEC was working vigorously to influence the exploration of oil of its member countries. Aside from the previously discussed issues associated with oil in Nigeria, oil spills and hydrocarbon contamination are also threats in Nigeria. As indicated by the official evaluations of the NNPC, an average of 2300 m^3 of oil is spilled in 300 separate incidents annually between 1976 and 1996 (Twumasi & Merem, 2006). However, some independent assessments have put the quantity at approximately 10 times higher than the reported values. The destruction caused by spills, especially in the dry land or freshwater swamp regions, is usually worrisome, as it results in negative impacts such as contamination of fishponds and destruction of crops and other sources of livelihood, and these effects remain in the areas for a long period of time. The effects of such loss of business can hinder children's education because their parents may no longer be able to afford the cost of tuition, which could lead to virtual desperation (Twumasi & Merem, 2006).

6.3 SOCIO-POLITICAL CONSEQUENCES

6.3.1 Oil Boom and Corruption in Nigeria

Since oil was discovered in Nigeria c. the 1950s, the Nigerian GDP has been doing pretty well. Approximately 90% of Nigeria's national revenue is generated from the oil sector, and that is the reason government screens this sector in order to keep up their standard in the world market. Because of the reliance of the GDP on oil export, it affects government spending and at the same time makes the government unable to apply exact policies that limit the level of corruption inside the economic framework. Additionally, the communities where the oil wells are dug are experiencing annihilation and low economic viability because the national government has refused to sanction and enforce laws relating to ecological protection against organizations such as Shell Petroleum Development Company (SPDC), Chevron, Mobil, and others that drill oil in the Niger Delta. As reported by Amnesty Program International, 70% of the population in the Niger Delta lives under one US dollar (USD1) per day because of environmental degradation and corrupt leadership, among other factors. These same leaders support the youths in their militant activities, which involves capturing innocent individuals and destroying government infrastructures (Eze, 2015). Corruption in Nigeria is affecting the entire system. Due to the tremendous resources available, government authorities exploit state apparatus to control the national treasury for their own use (Fagbadebo, 2007). Corruption is a crucial issue in Nigeria, particularly in the oil sector. There are cases of oil firms declining to publish their financial statement and reporting the correct data required by the government for appropriate monitoring. The government circulates the figure given to them and persuades the public to trust their financial statement despite the fact that they know that those figures are not valid. Every citizen of this great nation needs to know how much income accrues to the country from crude oil export and how the money is been utilized. Corruption, according to Eze (2015), is the ability to secure influence or riches wrongfully, with the sole aim of gaining secretly at public expense. It has been reported by some economic scientists that corruption has negative impacts on the Nigerian economy. Corruption has infiltrated government parastatals as well as the oil sectors. This practice has spread so much that it looks as if it has been legalized in the nation.

The spread of corruption in the nation can be credited to the prosperity gotten from oil, or how else can one explain that Nigeria, who is a major oil producer, still has a considerably large part of her population living in abject poverty? Nigeria still relies on foreign aids to meet essential obligation (Akinyetun, 2016). Government officials, directors of oil firms, and community chiefs are some of the money embezzlers in the nation. Income from crude oil is not appropriately distributed and checked, and payment to government for the mining of mineral resources are not properly accounted for. Corruption shows in all facets of government activities, for example, in assigning contracts, distribution of benefits, assemblage of public revenues, and legal declarations. Authorities required to perform these obligations partake at one point or the other in abusing the process.

Corruption is self-evident. Nigeria displays a traceable case for comprehending the link that joined political disquietude and corruption. The Nigerian political and economic systems are noted for misuse of office and corruption. According to the National Planning Commission, corruption in Nigeria has varying structures. Notable among these structures are collection of unlawful levy and dues, deceitful exchange, misappropriation of public funds and assets, manipulation of records, bribery, and false oath. Globally, Nigeria is ranked as one of the leading corrupt nations, and this position has belittled her in the comity of nations. Corruption as reported by scholars is worse in comparison with drug trafficking. Corrupt and awful administrations were the two main alibi frequently used by the military to forcefully overthrow the government in Nigeria (Fagbadebo, 2007).

Under normal circumstances, contracts of oil production are supposed to be awarded only to firms that would deliver the job efficiently and effectively, and at the right cost. On the contrary, favorite winners of oil contracts in Nigeria, regardless of their capacity and expertise, are companies that can give awesome reward to government officers. At present, oil production in Nigeria has taken aggregate control of the economy, and most legislative issues in Nigeria are centered on oil. Therefore, oil has become the dictator of the political economy of Nigeria (Akinyetun, 2016).

According to literature, Nigeria is the tenth nation in the world with the largest hydro-carbon reserve and is contributing approximately 4% to oil production in the world. However, despite Nigeria's oil richness and vast income generated from the business, there is no significant difference in the infrastructural development and standard of living of most Nigerians. The major factor responsible for the incomprehensible suffering amidst wealth in Nigeria is the corruption of the Nigerian government.

6.3.2 Impacts of the Oil Boom on Policy Formulation in Nigeria

In 1970, the Nigerian government moved the concentration of its agrarian policies that earlier gave total control of production to state, farmers, and elites such as serving and retired top military officers, the affluent, and bureaucrats, and began to direct production via public proprietorship, indigenization of foreign endeavors, price, interest, and conversion rate controls. Policy that stipulates limitation of specific capitals from certain subsectors is also included. However, in 1986, the policies were changed in consonance with a basic change program forced by the International Monetary Fund (IMF)/World Bank. Thus, appropriations, price, interest, and conversion rate controls, as well as the government's immediate inclusion in production, were reduced (Okolie, 1995). These changes in policies not only resulted from the effects of internal or external powers alone, but also from an intricate connection between them.

The post-1970 agrarian policies must be understood because of the adjustments in the global system of credits. Military rule, centralization, and immense oil incomes, consolidated with a universal credit administration controlled by private banks, created the statist strategies of the oil boom period. In the same vein, the decrease in oil incomes and shifts in the guidelines overseeing access to global credit from 1982 to 83 prompted the movement from the statist policies into management by private production (Okolie, 1995).

Moreover, the Nigerian state is conceded as having little control over the price of its major oil exports and imports, as she is being obliged by universal powers to invert her policies. As farming exports fell while oil production ascended in the 1970s, oil exports developed and accounted for more than 90% of export profit and 80% of the government's incomes. However, the military government mandates were centered on having more resources and power at the center (the federal government) to the detriment of the constituent state (Okolie, 1995). It is noteworthy that political initiative in the framework of this paper alludes not exclusively to the administration or to the authority of an organized state; rather, it embraces the totality of the political class that has impacted the system of government even from behind the scenes. Corruption has been the most despicable aspect of Nigeria's advancement. Unfortunately, the political class and military saddled with the duty of coordinating the affairs of the nation have been the significant leading party in executing this corruption (Ogbeide, 2012).

6.3.3 Oil Conflicts and Conglomerate Politics in Nigeria

More than 20 million individuals from different ethnicity and dialect are residents in the Niger Delta. The genesis of assault and conflict between communities and oil companies in the Niger Delta is dated back to the early 1990s. The history and implications of the oil strife in Nigeria and its effects on the Niger Delta biological systems as well as the national economy are enormous. In 2006, Nigeria was aglow by a blast of assault masterminded by militants against the foreign-owned companies in the oil-rich Niger Delta. Four foreign SPDC representatives were kidnapped in January 2006 by aggressors known as the Movement for the Emancipation of the Niger Delta (MEND). The hostility continued throughout the year, as the activists exploded pipelines, overran offshore apparatus, executed Nigerian officers, and hijacked more than 50 oil laborers. In a bid to stop its war on the oil companies, MENDs requested compensation for the ecological harm caused by the oil business, more control over oil incomes by state government, and enhancement of living conditions in the delta.

Since the oil boom in the mid-1970s, Nigeria's economy has been largely controlled by oil (Omeje, 2012). Oil assets represent approximately 40% of the GDP and more than 90% of foreign trade income. Reports have shown that oil reserves in Nigeria are evaluated at 36 billion barrels, while natural gas reserves are more than 150 trillion cubic feet. Crude oil production in the upstream area in Nigeria is ruled largely by western transnational oil corporations (TNOCs), such as Royal Dutch Shell, Exxon Mobil, Chevron Texaco, Agip, and Total. These oil companies run JV partnerships with the Nigerian federal government, represented mainly by the NNPC and its auxiliaries. The administration holds a 60% value share while the rest are held by multinational oil companies (Omeje, 2012).

The emergence of an oil-dependent economy in Nigeria in the 1970s prompted the disregard of different areas of the economy, particularly the agrarian sector, which used to be the mainstay of the economy. Despite Nigeria's huge oil resources, the World Bank reported that as a result of corruption, 80% of the oil incomes that accrued to the domestic front (i.e., the state and indigenous financier) benefit just 1% of the populace. Since the state has neglected the people and the region, the greater part of the Niger Delta populace, including those who do not have any oil company operating in their locality, seek the oil business for improvement of well-being. Ethnicity and provincialism are critical elements in Nigeria's legislative issues. The Nigerian government and the extensively nationalized oil sector have been overwhelmed by a lax coalition of an ethnic plurality of elites, to the detriment of the majority of the ethnic minorities, including those of the oil-bearing Niger Delta region. Statutorily, responsibility for every mineral resource in Nigeria is vested in the government (Omeje, 2012). All lands as specified by law are state property. Yet, this dubious law is just initiated when the vested economic or political interests of the nation are questionable. The national government appropriates and holds a larger piece of the oil incomes and rents while 13% of the incomes gotten from onshore and offshore oil resources are paid to oil-bearing states in the Niger Delta. Oil-related rents, for example, sovereignties, charges, oil export income, and premiums on JV investments, among others, are the soul of Nigeria's economy. The household spending plan and the enormous import exchange are primarily financed by oil incomes. Groups of the nation's world class, with solid interests in the assignment, allocation, and utilization of oil incomes, command all levels of government (Omeje, 2012). The Niger Delta experiences extreme ecological deterioration because of oil spills, loose environmental governance, government complicity, and continuous marine and air contamination. The nature and ways of accretion among the rentier space—a term used to express the obtaining and controlling of all manner of oil and oil-related assets, including the monetary advantages gotten from them, are the major elements of oil conflict in Nigeria (Omeje, 2012).

The details of an economic policy include the accumulation, adjustments, investigation, outline, and understanding of economic information. The nature of information contribution to policy formation becomes more important to policies that will affect the macro-economy in the most wanted sector for greater benefit to the economy—whether national, state, or local government. Economic policies are typically detailed to take care of recognized and examined issues that remain between the economy and its objectives over a given timeframe. This obviously differs from the abnormality of self-intrigue and propelled economic policies that are common with maverick governments and pioneers. The enormous increase in oil income as a gain of the Middle East war of 1973 made uncommon and spontaneous riches for Nigeria (Adedipe, 2004). Presently, with a specific goal to make the business condition favorable for new investments, the legislature has included the need to discover riches in socio-economic foundation, particularly in the urban regions. A few policy activities to address the inadequate structure and wasteful aspects were taken, but it is ineffectively actualized and occasionally opposing. Generally, the policy plan seems to react to oil circumstance or endeavor to exploit it. This ordinarily takes the form of "increase expense when oil profit increases so as to keep up with the position, but when there is a plunge in income, search is undertaken for an urgent way out of the crisis" (Adedipe, 2004).

6.3.4 Social Crisis, Governance, and Political Instability

A few variables such as scarce resources, feeble authenticity, and patron-client—or what is regularly referred to in Nigeria as "godfather"—have been identified as some of the causes of political crisis in African states and the third world. Scare resources incite inequality and a frail position in the world economic framework. State control of the finite resources gives room to government officials and politicians to control government spending in order to acquire fortunes. This prompted powerless authenticity, as the citizens lack confidence in their political leaders and to an extent, in the political structure. Engagement in government is low because the subjects see it as irrelevant to their lives. These administration issues have created an evil structure in a domain that induces precariousness in the political system as the people long for the subtle profits of good administration. The National Planning Commission identified systemic corruption as a main hindrance to development. As it has been noticed, the underprivileged are the victims of this deterioration while the ruling class deceptively makes decisions that give them undue advantage over others (Fagbadebo, 2007).

There is broad understanding in and outside government that the IMF and World Bank sold the idea called the Structural Adjustment Program (SAP) to the Nigerian government, and this created severe political and social strains. This unanimity stems, not from concurrence on the requirement for change, but rather from a general impression of developing pressures, estrangement, underestimation of poor people, social rot, a rising remote obligation profile, and flimsiness in the nation to specify a couple of the extending disagreements and weights, which have been produced or emphasized by the execution of the modified program since 1986. Life has turned out to be more difficult, mostly for the poor. Unemployment and inflation have increased to an exceptional level, and pressure within and between classes has been enormous. Several enterprises have shut down. Wastefulness and inefficiency in public organizations have become higher than ever, thus taking corruption in Nigeria to a level that has never been experienced in the nation's history. The foreign obligation profiles are higher than ever, practically quadrupling with alteration at the expense of a general policy of subsidization and the mass drop of workers. The affected regions remain neglected while human rights are infringed upon by political pressure amongst state and society. These are intense issues regularly dismissed by the IMF and the World Bank. However, unlike the IMF, the World Bank in its 1989 report on the sub-Saharan Africa crisis to sustainable growth, revealed that the necessity to guard the poor, legitimize popular associations, and advance regard for human rights, and the democratization of society within the setting of change is important, if any projects are to stay on course or benefit the people (Ihonvbere, 1993). In the opinions of the public and private segments of the Nigerian population, there is an acknowledgment that the boom days are over and that there is an earnest desire to be judicious in the utilization of resources. Some local businesses now depend less on imported inputs for their production processes. Unfortunately, this development is disrupted by the irregularity, untruthfulness, corruption, and wastefulness of the government officials. Moreover, the unmediated usage of the change program with little regard for the specificities and imbalances of the nation has the tendency to dissolve gains and create profound established disagreements and conflicts. Furthermore, the unequal circulation of the gains and agonies of change, the clear disregard of the living standards of the majority, and the suppression of opposition forces have potential to create problems if not properly managed. One of Nigeria's renown economic specialists reported that the disappointment of the SAP program to adequately react to the nation's emergency happened because it failed to consider the sociopolitical and economic substances of the nation. As indicated by him, government authorities did not comprehend what truly matters to SAP. They replicated the economic policies from Europe and America rather than using national policy. This act is the second line of bondage (Ihonvbere, 1993).

With tremendous riches from oil resources andsocio-economic and political prowess, Nigeria is considered as the giant of Africa. However, poor leadership in Nigeria has prompted stagnation, estrangement of the citizenry, and creation of a low level of system influence. It is informative to note that the initiative issue in the Nigerian commonwealth was a sign of the useless example of the times of the military era. The initiative example in Nigeria does not have the important concentration that fits for imparting national advancement or to advance political strength. On the other hand, Nigerian leaders are occupied with their cravings for the allotment and privatization of the Nigerian state. The fall of the Second Republic, for example, was facilitated by corruption and the orderly political wickedness that was rampant in governance. Several authorities likewise tormented the nation with incessant military overthrows and counter-upsets whenever their successors aim to improve the status of the nation (Fagbadebo, 2007).

Great administration could be expert when the operation of government is in accordance with the predominant legitimate and moral standards of the political group. At the point when this is the circumstance, framework influence will be high, and the public would seek to partake in the activities of the state with a conviction that the methods of governance would serve the best interests of the populace. All previous atrocities would be reduced drastically as people's rights would be secured under the rule of law. Political leaders would uphold probity and accountability in governance. However, the case is different in Nigeria. The populace lost confidence in the system because of reckless and careless administration. This problem is more obvious in the Niger Delta where oil exploration has devastated the environment and the people. This region produces the bulk of the resources that sustains the country, yet the people are exceedingly undeveloped and poor (Fagbadebo, 2007). The frustration in Nigeria has led to the emergence of several illegal businesses. For instance, prostitution has increased as young, skantily-clad school girls line the streets or wander about searching for clients. It is not a surprise that more than half a million Nigerians are now HIV infected. High school pregnancies have turned out to be uncontrollable, and quack abortion centers have doubled in the main cities. Child abuse, divorce, matrimonial savagery, murder, and stealing of public properties have all increased in recent time. Nigerians began to crave for junks for nourishment and second-hand apparel and cars among others, in order to adapt to the torments of change. Public transportation turned out to be extremely expensive because of the shortage of spare parts and the irregular supply of fuel. The elites abandoned their official vehicles because of cost of maintenance while the clinics and hospitals are in a pitiable condition (Ihonvbere, 1993).

The Niger Delta's predicaments have constituted an overwhelming issue in Nigerian political, social, and economic discussions. Before independence, the minority groups of Nigeria have expressed fears of the possibility of being marginalized by the dominant ethnic groups in Nigeria. It was this agitation that prompted the setting up of the Sir Henry Willinck Commission of 1957. However, the outcry then was silenced after being guaranteed that there would be consideration for them in the constitution. A couple of years after the nation's freedom, it became obvious that the fears earlier expressed by the minorities, particularly those of the oil-producing Niger Delta, were genuine. They have protested that the entitlement that should accrue to them as the oil-producing region have not been fully allotted to them.

6.4 ECONOMIC CONSEQUENCES

6.4.1 Effect of the Oil Boom on the Nigerian Economy

Crude oil provides a major source of economic rent to a nation, since oil products could be sold at a value that far surpasses its cost of extraction. Some researchers have disclosed that an economy with abundant resources will seek more rents than those with little resources. In Nigeria, for instance, oil riches have been identified to be one of the primary drivers of the rent-seeking activities and corruption. The oil boom of the 1970s was responsible for the emergence of disorderliness in Nigeria. There was a remarkable increase in the foreign exchange earnings during the oil boom era in Nigeria. However, the utilization of oil boom resources in funding some laudable public projects in Nigeria was characterized by political corruption. Also, there was a drastic reduction in the implementation of numerous investment projects, which led to an increase in negative rates of return (Mohammed, 2006). The nation's major productive base includes the farming, production of crude oil, and different hydrocarbons and is said to represent more than 90% of foreign trade and 75% of employment. Five years ago, Nigeria's economy improved by an average of 7%, and this improvement was facilitated by the oil sector, which represents more than 30% of the total national output and 70% of all export. Since oil price began to fluctuate, there has been a great deal of focus on what should be done to guarantee consistent development in the oil sector in Nigeria regardless of the status of the world market. Due to instability in oil prices and Nigeria's over-reliance on oil, numerous economic analysts have raised concern about the fate of the economy of Nigeria in the nearest future. With more than 65% of Nigeria's income originating from oil since the last decade, Nigeria's monetary policy remains intensely impacted by the oil business and its unpredictable development (Igberaese, 2013).

Subsequently, Nigeria's income decreased when oil prices fell after the boom. The increase in oil prices from 2005 to 2008 reflected an increase in revenue and expenditure for the Nigerian government. However, the world financial crisis in 2009 prompted a fall in world oil prices, and this development also caused Nigeria's income to fall. The impacts of oil on Nigeria's GDP and export are influenced by the world oil prices. The principle of comparative advantage became irrelevant because adjustments in production did not influence the economic development. However, the values of oil exports often dictated by oil prices have impacted the economic development in Nigeria (Igberaese, 2013).

6.4.2 The Global Fall in Oil Price

Globally, the oil price fell by more than 40% at $115 a barrel. It is presently below $70 per barrel. Several oil-exporting nations such as Russia, Nigeria, Iran, and Venezuela are affected by this price declination. *The Economist* magazine reported that oil-producing nations whose financial plans rely on high costs of crude oil are stuck by an unfortunate situation. As a result of the current oil crisis, Nigeria has been compelled to raise interest rates and devalue the Naira, while Venezuela looks nearer and nearer to defaulting on its obligation (Adamu, 2015). The oil crisis has re-occurred over time. For instance, the two oil crises in 1973 and 1979 are confirmations that it should not be the only dependent sector for Nigeria. In a bid to alleviate the oil crisis, the OPEC was formed in September 1960 to control oil transactions between member states. Nigeria became part of OPEC in 1971. However, currently, there has been a whirl of contentions concerning the conceivable causes of the fall in oil value. A few analysts have noticed that the fall in oil value emerged from US refinery support, OPEC inaction, penetration of world oil production by oil nations that are not part of OPEC, and the irregular actions of some members of OPEC. In addition, financial crimes are also recognized as one of the causes of the oil price fall in

Nigeria. This is so because organizations would still maintain their extravagant way of life even during the crash period of oil price in the world. And by so doing, the directors and executives of such organizations are forced to carry out activities contrary to the organizations' policies (Ogochukwu, 2016). This fall in oil price has influenced the Nigerian economy greatly. To reduce the effects of the fall in oil price on the Nigeria economy, the present Central Bank of Nigeria's Governor, Godwin Emefiele, commenced a severe change in the banking sector in order to stabilize the economy. For instance, to guarantee a proper control of the financial sector in Nigeria, the Central Bank of Nigeria, in a report delivered on the November 21, 2015, advised three commercial banks to recapitalize when they could not meet the base capital ampleness rate of 10% before June 2016. The oil price fall crisis occurrence had in reality influenced the Nigerian economy by causing lack of funds for financial services. Currently, it has been revealed by the banks that few oil marketers owed some Nigerian banks approximately 5 trillion Naira. This is harmful to the financial institutions in Nigeria and could subject them to liquidity and the inability to pay their staff's salaries. It could also lead to massive retrenchment of staffs and thus contribute to the already outrageous unemployment situation in Nigeria. The fall in oil price also affects the budgetary issue for the oil sector and the capital market because of their connection to the financial world and the Nigerian banks. Moreover, another effect of the oil price fall is currency devaluation, which causes inflation. Costs of services, merchandise, and commodities have increased and are still increasing. Importation also turns out to be more costly because more Naira will pursue the few accessible Dollars. This situation aggravated the problems faced by the Nigerian economy. To control the effect of this oil price fall, administrative measures to alleviate the impacts on the economy are a necessity because only a preserved and stable macroeconomic environment and a sound and lively financial framework can drive the economy to accomplish its national desire to become one of the 20 biggest economies in the world by the year 2020 (Ogochukwu, 2016).

6.4.3 Impacts of Oil Shock Susceptibility in Nigeria

The easiest measure of any economy is the GDP, which comprises the estimation of all merchandise and enterprises recorded in a given year. The economy comprises not only the sectors in a country but also the organizations and administrators of both sectors and the resources (both natural and human). The federal, state, and local governments through their yearly spending plans, run different sectors of the economy. The success of the operation of the petroleum sector depends on the economy and vice-visa. Every sector is financed in the national budget, and the commitment of the sectors in any given year makes up the annual spending plan. The oil business has had both positive and negative effects on the Nigerian economy. However, there has been more negative than positive effects because of poor execution of government policies (Ijeh, 2010). From the economic point of view, the everyday prices of oil might be dictated by free market strengths; however, sharp shifts in price level are basically influenced by political factors. An example of this is the politically induced chaos in the Middle East where stealing of the crude oil supply is common. Oil price shocks occur due to low price elasticity of demand and supply. The consequence of this is that price variety is required to clear the market, that is, to shift the market towards an equilibrium point. Crude oil price fluctuation is usually low, particularly in the short run. This is because it takes some time before the expending apparatus or capital stocks are supplanted with more-productive substitutes. Nonetheless, substitution happens in the long run, and price fluctuation is substantially bigger. In any case, price elasticity is still less than one (Ibrahim, Asekomeh, Mobolaji, & Adeniran, 2014). Generally, oil prices have been more unpredictable than any other products or resource since World War II. The pattern of interest and supply in the world economy, coupled with the activities of OPEC, influences the price of oil. The current changes in oil prices in the world economy are very fast and unusual. This is mostly because of increased request of oil by China and India. The significant fluctuation of oil prices has induced incredible difficulties for policymaking. The transmission instruments through which oil price have effected genuine economic activity are traceable to both demand and supply channels. The supply reactions are identified by the fact that crude oil is a fundamental contributor to production, and therefore an increase in oil price prompts a corresponding increase in production costs that would compel firms to lower yield. Oil price changes also involve demand reactions on utilization and investment. On average, crude oil prices have increased from USD25 per barrel in 2002 to USD55 per barrel in 2005. The increase in oil prices had a temporal potential of conflicting effects on world demand and development (Oyeyemi, 2013).

The assessment of the outcomes of oil price shock on development in Nigeria is pertinent. As a small open economy, Nigeria has no genuine impact on the world price of oil, although it is incredibly affected by the impact of oil price instability both as an exporter of unrefined petroleum and shipper of refined oil-based goods. Therefore, it implies that change (increase or fall) of oil price in the world benefits or hurts Nigerian economy. Fundamentally, the essence of the issue lies in the fact that Nigeria has solely depended on this commodity throughout the years, thus making her a mono-commodity economy. In 2008, when oil price tumbled from a pinnacle of $147 to approximately $37.81 per barrel, the Nigerian budget was reduced in terms of income and expenditure. These cuts had consequent impacts on all sectors of the Nigerian economy. Oil price has been found to have had a more blunt impact on the conversion rate of the Naira than any other economic variable. This is due to the fact that profit from crude oil sale represents a substantial piece of Nigeria's foreign exchange, estimated at approximately 90% (Oriakhi & Iyoha, 2013). Over-dependence on crude oil by the Nigerian government has led to complete abandonment of nonoil sectors, and this has made her defenseless against the impulses of the global oil market.

The Nigerian government and private individuals prefer to go for the quick return on investment obtainable in the petroleum sector instead of moderate income in the agricultural sector and solid minerals. The business sectors for oil and oil-based commodities were more appealing worldwide than the market for nonoil-based goods. Research conducted by the Geological Survey Department of Ministry of Solid Minerals, the Raw Materials and Development Council, the Nigerian Mining Council, and the National Steel Council, revealed that Nigeria has several mineral assets of commercial quantity that could be exploited for export and for local uses as an approach to diversify the economy from total dependence on crude oil. Policy thrust in the past centered on crude oil until the Ministry of Solid Minerals was established in 1995. In spite of this, financial constraints are responsible for the very little progress that has been made by the ministry in exploring the substantial stock of bitumen found in four states of the federation (Ijeh, 2010). The Niger Delta states, namely Rivers, Edo, Delta, Bayelsa, and Akwa Ibom, have been identified as the heart of crude oil production in Nigeria. The effects of oil on the Nigerian economy cannot be isolated from the effects of the sector on the Niger Delta and vice-visa. Various crises have occurred among the stakeholders, the Nigerian government, oil companies and oil-bearing communities in the Niger Delta over oil-related issues. In order to have a peaceful environment, government over the years has set up boards and commissions to investigate the activities of oil companies in the Niger Delta and the challenges faced by the indigenes. Some of these bodies are: the Oil and Mineral Producing Areas Development Commission (OMPADEC), the Clean Nigeria Association (CAN), the Niger Delta Development Commission (NDDC), among others, to handle some key issues affecting the Niger Delta. However, the problem of the Niger Delta is not a lack of policy but poor policy execution.

6.4.4 Oil Boom in Nigeria: Dutch Disease or Debt

The resource curse occurs when nations with more natural resource endowment develop slower than nations that are poor in resources. One impact could be an increase in the real exchange rate of an economy brought about by an ascent in exports taking after a resource boom, otherwise called Dutch disease. Despite Nigeria's natural resource wealth, there is no tangible economic development. Nigeria's real GDP development rate is estimated at 1% per annum since 1960 as compared to Botswana, another resource-rich sub-Saharan African economy, which has maintained an economic development of 7% yearly since 1960. Some scholars are of the opinion that resource waste and corruption in Nigeria are responsible for such low developmental rates and therefore suggested that instead of Dutch illness, the most appropriate term for Nigeria's predicament is "resource curse." Another school of thought identified the instability of spending as the variable obstructing development. Instability can be viewed as an expense on investment. Investment requires irreversible choices; just like capital, once introduced, cannot be moved to a different sector. To clarify high instability in oil-rich countries, it has been reported that nations with many resource rents but with less savings are probably going to overspend in great years, and under-alter in awful years. Governments such as that in Nigeria are thought to be particularly defenseless against what amounts to overgrazing the commons (Budina et al., 2007). However, the GDP development rate in Nigeria has changed from 1% to 5% per annum from year 2000 to 2008 (de Wit & Crookes, 2013).

According to Harm (2008), a real appreciation in the domestic currency may reflect an increase in the relative price of non-tradable goods. The reason for this price increase may be an exogenous increase in the available gross national income, which is the outcome of raw material findings. Although such temporal gains are generally regarded as a blessing, they can have a negative impact on the economy in the medium and long term, and this is referred to as the Dutch disease. The term *Dutch disease* was first used by economists in describing the problematic structural consequences of the reduction of natural gas in the Netherlands during the 1960s. The term *Dutch disease* suggests that such a development is to be assessed negatively. It should not be forgotten that the real cause of the changed production pattern is an income gain (Harm, 2008). There has been an argument that the so-called "resource curse" causes poor growth and raises the intensity of incidence and duration of conflict in oil-rich countries. While oil resources have for quite some time been viewed as beneficial to economic and political advancement in a nation such as Nigeria, the current poor economic performance of oil exporters and the frequent occurrence of civil wars in mineral-rich economies have brought about global rethink that the resource wealth earlier attained might be to a greater degree a curse than a blessing. A few analysts have evaluated that oil exports are positively correlated with the onset of civil war whereas others reported that oil export is fundamentally connected with a subset of civil wars, precisely, secessionist wars. Since oil economies have the most elevated amounts of rents that are accessible in the economy, the contention for the legitimacy of the resource control and political brutality interfacing in such economies is significant to cases that mineral resource rents promote armed rebellion in less-developed nations. The thoughts that rebels can do well out of war was proposed as a more persuading clarification as regards the onset of contention than sociopolitical grievances, salary, resource imbalance, and ethnic competition or the truancy of a vote-based system. Natural resources offer revolt groups a subsidizing opportunity since they could create rents within their area and easily plundered resources on a continuous basis. The thought that the presence of oil rents produces more conflict agrees with the prevailing theories of rent seeking (Deacon & Rode, 2012). Rent seeking can extensively be related to activities that try to make, keep up, or change the rights and foundations on which specific rents are based. Since oil rents often give them salaries that are higher than what would have been earned in an ideal enterprise, more motivators are added to make and keep up these rents. Rent seeking can be conceptualized as actions ranging from bribery, political campaigning, and taking up of arms. Actually, a critical issue is the degree to which a non-booty resource such as oil exposed oil-subordinate states to based revolt, and therefore results into destructive types of rent seeking (Deacon & Rode, 2012). A common reference point of poor management of wealth in oil-rich countries is the purported Dutch disease, named after Holland's poor record in dealing with its gaseous petrol riches in the 1960s. Nevertheless, there may be an error in naming Nigeria at present as another kind of Dutch disease. This is simply because for the past 20 years, there have been serious unemployment issues in Nigeria—a characteristics that cannot be associated with a Dutch disease. Nigeria has had times of both disproportionate spending and underspending and recently began execution of a smooth policy. As long as that policy is kept, there will be no real Dutch disease problem now or in the nearest future (Budina et al., 2007). Besides the above-mentioned issues, there are other challenges that could prompt the description of Nigeria as another case of an oil-rich country surrendering to the Dutch disease. Specifically, high spending out of what is a fundamentally tradable resource has led to low development and battles for rare resources, thus drawing labor and capital out of the innovative sectors. The fact in this clarification is that there is a resource shortage that outweighs variable prices. Generally, increments in non-innovative activities can be met without a decrease in the innovative sectors. So, that is where the description of Nigeria as a Dutch disease victim came from (Budina et al., 2007).

I. BACKGROUND INFORMATION ON PETROLEUM INDUSTRY ACTIVITIES AND THE NIGERIAN ENVIRONMENT

6.5 CONCLUSION

The issues about poor growth of the national economy and political insecurity in Nigeria were aggravated by poor administration and corruption. Corrupt and clumsy administrations as well as a severe domestic sociopolitical environment have done great harm to Nigeria. Many times, decisions are made by the government, not according to what the people desire, but rather on what the leaser countries and world financial organizations prescribed. An educated common society is important to adjust the power of the Nigerian State. This could be a panacea for the wrong use of power by public authorities and therefore would engender mental re-orientation towards making a significant positive advancement (Fagbadebo, 2007). It is our conviction that structural change is critically required in Nigeria in order to cut waste, revive the economy and production, invigorate reserve funds, and rebuild economic relationship with foreign markets. However, for the above to be achieved, certain internal and external policies must be set up. For example, the Bank must understand that all African nations are not the same (Ihonvbere, 1993).

6.6 RECOMMENDATIONS

* There is a need to diversify the productive base of the present economy by giving appropriate concentration to other sectors, which include agriculture, manufacturing, and tourism, in order to open up an expansive range for inflow of revenue to the government treasury.
* There is an urgent need for the Nigerian government to begin to correct her past policies so as to eliminate hazard impacted on the system by oil and thus save the Nigerian economy.
* Finally, adequate attention should be directed towards engaging in laudable projects that will meet the basic needs of the people as well as guarantee their well-being.

References

Adamu, A. (2015). The impact of global fall in oil prices on the Nigerian crude oil revenue and its prices. In: *Proceedings of the second middle east conference on global business, economics, finance and banking*, Dubai-UAE. Paper ID: D508; pp. 22−24.

Adedipe, B. (2004). The impact of oil on Nigeria's economic policy formulation. In: *Paper presented at the conference on Nigeria: Maximizing pro-poor growth: Regenerating the socio-economic database, organized by overseas development institute in collaboration with the Nigerian Economic Summit Group*, 16th−17th June, 2004.

Akinyetun, T. S. (2016). Nigeria and oil production: Lessons for future. *International Journal of Multidisciplinary Research and Development*, 3(5), 19−24.

Brock, S. (2014). *Dutch disease and the oil and boom and bust*. OxCarre Research Paper 133. Department of Economics OxCarre.

Budina, N., Pang G., & van Wijnbergen, S. (2007). *Nigeria's growth record: Dutch disease or debt overhang?* World Bank Policy Research Working Paper 4256.

Deacon, R.T., & Rode, A. (2012). Rent seeking and the resource curse; Department of Economics. http://econ.ucsb.edu/~deacon/RentSeekingResourceCurse%20Sept%2026.pdf

de Wit, M., & Crookes, D. (2013). *Oil shock vulnerabilities and impacts: Nigeria case study*. South Africa: Faculty of Economic and Management Sciences, Stellenbosch University, 25 pp.

Eze, O. J. (2015). Analysis of oil export and corruption in Nigeria economy. *International Journal of Economics, Commerce and Management*, 3(7), 112.

Fagbadebo, O. (2007). Corruption, governance and political instability in Nigeria. *African Journal of Political Science and International Relations*, 1(2), 028−037.

Harm, P. (2008). *Internationale Makroökonomik; Deutsche Nationalbibliograhie* (pp. 276−278). Mohr Siebeck Tübingen, ISBN 978-3-16-148775-0.

Ibrahim, A., Asekomeh, A., Mobolaji, H., & Adeniran, Y. A. (2014). Oil price shocks and Nigerian economic growth. *European Scientific Journal*, 10(19), 375−391.

Igberaese, T. (2013). *The effect of oil dependency on Nigeria's economic growth*. The Hague, Netherlands: Master of Arts Degree in Development Studies, International Institute of Social Studies.

Ihonvbere, J. O. (1993). Economic crisis, structural adjustment and social crisis in Nigeria. *World development*, 21(1), 141−153.

Ijeh, C. A. (2010). *Assessing the impact of over dependence on oil revenue to Nigeria economy*. Master of Business Administration Thesis. Enugu Campus, Enugu, Nigeria: Department of Management, Faculty of Business Administration, University of Nigeria, 117 pp.

Mohammed, S. (2006). *Corruption in Nigeria*. Lancaster: Lancaster University Management School, 27 pp.

Odularu, G. O. (2008). Crude oil and the Nigerian economic performance; oil and gas business. Department of Economics & Development Studies: Covenant University (The electronic Scientific Journal).

Ogbeide, M. M. (2012). Political leadership and corruption in Nigeria since 1960: A socio economic analysis. *Journal of Nigeria Studies*, 1(2), Fall 2012.

Ogochukwu, N. O. (2016). The oil price fall and the impact on the Nigerian economy: A call for diversification. *Journal of Law, Policy and Globalization*, 48, 84−93.

Okolie, A. C. (1995). Oil rents, international loans and agrarian policies in Nigeria, 1970−1992. *Review of African Political Economy*, 64, 199−212.

Omeje, K. (2012). Oil conflict and accumulation politics in Nigeria. ECSP Report, issue 12.

Oriakhi, D. E., & Iyoha, D. O. (2013). Oil price volatility and its consequences on the growth of the Nigerian economy: An examination (1970−2010). *Asian Economic and Financial Review*, 3(5), 683−702.

Oyeyemi, A. M. (2013). The growth implications of oil price shock in Nigeria. *Journal of Emerging Trends in Economics and Management Sciences (JETEMS)*, 4(3), 343−349.

Pinto, B. (1987). Nigeria during and after the oil boom: A policy comparison with Indonesia. *The World Bank Economic Review*, 1(3), 419−445.

Twumasi, Y., & Merem, E. (2006). GIS and remote sensing applications in the assessment of change within a coastal environment in the Niger Delta Region of Nigeria. *International Journal of Environmental Research & Public Health*, 3(1), 98−106.

I. BACKGROUND INFORMATION ON PETROLEUM INDUSTRY ACTIVITIES AND THE NIGERIAN ENVIRONMENT

THE EFFECTS OF CRUDE OIL EXPLORATION ON THE SOCIO-CULTURAL AND ECO-ECONOMICS OF NIGERIAN ENVIRONMENT

The Impacts of Seismic Activities on the Geology of Oil-Producing Regions of Nigeria

Oluwaseun Omolaja Fadeyi[1] and John Olurotimi Amigun[2]

[1]University of Trier, Trier, Germany [2]Federal University of Technology, Akure, Nigeria

7.1 INTRODUCTION

Nigeria is comprised of different agro-ecological zones, one of which is the freshwater swamp forest where the popular oil-producing "Niger Delta" region is located. This distinctive land mass covers an area of approximately 70,000 km^2, a figure that accounts for approximately 8% of the total land area of the country (Mmom & Igbuku, 2015). Politically, oil-producing areas in Nigeria are comprised of 10 states, with Anambra State being the most recent inclusion. Mmom and Igbuku (2015) further reiterated that the population in the area is approximately 31 million spreading across 40 ethnic backgrounds with more than 200 spoken dialects that differ significantly from one another. Major groups in the region include the Ijaws, Itsekiris, Isokos, and Urhobos, as well as the Ogoni people. Economic analysts according to Mmom (2003) have estimated that close to 70% of the region is rural, whereas the rest is urban. Hence, the poverty level is quite high as about 66% of the populace live on an average monthly earning of USD75. However, the region accounts for approximately 90% of Nigeria's export earnings and 70% of national revenue, the majority of which are derived from crude oil (Akinyetun, 2016). With the many benefits derived nationally from the Niger Delta region, it is rather surprising that this oil-rich region suffers the direct impacts of exploitation and exploration of crude oil, so much so that the area still languishes in the pains and poverty due to drawbacks in socioeconomic and infrastructural development. Furthermore, some of the impacts of crude oil exploration have in the past resulted in loss of lives. It is also documented that destruction of flora and fauna seem to result in unemployment, among other notable impacts of crude oil exploration and exploitation (Ogwu, Badamasuiy, & Joseph, 2015). As a result, oil, which should be a (re)source of blessing, is now seen as a curse due to the many challenges emanating from its exploration and production.

Seismic activities within oil-producing areas of Nigeria are mainly associated with hydrocarbon drilling programs. Seismic surveys provide detailed information both onshore and offshore as to the depth and extent of the target (oil or gas), depending on the aim of the survey. Seismic *reflection* and seismic *refraction* are two geophysical methods mainly employed for hydrocarbon exploration. Generally, seismic surveys involve sending energy into the ground with the use of wave-generating devices or explosives (Ifiok & Igboekwe, 2011). As the wave travels within the subsurface, it hits several rock layers. Hence, depending on the density of each layer hit, the waves are either reflected or refracted. Waves that succeed in getting back to the surface are recorded by a device known as a *geophone* connected to impulse-reading computer tapes and a computer for detailed geophysical interpretation. On land, *vibreosis* can be used to generate sound waves, whereas *airguns* are used offshore. Rather than geophones, *hydrophones* are used to detect waves returning to the surface in offshore seismic surveys (Fig. 7.1).

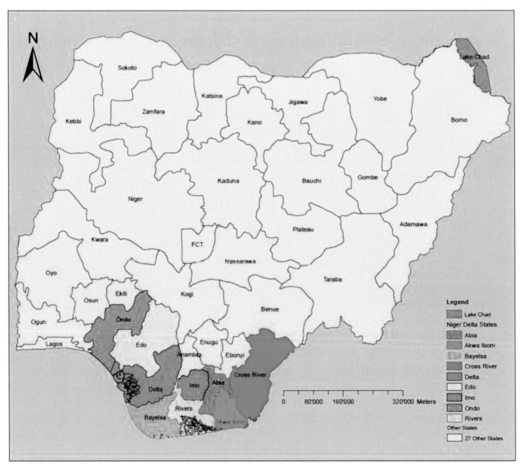

FIGURE 7.1   Map of Nigeria showing the oil-producing states in the Niger Delta region. *Source: Ite, A.E., Ibok, U.J., Ite, M.U. and Petters, S.W. (2013). Petroleum exploration and production: Past and present environmental issues in the Nigeria's Niger Delta. American Journal of Environmental Protection, 1(4): 78–90.*

II. THE EFFECTS OF CRUDE OIL EXPLORATION ON THE SOCIO-CULTURAL AND ECO-ECONOMICS OF
NIGERIAN ENVIRONMENT

7.2 SEISMIC ACTIVITIES: OFFSHORE/ONSHORE IMPACTS

The impacts of seismic activities are enormous, depending on whether the survey in question is onshore or offshore. Several literature exist with respect to the impacts of hydrocarbon exploration and exploitation, particularly within the marine environment of the Niger Delta. Land-based impacts of seismic activities vary from place to place. Some of the impacts so far experienced since exploration and exploitation commenced within the Niger Delta are discussed in subsequent sections of this paper. To better understand the impacts associated with spills, there is need to understand the occurrence of hydrocarbon within marine environments and in soils. In marine environments, oil may be seen to float or remain at the deeper part of the water surface. This depends on the density and type of oil. In soils, oil occurs in different states; in a gaseous states through volatilization, in a solid state through adsorption to soil particles, in a liquid state through dissolution in soil water, and in an immiscible state (a state in which fuels and other chlorinated hydrocarbons may be present in individual phases) (Konečný, Boháček, Müller, Kovářová, & Sedláčková, 2003).

7.2.1 Spills

Most environmental concerns associated with seismic activities are mostly the results of spill and in some cases noise . Crude oil, which has its merit as the main source of Nigeria's economy, has also been described as one of the major environmental pollutants within the Niger Delta region (Imasuen & Omorogieva, 2013). This is due to the reoccurrence of oil spill events that give rise to environmental degradation, soil depletion, water contamination, and atmospheric pollution, thus adversely affecting the inhabitants and the habitats where exploration is taking place. Civil unrest against environmental degradation has also been witnessed in the Niger Delta (Njoku, Akinola, & Oboh, 2009). There are many ways by which oil spills may occur. Some common ways include human error or deliberate acts, disasters, and equipment failure, among other means (Anderson & LaBelle, 2000). Spills are not peculiar to offshore seismic drilling events only. In fact, they tend to occur onshore as much as they occur at sea. When a spill occurs offshore, it takes a great deal of remediation efforts to contain and subsequently clean up such spill site. Shekwolo (2005) identified major causes of oil spill within the Niger Delta region between 2000 and 2005 (Fig. 7.2).

Annually, there are several reports of oil spills in the Niger Delta. These events are often caused by aging equipment as well as by poorly maintained pipelines and/or activities of vandals and thieves. Generally, the spills have a serious negative impact on the aquatic environment and farmlands. Since livelihood of some of the persons in the region depends on the affected environmental compartments, it eventually leads to poverty in the long run as well as serious health hazards (Amnesty International, 2015). Recently, Shell petroleum, the main crude oil exploration company, made a compilation of oil spill events between 2007 and 2013 (Table 7.1). The result showed the amount (in volumes) of oil spilled during the period. Although these figures have been debated and contested by some groups, what remains to be seen is the amount of remediation procedures commissioned to clean up spilled oil.

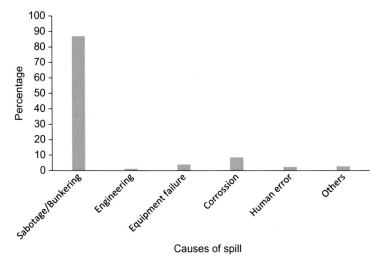

FIGURE 7.2   Causes (in %) of oil spills in Nigeria. *Source: Modified after Shekwolo, P.D. (2005). Oil production and sustainable environmental management: SPDC experience. In I.N. Mogbo, A.A. Oladimeji, S. Jonah, & G.A. Okoli (Eds.), Current trends in science and environmental management (pp. 35–57). Minna, Niger State: Proceedings of School of Science and Science Education, Federal University of Technology, Minna.*

TABLE 7.1   Details of Recent Oil Spill Between 2007 and 2013

Year	2007	2008	2009	2010	2011	2012	2013	Total
Total number of spills	320	210	190	170	201	192	200	1489
Approx. volume in barrels	26,000	100,000	120,000	23,000	18,000	22,000	20,000	329,000

Shell Production and Development Company (SPDC). (2013). Annual newsletter. Warri: SPDC

Aside the estimate carried out by Shell between 2007 and 2013, one can categorically point to certain oil spills that actually caused pronounced environmental pollution within the Niger Delta. Since oil exploration started in the area, approximately 1.5 million tons of crude oil have so far been spilled (Ogwu et al., 2015). Notably amongst oil spill events include Forcados tank 6 Terminal in Delta state where approximately 570,000 barrels' worth was spilled, and 421,000 was spilled at Funiwa 5 well between January 17 and January 30, 1980 (Gabriel, 2014; Tolulope, 2004; Ukoli, 2005). There are several other oil spill events mentioned in the literature, with each spilling several thousand in barrels as reported by Nigeria's apex petroleum regulating agency, Nigerian National Petroleum Cooperation (Twumasi & Merem, 2006).

Often times, exploration companies claim to pay compensation and even award remediation projects to indigenous firms (Amnesty International, 2015). However, no activity is seen in most contaminated areas after a drilling program. This trend often annoys the populace, who sometimes sabotage the work of the exploration firms. In cases where remedial projects are truly awarded, firms saddled with this responsibility often boycott the projects after receiving initial payments from exploration companies. As a result, there is often a conflict of interest between project owners and the contractor as well as communities where oil wells have been dug (Amnesty International, 2015).

7.2.2 Pollution Resulting From Blowout

Soil and water pollution might arise when there is a blowout on a rig or when oil is spilled through other means. This may have serious consequences on workers. Equipment that cost millions of dollars may also be damaged in the process. Blowouts are consequent upon a combination of gas and oil mixtures; each component can, however, result in individual blowout. Gas-caused blowouts can pose extreme danger but will often cause minimal environmental hazards (Anderson & LaBelle, 2000). Other than the human causes of oil spill, other factors also contribute one way or another to spillage around the Niger Delta. As a result, preventing oil spills must be a priority, so that once they occur, swift and effective clean-up and rehabilitation of pollution and environmental damage must be embarked upon in order to protect human rights and lives. If pollution and environmental damage persist, this is then termed as the violation of human rights, as people are driven into poverty through long-term damage to livelihoods and health (Amnesty International, 2015).

7.2.3 Biodiversity Disturbance

The use of an air gun for generation of wave in offshore seismic surveys produce noise in the range of 259 decibels (Ifiok & Igboekwe, 2011). Acoustic harnessing devices used in seismic surveys to gather sound energy at a point also produce noise that has adverse effects on marine animals that depend on sound as part of their adaptive features (Ogwu et al., 2015). Rig construction and subsequent drilling of hydrocarbon can also cause aquatic life to migrate far away. Prolonged activities at sea may lead to disappearance or complete extinction of some endangered species (Oloruntegbe, Akinsete, & Odutuyi, 2009; Ogwu et al., 2015). The same effect of noise is witnessed by humans in onshore seismic surveys where approximately 120 decibels of sound is generated using very large hammers (Ifiok & Igboekwe, 2011). The result of constant contact with such a huge amount of noise may cause hearing loss. Subsequent drilling renders land unfit for other purposes such as agriculture, which was the main source of community wealth before oil discovery.

7.2.4 Gas Venting and Flaring

Atmospheric exposure or burning of gases associated with exploration and exploitation processes have resulted in approximately 250 toxic substances being released into the environment within the Niger Delta (Ogwu et al., 2015). This is largely due to the ineffective implementation of gas venting and flaring laws (Ita & Ibok, 2013). Christen (2004) explained that the situation is different in most developed oil-producing nations where sustainable disposal methods are adopted. Volatile organic compounds (VOCs), carbon monoxide, carbon dioxide (CO_2), NO_x and SO_2 are some of the gases emitted during venting and flaring events (Oloruntegbe et al., 2009). Ite et al. (2013) reiterated that despite having a strong law in place that frowns upon gas flaring, Nigeria still generates more than 40 million tons of CO_2 daily as a result of flaring events from approximately 123 sites (Uyigbe & Agho, 2007). In an earlier study, Oyinlola (1995) explained that emission from gas flaring completely impacted crops planted a few meters away, while 45% of yields were affected at farther distances from a gas flaring station. Some authors have stressed that gas flaring has an impact on human health (Ekpoh & Obia, 2010). Rapid corrosion of zinc roofs has also been associated with flaring (Ekpoh & Obia, 2010; Obia, 2010). As a result of these challenges, rapid mitigation procedures needs to be implemented within the next few years in order to salvage the situation.

7.2.5 Ground Movement

Continuous puncturing of the ground during oil well drilling might have future implications in terms of movement within the subsurface. Although there has not been any record of a dangerous ground movement within the delta, there could be a possibility of future ground movement if the current methods adopted for oil prospecting continue. Seismic blasting and drilling of holes are some of the key techniques adopted during the process of oil exploration. These methods both have the tendency of disturbing the stratigraphy of subsurface soil/rock layers, thereby causing a major realignment of the disturbed underground rock strata. As the drilling proceeds, the subsurface rocks are fractured. This could result in some forms of faulting within the subsurface.

7.3 MITIGATING THE IMPACTS

There are so many techniques in practice available to clean up contaminated environmental components. These remediation methods depend largely on the type of contaminated media (soil or water) as well as the extent of contamination. For contaminated water bodies, eventual pollution is the result of the activities of pipeline vandals (sabotage), pipeline corrosion, and engineering problems, among other challenges (Shekwolo, 2005) within the marine environment. Larson (2010) discussed some of the most common techniques employed for cleanup. The methods include physical, chemical, thermal, and biological remedial techniques. Certain techniques, mostly physical methods, are peculiar only to remediation in the marine environment, whereas phytoremediation is peculiar to soil resuscitation (cleanup). Specific methods such as the biological remediation can be used in both cases (Table 7.2).

7.4 CONCLUSION

It is no longer news that seismic activities, the majority of which are from hydrocarbon exploration and exploitation, have wreaked havoc within the Niger Delta region. Activities of pipeline vandals as well as the unrest within the region have further

TABLE 7.2   Some Techniques for Cleaning Up Oil in Environmental Compartments

Biological	Physical	Adsorption	Chemical
Bioremediation	Use of skimmers (for mechanical collection)	Natural organic sorbents	Use of solidifiers
Biostimulation of native organisms	Use of booms	Mineral sorbent	Use of dispersants
Bioaugmentation—introduction of microorganisms		Synthetic organic/ polymer	
Phytoremediation			

Modified after Bandura, L., Woszuk, A., Kołodý nska, D. and Franus, W. (2017). Application of mineral sorbents for removal of petroleum substances: A review. Minerals, 7(37): 1–25.

compounded this problem. In view of this, there is need for continuous remediation efforts and actions towards reducing additional environmental hazards resulting from crude oil exploration. The Niger Delta, although seriously contaminated, can witness a tremendous turn-around if indigenous clean-up firms are encouraged and empowered. Such firms must also invest in the training and retraining of employees. Exploration firms should also be periodically checked in order to ensure that they follow environmental safety standards. This is the way out of the menace of environmental pollution within the Delta (Amnesty International, 2015).

References

Akinyetun, T. S. (2016). Nigeria and oil production: Lessons for future. *International Journal of Multidisciplinary Research and Development*, 3(5), 19–24.

Amnesty International (2015). *CLEAN IT UP: Shell's false claims about oil spill response in the Niger delta*. London: Amnesty International.

Anderson, C. M., & LaBelle, R. P. (2000). Update of comparative occurrence rates for offshore oil spills. *Spill Science Technology Bulletin*, 6(5), 303–321.

Bandura, L., Woszuk, A., Kołodý nska, D., & Franus, W. (2017). Application of mineral sorbents for removal of petroleum substances: A review. *Minerals*, 7(37), 1–25.

Christen, K. (2004). Environmental impacts of gas flaring, venting add up. *Environmental Science & Technology, 38* (24), 480A.

Ekpoh, I., & Obia, A. (2010). The role of gas flaring in the rapid corrosion of zinc roofs in the Niger Delta Region of Nigeria. *The Environmentalist*, 30(4), 347–352.

Gabriel, A. (2014). *The Nigerian petroleum industry and the Nigerian economy*. Abuja: Nigerian Forum.

Ifiok, U., & Igboekwe, M. U. (2011). The impacts of seismic activities on marine life and its environment. *International Archive of Applied Sciences and Technology*, 2(2), 1–10.

Imasuen, O. I., & Omorogieva, O. M. (2013). Comparative study of heavy metals distribution in mechanic workshop and refuse dumpsite at Oluku and Otofure, Edo State, South-Western Nigeria. *Journal of Applied Science and Environmental Management*, 17(3), 425–430.

Ita, A. E., & Ibok, U. J. (2013). Gas flaring and venting associated with petroleum exploration and production in the Nigeria's Niger Delta. *American Journal of Environmental Protection*, 1(4), 70–77.

Ite, A. E., Ibok, U. J., Ite, M. U., & Petters, S. W. (2013). Petroleum exploration and production: Past and present environmental issues in the Nigeria's Niger Delta. *American Journal of Environmental Protection*, 1(4), 78–90.

II. THE EFFECTS OF CRUDE OIL EXPLORATION ON THE SOCIO-CULTURAL AND ECO-ECONOMICS OF NIGERIAN ENVIRONMENT

Konečný, F., Boháček, Z., Müller, P., Kovářová, M., & Sedláčková, I. (2003). Contamination of soils and groundwater by petroleum hydrocarbons and volatile organic compounds − Case study: ELSLAV BRNO. *Bulletin of Geosciences, 78*(3), 225−239.

Larson, H. (2010). *Responding to oil spill disasters: The regulations that govern their response,* Ontario: s.n. Retrieved on 26 February, 2017 from http://www.wiseintern.org/journal/2010/HattieLarson_Presentation.pdf.

Mmom, P. C. (2003). *The Niger Delta: A spatial perspective to its Development.* Port-Harcourt: Zelon Enterprises.

Mmom, P. C., & Igbuku, A. (2015). Challenges and prospect of environmental remediation/restoration in Niger Delta of Nigeria: The case of ogoniland. *Journal of Energy Technologies and Policy, 5*(1), 5−10.

Njoku, K. L., Akinola, M. O., & Oboh, B. O. (2009). Phyto-remediation of crude-oil contaminated soil: The effect of growth of glycine max on the physico-chemical and crude oil content of soil. *Nature and Science, 7*(10), 79−85.

Obia, A. (2010). The effect of industrial air−borne pollutants on the durability of galvanized iron roofs in the tropical humid region of Nigeria. *Global Journal of Environmental Sciences, 8*(2), 89−93.

Ogwu, F. A., Badamasuiy, S., & Joseph, C. (2015). Environmental risk assessment of petroleum industry in Nigeria. *International Journal of Scientific Research and Innovative Technology, 2*(4), 61−71.

Oloruntegbe, K. O., Akinsete, M. A., & Odutuyi, M. O. (2009). Fifty years of oil exploration in Nigeria: Physico-chemical impacts and implication for enviromental accounting and development. *Journal of Applied Sciences Research, 5*(12), 2131−2137.

Oyinlola, O. (1995). External capital and economic development in Nigeria (1970−1991). *The Nigerian Journal of Economic and Social Studies, 37*(2&3), 205−222.

Shekwolo, P.D. (2005). Oil production and sustainable environmental management: SPDC experience. In I.N. Mogbo, A.A. Oladimeji, S. Jonah, & G.A. Okoli (Eds.), *Current trends in science and environmental management* (pp. 35−57). Minna, Niger State: Proceedings of School of Science and Science Education, Federal University of Technology, Minna.

Shell Production and Development Company (SPDC) (2013). *Annual newsletter.* Warri: SPDC.

Tolulope, C. (2004). Poverty and social challenges of climate in Nigeria. *Wome Issues, III*(1), 1−35.

Twumasi, J., & Merem, K. (2006). *Pouring oil on troubled waters.* Lagos: Times Magazine.

Ukoli, M. (2005). *Environmental factors in the management of the oil and gas industry in Nigeria* [Online]. Available at: www.cenbank.org [Accessed 12 May 2017].

Uyigbe, E., & Agho, M. (2007). *Coping with climate change and environmental degradation in the Niger Delta of southern Nigeria.* Benin City: Community Research and Development, Centre.

Further Reading

Fingas, M., & Fieldhouse, B. (2011). Review of solidifiers. *Oil Spill Science Technology,* 713−733. Available from http://dx.doi.org/10.1016/B978-1-85617-943-0.10014-0.

8

The Effects of Crude Oil Exploration on Fish and Fisheries of Nigerian Aquatic Ecosystems

Oluwaseyi Y. Mogaji[1], Akeem O. Sotolu[2], Peace C. Wilfred-Ekprikpo[3] and Bidemi M. Green[4]

[1]National Institute for Freshwater Fisheries Research, New-Bussa, Niger State, Nigeria
[2]Nasarawa State University, Keffi, Nigeria [3]Nigerian Institute for Oceanography and Marine Research, Lagos, Nigeria [4]University of Lagos, Lagos, Nigeria

8.1 HISTORY OF NIGERIAN AGRICULTURAL ECONOMY

Nigeria is one of the West African countries with motley climatic conditions. Nigerialies between 3°E and 15°E, and between 4°N and 14°N. It is centrally tropical, equatorial in the South, and arid in the North. As a country, it borders with Niger and Chad in the northern part, Benin in the West, Cameroun in the East, and the Atlantic Ocean in the South. Nigeria is divided into six geo-political zones (Fig. 8.1) embedded on a total land mass of 923,768 km that is made up of 910,768 km of land and 13,000 km water; endowed with a great deal of mineral resources such as fossil fuel (crude oil, natural gas, coal, and lignite), radioactive minerals (uranium, monazite, and zircon), metallic minerals (tin, columbite, iron, lead, zinc, gold), nonmetallic minerals (limestone, marble, gravel, clay, shale, feldspar), and arable land.

Nigeria is incipiently an agriculture-dependent country that contributes 40% of the Gross Domestic Product (GDP) and employs approximately 70% of the working population in Nigeria (CIA, 2013). Nigeria repositioned her focus to crude oil export in the 1970s and decades of slow economic growth later; a bane to agriculture that is also the largest economic activity in the rural area inhabited by almost 50% of the population. Given the tremendous resource both in human capital and natural resources in the most populous nation in Africa with a population of more than 150 million and a labor force of 53.83 million (CIA World Fact Book, 2013), the performance of the economy has been

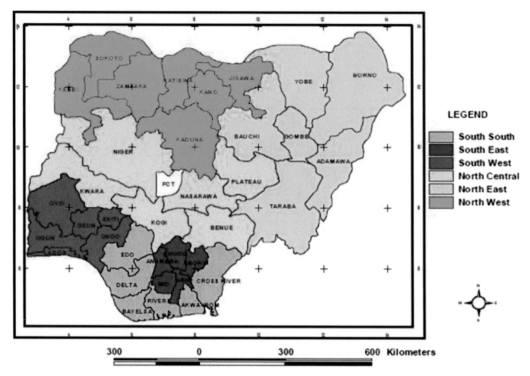

FIGURE 8.1   Nigeria's geo-political zones.

staggering and disappointing. Nigeria is said to be suffering from resource curse (Aluko, 2004; Otaha, 2012). Besides being Africa's largest producer of crude oil, Nigeria's gas reserves ranks sixth globally, and it has the eighth-largest crude oil reserve in the world, concentrated mainly in the Niger Delta (Fig. 8.2) (Sanusi, 2010). Despite these endowments, the nation ranks among the world's poorest economies. With untapped potential for growth and development in the availability of land, water, labor, and its large internal markets, it is estimated that approximately 84 million hectares of Nigeria's total land area has potential for agriculture. However, only approximately 40% of this is land under cultivation. This results in heavy dependence on food imports even with the potential to harness her rich vegetation that can support heavy livestock population, irrigation with a surface, and underground water of approximately 267.7 billion cubic meters and 57.9 billion cubic meters, respectively (Chauvin, Mulangu, & Porto, 2012; Lipton, 2012). Indeed, Nigeria's large and growing population provides an inherent capacity for a vibrant internal market for increased agricultural productivity.

In spite of these opportunities, the state of agriculture in Nigeria remains poor and largely underdeveloped as the sector continues to rely on primitive methods to sustain a growing population without efforts to add value. This has reflected negatively on the productivity of the sector, its contributions to economic growth, as well as its ability to perform its traditional role of food production, among others. This situation of the sector has

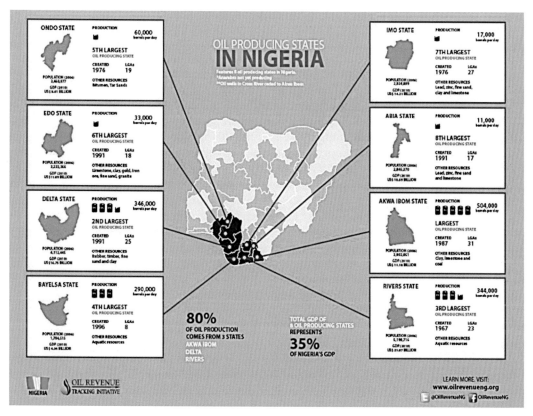

FIGURE 8.2   Nigeria's oil-producing states.

been blamed on oil glut and its consequences on several occasions (Falola & Haton, 2008). The contribution of petroleum to GDP stood at 0.6% in 1960, whereas agriculture's contribution was 67%. By 1974, the contribution of petroleum had increased to 45.5%, almost doubling that of agriculture, which had decreased to 23.4% by 1974 (Yakub, 2008). However, this pattern was not as a result of increased productivity in the nonagricultural sectors as expected in an industrial revolution; rather, it was the outcome of low productivity due to negligence of the agriculture sector.

The subsectors of the agriculture sector in Nigeria have potentials for growth. In the fishery subsector, local production is inadequate for domestic demand and consumption. Nigeria imports 700,000 MT of fish annually, which is 60,000 MT more than the total domestic production (Essien & Effiong, 2010). However, the subsector has recorded the highest average growth rate of 10.3% (1961−2011) compared to the 6% recorded in crop production in the same period. With an average contribution of 4.3% to total agriculture GDP between 1960 and 2011 and provision of at least 50% animal protein, fisheries contribute to economic growth by enhancing food security and improving livelihood of fish farmers and their households (Essien & Effiong, 2010; Gabriel, Akinrotimi, Bekibele, Onunkwo, & Anyanwu, 2007).

II. THE EFFECTS OF CRUDE OIL EXPLORATION ON THE SOCIO-CULTURAL AND ECO-ECONOMICS OF NIGERIAN ENVIRONMENT

8.2 CRUDE OIL EXPLORATION IN NIGERIA

8.2.1 History

Exploration is the mining or excavation of mineral resources from the terrestrial and aquatic environments using technological procedures. The Nigerian Environmental Study/Action Team as highlighted by Mba (1995) identified three (3) categories of mineral resources, which are fuel, metallic, and industrial minerals, which differ in their exploration processes (Bayode, Adewunmi, & Odunwole, 2011). The exploitation of fuel mineral involves exploration, extraction, processing, and transportation, as well as storage and consumption of petroleum, natural gas, coal, lignite, and uranium (Bayode et al., 2011).

The ethnology of the Niger Delta region consists of the Rivers, Bayelsa, Akwa-lbom, Delta, Ondo, Edo, and the Igbo in Abia and Imo States (Idowu, 2012). Fishing and agriculture were the two major traditional occupations of the Niger Delta people. During the colonial era, forestry was introduced as a third major economic activity in the region. Today, although agriculture, fishing, and forestry still account for approximately 44% of employment, all three economic activities have declined since the discovery and production of crude oil in commercial quantity.

Nigeria has a coastline of approximately 853 km bordering the Atlantic Ocean with a continental shelf area of 37,934 km^2 (Oyase & Jemerigbe, 2016). The coastal area is humid, and the average temperature is between 24°C and 32°C, with an average annual rainfall between 1500 and 4000 m (Kuruk, 2004).

Also, Nigeria is blessed with a total land mass of 923,768 km^2 made up of 918,768 km^2 of terrestrial land and 13,000 km^2 of water (CIA World Fact Book, 2013). The Niger Delta is located in the Atlantic coast of Southern Nigeria and is the world's second largest delta, with a coastline of approximately 450 km (Awosika, 1995); an extended spatial location of approximately 20,000 km^2, Africa's largest wetland and the third largest in the world (Anifowose, 2008; Chinweze & Abiola-Oloke, 2009; CLO, 2002; Powell et al., 1985). A report shows that 2370 km^2 of the Niger Delta area consists of rivers, creeks, and estuaries. Stagnant swamps occupy approximately 8600 km^2, and mangrove swamps extend approximately 1900 km^2 as the largest mangrove swamp in Africa (Awosika, 1995). The Niger Delta is categorized as a tropical rainforest with ecosystems that constitute different species of plants and animals in both an aquatic environment and a terrestrial environment. Thus, the region can be grouped into four ecological zones, which are the coastal inland zone, freshwater zone, lowland rainforest zone, and mangrove swamp zone. The Niger Delta is regarded as one of the ten most important wetlands and marine ecosystems in the world (ANEEJ, 2004). In 1991, approximately 25% of the Nigerian population lived in the Niger Delta area. By 2005, the population of the area rose to approximately 30 million, which represent approximately 23% of Nigeria's total population (Uyigue & Agho, 2007).

8.2.2 Crude Oil Pollution in Niger Delta

Approximately 1.5 million tons of crude oil has been spilled into the Niger Delta ecosystem over the past 50 years. This volume is 50 times the estimated quantity of crude oil spilled in the popular Exxon Valdez spill in Alaska in 1989 (Niger Delta Resource Damage

Assessment and Restoration Project, 2006). The first case of oil spill in Nigeria was at Araromi, Ondo State in 1908 (Tolulope, 2004). Since then, oil spill has continued in the country especially in the oil-producing Niger Delta and became more regular after commercial production started in the late 1950s. Some popular oil spill scenarios are: the Forcados tank 6 Terminal in Delta State, which spilled 570,000 barrels of oil into the Forcados estuary, thereby contaminating the aquatic and swamp forest in July 1979 (Tolulope, 2004), Oyakama oil spill of approximately 30,000 barrels in May 1980, and spillage of 500 barrels and 5000 barrels in the year 1979 and 1983, respectively, at Oshika village, River State from the Ebocha-Brass (Ogada-Brass 24) pipeline, which resulted in loss of organisms such as fish, crabs, and shrimp. Some of the effects of these spills in less than a year were an increase in mortality of embryonic shrimp and decreased reproduction (Tolulope, 2004).

Also, the Ogada-Brass pipeline oil splatter near Etiama Nembe in February 1995 was approximately 24,000 barrels of crude oil, which dispersed over freshwater forest and brackish mangrove swamp (Onwuteaka, 2016). The Eket oil spill incident of approximately 40,000 barrels by Mobil in 1998 and one of the largest spill recorded in Nigeria of approximately 200,000 barrels as a result of an offshore well uprush in January 1998 into the Atlantic, which damaged 340 hectares of mangrove forest (Nwilo & Badejo, 2004, 2005a, 2005b). An average of 221 spills per year has been documented since 1989 by the Shell Petroleum Development Company (SPDC) in its field operation constituting 7350 barrels annually (SPDC Nigeria Brief, 1995). Crude oil spill cases in Nigeria from 1976 to 1983 are shown in Table 8.1. From 1976 to 1996, a total of 4647 oil spill incidents discharging approximately 2,369,470 barrels of crude oil into the environment of which 1,820,410.5 (77%) were not recovered (Onwuteaka, 2016). In many other countries, this would be a national emergency.

TABLE 8.1   Statistical Frequency of Oil Spills from 1976 to 1983 in Nigeria

Year	Oil Spills Count	Net Volume (Barrels)
1976	128	20,023
1977	104	31,144
1978	154	97,250
1979	157	630,405
1980	241	558,053
1981	233	22,840
1982	213	33,612
1983	130	32,467
Total	**1360**	**1,425,794**

Source: Nwankwo, J. N. (1984). Oil and environmental pollution. In Paper presented at the conference on strategies for the fifth national development plan: 1986–1990, November 25–29. Ibadan: NISER.

II. THE EFFECTS OF CRUDE OIL EXPLORATION ON THE SOCIO-CULTURAL AND ECO-ECONOMICS OF
NIGERIAN ENVIRONMENT

The spills from the Niger Delta may be attributed to material imperfection, pipeline corrosion, and ground erosion because of the extensive system of pipelines laid across the region. However, in a trade of blame, the oil companies opine that most of the spills were caused by sabotage. However, the Department of Petroleum Resources (DPR) contends that 88% of oil spill incidents investigated were due to equipment lapses, vandalism, oil blowouts from flow stations, and accidental and deliberate releases from oil tankers at sea (Nwilo & Badejo, 2004, 2005a). In any case, the effects of oil exploration or spill in an aquatic ecosystem are perilous (Table 8.2). Crude oil pollution distorts the beach aesthetics and depletes fishery potentials, impacts on water productivity, and human health. As with terrestrial oil pollution, the initial impact of oil on an aquatic system manifests in the suffocation of native biotic communities due to the covering of water surfaces by oil in such a way that primary producers and higher species in the food chain become impacted.

8.3 NIGERIAN FISH AND FISHERY SECTOR

Aquaculture is still being practiced at the subsistence level in most countries of the world. Because of this, statistics of production are difficult to collect, and this makes proper assessment difficult. Nigeria has been cited as an area with high aquaculture potential. Production in Nigeria is characterized by small-scale rural activities. Some people engage in aquaculture as a secondary source of livelihood or on a part-time basis in small farms or small freshwater ecosystems. Fish are very prolific. A fish can lay up to one million eggs, and under culture systems, all the eggs may hatch, become fry, and grow to maturity or table size within 3 months without losses (Nicholas, 2016).

Aquaculture is the fastest-growing animal-based food producing sector, particularly in developing countries. The sector alone contributes nearly a third of the world's supply of fish products. Aquaculture produces more than 220 species, unlike terrestrial farming, where the bulk of the production is based on a limited number of species. Aquaculture in Nigeria is based principally on cichlids, catfish, and cyprinids, which contribute 23%, 43%, and 15%, respectively, to the total production. Nigeria is responsible for more than half the production in sub-Saharan Africa and only six countries (Nigeria, Zambia, Madagascar, Togo, Kenya, and South Africa) account for 89% of production (Nicholas, 2016). Fishing News International (FNI, 1980) reported that production of fish in Nigeria stands at 11.3 tons, which represent 2.3% of the world catch. Recently, a geographical information system (GIS) was used to evaluate the potential of fish farming in Africa, with the outcome that nearly half of the continents' surfaces are ajudged as having a positive aquaculture potential (Central Institute for Freshwater Aquaculture (CIFA), 1993). With this revelation, the question that readily comes to mind is, "if most of Nigeria's potential fisheries resources are subjected to oil exploration threats, what would be the fate of this sector and its contribution to national development in terms of foreign exchange earnings, food security, employment generation, unifying factor in rural communities, cultural identity, aesthetic value, and scientific study of aquatic life?"

TABLE 8.2   Some Oil-Polluted Sites in the Niger Delta

Location	Environment	Impacted Area (ha)	Nature of Incident
BAYELSA STATE			
Biseni	Freshwater swamp forest	20	Oil spillage
Etiama/Nembe	Freshwater swamp forest	20	Oil spillage & fire outbreak
Etelebu	Freshwater swamp forest	30	Oil spill incident
Peremabiri	Freshwater swamp forest	30	Oil spill incident
Adebawa	Freshwater swamp forest	10	Oil spill incident
Diebu	Freshwater swamp forest	20	Oil spill incident
Tebidaba	Freshwater swamp forest Mangrove	30	Oil spill incident
Nembe Creek	Mangrove forest	10	Oil spill incident
Azuzuama	Mangrove	50	Oil spill incident
9 sites			
DELTA STATE			
Opuekebe	Barrier forest island	50	Salt water intrusion
Jones Creek	Mangrove forest	35	Spillage & burning
Ugbeji	Mangrove	2	Refinery waste
Ughelli	Freshwater swamp forest	10	Oil spillage-well head leak
Jesse	Freshwater swamp forest	8	Product leak/burning
Ajato	Mangrove		Oil spillage incident
Ajala	Freshwater swamp forest		Oil spillage incident
Uzere	Freshwater swamp forest		Oil spillage incident
Afiesere	Freshwater swamp forest		Oil spillage incident
Kwale	Freshwater swamp forest		Oil spillage incident
Olomoro	Freshwater Swamp forest		QC
Ughelli	Freshwater swamp forest		Oil spillage incident
Ekakpare	Freshwater swamp forest		Oil spillage incident
Ughuvwughe	Freshwater swamp forest		Oil spillage incident
Ekerejegbe	Freshwater swamp forest		Oil spillage incident
Ozoro	Freshwater swamp forest		Oil spillage incident
Odimodi	Mangrove forest		Oil spillage incident
Ogulagha	Mangrove forest		Oil spillage incident

(Continued)

II. THE EFFECTS OF CRUDE OIL EXPLORATION ON THE SOCIO-CULTURAL AND ECO-ECONOMICS OF NIGERIAN ENVIRONMENT

TABLE 8.2   (Continued)

Location	Environment	Impacted Area (ha)	Nature of Incident
Otorogu	Mangrove forest		Oil spillage incident
Macraba	Mangrove forest		Oil spillage incident
20 sites			Oil spillage incident
RIVERS STATE			
Rumuokwurusi	Freshwater swamp	20	Oil spillage
Rukpoku	Freshwater swamp	10	Oil spillage

Niger Delta Resource Damage Assessment and Restoration Project. (2006). Federal Ministry of Environment, Abuja, Nigerian Conservation Foundation Lagos. WWF UK and CEESP-IUCN Commission on Environmental, Economic, and Social Policy.

TABLE 8.3   Degree of Flaring in Major Gas-Producing Nations (1991)

Nation	Proportion (%)
United State of America	0.6
Holland	0.0
Britain	4.3
Former Union of Soviet Socialist Republic (USSR)	1.5
Mexico	5.0
OPEC NATIONS	
Nigeria	**76.0**
Libya	21.0
Saudi Arabia	20.0
Algeria	19.0

The World Bank (1995). Defining an environmental development strategy for the Niger Delta (Vols. 1 and 2). Washington, DC: World Bank.

8.4 IMPACTS OF ACID RAIN AND GAS FLARING ON THE NIGER DELTA ECOSYSTEM

Another problem within the Niger Delta area is the presence of acid rain caused by gas flaring, which has resulted in biodiversity loss, especially the destruction of forest and economic crops. The dominance of grasses and shrubs in some parts of the region is due to acid rain, which has caused loss of natural forest, although other factors could equally have been responsible, such as agricultural activities and the exploration and exploitation of crude oil (Opukri & Ibaba, 2008; Uyigue & Agho, 2007). Studies have shown that the concentration of acid in rainwater appears to be higher in the Niger Delta region and decreases further away from the region (Uyigue & Agho, 2007). Moreover, as Nigeria remains top among countries that flare gas (Table 8.3), the heat from gas flaring

TABLE 8.4   Impacts of Crude Oil Exploration on Habitats

Reference	Period	Response
Lewis (1981)	Acute	
	Less than15 days	Casualties of birds, turtles, fish, and invertebrates
	10–30 days	Defoliation and casualties of small (<1 m) mangroves; loss of aerial root community
	Chronic	
	30 days–12 months	Defoliation and casualties of medium (<3 m) mangroves; tissues damage to aerial roots
	1–5 years	Death of larger (>3 m) mangroves; loss of oiled aerial roots and regrowth of new ones (sometimes deformed); decolonization of oil-damaged areas by new seedlings
	1–10 years	Reduction in litter fall, reduced reproduction, and reduced survival of seedlings; death or reduced growth of young trees colonizing oiled sites? Increased insect damage?
	10–15 years	Complete recovery
Lamparelli et al. (1997)	Initial effect	
	0–12 months	Seedling and saplings die; no structural alterations can be measured
	Structural damage	
	1–4 years	High mortality is observed, and the oil impact can be measured in terms of major structural alterations
	Stabilization	
	4–9 years	No or few additional alterations to the structural parameters; sapling growth is observed
	Recovery	
	> 9 years	It is possible to measure improvements in the structural tree parameters; ecosystem may not recover fully to its original state

National Research Council, Oil in the Sea 3: Inputs, Fates, and Effects. National Academy Press, Washington, D.C.

exterminates flora around the flaring area, destroys mangrove swamps and salt marshes, declines the growth of some plants, and induces soil degradation and atrophied agricultural productivity (Mba, 2000; UNDP, 2006).

Oil exploration and exploitation activities have significantly contributed to environmental degradation of the Niger Delta region (Table 8.4) (Bayode et al, 2011). Gas flaring in the area is a major source of CO_x, NO_x, SO_x, and particulate matter, and the cumulative environmental impacts of these flaring activities result in contaminant build-up on land and shallow groundwater. It also contributes to the greenhouse effect and general global warming and has also caused high concentration of acid rain within the region (Bayode et al, 2011; FGN, 2012; Irhivben & Omonona, 2013). In terms of organisms, they vary greatly in their sensitivity to petroleum hydrocarbons, and predicting the environmental impacts of specific releases of a quantity of petroleum hydrocarbon requires much site-specific information about the nature of the receiving body (Bayode et al., 2011). In an aquatic environment, oil film floating on the surface of water deprives natural aeration and leads to death of freshwater or marine life. On land, it leads to declined vegetation growth and soil infertility for a long period.

8.5 IMPACTS OF PETROLEUM-RELATED ACTIVITIES ON FISHERIES

There is no doubt that oil pollution caused by spillages from the oil industry located primarily in the Niger Delta region has done a great deal of harm to the region. Some past spills have necessitated the complete relocation of some communities, loss of ancestral homes, pollution of freshwater, loss of forest and agricultural land, destruction of fishing grounds, and reduction of fish population, which is the major source of income for the Niger Delta people (Tolulope, 2004). It is a complex and problematic issues to achieve remediation of oil-contaminated areas, as it takes a few years to effect recovery, and this depends on the geology and type of soil (Salako, Sholeye, & Ayankoya., 2012). Thus, laws to prevent unnecessary destruction of the environment as well as regulate the activities of the oil companies were enacted. But whether these laws have been strictly enforced by the agencies empowered by law to do so, and whether punishments or penalties stipulated by these laws to deter potential offenders are adequate are debatable. For instance, there is a fineable sum of 500,000 Naira (USD1,430) "for each day of failure to report the occurrence" to the National Oil Spill Detection and Response Agency (NOSDRA, 2006). For neglecting a cleanup "to all practical extent, including remediation", a fine of up to one million Naira (USD2,857) is levied on the offender (Onwuteaka, 2016) (Fig. 8.3).

The use of fossil fuel such as petroleum affects the earth biosphere, as it releases pollutants and greenhouse gases into the air through events such as oil spills (Oteh & Eze, 2012). Crude oil exploration has brought numerous devastating environmental incidents that upset ecological balance (e.g., distorted food chain), thereby destabilizing the biology and biodiversity of the ecosystem. All stages of crude oil exploration cause grave consequences on aquatic and terrestrial biodiversity. When crude oil is discharged on the surface of water, whether marine or fresh, it undergoes physical, chemical, and biological alteration, is moved by wind and currents, and is injected into the air. Crude oil impact in an ecosystem may cause physical smothering, which affects the physiological functions of fish exposed to it. The organisms at the sea surface or at the water's edge are more susceptible to oil contact. Baker (1971) reported that research carried out on the toxicity of Bonny Light Crude Oil (BLCO) Premium Motor Spirit (PMS), Qua Iboe Light Crude Oil (QILCO), and Dual Purpose Kerosene (DPK) on aquatic organisms has revealed their lethal, acute, short- and long-term effects on organisms.

FIGURE 8.3 Crude oil spill in the Niger Delta.

Baker (1971) also reported that as little as 0.1 ppm of crude oil can seriously affect fish, amphibians, crustaceans, and plankton. Crude oil has the potential to kill quickly by coating aquatic lives and interfering with gas exchange necessary for life. The contaminating impact has longer and perhaps more severe effects on benthic organisms (Baker, 1971). Sloman and Wood (2004) reported that physical environmental disturbances directly affect fish at the ecological level by rupturing fish assemblages and causing fragmentation and loss of genetic variability in fish populations. Thus, crude oil exploration poses grave danger to the biodiversity of the aquatic ecosystem. Oil toxicity also poses danger within 48 hours after a spill on different species of birds by reducing their fur insulating ability and damaging their water repellent nature (Salako et al., 2012).

The fish diversity in the Niger Delta is enormous, and it includes both freshwater and marine species (Ekpo, 2012, 2013). Fishes of economic importance include catfish (*Chrysichthys nigrodigitatus, Clarias gariepinus*), African bony tongue (*Heterotis niloticus*), among others. According to the list from the Statistics Unit of the Federal Department of Fisheries (FDF), landings by the artisanal fishers in the different fishing settlements also include crayfish, sole, tilapia, and snakehead. A serious threat posed by oil-related pollution is the impact on underground waters (Agbonifo, 2016; Raji & Abejide, 2013) through seepage of crude oil, which results in underground water pollution. The polluted groundwater recharges aquatic ecosystems (streams, rivers, well, etc.) contaminating these water bodies in the process as well as increasing the cases of water-borne diseases (UNDP, 2006). The fish species in these environments are forced to migrate to safer areas that are free of crude oil pollution. This results in dwindling catch and loss of income to fisherfolks (Fig. 8.4).

FIGURE 8.4   Fisherman inspecting his trap, which is empty due to oil spill. *Adapted from Brown, I., & Tari, E. (2015). An Evaluation of the effects of petroleum exploration and production activities on the social environment in Ogoni Land, Nigeria*. International Journal of Science and Research, 4 *(4): 273–282.*

8.6 CONCLUSION

The exploration of crude oil has had debilitating effects on fisheries and its aquatic ecosystem. Various studies showed the inadequate management gap and insensitivity to approach the colossal damage of oil spill in the Niger Delta region of Nigeria. It is against this backdrop that arise many fragments of economic woes, suffering and uprising in the region. The fate of freshwater fisheries and its aquatic ecosystem, socio-cultural activities of the people, preservation and sustainability of aquatic life, environmental resuscitation, and economic diversity, among other relating factors in the Niger Delta region of Nigeria remain bleak. However, there could be hope if stringent bioremediation policies are implemented, which will involve monitoring, evaluation, and rehabilitation of the affected region by all stakeholders.

References

Agbonifo, P. (2016). Oil spills injustices in the Niger Delta region: Reflections on oil industry failure in relation to the United Nations Environment Programme (UNEP) report. *International Journal of Petroleum and Gas Exploration Management, 2*(1), 26–37.

Aluko, S. (2004). *Making natural resources into a blessing rather than a curse: In covering oil. A reporter's guide to energy and development*. New York: Open Society Institute.

ANEEJ. (2004). *Oil of poverty in the Niger Delta. A publication of the African Network for Environment and Economic Justice.*

Anifowose, B. (2008). Assessing the impact of oil and gas transport on Nigeria's environment. U21. In *Postgraduate research conference proceedings 1*. UK: University of Birmingham.

Awosika, L. F. (1995). Impacts of global climate change and sea level rise on coastal resources and energy development in Nigeria. In J. C. Umolu (Ed.), *Global climate change: Impact on energy development*. Nigeria: DAMTECH Nigeria Limited.

Baker, J. M. (1971). *Seasonal effect. The Environmental effects of oil pollution on littoral community* (pp. 44—45). London: Institute of Petroleum.

Bayode, O. J. A., Adewunmi, E. A., & Odunwole, S. (2011). Environmental implications of oil exploration and exploitation in the coastal region of Ondo State, Nigeria: A regional planning appraisal. *Journal of Geography and Regional Planning*, 4(3), 110—121.

Brown, I., & Tari, E. (2015). An Evaluation of the effects of petroleum exploration and production activities on the social environment in Ogoni Land, Nigeria. *International Journal of Science and Research*, 4(4), 273—282.

Central Institute for Freshwater Aquaculture (CIFA). (1993). *A strategic reassessment of fish farming potential in Africa*. CIFA Technical Paper. No. 32. Rome: FAO. 1998. 170 pp.

Chauvin, N., Mulangu, F., & Porto, G. (2012). *Food production and consumption trends in Sub-Saharan Africa: Prospects for the transformation of the agricultural sector* (pp. 1—74). UNDP working paper for African human development report.

Chinweze, C., & Abiola-Oloke, G. (2009). Women issues, poverty and social challenges of climate changes in the Nigerian Niger Delta context. In *7th International conference on the human dimension of global environmental changes*. Bonne, Germany: UN Campus.

CIA. (2013). *The world fact book*. Retrieved 03.03.13, from <https://www.cia.gov/library/publications/the-world-factbook/geos/ni.html>.

Civil Liberties Organization. (2002). *Blood trail: Repression and resistance in the Niger Delta*. Ikeja: CLO.

Ekpo, I. E. (2012). Diversity and condition factor of fish species of Ikpa River at Nwaniba in Niger Delta, Nigeria. *FUTA Journal Research in Science*, 1(1), 36—47.

Ekpo, I. E. (2013). Effects of physico-chemical parameters on zooplankton species and density of a tropical rainforest river in Niger Delta, Nigeria using Canonical Cluster Analysis. *The International Journal of Engineering and Science*, 2(4), 13—21.

Essien, A. I., & Effiong, J. O. (2010). *Economic implications of fish landings in Nigeria*: A case study of Ayadehe and Oku Iboku fishing communities in Itu Local Government Area of Akwa Ibom state, Nigeria. *International Journal of Economic Development Research and Investment*, 1(2&3), 10—18.

Falola, T., & Heaton, M. (2008). *A history of Nigeria*. (p. 183). New York: Cambrige University Press.

Federal Government of Nigeria. (2012). *Nigeria's path to sustainable development through green economy: Country report to the Rio +20 summit*.

Fishing News International (FNI). (1980). *Fishing news International Fisheries statistics of Nigeria* (1st ed.). FNI 19(5): 1980.

Gabriel, U. U., Akinrotimi, O. A., Bekibele, D. O., Onunkwo, D. N., & Anyanwu, P. E. (2007). Locally produced fish feed, potentials for aquaculture development in Sub-Saharan. *Africa, African Journal of Agricultural Research*, 2(7), 287—295.

Idowu, O. F. (2012). Niger Delta crises: Implication for society and organizational effectiveness. *British Journal Art and Socila Sciences*, 7(2), 100—112.

Irhivben, B. O., & Omonona, B. T. (2013). Implications of oil exploration on agricultural development in Delta State, Nigeria. *International Journal of Humanities and Social Science Invention*, 2(4), 59—63.

Kuruk, P. (2004). Customary water laws and practices: Nigeria. <http://www.fao.org/legal/advserv/FOA/UCNCS.Nigeria.pdf>.

Lamparelli, C. C., Rodeigues, F. O., & Orgler de Moura, D. (1997). Long-term assessment of an oil spill in a mangrove forest in Sao Paulo, Brazil. In B. Kjerfve, L. Drude de Lacerda, & W. H. SalifDiop (Eds.), *Mangrove ecosystem studies in Latin America and Africa* (pp. 191—203). Paris, France: UNESCO.

Lewis, E. L., & Perkin, R. G. (1981). The practical salinity scale 1978. *Deep Sea Research*, 28A(4), 307—328.

Lipton, M. (2012). *Learning from others: Increasing agricultural productivity for human development in Sub-Saharan*. United Nations Development Programme, Regional Bureau for Africa. <http://web.undp.org/africa/knowledge/WP-2012-007-Lipton-Agriculture-Productivity.pdf>.

Mba, C.H. (1995). Mineral resources exploitation in Nigeria: The need for effective physical and environmental planning. In *Paper presented at the 26th annual conference of Nigerian Institute of Town Planners*, held at Hill-Station Hotel, Jos, Nigeria, October 1995.

II. THE EFFECTS OF CRUDE OIL EXPLORATION ON THE SOCIO-CULTURAL AND ECO-ECONOMICS OF NIGERIAN ENVIRONMENT

Mba, C. H. (2000). Environmental protection and national development: Towards optimal resolution of conflicting measures and strategies. In A. Akpuru-Aja, & A. C. Emeribe (Eds.), *Policy and contending issues in Nigerian national development strategy*. Enugu, Nigeria: John Jacob's Publishers, Ltd.

National Oil Spill Detection and Response Agency (NOSDRA). (2006). Oil spill recovery, clean-up, remediation and damage assessment regulations, 2011. Section 17; 6.

Nicholas, O. E. (2016). Aquaculture for sustainable development in Nigeria. *World Scientific News, 47*(2), 151–163.

Niger Delta Resource Damage Assessment and Restoration Project. (2006). Federal Ministry of Environment, Abuja, Nigerian Conservation Foundation Lagos. WWF UK and CEESP-IUCN Commission on Environmental, Economic, and Social Policy.

Nwankwo, J. N. (1984). Oil and environmental pollution. In *Paper presented at the conference on strategies for the fifth national development plan: 1986–1990*, November 25–29. Ibadan: NISER.

Nwilo, C. P., & Badejo, T. O. (2004). *Management of oil dispersal along the Nigerian coastal areas*. Lagos, Nigeria: Department of Survey & Geoinformatics, University of Lagos.

Nwilo, C. P., & Badejo, T. O. (2005a). *Impacts and management of oil spill pollution along the Nigerian coastal areas*. Lagos, Nigeria: Department of Survey & Geoinformatics, University of Lagos. <www.fig.net/pub.figpub36/chapters/chapter_8.pdf>.

Nwilo, C. P., & Badejo, T. O. (2005b). Oil spill problems and management in the Niger Delta. In *International oil spill conference* (pp. 567–570), Miami, FL, USA.

Onwuteaka, J. (2016). Hydrocarbon oil spill clean-up and remediation in the Niger Delta. *Journal of Geography, Environment and Earth Science International, 4*(3), 1–18.

Opukri, O. C. O., & Ibaba, I. S. (2008). Oil induced environmental degradation and internal population displacement in the Nigeria's Niger Delta. *Journal of Sustainable Development in Africa, 10*(1), 173–193.

Otaha, J. (2012). Dutch disease and Nigeria oil economy. *African Research Review, 6*(1), 82–90.

Oteh, C. O., & Eze, R. C. (2012). Vandalization of oil pipelines in the Niger Delta Region of Nigeria and poverty: An overview. *Studies in Sociology of Science, 3*(2), 13–21.

Oyase, A., & Jemerigbe, R. (2016). Contribution of aquaculture to poverty reduction and food security in Nigeria. *International Journal of Applied Microbiology and Biotechnology Research, 4*, 26–31.

Powell, C. B., White, S. A., Ibiebele, D. O., Bara, M., DutKwicz, B., Isoun, M., & Oteogbu, F. U. (1985). Oshikaoil spill environmental impact; effect on Aquatic biology. In *Paper presented at NNPC/FMHE international seminar on petroleum industry and the Nigerian environment* (pp. 168–178), Kaduna, Nigeria.

Raji, A. O. Y., & Abejide, T. S. (2013). An assessment of environmental problems associated with oil pollution and gas flaring in the Niger Delta region Nigeria, C. 1960s–2000. *Arabian Journal of Business and Management Review, 3*(3), 46–62.

Salako, A., Sholeye, O., & Ayankoya, S. (2012). Oil spills and community health: Implications for resource limited settings. *Journal of Toxicology and Environmental Health Sciences, 4*(9), 145–150.

Sanusi, L. S. (2010). *Growth prospects for the Nigerian economy*. Convocation lecture delivered at the Igbinedion University, Benin City, Edo State, Nigeria.

Shell Petroleum Development Company. (1995). *People and the environment*. Annual Report.

Sloman, K. A., & Wood, C. M. (2004). Fish social behaviour and the effects of trace metal contaminants. In K. Sloman, C. Wood, & D. MacKinlay (Eds.), *Behaviour, physiology and toxicology in fish* (pp. 121–126).

The World Bank (1995). *Defining an environmental development strategy for the Niger Delta* (Vols. 1 and 2). Washington, DC: World Bank.

Tolulope, A. O. (2004). *Oil exploration and environmental degradation: The Nigerian experience. International information archives* (2, pp. 387–393). International Society for Environmental Information Science EIA04-039.

United Nation Development Programme. (2006). Niger delta human development report, 218 pp.

Uyigue, E., & Agho, M. (2007). *Coping with climate change and environmental degradation in the Niger Delta of Southern Nigeria*. Community Research and Development Centre Nigeria (CREDC), 31 pp.

Yakub, M. (2008). The impact of oil on Nigeria's economy: The boom and bust cycles. *Bullion-Central Bank of Nigeria, 32*(2), 41–50.

II. THE EFFECTS OF CRUDE OIL EXPLORATION ON THE SOCIO-CULTURAL AND ECO-ECONOMICS OF NIGERIAN ENVIRONMENT

The Impacts of Petroleum Production on Terrestrial Fauna and Flora in the Oil-Producing Region of Nigeria

K.S. Chukwuka[1], C.G. Alimba[1], G.A. Ataguba[2] and W.A. Jimoh[3]

[1]University of Ibadan, Ibadan, Nigeria [2]University of Agriculture, Makurdi, Nigeria
[3]Federal College of Animal Health and Production Technology, Ibadan, Nigeria

9.1 INTRODUCTION

The Niger Delta, the oil-producing region of Nigeria, extends over approximately 70,000 km[2] of Nigeria's land mass. This region, which is comprised of Abia, Akwa-Ibom, Cross River, Delta, Bayelsa, Rivers, Edo, Imo, and Ondo States, separates the Bight of Benin from the Bight of Biafra within the larger Gulf of Guinea. The Niger Delta, the largest wetland in Africa and the third largest in the world (Anifowose, 2008) is characterized with highly diverse ecosystems that support numerous species of terrestrial and aquatic fauna and flora before the discovery of crude oil in 1956. Following crude oil discovery and its subsequent exploitation and exploration, the flora- and fauna-rich biodiversity region has been greatly affected ,with most populations of plant and animal either destroyed or their dynamics altered.

9.2 ENVIRONMENTAL CHARACTERISTICS OF THE NIGER DELTA

The Niger Delta region is characterized with high rainfall, 2500 mm/year, with a temperature range of 18–33°C and humidity that is above 95% all through the year (UNDP, 2016). This second largest delta in the world encompasses rivers, creeks, estuaries, and swamps (Awosika, 1995). Its ecosystem is highly diverse and serves as habitats to numerous species of terrestrial fauna and flora (Uyigue & Ogbeibu, 2007). The biodiversity richness of the Niger Delta region was described as being "extraordinary" and comprised of diverse species of flora and fauna, both aquatic and terrestrial forms (Zabbey, 2004). Ecologically, the Niger Delta region is considered one among the ten most important wetland and marine ecosystems in the world. It was earlier reported to be characterized by two ecological zones: a tropical rainforest in the northern part and an extensive mangrove forest in the southern reaches (Hutchful, 1985). Singh, Moffat, and Linden (1995) later identified and classified the region into four distinct ecological regions: freshwater swamp forests, lowland rainforests, barrier island forests, and mangrove forests. The lowland rainforest zone of the Niger Delta consists of wildlife diversities that include animals from various classes of vertebrates and invertebrates with a good number of these species indigenous to the region (Ebeku, 2005). The mangrove zone covers a total area of more than $150,000 \text{ km}^2$ globally (Spalding, Kainuma, & Collins, 2010) and are found in about 124 countries with tropical climates as well as subtropical regions of temperate countries (FAO, 2007). Africa has a total mangrove cover of $27,945.72 \text{ km}^2$ with West Africa accounting for more than half of this total area (Spalding, Kainuma & Collins, 2010). Nigeria has a total cover of 7355.57 km^2 (Spalding, Kainuma &Collins 2010), which makes it the largest mangrove cover in Africa.

9.3 BIOLOGICAL DIVERSITY OF THE NIGER DELTA REGION

The ecosystem of the area is highly diverse and supports numerous species of terrestrial and aquatic fauna and flora as well as humans (Uyigue & Ogbeibu, 2007). Hence, it was declared the key zone for conservation of the western coast of Africa due to its extraordinary biodiversity (Zabbey, 2004). The composition of diverse plant and animal species in the Niger Delta region compared to any other regions in Nigeria has been well described as the uniqueness of the area (Singh et al., 1995; Ubom, 2010; Ugochukwu & Ertel, 2008). UNDP (2016) reported approximately 148 species of birds from 38 families and high diversity of primates and many other plant and animal species inhabiting the regions. Spalding et al. (2010) reported that six species of mangrove plants were very prominent in the Niger Delta: *Rhizophora racemosa*, *Rhizophora mangle*, *Rhizophora harrisonii*, *Avicennia germinans*, *Laguncularia racemosa*, and *Conocarpus erectus*. In addition to these, *Nypa fruticans* and *Acrotichum aureum* are introduced species that are common in the region (Corcoran, Ravilious & Skuja, 2007; James et al., 2007). It is well known that the mangrove is one of the most productive ecosystems in the world due to its rich community of fauna and flora (Ugochukwu & Ertel, 2008). Other important terrestrial flora prominently reported in the region include *Symphonia globuli*, *Raphia vinifera* (Raphia Palm), and the oil palm (*Elaeis guineensis*) (McGinley & Duffy, 2007). These species are conspicuously seen along the creek of the Niger Delta. Many plant species that are locally and globally labeled as endangered species can be found in the Niger Delta region (UNDP, 2016). This suggests the importance of the Niger Delta in conserving many plant and animal species. Also, the plant diversity is of great local and regional importance (IUCN, 1992).

There are similarly diverse animal species in the Niger Delta region. For instance, the Niger Delta region serves as home to *Loxodonta africana cyclotis* (Forest Elephants), *Trichechus senegalensis* (West African Manatee), *Procolobus epieni* (red colobus monkey),

Choeropsis liberensis heslopi (Pigmy hippopotamus), *Cercopithecus erythrogaster* (white-throated guenon) and *Cercopithecus sclateri* (Sclater's guenon). These six mammals are in the IUCN (International Union for Conservation of Nature) red list as endangered species (UNDP, 2016). Also, a high diversity of primates is found in the region. For instance, the endangered *Pan troglodytes vellerosus* (Nigeria-Cameroun chimpanzee) can be found in the Niger Delta region. Also, *Limnotragus spekei* (aquatic antelope) and *Kobus ellipsyprimnus* (water buck) are present in the tangles of the swamp forest of the region. No fewer than 148 water-related bird species from 38 families can be sighted in the Niger Delta (World Bank, 1995). One hundred and fifty eight genera and 64 families with 20 endemic species of fish have been reported in the region (Powell, 1993). The Niger Delta region remains the stronghold for *Osteolaemus tetraspis* (West African dwarf crocodiles) and much molluscan diversity, including snails and bivalves. In summary, it was estimated that more than 46,000 plant species of which about 205 are endemic, and approximately 484 species in 112 families threatened with extinction exist in the Niger Delta region (Salau, 1993). Also, approximately 24 out of 274 mammals, 10 out of 831 birds, and 2 out of 114 reptiles that are reported to be in the region are endangered (WRI, 1992).

9.4 IMPORTANCE OF BIODIVERSITY TO THE TERRESTRIAL ECOSYSTEMS OF THE NIGER DELTA REGION

Biodiversity is the key factors in an ecosystem: functioning, stability and resilience (Chaplin et al., 1997). The larger population of the Niger Delta survives on services provided by this ecosystem: agriculture, industry, fishing, food, drinking water, wood, shelter, medicine, employment and esthetics. Genetic and species diversities are the stabilizing factors in any ecosystem diversity. These factors are closely related, with each interacting and influencing the other. If one of the factors is altered, it will affect the other, which may lead to ecosystem instability, an indicator of degradation. This can be exemplified using a well-known biological process: during photosynthesis, plants produce and release oxygen as a by-product. This helps to maintain the quality of the atmosphere by providing oxygen for respiring organisms. The vegetation of the wetlands absorbs carbon dioxide, the by-product of respiration for food synthesis, hence purifying the atmosphere. This suggests that forests in the Niger Delta play the role of carbon sequestration by using carbon dioxide produced by respiring animals for photosynthesis, a natural system of cleaning the atmosphere. Wetlands of the Niger Delta provide shelter and food for numerous animal species in the region (Dupont, Jahns, Marret, & Ning, 2000). The decomposers in the Niger Delta ecosystems play a key role in waste decomposition via organic waste decay, hence helping in solid waste disposal and subsequently increasing soil nutrient renewal and soil formation (Abam, 2001; Kokwono, 1994; Uluocha & Okeke, 2004). The biological activities in the Niger Delta region due to great plant and animal diversities include but are not limited to support of parasite-host, hunt-hunted, symbiotic and other relationship among organisms; manufacture of food for various organisms by green plants; and hydrological, mineral, and gaseous exchange between the soil, water and the air (Alonso, Dallmeier, Granek, & Raven, 2001). This accounts for interdependence existing among the biotic and abiotic components of the ecosystem. Natural processes in the system enhance the self-supporting nature of the system by regulating the population of organisms to accommodate the carrying capacity and functioning of the ecosystem. Also, the flow of energy and recycling of organic nutrients are maintained to enhance the ecosystem integrity (Iwegbue, Ekakitie, & Egun, 2006). The ecosystem integrity provided by the biodiversity in the region is very important and needs to be maintained sustainably, as direct loss of habitat by human activities may impact adversely on the diversity of the species. Human activities that will greatly impact on the adaptability and survival of plant and animal species in the Niger Delta region are ecosystem degradation, loss of species diversity, and environmental contamination.

9.5 IMPACTS OF OIL EXPLORATION AND SPILLAGE IN THE NIGER DELTA REGION OF NIGERIA

Oil was first discovered in 1956 by Shell British Petroleum (now Royal Dutch Shell) at Olobiri in the eastern Niger Delta of Nigeria. However, petroleum exploitation started in 1958 (NNPC, 2016). Currently, there are nine oil-producing states in Nigeria, with 185 local government areas. There are more than 800 oil-producing communities in the Niger Delta, with more than 900 producing oil wells and several petroleum production—related facilities (Osuji & Onojake, 2004). Oil and gas pipelines cover approximately 7000 km (Anifowose, 2008; Onuoha, 2007). The Niger Delta has the largest natural gas reserve and the second largest oil reserve in Africa (Kadafa, 2012). Oil exploration and exploitation, which was discovered several decades ago, is still on-going in the Niger Delta. This has increased oil spillage and gas flaring in the region, leading to environmental pollution.

9.6 IMPACTS OF OIL POLLUTION IN THE NIGER DELTA ENVIRONMENT

The processes involved in oil exploration and exploitation have numerous negative impacts on biodiversity and ecosystem functioning, especially when there is little or no remediation process at the postexploration or the postproduction process. Starting from the clearing of seismic lines for oil exploration to the installation of pipelines requires destruction of a larger area of the terrestrial habitat to make pipeline tracks. These pipelines are laid across the rainforest and mangrove regions. With incessant vandalism and failure of pipeline integrity due to aging and defects in material, high incidences of oil leakages and accidental discharges into the environment are enormous. Most oil leakages from pipelines and incidences of oil discharge occur in the mangrove swamp forest, the most reproductive ecosystems (Zabbey, 2004). Therefore, the exploration and production of petroleum and its subsequent transportation and distribution in the Niger Delta have led to pollution of terrestrial and aquatic habitats, with serious threats to associated flora and fauna (Uluocha & Okeke, 2004). Current findings have shown that more than 70 Protected Areas in the Niger Delta of Nigeria have lost substantial portions of their areas, including the biodiversity (Phil-Eze & Okoro, 2009). Oil exploitation in the Niger Delta of Nigeria has similarly increased deforestation and illegal logging activities. This led to alterations in animal habitats, loss of biodiversity, and vegetation fragmentation. These occurrences led the United Nations' Human Development to make the following statement about the Niger Delta region of Nigeria, "there is a strong feeling in the region that the degree and rate of degradation are pushing the delta towards ecological disaster" (UNDP, 2006). The oil fraction emissions and associated pollution resulting from the activities of oil industries in the Niger Delta have strong harmful impacts in the region. The oil fractions (petroleum hydrocarbon) can bioaccumulate and bioconcentrate in lower forms and can be passed on through the food chain to higher trophic level organisms, including humans (Fu et al., 2003).

9.7 EFFECTS OF OIL EXPLORATION ON SOIL COMPACTION AND ORGANISMS

Soil compaction may be beneficial or detrimental to the growth of plants; however, the detrimental effects are much more obvious than the benefits. The detrimental effects of soil compaction on the growth and yield of plants and field crops cannot be overlooked. High levels of soil compaction are usually encountered in oil exploration sites and timber harvesting sites (with high forest and vegetation destruction). During oil exploration, compaction can occur via the use of tillage tools or heavy machinery or may be due to constant and heavy pedestrian traffic. Compaction obviously changes the soil structure and hydrology by altering the soil bulk density and fragmenting the soil aggregates, decreasing soil porosity and aeration as well as the infiltration capacity of the soil. Consequently, this leads to increased water runoff and soil erosion. Severe soil compaction affects plants physiology (Kozlowski, 1999). This will similarly affect plant root structure with reduced water absorption and subsequent plant-water balance and relations, and thus lead to reduced plant leaf development, which will affect photosynthesis due to smaller leaf sizes. Soil compaction also affects the functioning of plant growth hormones, especially abscisic acid and ethylene. The compaction of surface and subsoils adversely affects mineral absorption in plants. Serious soil compaction affects forest regeneration negatively by inhibiting seed germination and seedling growth and leading to the death of plants.

9.8 IMPACTS OF VEGETATION CUTTING/CLEARING ON THE ANTHROPOLOGICAL/CULTURAL HERITAGE OF THE NIGER DELTA PEOPLE

The Niger Delta region of Nigeria is synonymous to the goose that lay the "golden egg" from the standpoint of income and revenue generation to the country and as such can be compared with the biblical "Aaron" in that she has nothing to show for her natural endowment. Oil exploration and production activities in this region ushered many economic, social, anthropological, and/or cultural challenges into the traditional lifestyle of the indigenous people of the Niger Delta region. The most obvious of them involved the land-use patterns, emigration (due to increased access and opportunities for job placement), sociocultural influences from foreigners and new entrants into the region, and access to goods and services (education, healthcare, social amenities, etc.). Be that as it may, oil exploration and production in the area, instead of benefiting the indigenes, made them victims of environmental mismanagement.

Indeed, the Niger Delta region experiences both environmental and cultural challenges. However, the environmental challenges are faced on a daily basis, whereas the cultural challenges are better imagined than seen. The shrines where gods and goddesses are worshiped have been desecrated, cleared, and no longer exist and are replaced with flow stations (Fentiman, 1996). On the most obvious note, native people attach importance to the conservation of their cultural assets and therefore destruction of these cultural values is undesirable. It is important to emphasize that these irreplaceable heritages cannot be quantified in monetary terms. Cultural heritage in this region is considered as the most visible sign of collective identity and therefore when threatened, the perpetrators are conceived and perceived as enemies of such heritage. Hence, one of the many causes of anxiety today is the concern about the loss of cultural heritage.

9.9 EFFECTS OF DRILLING AND NOISE ON GERMINATION, GROWTH, AND DEVELOPMENT OF PLANTS

Drilling activities involve vibrations from blasting and forceful compaction of soils. The forceful effects of these vibrations may create substantial problems for surrounding flora, influencing their structures and physiology (Svinkin, 2008). These drilling activities are accompanied by noise and/or sound, which may affect the growth and development of living organisms in varying degrees. Sound has both positive and negative effects on the growth and development of plants; however, no known effect has been reported on the germination of seeds. Few records on the adverse effects of noise show that it has negative effects on seedling recruitment, survival, and establishment (Francis, Kleist, Ortega, & Cruz, 2012). Random noise has been reported to affect plant growth negatively (Collins & Foreman, 2001). Similarly, in a recent study, noise has been reported to have profound effects on seed germination, plant height, and number of leaves of *Cyamopsis tetragonoloba* grown in India (Vanol & Vaidya, 2014).

9.10 IMPACTS OF OIL SPILLAGE ON SOIL PROPERTIES

Generally, in most oil-producing countries of the world, the major causes of oil spillages include pipeline leakages, hose failures of loading tankers, malfunctioning of the vehicular-hose manifold, blowouts, oil wells and flow-lines sabotage, and over-pressure failures/overflow of process equipment components (Awobajo, 1981; Okpokwasili & Amanchukwu, 1988). Among the abiotic components of the ecosystem, the soil, the recipient and major reservoir of the spilled oil, is the most affected by oil spillage. Petroleum products in the form of spills in the soil are readily absorbed and bioaccumulated in plants from where they are taken up by other organisms of higher trophic levels in the food chain. These spills are known to contain heavy metals, which are hazardous, toxic, and carcinogenic and therefore constitute health risks to humans and other living organisms. Oil spills affect the physico-chemical properties of the soil in no small measures. Benka-Coker and Ekundayo (1995) reported that there were heavy metal concentrations (lead, iron, and zinc) in oil-spilled zones of the Niger Delta. They also reported that electrical conductivity, exchangeable cations, available phosphorus, and total nitrogen of these soils were comparatively low, whereas the total organic carbon was high.

9.11 IMPACTS OF OIL SPILLAGE ON PLANTS AND AGRICULTURAL FARMLANDS

The degradation of agricultural land due to oil prospecting, exploration, and production has caused untold hardship to the growing population of the Niger Delta region in Nigeria. This degradation is commonly observed as shoreline erosion and landscape destruction, thereby reducing the available land space for agriculture and food production. Resulting from exploitation of the petroleum resources, the remaining available agricultural lands have been made to become uncropable due mainly to laying of pipelines and associated oil spillage. Crude and refined oil spillage has deleterious effects on species diversity and primary productivity of a terrestrial ecosystem. It slows down vegetation recolonization and accelerates habitat deterioration via sheet erosion. Generally, vegetation recovery following an oil spill is abnormally slow (Kinako, 1981). Different studies independently reported the effects of crude and refined oils, including spent lubricating oil, on soils and soil organisms. They concluded that oil derivatives and/or components suppress seed germination and evapo-transpiration in plants, leading to stomatal abnormalities and death in diverse food crops (Agbogidi, Eruotor, & Akparobi, 2006; Anoliefo & Vwioko, 1994; Anoliefo, 1991; Atuanya, 1987; Gill, Nyawuame, & Eruikhametalor, 1992; Rowell, 1977). This will increase or constitute a threat to food security in the Niger Delta region and Nigeria in general.

9.11.1 Germination of Seeds

Environmental pollution and degradation resulting from oil spillage in major oil-producing regions of the world as in the Niger Delta cannot be eschewed. The effects of oil spillage on the germination of seeds have been reported by several researchers. Udo and Fayemi (1974) reported that oil spillage accounted for a 50% reduction in the germination of *Zea mays* L. seeds and seedling development. The poor growth of the seedlings was attributed to suffocation of the plants due to exclusion of air by the oil, which interfered with the plants' soil-water relationships. Similarly, Agbogidi and Eshegbeyi (2006) showed that the germination and seedling growth of a cash crop, *Dacryodes edulis* (African pear) (an endangered species) in oil-producing areas of Delta State was affected by oil spillage. Amadi, Dickson, and Moate (1993) and Amadi, Abbey, and Nma (1996) reported that crude oil concentration above 3% in the soil will reduce germination by suffocating seeds, thereby affecting their physiological activities. Anoliefo and Vwioko (1994) similarly reported that contamination of soil with 4% and

5% spent oil consistently inhibited germination of hot pepper and tomatoes seeds. Evidence on the toxic effects of crude and spent oil on plant seed embryos abounds (Baker, 1970; Gill, Nyawuame & Eruikhametalor, 1992; Kolattukudy, 1979). Their findings show that penetration of crude and/or spent oil into seed embryos, plant tissues, and cells is lethal and can lead to plant injury and/or death.

9.11.2 Seedling Growth, Development and Yield

Oil exploration at all stages has adverse consequences on the environment. The most important and persistent environmental problem in the oil-producing areas of Nigeria is oil spillage. More than 6000 spills had been recorded within the four decades of oil exploitation in Nigeria (Department of Petroleum Resources, 1997), with an average of 150 spills per annum. Environmental impacts of oil spill on the teaming population of Nigeria, especially those living within oil-producing areas, are enormous. Oil spills have impacted negatively on most agricultural lands in the oil-producing areas of Nigeria and have turned productive lands to wastelands; as a result, farmers have been forced to abandon their agricultural fields. Chindah and Braide (2000) reported that oil spill caused great damage on crops grown in the Niger Delta communities due to high retention time of oil occasioned by limited flow. Similarly, Onyegeme-Okerenta, Alozie, and Wegwu (2015) reported significant decrease in chlorophyll contents of *Abelmoschus esculentus* L. Moench (Okra) in a polycyclic aromatic hydrocarbon (PAH)—polluted agricultural land in an oil-producing region of Nigeria. The result of this is reduction in land productivity, crop yield, and farmers' income, with consequent food insecurity. In the same vein, Inoni, Omotor, and Adun (2006) reported that oil spill reduced land productivity and crop yield by 10% while farmers' income dropped by 5%.

9.12 IMPACTS OF GAS FLARING ON SOIL PROPERTIES

Gas flaring in the Niger Delta region of Nigeria dates back to 1956 with the discovery and exploitation of petroleum products. The Niger Delta region is one of the most oil-impacted and polluted areas on a global scale (Ikelegbe, 2005). This is attributed to the uncontrolled and wasteful flaring of gas during oil production, which has impacted negatively on the terrestrial ecosystem in the region. The negative impacts of gas flaring on soil and terrestrial ecosystems in the Niger Delta, Nigeria, have been well studied. Ukegbu and Okeke (1987) reported that gas flaring increased air/soil temperature, which led to a 20%−61% reduction in the soil microbial load in Imo State, one of the oil-producing zones of the Niger Delta region. Alterations in soil temperature subsequently affected the various soil properties, including important nutrient elements (N, P, K, and Na) and electrical conductivity (Atuma & Ojeh, 2013). This will invariably lead to reduced soil fertility and life forms in the area. Also, gas flaring has been reported to increase soil acidity, which simultaneously depleted soil organic matter and total nitrogen content of the affected soil (Alakpodia, 2000; Ogidiolu, 2003). This will make the soil unproductive as most soil animals and microbes will be destroyed (Table 9.1).

TABLE 9.1 Gas Production and Utilization Between 1970 and 2010 (Million m^3) in the Niger Delta Region, Nigeria

Year	Production	Utilization	% Utilized	Flaring	% Flared
1970	8039	82	0.89	7957	98.98
1971	12,975	185	1.43	2790	98.57
1972	17,122	274	1.60	16,848	98.40
1973	21,882	395	1.81	21,487	98.19
1974	27,170	394	1.45	26,776	98.55
1975	18,656	323	1.73	18,333	98.27
1976	21,276	659	3.10	20,617	96.90
1977	21,924	972	4.43	20,952	95.57
1978	21,306	1866	8.76	19,440	91.24
1979	27,619	1546	5.60	26,073	94.40
1980	24,551	1647	6.71	22,904	93.29
1981	17,113	2951	17.24	14,162	82.76
1982	15,382	3442	22.38	11,940	77.62
1983	15,192	3244	21.35	11,948	78.65
1984	16,255	3438	21.15	12,817	78.85
1985	18,569	3723	20.05	14,846	79.95
1986	18,739	4822	25.73	13,917	74.27
1987	17,085	4794	28.06	12,291	71.94
1988	20,253	5516	27.23	14,737	72.76
1989	25,053	6323	25.24	18,730	74.76
1990	28,163	6343	22.52	21,820	77.48
1991	31,587	7000	22.16	24,588	77.84
1992	32,465	7058	21.74	25,406	78.26
1993	33,445	7536	22.50	25,908	77.50
1994	32,793	6577	20.06	26,216	79.94
1995	32,980	6910	20.95	26,070	79.05
1996	35,450	8860	24.99	26,590	75.01
1997	37,150	12,916	34.76	24,234	65.23
1998	37,039	13,407	36.19	23,632	63.80

(Continued)

II. THE EFFECTS OF CRUDE OIL EXPLORATION ON THE SOCIO-CULTURAL AND ECO-ECONOMICS OF NIGERIAN ENVIRONMENT

TABLE 9.1   (Continued)

Year	Production	Utilization	% Utilized	Flaring	% Flared
1999	43,636	21,274	48.75	22,362	51.25
2000	42,732	18,477	43.23	24,255	56.76
2001	52,453	25,694	48.98	26,759	51.02
2002	48,192	23,357	48.46	24,835	51.53
2003	51,766	27,823	53.74	23,943	46.25
2004	58,964	33,873	57.44	25,091	42.55
2005	59,285	36,282	61.19	23,003	38.80
2006	82,037	53,453	65.15	28,584	34.84
2007	84,707	57,400	67.76	27,307	32.24
2008	80,604	58,793	72.94	21,811	27.06
2009	64,883	46,896	72.27	17,987	27.72
2010	67,758	51,290	75.69	16,468	24.30

Sources: Combined adaptation from Central Bank of Nigeria. (1996). Statistical Bulletin 7 (1), 138 (CBN, 1996) and Adole, T. (2011). A Geographic Information System (GIS) based assessment of the impacts of gas flaring on vegetation cover in Delta State, Nigeria. M.Sc. Dissertation, Environmental Sciences of the University of East Anglia, Norwich (Adole, 2011).

9.13 IMPACTS OF GAS FLARING ON STANDING PLANTS AND CROPS

Gas flaring impacts negatively on the terrestrial ecosystems with, particular emphasis on plant growth and development. According to Isichei and Sandford (1976), flare sites had higher temperatures, and this had profound effects on soil fertility, standing vegetation, and crop growth. A survey of vegetations within and around 100 m from flare sites revealed that these vegetations were charred. High temperature does not enhance plant physiology, therefore impairing a plant's normal growth, photosynthesis, and flowering. Orimoogunje, Ayanlade, Akinkuolie, and Odiong (2010) similarly observed that high temperature changes in a gas-flared site led to stunted crop growth, scorched plants, and withering of young crops. This report was in support of the findings that cassava yield was negatively affected at gas-flared areas (Atuma & Ojeh, 2013). In general, oil spills have been shown to cause various damages to marsh vegetations. These damages result in the reduction of the following parameters: growth, photosynthetic rate, stem height, density, and above-ground biomass of *Spartina alterniflora* and *Spartina patens* and eventual death of the plants (Krebs & Tamer, 1981).

9.14 TOXIC IMPACTS OF CRUDE OIL POLLUTION ON MANGROVE PHYSIOLOGY

Apart from the aforementioned impacts of crude oil on soil and plants, its exploration activities and associated discharges on mangrove ecosystems in the Niger Delta severely affects mangrove anatomy and physiology. The mechanism of toxicity depends on the type of oil as well as the period of exposure. According to Shigenaka (2014), light oils impact more toxicity at the acute toxicity stage to mangroves than do heavy oils. The Nigerian Bonny light crude would therefore be acutely toxic to mangroves if spilled or deliberately leaked into the environment. Although, specific accounts of the effects of crude oil exposure on Nigerian mangrove species are not completely documented; however, world history from oil spillage and leakages showed that mangrove ecosystems are the most severely impacted by the petroleum fractions. The water around mangroves, sediment, mangrove leaves, pneumatophores, and roots have been reported to be sensitive to oil fractions (Diab & Bolus, 2014). Oil coating on the water surface restricts free exchange of oxygen between water and air. Mangrove pneumatophores and accompanying lenticels normally reside within the intertidal region that is also affected by oil pollution, and this affects gaseous exchange.

The combination of physical impacts and toxicity of the oil creates additional stress on the mangroves; however, there is no cumulative effect of exposure of seedlings to different oils at different durations of exposure (Proffitt & Devlin, 1998). It is possible that the light fraction of oil and lightweight hydrocarbons released during gas flaring, which find their way into the surrounding water, will reduce the ability of the mangroves to exude salts. Adsorption of oil by mangroves through the pneumatophores is responsible for toxicity of crude oil to mangroves, with greater adsorption occurring at periods of low atmospheric temperatures regardless of diurnal changes (Partani, Ghiassi, Darban, & Saeedi, 2015). Decomposition of mangrove leaves has been shown to be exacerbated by hydrocarbons, as reported during an oil spill in Tacarigua, Venezuela (Bastardo, 1993). Defoliation is the first response expressed from mangroves as a result of oil toxicity, followed by mortality, which in fact is a function of the amount of leaves lost, while leaf deformation is also observed, but seedling mortalities occur due to prolonged exposure (Lacerda et al., 2013).

Oil fractions have been shown to induce genetic toxicity in mangroves species. Mutagenesis in mangroves as a result of exposure to PAHs from crude oil was first reported in the form of loss of chlorophyll-synthesizing genes due to heterozygosity for the gene (Klekowski, Corredor, Morell, & Del Castillo, 1994). This was further supported by the subsequent studies of Duke and Watkinson (2002) and Profit and Travis (2005) in which they concluded that the presence of PAHs in the sediments is the major reservoir that unleashes the mutagenic effects on the mangroves.

Mangroves are affected by changes in soil conditions (McKee, 1993, 1995), hence source operations and recordings during seismic operations (Van Dessel & Omoku, 1994), dredging (Corcoran, Ravilious & Skuja, 2007), and oil spillage (Kadafa, 2012) affect mangroves since soil conditions are altered in each case. Seedlings of *Avicenia marina* were reported to be killed massively by presence of crude oil within 14 days, with only 0.6% survival, while propagules were seemingly unaffected by shading and canopy but were inhibited from establishment by the presence of crude oil in soil sediment (Grant, Clarke, & Allaway, 1993). Similarly, exfoliation, total biomass, and chlorophyll content of *R. racemosa* is impacted negatively by application of crude oil in the soil at chronic, acute, and moderate levels (Mensah, Okonwu, & Yabrade, 2013). Acute and chronic exposure of *A. germinans* to crude oil was reported to affect growth and survival, with high response scores being recorded for treatment (Chindah, Braide, Amakiri, & Onokurhefe, 2008). Sodre et al. (2013) reported that although some key physiological indices of *L. racemosa* subjected to oil contamination in their growth substrate was relatively unaffected by the presence of the oil as a contaminant, the shoot dry weight was only about 80% of normal weight recorded for shoots after 128 days of growth. Proffitt and Devlin (1998), reported a contrary result from treatment of *R. mangle* seedlings and saplings with crude oil after the initial treatment and concluded that oiling episodes have no cumulative effect on the mangrove regardless of life stage, oil type, and method of oil application. This conclusion was, however, not in conformity with the findings of Ellison (1999), who suggested that several factors, including experimental design during the trials and nurturing of the plants, as well as deviations from real field conditions, may have been responsible for the variation in results.

Chronic exposure of mangroves is difficult to control since the nature of mangroves make the cleaning of spills difficult. However, Shigenaka (2014) suggested that chronic exposure to crude oil can be determined via several symptoms presented by the mangrove species. These include:

1. Differential growth, changes in reproduction periods as well as approach
2. Development of anatomical features to help cope with the stress
3. Reduced resilience to other stressors

Mangrove roots tend to accumulate more oil than other parts, hence they are affected by chronic toxicity. According to Diab and Bolus (2014), roots of *A. marina* exposed to petroleum pollutants accumulated more than 500% of the oil in the roots than the leaves. Similarly, the use of Geographical Information System in studying the effects of crude oil spill on mangroves in Southeast Brazil showed that trees had structural defects several years after exposure (Santos, Cunha-Lignon, Schaeffer-Novelli, & Cintrón-Molero, 2012).

9.15 IMPACTS OF CRUDE OIL DISPERSION, EXPLORATION, AND GAS FLARING ON TERRESTRIAL FAUNA

The biological effects of oil pollution on terrestrial fauna have been well studied based on the composition of crude oil and the relative toxicity of its fractions. Crude oil is complex mixtures of natural products, with a wide range of molecular weights and structures. The low-boiling saturated hydrocarbons (gasoline range) at low concentrations produce anesthesia and narcosis and at greater concentration, cell damage, and death in a wide variety of soil invertebrates and lower vertebrates (Onwurah, 2000). The higher-boiling saturated hydrocarbons (kerosene and lube oil range) are probably not directly toxic, although they might interfere with nutrition and the reception of chemical clues that are necessary for communication between many animals (Onwurah, 2000). Aromatic hydrocarbons are abundant in crude oil and represent the most dangerous fraction. Low-boiling aromatics (benzene, toluene, xylene) can cause acute poisoning to many lung-breathing invertebrates and vertebrates, including man.

There is more information on field studies with plants due to their inability to move, whereas scanty information has been documented with their fauna counterpart. However, experimental studies abound and have shown that crude oil altered hematology, biochemical parameters, and histopathology of the liver and kidney, attesting to its systemic toxicity in rodents (Leighton, Lee, Rahimtula, O'brien, & Peakall, 1985; Orisakwe et al., 1989). In Nigeria, crude oil is predominantly found in the Niger Delta areas and over the years, the population residing in this region have continuously used this crude oil as an antidote for various ailments, including gastrointestinal disorders, burns, foot rot, leg ulcers, and poisoning (Orisakwe et al., 1989). However, no study was found that has reported the incidence of cancer, despite the high level of exposure among the inhabitants of the Niger Delta in Nigeria. There is information from other countries showing incidence of tumors among workers occupationally exposed to various tars and oils in the Scottish oil shale industry (Leitch, 1922, 1923, 1924; Twort & Ing, 1928). In a laboratory study, Clark, Walter, Ferguson, and Katchen (1988) exposed mice to Wilmington crude oil via the skin and observed skin tumor development on the site of application. This suggests cancer-forming potentials of crude oil in mice. Similarly, an excess of lung cancer was reported in a large cohort of Japanese workers exposed to kerosene, diesel oil, crude petroleum, and mineral oil. Although the direct causation of cancer by crude oil and crude oil residues has not yet been conclusively demonstrated, a wider range of hydrocarbons in the crude oil can act as potent tumor initiators. This will spell doom for most animals during exposure to crude oil and its fractions.

Birds are easily affected by oil on the surface of a mangrove ecosystem because they spend long periods sitting on the water. When feathers are soaked by crude oil this will cause the birds to reduce or stop flying; hence, feeding or preening will be equally affected. Incidental ingestion of oil via drinking may damages their gut, leading to starvation and subsequently death. In some cases, cold may develop or poisoning through shock. The impact on the local population of birds is serious and has been shown to lead to reduction of bird population for many years (SPDC, Nigeria, 1997). Crude oil has been shown to be embryotoxic, and its ingestion during egg formation and the laying period may affect the development of progeny (Gorsline & Holmes, 1981).

In conclusion, crude oil exploration and gas flaring have made a great impact on soil, flora, and fauna of the mangrove and terrestrial ecosystems in the Niger Delta region of Nigeria. Organisms in this region have shown differential response to the surge of exposure to crude oil and crude oil fractions. These range from cellular, systemic, and organismal, culminating in the death of the organisms in most chronic exposures.

References

Abam, T. K. S. (2001). Regional hydrological research perspectives in the Niger Delta. *Hydrological Science, 46,* 13–25.

Adole, T. (2011). *A Geographic Information System (GIS) based assessment of the impacts of gas flaring on vegetation cover in Delta State, Nigeria.* Norwich: M.Sc. Dissertation, Environmental Sciences of the University of East Anglia.

II. THE EFFECTS OF CRUDE OIL EXPLORATION ON THE SOCIO-CULTURAL AND ECO-ECONOMICS OF NIGERIAN ENVIRONMENT

Agbogidi, O. M., & Eshegbeyi, O. F. (2006). Performance of *Dacryodes edulis* (Don, G. Lam, H. J.) seeds and seedlings in a crude oil contaminated soil. *Journal of Sustainable Forestry, 22*(3-4), 1–13.

Agbogidi, O. M., Eruotor, P. G., & Akparobi, S. O. (2006). Effects of soil contaminated with crude oil on the germination of maize (*Zea Mays* L.). *Nigerian Journal of Science and Environment, 5,* 1–10.

Alakpodia, I. J. (2000). Soil characteristics under gas flare in Niger Delta, Southern Nigeria. *International Journal of Environmental and Policy Issues, 1*(2), 1–10.

Alonso, A., Dallmeier, F., Granek, E., & Raven, P. (2001). *Biodiversity: Assessment of biodiversity program and president's committee of advisor on science and technology* (p. 31). USA: Washington DC.

Amadi, A. A., Dickson, A., & Moate, G. O. (1993). Remediation of oil polluted soils: Effect of organic nutrient supplements on the performance of maize *Zea mays* L. *Water Air Soil Pollution, 66,* 59–76.

Amadi, A., Abbey, S. O., & Nma, A. (1996). Chronic effects of oil spill on soil properties and microflora of a rainforest ecosystem in Nigeria. *Water, Air Soil Pollution, 86,* 1–11.

Anifowose, B. (2008). Assessing the impact of oil & gas transport on Nigeria's environment. *U21. Postgraduate Research Conference Proceedings 1, University of Birmingham UK.*

Anoliefo, G. O., & Vwioko, D. E. (1994). Effects of spent lubricating oil on growth of *Capsicum annum* L. and *Lycopersicon esculentum* (Miller). *Environmental Pollution, 88,* 361–364.

Anoliefo, G.O. (1991). Forcados blend crude oil effects on respiratory mechanism, mineral element composition and growth of *Citrullus vulgaris Schead.* Unpublished Doctoral Thesis, University of Benin, Benin City, Nigeria.

Atuanya, E. I. (1987). Effects of waste engine oil pollution on physical and chemical properties of soil: A case study of waste oil contaminated Delta soil in Bendel State. *Nigerian Journal of Applied Science, 5,* 155–175.

Atuma, M. I., & Ojeh, V. N. (2013). Effect of gas flaring on soil and cassava productivity in Ebedei, Ukwuani Local Government Area, Delta State, Nigeria. *Journal of Environmental Protection, 4,* 1054–1066.

Awobajo, A. O. (1981). *An analysis of oil spill incidents in Nigeria.* In *Proceeding of International Seminar on Petroleum Industry and the Nigerian Environment* (pp. 7–63). Nigeria: NNPC, Lagos.

Awosika, L. F. (1995). Impacts of global climate change and sea level rise on coastal resources and energy development in Nigeria. In J. C. Umolu (Ed.), *Global climate change: Impact on energy development. Nigeria: DAMTECH Nigeria Limited.*

Baker, J. M. (1970). The effect of oil on plants physiology. In E. B. Cowell (Ed.), *The ecological effect of oil pollution on littoral communities* (pp. 8–98). London: Applied Science Publishers.

Bastardo, H. (1993). Decomposition process in *Avicennia germinans, Rhizophora mangle* and *Laguncularia racemosa* under oil spill. *Acta Biol Venez, 14,* 53–60.

Benka-Coker, M. O., & Ekundayo, J. A. (1995). Effects of an oil spill on soil physico-chemical properties of a spill site in the Niger Delta Area of Nigeria. *Environmental Monitoring and Assessment, 36,* 93–104.

Central Bank of Nigeria (1996). *Statistical Bulletin, 7*(1), 138.

Chaplin, F. S., III, Walker, B. H., Hobbs, R. J., Hooper, D. V., Lawton, J. H., Sala, O. E., & Tilman, D. (1997). Biotic control over the functioning of ecosystems. *Science, 277,* 500–503.

Chindah, A.C., & Braide, S.A. (2000). The impact of oil spills on the ecology and economy of the Niger Delta. In *Proceedings of the Workshop on Sustainable Remediation Development Technology held at the Institute of Pollution Studies.* Rivers State University of Science and Technology, Port Harcourt.

Chindah, A. C., Braide, S. A., Amakiri, J. O., & Onokurhefe, J. (2008). Effect of crude oil on the development of white mangrove seedlings (*Avicennia germinans*) in the Niger delta, Nigeria. *Estudos de Biologia, 30*(70/71/72), 77–90.

Clark, C. R., Walter, M. K., Ferguson, P. W., & Katchen, M. (1988). Comparative dermal carcinogenesis of shale and petroleum-derived distillates. *Toxicology and Industrial Health, 4,* 11–22.

Collins, M. E., & Foreman, J. E. K. (2001). The effect of sound on the growth of plants. *Canadian Acoustics, 29*(2), 2–8.

Corcoran, E., Ravilious, C., & Skuja, M. (2007). *Mangroves of western and central Africa.* http://www.crc.uri.edu/download/GH2009D002_122713.pdf, http://books.google.com/books?hl = en&lr = &id = xdrMCcDQaaoC&oi = fnd&pg = PA3&dq = Mangroves + of + Western + and + Central + Africa&ots = 2PbPbcYvWl&sig = Uu4Qi ZuDQaphZqx-OfSZ8erl0IM.

Department of Petroleum Resources (DPR). (1997). Annual reports. Department of Petroleum Resources, Abuja, 191.

II. THE EFFECTS OF CRUDE OIL EXPLORATION ON THE SOCIO-CULTURAL AND ECO-ECONOMICS OF NIGERIAN ENVIRONMENT

Diab, E. A., & Bolus, S. T. (2014). The effect of petroleum pollutants on the anatomical features of *Avicennia marina*. (Forssk.) Vierh. *International Journal of Science and Research*, 3(12), 2503–2515.

Duke, N. C., & Watkinson, A. J. (2002). Chlorophyll-deficient propagules of *Avicennia marina* and apparent longer term deterioration of mangrove fitness in oil-polluted sediments. *Marine Pollution Bulletin*, 44(11), 1269–1276. doi:10.1016/S0025-326X(02)00221-7

Dupont, L. M., Jahns, S., Marret, F., & Ning, S. (2000). Vegetation change in equatorial West Africa: Time-slices for the last 150 ka. *Palaeogeography, Palaeoclimatology, Palaeoecology*, 155, 95–122.

Ebeku, K.S.A. (2005). Oil and the Niger Delta people in international law. Resource rights, environmental and equity issues. In *OGEL Special Study Vol. 5, published in November 2005 by Oil, Gas & Energy Law Intelligence* (OGEL).

Ellison, A. M. (1999). Cumulative effects of oil spills on Mangroves. *Ecological Applications*, 9(4), 1490–1492.

Fentiman, A. (1996). The anthropology of oil: The impact of the oil industry on a fishing community in the Niger Delta. *Social Justice*, 23(4), 87–99.

Food and Agricultural Organization (FAO) (2007). The world's mangroves 1980-2005. FAO Forestry Paper (vol. 153) Rome: FAO. Available from http://dx.doi.org/978-92-5-105856-5.

Francis, C. D., Kleist, N. J., Ortega, C. P., & Cruz, A. (2012). Noise pollution alters ecological services:enhanced pollination and disruptedseed dispersal. *Proceedings of the Royal Society Biology Science*, 279, 2727–2735.

Fu, J., Mai, B., Sheng, G., Zhang, G., Wang, X., Peng, P., ... Tang, U. W. (2003). Persistent organic pollutants in environment of the pearl river Delta, China: An overview. *Chemosphere*, 52, 1411–1422.

Gill, L. S., Nyawuame, H. G. K., & Eruikhametalor, A. O. (1992). Effect of crude oil on the growth and anatomical features of *Chromolaena odorata* (L). *Newsletter*, 5, 46–50.

Gorsline, J., & Holmes, W. N. (1981). Effects of petroleum and adreno-cortical activity and onhepatic naphthalene-metabolizingactivity in mallard ducks. *Archives of Environmental Contamination and Toxicology*, 10, 765–777.

Grant, D. L., Clarke, P. J., & Allaway, W. G. (1993). The response of grey mangrove (*Avicennia marina* (Forsk, Vierh). Seedlings to spills of crude oil. *Journal of Experimental Marine Biology and Ecology*, 171(2), 273–295. Available from http://dx.doi.org/10.1016/0022-0981(93)90009-D.

Hutchful, E. (1985). Oil companies and environmental pollution in Nigeria. In C. Ake (Ed.), *Political economy of Nigeria* (pp. 113–140). Lagos: Longman.

International Union for the Conservation of Nature and Natural Resources (IUCN) (1992). A guide to the convention to the biological diversity. *Environmental Policy and Law Paper No.30*. Gland, Switzerland: IUCN Environmental Law Centre.

Ikelegbe, A. (2005). The economic of conflict in the oil rich Niger Delta region of Nigeria. *Journal of Third World Studies*, 43(2), 24–50.

Inoni, O. E., Omotor, D. G., & Adun, F. N. (2006). The effect of oil spillage on crop yield and farm income in DeltaState, Nigeria. *Journal of Central European Agriculture*, 7(1), 41–48.

Isichei, A. O., & Sanford, W. W. (1976). The effects of waste gas flare on the surface vegetations in South- Eastern Nigeria. *Journal of Applied Ecology*, 13, 173.

Iwegbue, C. M. A., Ekakitie, A. O., & Egun, A. C. (2006). Mineralization of nitrogen in wetlands soils of the Niger Delta amended with water hyacinth (*Eichhornia* sp). *Waste International Journal of Soil Science*, 1, 258–263.

James, G. K., Adegoke, J. O., Saba, E., Nwilo, P., & Akinyede, J. (2007). Satellite-based assessment of the extent and changes in the mangrove ecosystem of the Niger Delta. *Marine Geodesy*, 30(3), 249–267. Available from http://dx.doi.org/10.1080/01490410701438224.

Kadafa, A. A. (2012). Environmental impactof oil exploration and exploitation in the Niger Delta of Nigeria. *Global Journal of Science Frontier Research environment and earth sciences*, 12(3), 2249–4646, online ISSN.

Kinako, P. D. S. (1981). Short-term effects of oil pollution on species numbers and productivity of a simple terrestrial ecosystem. *Environmental Pollution, (Series A)*, 26(1), 87–91.

Klekowski, E. J., Corredor, J. E., Morell, J. M., & Del Castillo, C. A. (1994). Petroleum pollution and mutation in mangroves. *Marine Pollution Bulletin*, 28(3), 166–169. Available from http://dx.doi.org/10.1016/0025-326X(94) 90393-X.

Kokwano, J. O. (1994). An overview of the current status of biodiversity in Africa. *Whydah Newsletter*, 3(9), 1–8.

Kolattukudy, P. E. (1979). Oxidation of paraffins by plant tissues. *Plant Physiology.*, 44, 315–317.

Kozlowski, T. T. (1999). Soil compaction and growth of woody plants. *Scandinavian Journal of Forest Research*, 14, 596–619.

II. THE EFFECTS OF CRUDE OIL EXPLORATION ON THE SOCIO-CULTURAL AND ECO-ECONOMICS OF NIGERIAN ENVIRONMENT

Krebs, C.T., & Tanner, C.E. (1981). Restoration of oiled marshes through sediment stripping and spartina propagation. In *Proceeding of the 1981 oil spill conference* (pp. 375–385). Washington, DC: American Petroleum Institute.

Lacerda, L. D., Conde, J. E., Kjerfve, B., Alvarez-Leon, R., Alarcon, C., & Polania, J. (2013). American mangroves. In L. D. de Lacerda (Ed.), *Mangrove ecosystems: Function and management* (pp. 1–62). New York: Springer Science and Business Media.

Leighton, F. A., Lee, Y. Z., Rahimtula, A. D., O'brien, P. J., & Peakall, D. B. (1985). Biochemical and functional disturbances in red blood cells of herring gulls ingesting Prudhoe bay crude oil. *Toxicology and Applied Pharmacology, 81*, 25–31.

Leitch, A. (1922). Parraffin cancer and its experimental production. *The British Medical Journal, 2*, 1104–1106, <http://www.jstor.org/pss/20421874>.

Leitch, A. (1923). The experimental inquiry into thecauses of cancer. *The British Medical Journal, 4*, 210–310, <http://www.pubmedcentral.nih.gov/articlerender.fcgi?artid = 2317302>.

Leitch, A. (1924). Mule spinners cancer and mineraloils. *The British Medical Journal, 2*, 941–943, <http://www.jstor.org/pss/20438450>.

McGinley, M., & Duffy, J.R. (2007). Species diversity. In J.C. Cutler (Ed.), *Encyclopedia of earth*. Washington, DC: Environmental Information Coalition, National Council for Science and the Environment (8 February 2007). Available at <http://www.eoearth.org/article/Species_diversity>, last accessed 5 February 2008.

McKee, K. L. (1993). Soil physicochemical patterns and Mangrove species distribution--reciprocal effects? *The Journal of Ecology, 81*(3), 477. Available from http://dx.doi.org/10.2307/2261526.

McKee, K. L. (1995). Interspecific variation in growth, biomass partitioning, and defensive characteristics of neotropical mangrove Seedlings: Response to light and nutrient availability. *American Journal of Botany, 82*(3), 299–307.

Mensah, S. I., Okonwu, K., & Yabrade, M. (2013). Effect of crude oil application on the growth of mangrove seedlings of *Rhizophora racemosa* G. Meyer. *Asian Journal of Biological Sciences, 6*, 138–141.

Nigerian National Petroleum Corporation, (NNPC). (2016). History of the Nigerian petroleum industry. History. aspx.

Ogidiolu, A. (2003). Effects of gas flaring on soil and vegetation characteristics in oil producing region of Niger Delta, Nigeria. *International Journal of Ecology and Environmental Dynamics, 1*(1), 47–53.

Okpokwasili, G. C., & Amanchukwu, S. C. (1988). Petroleum hydrocarbon degradation by candida species. *Environment International, 14*, 243–247.

Onuoha, F. (2007). Poverty, pipeline vandalisation/explosion and human security: Integratingdisaster management into poverty reduction in Nigeria. *African Security Review, 16*(2), 95–108.

Onwurah, I. N. E. (2000). *A perspective of industrialand environmental biotechnology* (p. 148). Nigeria: Snaap Press/ Publishers Enugu.

Onyegeme-Okerenta, B. M., Alozie, S. C., & Wegwu, M. O. (2015). Physico-chemical properties of soil polluted with petroleumcrankcase oil and chlorophyll concentration of *Abelmoschus esculentus* (okra). *Journal of Environment and Earth Science, 5*(20), 80–88.

Orimoogunje, O. I., Ayanlade, A., Akinkuolie, T. A., & Odiong, A. U. (2010). Perception on the effect of gas flaring on the environment. *Research Journal of Environmental and Earth Sciences, 2*(4), 188–193.

Orisakwe, O. E., Njan, A. A., Afonne, O. J., Akumka, D. D., Orish, V. N., & Udemezue, O. O. (1989). Investigation into the nephrotoxicity of Nigerian bonny light crude oil in albino rats. *International Journal of Environmental Research. Public Health, 1*, 106–110, <http://www.ncbi.nlm.nih.gov/pubmed/16696185>.

Osuji, L. C., & Onojake, C. M. (2004). Trace heavy metals associated with crude oil: A case study of ebocha-8 oil-spill-polluted site in Niger Delta, Nigeria. *Chemistry & Biodiversity, 1*(11), 1708–1715.

Partani, S., Ghiassi, R., Darban, K., & Saeedi, M. (2015). Investigating natural physical adsorption of oil content by mangroves, A field-scale study. *International Journal of Environmental Research, 9*(1), 373–384.

Phil-Eze, P. O., & Okoro, I. C. (2009). Sustainable biodiversity conservation in the Niger Delta: A practical approach to conservation site selection. *Biodiversity and Conservation, 18*, 1247, Retrieved from http://link.springer.com/article/10.1007/s10531-008-9451-z on 05/08/2016.

Powell, C.B. (1993). Sites and species of conservation interest in the central axis of the Niger Delta (Section C). In *Report submitted to the Natural Resources Conservation Council of Nigeria, biodiversity unit*. Rivers State University of Science and Technology, Port Harcourt, Nigeria.

II. THE EFFECTS OF CRUDE OIL EXPLORATION ON THE SOCIO-CULTURAL AND ECO-ECONOMICS OF NIGERIAN ENVIRONMENT

Proffitt, C. E., & Devlin, D. J. (1998). Are there cumulative effects in red mangroves from oil spills during seedling and sapling stages? *Ecological Applications, 8*(1), 121−127. Available from http://dx.doi.org/10.1890/1051-0761 (1998)008[0121:ATCEIR]2.0.CO;2.

Proffitt, C. E., & Travis, S. E. (2005). Albino mutation rates in red mangroves (*Rhizophora mangle* l.) As a bioassay of contamination history in Tampa bay, Florida, USA. *Wetlands, 25*(2), 326−334. Available from http://dx.doi.org/10.1672/9.

Rowell, M. J. (1977). The effect of crude oil on soils: A review of the literature. In J. A. Toogood (Ed.), *The reclamation of agricultural soils after spills, Part 1* (pp. 1−33). *Canada*: Edmonton Publishers.

Salau, A.T. (1993). Environmental Crisis and Development in Nigeria. Inaugural Lecture, delivered on Thursday, 11th February, 1993 at the University of Port Harcourt, Nigeria. Inaugural Lectures Series No. 13, 38.

Santos, L. C. M., Cunha-Lignon, M., Schaeffer-Novelli, Y., & Cintrón-Molero, G. (2012). Long-term effects of oil pollution in mangrove forests (Baixada Santista, Southeast Brazil) detected using a GIS-based multitemporal analysis of aerial photographs. *Brazilian Journal of Oceanography, 60*(2), 159−170.

Shell Petroleum Development Company, (SPDC). (1997). Oil spills in the Niger Delta - Monthly data for 2016. Retrieved July 1, 2016, from http://www.shell.com.ng/environment-society/environment-tpkg/oil-spills/data-2016.html#textwithimage_4.

Shigenaka, G. (2014). Oil toxicity and effects on mangroves. In R. Hoff, & J. Michel (Eds.), *Oil spills in mangroves-planning & response considerations* (pp. 30−46). US Department of Commerce.

Singh, J., Moffat, D., Linden, O., and Division, W.B.W.C.A.D.I., and E. O. (1995). *Defining an environmental development strategy for the Niger delta*. World Bank. Retrieved from https://books.google.co.th/books?id = 7Q0sAQAAIAAJ.

Sodre, V., Caetano, V. S., Rocha, R. M., Carmo, F. L., Medici, L. O., Peixoto, R. S., ... Reinert, F. (2013). Physiological aspects of mangrove (*Laguncularia racemosa*) grown in microcosms with oil-degrading bacteria and oil contaminated sediment. *Environmental Pollution, 172*(2), 243−249.

Spalding, M., Kainuma, M., & Collins, L. (2010). *World Atlas of Mangroves*. New York: Taylor and Francis Group.

Svinkin, M.R. (2008). Soil and structure vibrations from construction and industrial sources. Sixth international conference on case histories in geotechnical engineering (August 11, 2008). Paper8. Missouri University of Science and Technology, Scholars' Mine, 14.

Twort, C. C., & Ing, H. R. (1928). Mule spinners cancer and mineral oils. *Lancet, 214*, 752.

Ubom, R. M. (2010). Ethnobotany and biodiversity conservation in the Niger Delta, Nigeria. *International Journal of Botany, 6*, 310−322.

Udo, E. J., & Fayemi, A. A. A. (1974). The effect of oil pollution of soil on germination, growth and nutrient uptake of corn. *Journal of Environmental Quality, 4*(4), 537−540.

Ugochukwu, C. N. C., & Ertel, J. (2008). Negative impacts of oil exploration on biodiversity management in the Niger De area of Nigeria. *Impact Assessment and Project Appraisal, 26*(2), 139−147.

Ukegbu, D., & Okeke, A.O. (1987). Flaring of associated gas in oil and industry: Impact on growth, productivity and yield of selected farm crops. In *Petroleum Industry and the Nigerian Environment, Proceedings of 1987 Seminar*. NNPC, Lagos.

Uluocha, N., & Okeke, I. (2004). Implications of wetlands degradation for water resources management: Lessons from Nigeria. *Geographical Journal, 61*, 151.

United Nations Development Programme (UNDP). (2006). *Niger Delta human development report. Abuja: United Nations Development Programme*. Abuja: United Nations Development Programme, 229. http://hdr.undp.org/sites/default/files/nigeria_hdr_report.pdf.

United Nations Development Programme, (UNDP). (2016). Niger Delta biodiversity project. The GEF's strategic programme for West Africa (SPWA) − Sub-component biodiversity UNDP GEF PIMS no.: 2047 GEFSEC Project ID: 4090, 171. retrieved from http://www.undp.org/content/dam/undp/documents/projects/NGA/Niger%20Delta%20Biodiversity_Prodoc.pdf on 04/08/2016.

Van Dessel, J. P., & Omoku, P. S. (1994). Environmental impact of exploration and production operations on the Niger Delta Mangrove. *Health, safety & environment in oil & gas exploration & production*, , 437−445 . Jakarta. Retrieved from http://www.onepetro.org/mslib/app/Preview.do?paperNumber = 00027146&societyCode = SPE, http://www.onepetro.org/mslib/servlet/onepetropreview?id = 00027146&soc = SPE

Vanol, D., & Vaidya, R. (2014). Effect of types of sound (music and noise) and varying frequency on growth of guar or cluster bean (*Cyamopsistetragonoloba*) seed germination and growth of plants. *Quest, 2*(3), 9−14.

II. THE EFFECTS OF CRUDE OIL EXPLORATION ON THE SOCIO-CULTURAL AND ECO-ECONOMICS OF NIGERIAN ENVIRONMENT

Uyigue, E., & Ogbeibu A.E. (2007). Climate change and poverty: Sustainable approach in theNiger Delta region of Nigeria. Retrieved from http://2007amsterdamconference.org/Downloads/AC2007_UyigueOgbeibu.pdf 05/ 08/2016.

World Bank. (1995). Defining and environmental development strategy for the Niger Delta. 1:59.

World Resources Institute (WRI) (1992). *World resources. World Resources Institute, in collaboration with the United Nations Environment Programme and the United Nations Development Programme.* New York, NY: Oxford University Press.

Zabbey, N. (2004). Impacts of extractive industries on the biodiversity of the Niger Delta. National workshop on coastal and marine biodiversity management. *NNPC, Nigerian National Petroleum Co-operation, Monthly petroleum information* (1984, p. 53). September, Lagos, Nigeria.

10

Petroleum Industry Activities and Human Health

I. Lucky Briggs[1] and B. Chidinma Briggs[2]

[1]ESPAM-Formation University, Cotonou, Republic of Benin [2]Pre-natal Section
of the Maternal and Childcare Center, Amuwo-Odofin, Lagos, Nigeria

10.1 INTRODUCTION

Oil contamination, usually from drilling processes, creates problems that disrupt the lives of people living very close to oil camps, pumping stations, and pipelines. It also creates hazards to the local environment. Drinking water, topsoil, and livestock of people living in regions where oil is abundant are contaminated with petrochemicals due to the extraction process of oil. In many contaminated sites, acute and chronic illnesses that had resulted from exposure to petrochemicals have been recorded. Environments in the world, located around large oil wells, had supported healthy human life and vibrant ecosystems before the petrochemical industries were established. Oil contamination from drilling processes has, however, affected the people living in these areas to a large extent since their environments became polluted. The water, soil, and air have been severely plundered by petroleum pollutants. As a result, wildlife, livestock, and humans fall sick more frequently (Gay, Shepherd, Thyden, & Whitman, 2010).

With crude oil and production chemicals polluting water supplies, air, and surrounding plant and animal life, human health has suffered. Because a high percentage of those affected by oil contamination live off the land, their local economies have also been impacted by the destruction of the surrounding environment. In addition to these physical effects, economic stress affects the psychological health of the individuals living in polluted areas.

The Niger Delta covers approximately 70,000 km^2 and is home to more than 40 ethnic groups; the SPDC alone prospects on over 40% of this land and operates pipelines, wells, and flow stations that are located around homes, farms, and communities. Hundreds of thousands of these people are affected by the resulting oil contamination near their homes. About 60% of the region's inhabitants who have little money and depend on fishing and agriculture to make ends meet are severely affected (UNDP, 2006).

10.2 MEDIUM OF TRANSMISSION OF PETROLEUM PRODUCTS TO HUMANS

There are various media through which people in the Niger Delta and Nigeria at large are exposed to petroleum products, and thus become affected by its deleterious effects. Some of the common media of oil transmission to humans are discussed below.

10.2.1 Exhaust

The combustion of petroleum distillates is usually incomplete and the incompletely burned compounds produced, alongside water and carbon dioxide, are often toxic to life. These incompletely burned compounds include carbon monoxide, methanol, and fine particulates of soot (Ishisone, 2004).

10.2.2 Volatile Organic Compounds (VOCs)

Various solids and liquids including gasoline, diesel, and jet fuel emit VOCs. However, these often produce both short- and long-term negative impacts. Petroleum VOCs for example, are damaging to health and pollute the atmosphere with some VOCs (benzene in particular) exhibiting very high toxic and carcinogenic behaviors including damage to DNA. Approximately 1% of petroleum consists of benzene, which implies that it is a component of vehicle exhaust fumes. Aliphatic, volatile compounds are vapors obtained from spills of diesel and crude oil, although these compounds are less toxic than benzene. Regardless of their low toxicity, if their concentration in the air is high, they will still pose serious health problems even if the level of benzene is low. Collectively, these compounds are measured as total petroleum hydrocarbons (TPH). TPH levels may be high in indoor spaces as a result of gasoline, diesel, or jet fuel stored in underground storage tanks or brown field sites, which are not safe and can cause adverse health effects from inhalation (Ishisone, 2004). VOCs are also introduced into communities where gas is flared. Nitrogen oxides, carbon monoxide, carbon dioxide, sulfur oxides, hydrocarbons, and photochemical oxidants are pollutants associated with gas flaring. These pollutants usually possess pungent odor (Nriagu, Udofia, Ekong, & Godwin, 2016) and residents are usually exposed to large amounts of heat from the flare as well as soot and black carbon (Luginaah, Martin, Elliot, & Eyles, 2002).

10.2.3 Waste Oil

Oil is used in multiple parts of an automobile: from the transmission, hydraulics and brakes to the motor, crankcase and gearbox, making vehicles a particularly prevalent source of waste oil. Waste oil contains a range of contaminants and impurities from used oil, and when it drips from vehicle engines onto the ground, both the oil and contaminants quickly enter the water table. Once this waste oil and its associated toxins (including benzene) pollute the water table, soil and drinking water are subsequently affected, with the pollutants eventually being carried into rivers and oceans as a result of dispersal and runoff.

10.2.4 Oil Spills

Oil spill is a form of pollution described as the release of a liquid petroleum hydrocarbon into the environment, especially marine areas, due to human activities. Oil spills may be due to release of crude oil from tankers, pipelines, railcars, offshore platforms, drilling rigs and wells, as well as spills of refined petroleum products and their byproducts, heavy fuels used by large ships such as bunker fuel, or the spill of any oily refuse or waste oil. Spilled oil can penetrate into the structure of the plumage of birds and the fur of mammals, reducing their insulating ability and making them more vulnerable to temperature fluctuations and much less buoyant in the water (Ismail & Lewis, 2006; Jernelöv, 2010).

10.3 HEALTH EFFECTS OF PETROCHEMICAL EXPOSURE

Crude oil is a mixture of many different kinds of organic molecules, of which some are highly toxic and can cause cancer. Extraction of crude oil is a complicated process that has the potential for unfortunate consequences. In the Niger Delta, oil spills have extensively contaminated the landscape, damaging both the environment and the health of the people living in the area. Oil kills fish quickly at parts per million concentrations. The inhalation of smoke from natural gas flaring has been associated with a host of health problems in the Niger Delta region of Nigeria (Ishisone, 2004). Birth defects are caused by crude oil and petroleum distillates (Ibaba & Opukri, 2008; Mekuleyi, Ayorinde, Lawson, Ndimele, & Fakoya, 2015). Various psychological health problems have arisen in those living in the surrounding areas. The types of health effects that can occur from a given spill depend largely on the area's level of economic development. The relative speed with which the oil contamination is cleaned up, along with the geographic location of the spill, also impact the consequent health effects.

It is known that people residing in areas where the air is contaminated with volatile petroleum products due to oil exploration, experience shortness of breath, eye irritation, dizziness, cough, nose congestion, sore throat, phlegm, and weakness. These health effects were reported to be associated with the presence of large amounts of petrochemical toxicants in the blood of exposed individuals (Mark, 2009). The study by Nriagu et al. (2016) showed that oil contamination at the Niger Delta communities gave rise to psychological problems such as high level of worry, annoyance, and intolerance. These emotional distresses induce dysregulation of many inter-related physiological issues involving cardiovascular, endocrinological, and immunological systems.

10.3.1 Reproductive Health

Adverse effects on reproductive health have been associated with areas of oil contamination. Miscarriages, stillbirths, infertility, sterility, and birth defects have all been linked to oil contamination.

10.3.2 Cancer

Cancers of the stomach, rectum, skin, soft tissue, and kidneys were found to be more common among males, while cancers of the cervix and lymph nodes were more noted in females (Armstrong, Córdoba, Sebastián, & Stephens, 2001). Benzene (a carcinogen), that has the potential of causing leukemia in humans is found in both crude oil and gasoline. Reports have shown that people exposed to benzene are susceptible to infections, as the compound lowers the white blood cell count in humans. Chronic exposure to crude oil is known to lead to increased cancer incidence (Allison, Davies, & Uyi, 2006; Prasad & Kumari, 1987).

10.3.3 Respiratory Issues

The toxic cloud that resulted from the burning of oil wells in the Niger Delta region of Nigeria caused many respiratory health problems in those exposed. The smoke plume contained many heavy metals and other particulate matters along with hydrocarbons, which proved detrimental when inhaled (Husain, 1998). The fire particulates of soot blacken humans lungs and thereby cause heart problems or death.

10.3.4 Psychological Health

Many people living around crude oil contaminated sites in the Niger Delta experience psychological problems. It can be concluded that psychological symptoms such as stress, anxiety, and depression are present in all cases when oil contamination affects a community either directly or through other means. (Downs, Palinkas, Petterson, & Russell, 1993). In the Niger Delta region of Nigeria, problems with psychological health have resulted from oil that had contaminated crops and livestock as well as water supplies to the indigenous populations who are dependent on farming and the environment to survive. With water supplies and crops poisoned, and livestock such as chickens dying from ingestion of this water, depression would affect members of such communities (Cabrera, 2008).

10.4 CONCLUSION

Oil is one of the most important commodities in the world. Without it, most vehicles would not move, many homes cannot be heated, and national economies—especially in certain developing countries—would crumble. Unfortunately, with oil exploration and extraction practices comes an inherent risk of spills which ranges in magnitude from a few gallons to millions of gallons depending on the source of discharge. The various health problems associated with petrochemical exposure in the Niger Delta region of Nigeria include reproductive dysfunction, incidence of cancer, respiratory issues, and psychological health issues.

10.5 RECOMMENDATIONS

The following recommendations are suggested in order to reduce or eradicate the various health problems connected with oil contamination in the Niger Delta region of Nigeria.

1. People in this region should embrace the use of biodiesel instead of petroleum fractions.
2. Cellulose obtained from fibrous plant material should be used to produce alternative products to many oil-based chemicals.
3. Since plants and animal fats can produce good lubricants whose qualities are similar to that of motor oil and grease, their production should be intensified.
4. False floors at gasoline stations should be created to catch gasoline and oil drips from escaping into the water table.

References

Allison, M. E., Davies, O. A., & Uyi, H. S. (2006). Bioaccumulation of heavy metals in water, sediment, and periwinkle (*Tympanotonus fuscatus* var radula) from the Elechi Creek, Niger Delta. *African Journal of Biotechnology, 5*(10), 968–973.

Armstrong, B., Córdoba, J. A., Sebastián, M. S., & Stephens, C. (2001). Exposure and cancer incidence near oil fields in the Amazon basin of Ecuador. *Journal of Occupational and Environmental Medicine, 58,* 517–522.

Cabrera, R. (2008). *Technical summary report.* ChevronToxico website: <http://chevrontoxico.com> Retrieved 09.11.10.

Downs, M. A., Palinkas, L. A., Petterson, J. S., & Russell, J. (1993). Social, cultural, and psychological impacts of the, *Exxon-Valdez* oil spill. *Human Organization, 52*(1), 1–13.

Gay, J., Shepherd, O., Thyden, M., & Whitman, M. (2010). *The health effects of oil contamination: a compilation of research.*

Husain, T. (1998). Terrestrial and atmospheric environment during and after the Gulf War. *Environment International, 24*(1), 189–198.

Ibaba, I. S., & Opukri, C. O. (2008). Oil induced environmental degradation and internal population displacement in the Nigeria's Niger Delta. *Journal of Sustainable Development in Africa, 10*(1), 173–193.

Ishisone, M. (2004). *Gas flaring in the Niger Delta: the potential benefits of its reduction on the local economy and environment.* Berkeley: Environmental Sciences Group Major, University of California.

Ismail, K., & Lewis, G. (2006). Multi-symptom illnesses, unexplained illness and Gulf War syndrome. *Philosophical Transactions of the Royal Society B, 361,* 543–551.

Jernelöv, A. (2010). The threats from oil spills: Now, then, and in the future. *AMBIO: A Journal of the Human Environment, 39*(5), 1–14.

Luginaah, L. N., Martin, T. S., Elliot, S. J., & Eyles, J. D. (2002). Community responses and coping strategies in the vicinity of a petroleum refinery in Oakville, Ontario. *Health and Place, 8,* 177–190.

Mark, A. S. (2009). The effects of petrochemical and related toxins on human health. *International Journal of Public Health, 1,* 1–20.

Mekuleyi, G.O.; Ayorinde, O.A.; Lawson, E.O.; Ndimele, P.E. and Fakoya, K.A. (2015). Impacts of endocrine disruptors on economically viable crustaceans in Nigeria: Overview and recommendations. In *Proceedings of 30th FISON annual conference held between 22−27 at Asaba, Delta State, Nigeria* (pp. 208–211).

Nriagu, J., Udofia, E. A., Ekong, I., & Godwin, E. (2016). Health risks associated with oil pollution in the Niger Delta, Nigeria. *International Journal of Environmental Research and Public Health, 13*(3), 346. <www.mdpi.com/journal/ijerph>.

Prasad, M. S., & Kumari, K. (1987). Toxicity of crude oil to the survival of the Fresh Water Fish *Puntius Sophore* (HAM). *Acta Hydrocheimica et Hydrobiologica, 15,* 29.

United Nations Development Programme (UNDP) (2006). *Niger Delta human development report.* Garki, Abuja, Nigeria: United Nations Development Programme.

Occurrence of Radioactive Elements in Oil-Producing Region of Nigeria

Ijeoma Favour Vincent-Akpu[1], *Bolaji Benard Babatunde*[1] *and Prince Emeka Ndimele*[2]

[1]University of Port Harcourt, Port Harcourt, Rivers State, Nigeria [2]Lagos State University, Ojo, Lagos State, Nigeria

11.1 INTRODUCTION

Agriculture, which was the mainstay of the Nigerian economy for decades, has been replaced with oil. The latter extracted from the Niger Delta region accounts for over 90% of Nigeria's foreign earnings. The Niger Delta's geographical areas as defined by the natural boundaries include Bayelsa, Delta, and Rivers with about 25,640 km^2 (Ashton-Jones, 1998). However, NDDC (2004) geographical classification of the Niger Delta identified nine states (Cross River, Abia, Imo, Delta, Bayelsa, Rivers, Edo, Ondo and Akwa Ibom) as constituents of the region with over 70,000 km^2.

This incredible ecosystem is, however, vulnerable to destruction by petroleum and its products due to oil industry activities within the area. The exploitation and exploration of crude oil and gas may bring economic benefits to a country but its activities are destructive to the environment even at the safest and best operating practices. These problems are exacerbated by frequent oil spills caused by equipment failure, vandalism, general negligence, and noncompliance to environmental regulations. Since the discovery of crude oil in 1956, millions of oil barrels have been discharged into the Niger Delta's ecosystems and this has been characterized by devastating effects on its natural resources and inhabitants (Zabbey, Hart, & Wolff, 2010).

Oil pollution occurs in the Niger Delta almost on a daily basis (Odiete, 1999), and this event has positioned the region as one of the most polluted areas in the world (Kadafa, 2012; UNDP, 2006; Zabbey, 2009). Over 2000 m^3 of oil are allegedly reported as annual oil effluent discharged into the Niger Delta ecosystems (Anderson, 2005). The quantity of oil spills (Table 11.1) in the Niger Delta between 1976 and 2010 is estimated to be about three million barrels (Amnesty International, 2011).

The Political Ecology of Oil and Gas Activities in the Nigerian Aquatic Ecosystem
DOI: http://dx.doi.org/10.1016/B978-0-12-809399-3.00011-2

TABLE 11.1　List of Oil Spills in the Niger Delta Between 1976 and 2010

Period	Total no. of spills	Quantity (Barrels)
1976–1986	1945	2,073,730.55
1987–1996	1688	371,632.05
1997–2005	3986	334,811.30
2006–2010	930	288,000
Total	8549	2,748,595.90

Adapted from Odiete, W.O. (1999). Environmental physiology of animals and pollution *(p. 185). Lagos: Diversified Resources Ltd.; Amnesty International. (2011).* UN confirms massive oil pollution in Niger Delta. *Available online at: <http://www.amnestyusa.org/news/news-item/un-confirms-massive-oilpollution-in-niger-delta> Accessed July 2015.*

A number of processes are involved in bringing crude oil trapped in the earth to the surface. Crude oil is brought to the surface under pressure during drilling. Lubricants and other chemicals such as drilling muds are used to enhance the process. During drillings, the cuttings from the hole are brought to the surface too. Ultimately, drilling fluids and drill cuttings become wastes at different stages of the drilling process with wastewater from oil production arising from the separation of oil and production water. An estimated 25% of production water represents the total oil production in River State and 42.2% in Delta State (CBI, 1995). The separation of oil and production water leaves a heavy oily hazardous discharge originating from impurities in crude oil being settled in separation tanks. Drilling fluids and the resultant muds and cuttings also represent a considerable waste load in the oil recovery zones. For a total of 4000 wells drilled in Delta and Rivers State up to 1994, it has been estimated that the total amount of cuttings and drilling muds should be about 7,000,000 m^3 since an average well is estimated to produce 2500 m^3 cuttings and drilling mud (World Bank, 1995). The petroleum industry in Nigeria remains the largest land-based source of pollutants to the aquatic ecosystem. Toxic wastes such as radioactive materials are associated with oil production. These wastes, often carcinogenic, mutagenic, or teratogenic in nature, are highly harmful to both human and animal health and also have the potential of eroding the environment of its biodiversities.

11.2　RADIOACTIVE ELEMENTS OF NATURAL ORIGIN AND ARTIFICIAL BOMBARDMENT

Some radionuclides such as uranium and thorium occur freely in nature in the earth's crust while others could be produced in the processes of crude oil extraction through artificial bombardment, or what is called natural transmutation. Artificially bombarded radioactive elements can also occur as byproducts or waste products of geothermal energy production.

During the production process, natural radioactive elements such as Radium-226 flow with crude oil waste and water mixture, and accumulate in separation plants, wellheads, muds, sludge, and pipes. The act of using the same equipment at various sites also promotes the spread of radioactive materials. When a natural radioactive element decays, it produces another radionuclide. For instance, decay of uranium and thorium breeds Radium-226, Lead-210, and Radon-222, all of which are easily transported amid the oil and gas products during recovery process.

11.3 TYPES OF COMMON IONIZING RADIATIONS

Radioactive elements of natural origin have three major kinds of particles or radiation, namely: alpha (α), beta (β) and gamma (γ). Alpha particles are represented by the helium atom ($_2^4$He). It is usually emitted as a decay product of radioactive elements of specific mass number but with proton number larger than 82. They have a positive charge with lesser energy when compared to beta particles. However, their inhalation or ingestion by humans is deleterious to health. A typical example of α radiation is shown below (Fig. 11.1). Uranium (U) loses two protons and two neutrons from its nucleus, which causes a reduction in mass of 4 and a reduction in charge of 2 to form thorium (Th) (Fig. 11.1).

β particles are represented by electrons. Their penetrating power is higher than alpha particles. Beta disintegration involves only a change of one unit of positive charge without any change in mass number. Two types of beta radiation (beta minus and beta plus) are common. While beta minus ($\beta-$) occurs during electron emission of beta disintegration, beta plus ($\beta+$) is identified by positron emission. Also, radiations of positron and electron emissions respectively are usually accompanied by an electron neutrino and antineutrino. Fig. 11.2 shows how thorium gained one electron to form another element. Like the alpha particle, it is dangerous to consume or inhale beta emission.

Gamma radiation is composed of both alpha and beta radiations. Examples are found in radio wave, light wave and X-rays. Gamma ray is represented by a neutron that has no charge. Due to its great penetrating power, long exposure to this radiation is harmful to health.

11.4 PRESENCE OF NATURAL RADIOACTIVE ELEMENTS IN CRUDE OIL FRACTIONS

11.4.1 Concentration of Natural Radioactivity in Crude Oil Field

Several factors such as condition of operation and geological formation have been indicated to cause variation in the concentration of natural radioactive elements from one production site to another. Figs. 11.3 and 11.4 show the sequence of natural disintegration of

$$_{92}^{234}U \longrightarrow {}_{90}^{230}Th + {}_2^4He$$

FIGURE 11.1   Decay of Uranium.

$$_{90}^{234}Th \longrightarrow {}_{91}^{234}Pa + {}_{-1}^0e$$

FIGURE 11.2   Decay of thorium.

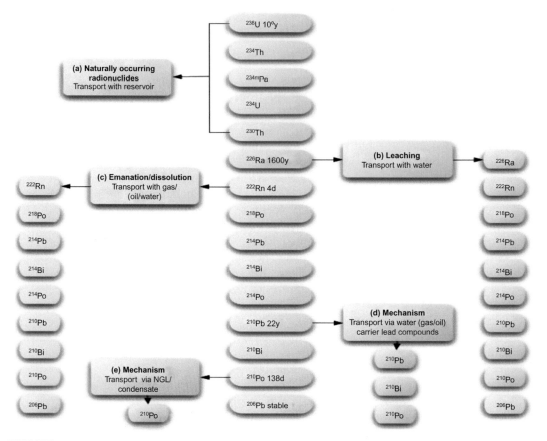

FIGURE 11.3   Uranium-238 decay series showing radionuclides associated with oil production. *OGP. (2008). Guidelines for the management of Naturally Occurring Radioactive Material (NORM) in the oil & gas industry. OGP Report No. 412 <www.ogp.org.uk>.*

Uranium-238 and Thorium-232, respectively. If these elements were left without any disturbance, both the original and the new radioactive elements would produce the same effects; this is called secular equilibrium. On the other hand, there would be no equilibrium if the activity of succeeding radioactive elements differs from that of its precursor (Paranhos, de Araújo, Brandão, Hazinb, & de O Godoy, 2005). The concentration of uranium and thorium in the earth's crust is estimated to be 4.2 and 12.5 ppm, respectively (Shawky, Amer, Nada, El-Maksoud, & Ibrahiem, 2001). While uranium and thorium remain in oil formation for a long period due to their insolubility (OGP, 2008), radioactive elements like radium can easily be pumped to the surface with the oil because of their solubility (Rajaretnam & Spitz, 2000; Shawky et al., 2001). Often, operators ignore this element, which further accumulates and becomes a threat to human health especially through direct exposure to it.

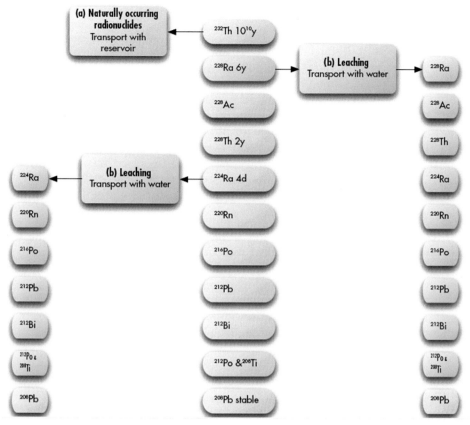

FIGURE 11.4   Thorium-232 decay series showing radionuclides associated with oil and gas production. *OGP. (2008). Guidelines for the management of Naturally Occurring Radioactive Material (NORM) in the oil & gas industry. OGP Report No. 412 <www.ogp.org.uk>.*

11.4.2 Natural Radioactive Elements in Abandoned Pipes From Oil Refining

Scales are often built up in the oil pipes as a result of brine. The pipes thereafter become blocked and eventually neglected. In the abandoned pipe, they persist to release radiation that pollute groundwater and the environment, and are potential health risks for the inhabitants. Different kinds of built-up scale have been found in the facilities used for oil exploration. Among these are sulfate scales ($SrSO_4$, $BaSO_4$) and carbonate scales ($CaCO_3$). Radium, which has a similar chemical structure to strontium (Sr), barium (Ba), and calcium (Ca), usually collides with these compounds to form radium sulfate and radium carbonate (Awwad, Attallah, El-Afifi, Ibrahium, & Aly, 2015). However, the presence of natural radioactive elements in the abandoned pipes in the Niger Delta has not been documented. However, in Brazil, radioactivity in scales and sludge from oil production has been reported (Godoy & Cruz, 2003).

11.4.3 Natural Radioactive Element in Gas Processing Facilities and Drilling Muds

Another site where radionuclides such as Radon-222 and its progeny, Lead-210, and Polonium-210 are located includes vessel for dehydration, processed waters, drilling muds and cuttings, and sludge. Some of the products of Radon-222 through natural disintegration include radon progeny, most of which are often short lived except Lead-210 and Polonium-210 whose half-lives are 22.6 years and 138 days, respectively. Most of the radon decay products (above 90%) are attached to airborne particulates or surfaces of the oil equipment and ambient aerosols.

11.4.4 Presence of Natural Radioactive Materials in Seawater Injection Systems

The oil recovery process where seawater is injected to assist recovery into the reservoir has the potential of introducing additional radioactive materials. In an environment with inadequate oxygen such as in the bedrocks, bacteria that can reduce sulfate have enhanced the uptake of more uranium from bio-fouling deposits. Therefore, in injection systems that use large quantities of seawater, concentration of uranium in the bio-fouling units poses a great threat to the health of the workers. Concentrations of uranium of about 2% by weight have been identified in seawater injection systems (OGP, 2005).

11.4.5 Radioactive Elements Associated With Cement and Building Materials Used in Crude Oil Processes

The naturally occurring radioactive elements (NRE) have an affinity for certain natural minerals, materials, and other resources. Therefore, the exploitation of these resources, as well as consumer products discharged into the environment, may lead to further increase of radioactivity in the products. A potential result is an increase in public and occupational exposures to radiation. Cement is extracted from solid minerals (limestone) while other minerals such as phosphogypsum, granite, and naturally occurring radioactive materials are usually associated with limestone. NRE increase during mining and production of cement resulting in the emission of ionization radiation, which is a problem to humans who work and live in houses built with cement blocks and other building materials (Cooper, 2005; El-Tahel, 2012). Notorious radionuclides with these effects are the ^{232}Th and ^{238}U and their progenies such as ^{210}Pb, ^{226}Ra, and nondecay series ^{40}K.

Building materials processed from natural origins often reflect the geology of their site of origin. The average ^{40}K, ^{232}Th and ^{226}Ra activity concentration in the earth's crust is estimated at 400, 30 and 35 Bq/kg, respectively. It has been reported that byproducts from industries with high concentrations of NRE such as coal slag and fly ash form large components in building materials (Ahmad, Hussein, & Aslam, 1998; Avwiri, 2005; Gbadebo & Amos, 2010). The internal radiation exposure effects of ^{222}Rn exhaled from building materials into the room air could be quantified using environmental parameters such as the ventilation rate. However, the use of a low reference radon concentration for building materials would assist in accounting for the effects from other sources without the need to exceed the action level.

11.5 ARTIFICIAL RADIOACTIVE ELEMENTS USED IN CRUDE OIL FRACTIONATING PROCESS

Radionuclides such as ^{90}Sr, ^{137}Cs and ^{131}I are manufactured. They are usually derived from atmospheric fallout from nuclear accidents and testing. However, activity concentrations of ^{137}Cs have been reported in Nigeria from environments farther from nuclear discharges (Junge et al., 2010). Manufactured radioactive elements such as ^{60}Co and ^{137}Cs are used for different purposes in the oil and gas industry. For instance, they are used to study pipelines and reservoir, level of nuclear density gauges, nuclear well logging, and as a radiotracers in the management of oil wells. Also, the radiation from these radioactive elements are important in monitoring and controlling the density of material in vessels of petrochemical plants, density of mud and cement grout, and to examine important metal components for cracks and other problems. Artificial radioactive elements (ARE) are incorporated into instruments and used to improve and modify electronics components, models and algorithms calibration, as well as to interpret and inverse the nuclear spectral data. However, ARE have long half-lives, which make them linger in the environment if not properly expunged.

11.6 OCCURRENCE OF NRE IN NIGER DELTA REGION OF NIGERIA

The existence of gamma-emitting natural radioactive elements in the soils and sediments from the Niger Delta has been reported. Nevertheless, the concentration of these NRE does not exceed the natural background levels and the reported global averages by the United Nations Scientific Committee on the Effects of Atomic Radiation (UNSCEAR) (Ajayi, Torto, Chokossa, & Akinlua, 2009). The effective and absorbed doses are within permissible limits given by UNSCEAR and the International Commission on Radiological Protection (ICRP). Reports on alpha-emitting elements like ^{210}Po and ^{210}Pb are scarce in the Niger Delta region. However, studies elsewhere have shown that ^{210}Po, which is usually accumulated in the alimentary glands of marine crustaceans and mollusks, is one of the most radiotoxic NRE (Babatunde, Sikoki, & Hart, 2015; Theng, Ahmad, & Mohamed, 2004). Furthermore, through ingestion of seafood and inhalation, ^{210}Po and ^{10}Pb have be alleged of delivering about 83% of the yearly effective dose to humans (Chen, Hou, Dahlgaard, Nielsen, & Aarkrog, 2001; UNSCEAR, 2000). There is likelihood that lack of reports on the above radio-toxicants in the samples from the Niger Delta could be accounted for by absence of appropriate equipment such as the alpha spectrometer required for their measurement or lack of expertise.

The findings on NRE (^{40}K, ^{226}Ra and ^{232}Th) from Niger Delta soils have been documented (Agbalagba, Avwiri, & Chad-Umoreh, 2012). All the hazard index parameters of these NRE are well below their permissible limits. Similarly, the report of Tchokossa, Olomo, Balogun, and Adesanmi (2012) on NRE (^{137}Cs, ^{238}U, ^{232}Th and ^{40}K) indicated that the

TABLE 11.2 Ranges of Activity Concentration ($Bq\ kg^{-1}$) of Radionuclides in the Sediments of the Bonny Estuary

^{238}U	^{232}Th	^{40}K	^{226}Ra	^{137}Cs
<383–601	482–819	1150 3120	193–242	43–59
<461–575	415–840	300–9800	269–484	<39–67
<469–812	370–540	<1032–3760	175–327	<44–50

estimated dose equivalent obtained annually from the concentrations of these NRE is less than the recommended safe level of 1 mSv annually. However, the gross beta and alpha activity concentrations investigated in soils and water samples within and around Imirigin oil field revealed that the NRE in the water samples exceeded WHO-recommended maximum permissible limit for drinking water, but the soil from the sampled area is still safe to be used as construction material for buildings (Meindinyo & Agbalagba, 2011). Table 11.2 shows concentrations of some natural radioactive elements in sediment obtained from the Bonny estuary (Babatunde et al., 2015).

Activity concentrations of ^{226}Ra, ^{228}Ra and ^{40}K were measured in some sachet drinking water samples produced in Nigeria (Ajayi & Adesida, 2009). According to their findings, the annual effective dose in all samples was higher than the maximum permissible limit. The natural radionuclide concentration in some medicinal plants such as carpetgrass (*Axonopus fissifolius*), lemon grass (*Cymbopogon citratus*) and speargrass (*Heteropogon contortus*), around oil and gas facilities in Ughelli and its surrounding communities has been documented (Oni, Gbadebo, Oni, & Sowole, 2011). The findings showed that the measured levels of the radioactivity in the samples are lower than the recommended annual effective dose, indicating a safe level of radiological health consequences.

11.7 HEALTH EFFECTS OF RADIOACTIVITY

The decay of radioactive elements is characterized by the release of ionized radiation, which has adequate energy to strip off electrons from atoms. In like manner, any tissue of an organism can be damaged when exposed to ionizing radiation. The predominant impact of ionized radiation is the induction of cancer with a latency period of years after exposure. High doses can result into acute radiation syndrome and burns (USEPA, 2011). Two major categories of radiation effects have been identified. These are stochastic effects that increase with dose (e.g., heart disease, reduction in cognitive ability, and cancer), and deterministic effects, which occur beyond a threshold dose (e.g., radiation-induced thyroiditis, radiation burns, acute and chronic radiation syndromes).

Reports have shown that concentration of ^{210}Po as low as 0.1–0.3 GBq in the blood of an adult male is likely to cause death within a month. Higher doses of this radioactive element can cause failure of bone marrow, and destroy kidneys and liver (Edwards & Lloyd, 1996; Harrison, Legget, Lloyed, Phlipps, & Scott, 2007). Details on the bioaccumulation of ^{210}Po as well as the adverse effects of its exposure to animals and humans have been fully reported (ICRP, 1994a, 1994b, 1996, 2006).

Furthermore, the continuous decay of radium-226 and lead-210 deposited in bone also produces polonium-210. It has great effects on the spleen, kidney, liver, and the lymph nodes. Inhalation of polonium either through cigarette smoke or radon in the air can be deposited on the mucous lining of the respiratory tract; when alpha particles are then emitted within the lung, the cells lining the airways would be damaged, thus resulting in lung cancer (Yoon et al., 2010; Zaga, Lygidakis, Chaouachi, & Gattavecchia, 2011).

References

Agbalagba, E. O., Avwiri, G. O., & Chad-Umoreh, Y. E. (2012). Gamma spectroscopy measurement of natural radioactivity and assessment of radiation hazard indices in soil samples from oil fields environment of Delta State, Nigeria. *Journal of Environmental Radioactivity, 109*, 64−70.

Ahmad, N., Hussein, A. J. A., & Aslam (1998). Radioactivity in construction materials and soil. *Journal of Environmental Radioactivity, 41*(2), 127−136.

Ajayi, O. S., & Adesida, G. (2009). Radioactivity in some sachet drinking water samples produced in Nigeria. *Iranian Journal of Radioactivity Research, 7*(3), 151−158.

Ajayi, T. R., Torto, N., Chokossa, P. T., & Akinlua, A. (2009). Natural radioactivity and trace metals in crude oil: implication for health. *Environmental Geochemical Health, 31*, 61−69.

Amnesty International (2011): *UN confirms massive oil pollution in Niger Delta*. Available online at: <http://www. amnestyusa.org/news/news-item/un-confirms-massive-oilpollution-in-niger-delta> Accessed July 2015.

Anderson, I. (2005). *Niger River basin: A vision for sustainable development* (p. 131). Washington, DC: The World Bank.

Ashton-Jone, N. (1998). *The human ecosystems of the Niger Delta*. Nigeria: Krat Books, Ltd Lagos.

Avwiri, G. O. (2005). Determination of radionuclide levels in soil and water around cement companies in Port Harcourt, Nigeria. *Journal of Applied Science and Environmental Management, 9*(3), 27−29.

Awwad, N. S., Attallah, M., El-Afifi, E. M., Ibrahium, H. A., & Aly, H. F. (2015). Overview about different approaches of chemical treatment of NORM and TE-NORM produced from oil exploitation. In V. Patel (Ed.), *Advances in petrochemicals*. (Chapter 5) <http://www.intechopen.com/books/advances-in-petrochemicals/overview-about-different-approaches-of-chemical-treatment-of-norm-and-te-norm-produced-from-oil-expl>.

Babatunde, B. B., Sikoki, F. D., & Hart, I. (2015). Human health impact of natural and artificial radioactivity levels in the sediments and fish of Bonny Estuary, Niger Delta, Nigeria. *Challenges, 6*, 244−257.

Carl Brothers International (CBI) (1995). The Niger Delta, Nigeria. Pollution assessment study (p. 70). Glostrup, Denmark: Report to the World Bank, Carl Bro. Group.

Chen, Q. J., Hou, X. L., Dahlgaard, H., Nielsen, S. P., & Aarkrog, A. (2001). A rapid method for the separation of Po-210 from Pb-210 by TIOA extraction. *Journal of Radioanalytical and Nuclear Chemistry, 249*, 587−593.

Cooper, M. B. (2005). *Naturally Occurring Radioactive Materials (NORM) in Australian Industries—Review of current inventories and future generation*. A Report prepared for the Radiation Health and Safety Advisory Council, ERS-006.

Edwards, A. A., & Lloyd, D. C. (1996). *Risk from deterministic effects of ionising radiation*. Doc. NRPB 7, no. 3, Chilton, Oxon.

El-Tahel, A. (2012). Assessment of natural radioactivity levels and radiation hazards for building materials used in Qassim area, Saudi Arabia. *Romanian Journal of Physiology, 57*, 726−735.

Paranhos, G. M. H., de Araújo, A. A., Brandão, Y. B., Hazinb, C. A., & de O Godoy, J. M. (2005). Radioactivity concentration in liquid and solid phases of scale and sludge generated in the petroleum industry. *Journal of Environmental Radioactivity, 81*, 47−54.

Gbadebo, A. M., & Amos, A. J. (2010). Assessment of radionuclide pollutants in bedrocks and soils from Ewekoro Cement Factory, Southwest Nigeria. *Asian Journalof Applied Science, 3*, 135−144.

Godoy, J. M., & Cruz, R. P. (2003). [226]Ra and [228]Ra in scale and sludge samples and their correlation with the chemical composition. *Journal of Environmental Radioactivity, 70*, 199−206.

Harrison, J., Legget, R., Lloyed, D., Phlipps, A., & Scott, B. (2007). Polonium-210 as a poison. *Journal of Radiological Protection, 27*, 17−40.

ICRP (1994a). Human respiratory tract model for radiological protection. *ICRP Publication 66 Annals of the ICRP, 24*(1−3).

ICRP (1994b). Dose coefficients for intake of radionuclides by workers. *ICRP Publication 68 Annals of the ICRP*, 24(4).

ICRP (1996). Age-dependent doses to members of the public from intake of radionuclides. Part 5: compilation of ingestion and inhalation dose coefficients. *ICRP Publication 72 Annals of the ICRP*, 26(1).

ICRP (2006). Human alimentary tract model for radiological protection. *ICRP Publication 100 Annals of the ICRP*, 36(1/2).

International Association of oil and gas producers OGP (2005). *Fate and effects of naturally occurring substances in produced water on marine environment*. OGP report 364.

Junge, B., Mabit, L., Dercon, G., Walling, D. E., Abaidoo, R., Chikoye, D., et al. (2010). First use of the ^{137}Cs technique in Nigeria for estimating medium-term soil redistribution rates on cultivated farmland. *Soil and Tillage Research*, 110, 211−220.

Kadafa, A. A. (2012). *Environmental Impacts of oil exploration and exploitation in the Niger Delta of Nigeria. Global Journal of Science Frontier Research Environment & Earth Sciences*, 12(3), 1−11.

Meindinyo, R. K., & Agbalagba, E. O. (2011). Radioactivity concentration and heavy metal assessment of soil and water in and around Imirigin oil field, Bayelsa State, Nigeria. *Journal Environmental Chemistry and Ecotoxicology*, 4(2), 29−34.

Niger Delta Development Commission (NDDC). (2004). *Biodiversity of the Niger Delta environment*. Niger Delta development commission master plan project final report.

Odiete, W. O. (1999). *Environmental physiology of animals and pollution* (p. 185). Lagos: Diversified Resources Ltd.

OGP. (2008). *Guidelines for the management of Naturally Occurring Radioactive Material (NORM) in the oil & gas industry*. OGP Report No. 412 <www.ogp.org.uk>.

Oni, O. M., Gbadebo, A. I., Oni, F. G. O., & Sowole, O. (2011). Natural activity concentrations and assessment of radiological dose equivalents in medicinal plants around oil and gas facilities in Ughelli and Environs, Nigeria. *Environment and Natural Resources Research*, 1(1), 201−206.

Rajaretnam, G., & Spitz, H. B. (2000). Effect of leachability on environmental risk assessment for naturally occurring radioactive materials in petroleum oil fields. *Health Physics*, 78(2), 191−198.

Shawky, S., Amer, H., Nada, A. A., El-Maksoud, T. M. A., & Ibrahiem, N. M. (2001). Characteristics of NORM in the oil industry from Eastern and Western deserts of Egypt. *Applied Radiation and Isotopes*, 55, 135−139.

Tchokossa, P., Olomo, J. B., Balogun, F. A., & Adesanmi, C. A. (2012). Radiological study of soils in oil and gas producing areas in Delta State, Nigeria. In T. O. Ajayi, B. Ezenwa, A. A. Olaniawo, R. E. K. Udolisa, & P. A. Taggert (Eds.), *Radiation protection dosimetry* (pp. 1−6). Oxford: Oxford University Press.

Theng, T. L., Ahmad, Z., & Mohamed, C. A. R. (2004). Activity concentrations of [210]Po in the soft parts of cockle (*Anadara granosa*) at Kuala Selangor. *Malaysia Journal of Radioanalytical and Nuclear Chemistry*, 262, 485−488.

United Nations Development Programme UNDP. (2006). *Niger Delta human development report*, 74.

UNSCEAR (2000). *Sources, effects and risks of ionization radiation. Report to the general assembly*. New York: United Nations.

USEPA. (2011). *Inventory of U.S. greenhouse gas emissions and sinks: 1990−2009*. <https://www3.epa.gov/climate-change/Downloads/ghgemissions/US-GHG-Inventory-2011-Complete_Report.pdf>.

World Bank. (1995). *Defining an environmental development strategy for the Niger Delta* (Vol. 1).Washington, DC: West Central Africa Department.

Yoon, M., Ahn, S. H., Kim, J., Shin, D. H., Park, S. Y., Lee, S. B., et al. (2010). Radiation-induced cancers from modern radiotherapy techniques: intensity-modulated radiotherapy versus proton therapy. *International Journal of Radiation, Oncology, Biology and Physics.*, 77, 1477−1485.

Zabbey, N. (2009). Pollution and Poverty in the Niger Delta Region − What is the Responsibility of Oil Companies in Nigeria? In *Paper presented at the University of Stavanger, Stavanger, Norway on 29thOctober, 2009* <www.cerhd.org>.

Zabbey, N., Hart, A. I., & Wolff, W. J. (2010). Population structure, biomass and production of the West African lucind, *Keletistes rhizoecus* (Bivalvia, Mollusca) in Sivibilagbara swamp at Bodo Creek, Niger Delta, Nigeria. *Hydrobiologia*, 654, 193−203.

Zaga, V., Lygidakis, C., Chaouachi, K., & Gattavecchia, E. (2011). Polonium and Lung Cancer. *Journal of Oncology*, 11.

II. THE EFFECTS OF CRUDE OIL EXPLORATION ON THE SOCIO-CULTURAL AND ECO-ECONOMICS OF NIGERIAN ENVIRONMENT

Mapping and Modeling Ecosystem Services in Petroleum-Producing Areas in Nigeria

Oludare Hakeem Adedeji[1] and Isa Olalekan Elegbede[2]

[1]Federal University of Agriculture, Abeokuta, Ogun State, Nigeria [2]Brandenburg University of Technology, Cottbus-Senftenberg, Germany

12.1 INTRODUCTION

The Niger Delta region of Nigeria is very important to the economic development of the country because of the large deposit of crude oil and petroleum production in the area. However, despite the huge financial benefits of the business, commercial crude oil exploration in the Niger Delta region has since brought about several environmental changes that have altered ecological configuration of the ecosystem. Worldwide, rapid changes in our ecosystem have occurred in the last century compared to previous times, mostly due to increasing human activities (MA, 2005a). The Millennium Ecosystem Assessment indicated habitat loss, pollution, overexploitation, climate change, and invasive species as the main drivers of ecosystem changes (Schneiders, Van Daele, Van Landuyt, & Van Reeth, 2012). The Niger Delta is presently the major petroleum-producing area of Nigeria and one of the several areas in the country in which the ecosystem services (ES) from the environment are hampered due to human developmental activities causing ecosystem degradation and untold hardship to the people (UNEP, 2011). Environmental degradation in the petroleum-producing region of Nigeria is epitomized by oil spillage, deforestation, loss of biodiversity, deterioration of drainage pattern, water quality degradation, gas flaring, et cetera, which have in a number of ways increased the vulnerability of the people and exposed the land to further degradation. According to the Millennium Ecosystem Assessment (MA et al., 2005a), nature generally provides a number of indispensable benefits or services to people, which are referred to as ES. The MEA in 2005 identified 24 services, which are classified under four major categories. They are the provisioning services (the material that people extract directly from ecosystems such as food, water, and forest products), regulating services (which modulate changes in climate and regulate floods, drought, diseases, waste, and water quality), cultural services (which consist of recreational, esthetic, and spiritual benefits), and supporting services (such as soil formation, photosynthesis, and nutrient recycling). However, anthropogenic modification and simplification of the biosphere to increase the supply of services from the Niger Delta ecosystem have seriously affected the productivity of the area. For instance, increases in the provisioning services, such as extraction of crude oil, have resulted in a decline in biological diversity, loss of vegetal cover, and air and water pollution. These in turn lead to a decline in the ES, subsequently affecting the resources available to the people and their livelihood.

All aspects of oil exploration and exploitation have adverse effects on the ecosystem and the local biodiversity because most of these oil spill incidences occur on land, swamps, and the offshore environment (Nwilo & Badejo, 2005; Twumasi & Merem, 2006; Uyigue & Agho, 2007). The Niger Delta region is rated as the most oil-impacted environment and polluted area in the world, most especially by environmental experts from the United Kingdom, the United States, and Nigeria (Ikelegbe, 2005; Kia, 2009; Obi, 2000). Oil spillage resulting from crude oil exploration has disastrous impacts on the environment in the region and has adversely affected people and their means of livelihood with attendant poverty status. Studies have shown that the quantity of oil spilled over 50 years was at least 9−13 million barrels, which is equivalent to 50 Exxon Valdez spills (FME et al., 2006), making the region one of the five most severely petroleum-damaged ecosystems in the world (Gbadegesin, 2008; Kadafa, 2012). Two of the major oil corporations in Nigeria, the Royal Dutch Shell and the Italian ENI, admitted to more than 550 oil spills in the Niger Delta in 2010, which is more than the only 10 spills per year across the whole of Europe between 1971 and 2011 (Amnesty International, 2009, 2011). Since the first oil spill in Nigeria at Araromi in the present Ondo state in 1908 (Tolulope, 2004), several other cases of oil spillage have occurred in the Niger Delta region. For instance, over 570,000 barrels of oil were spilled into the Forcados estuary in July 1979, polluting the aquatic environment and surrounding swamp forest (Tolulope, 2004; Ukoli, 2005). Furthermore, the Oyakama oil spillage in May 1980 caused a spillage of approximately 30,000 bbl of crude oil (Ukoli, 2005). Within the period 1976−96, there were a total of 4647 oil spill incidences discharging approximately 2,369,470 barrels of oil into the Niger Delta environment, of which 1,820,410.5 (77%) were not recovered (Twumasi & Merem, 2006), causing enormous damage to the fragile communities and environment. Major causes of oil spills in the Niger Delta include blowout, pipeline corrosion, equipment failure, and sabotage (Aaron, 1996). Other minor causes of oil spills also include accidental spills, overflow of tanks, valve failure, overpressure, sand cut through erosion, and engineering error (HRW, 1999).

12.2 CONCEPT OF ECOSYSTEM SERVICES

The natural resources provide many ES, which humans, over thousands of years, learned how to use, exploit, and often misuse (Glavan, Pintar, & Urbanc, 2015; Small, Medcalf, Parker, Haines-Young, & Potschin, 2013). Over the past 50 years, global changes in ecosystems have occurred, causing significant decline and degradation in the ES provided by the natural ecosystems. The concept of ES is an important paradigm of ecosystem management (Potschin & Haines-Young, 2011; Volk, 2013, 2015) focusing on the benefits that humans obtain from ecosystems (Seppelt, Dormann, Eppink, Lautenbach, & Schmidt, 2011; UNEP, 2005; Volk, 2015). The concept has received increased attention in recent years, and is seen as a useful construct for the development of policy-relevant indicators and communication for science, policy, and practice (Guerra, Maes, Geijzendorffer, & Metzger, 2016). It is an effective communication tool to bridge knowledge between science and policy (Maes et al., 2012; Viglizzo, Paruelo, Laterra, & Jobbágy, 2012). Seminal reports such as the Millennium Ecosystem Assessment (MA, 2005b) and the economics of ecosystems and biodiversity (TEEB, 2010) made the concept popular, as reflected by a rapidly increasing number of publications and ES assessment tools (Bagstad, Semmens, Waage, & Winthrop, 2013; Schägner, Brander, Maes, & Hartje, 2013; Seppelt et al., 2011). The Millennium Ecosystem Assessment is the first global checkup of our ecosystems and their capacity to provide us with ES. The assessment has shown us that we can no longer afford to take nature's benefits for granted because 15 out of 24 ES assessed are already degraded, threatening our ability to build vibrant communities. This degradation is expected to grow significantly worse in the first half of this century, threatening human well-being and the goals of development. Thus, there is need for concrete efforts by communities at the local, regional, and international level to focus attention on sustaining the steady supply of vital ES. Focusing on ES allows a decision maker to view services of nature as an input into a strategy to achieve a goal, much like physical or human capital (Ash et al., 2008; Daily et al., 2009). The Millennium Ecosystem Assessment (MA et al., 2005a) has shown that human well-being is inextricably linked with the ES rendered by nature, and it is these links that policy measures seek to influence by addressing the drivers of ecosystem change (Ash et al., 2008). Although ES (processes and functions) relate to the benefits derived from nature, they are not synonymous (Boyd & Banzhaf, 2007; Fisher & Turner, 2008). Ecosystem benefits are the measurable outcomes that affect the well-being of people, which provide the basis of economic valuation studies (Fisher & Turner, 2008; Lele, Springate-Baginski, Lakerveld, Deb, & Dash, 2013).

12.3 THE NIGER DELTA REGION

The Niger Delta of Nigeria lies between 4.01°N and 7.90°N and between 4.50°E and 10.56°E in the West African section of the tropical rainforest belt and has a humid tropical climate. The area can be described by its vast expanse of wetlands, sand barriers, several creeks, and rivers (Twumasi & Merem, 2006). It is a rich mangrove swamp covering more than 20,000 km^2 within wetlands of 70,000 km^2 formed primarily by sediment deposition from an intricate network of rivers and creeks. It is regarded as the third largest wetland in the world and Africa's largest delta with significant biological diversity (CLO, 2002; Powell et al., 1985; Twumasi & Merem, 2006). Kadafa (2012) classified the Niger Delta as a tropical rainforest with ecosystems comprised of diverse species of flora and fauna, both aquatic and terrestrial species. It consists of four main ecological zones: coastal inland zone, freshwater zone, lowland rainforest zone, and mangrove swamp zone and is

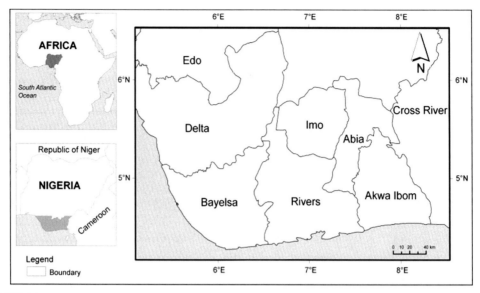

FIGURE 12.1   Niger Delta region of Nigeria .(Source: Onojeghuo and Blackburn, 2011)

considered one of the ten most important wetlands and marine ecosystems in the world (FME et al., 2006). The ecosystem is highly diverse, with numerous species of terrestrial and aquatic fauna and flora as well as human life (Uyigue & Agho, 2007). It harbors many locally and globally endangered species, and approximately 60%—80% of all plant and animal species found in Nigeria (Salau, 1993; WRI, 1992; Zabbey, 2004). It stretches approximately 240 km from Onitsha in the north to the outer barrier islands in the south and 480 km from the Benin River on the east to the Imo River on the west. There are nine states, namely: Abia, Akwa Ibom, Bayelsa, Cross River, Delta, Edo, Imo, Ondo, and Rivers that formed the region called the Niger Delta. Its large expanse of land and the many resources in the area have attracted a high human population for several decades. According to the Nigerian 2006 Census, approximately 33 million people lived in the Niger Delta Region (>265 people per km^2), making it one of the most densely populated regions in Africa and the world (Fig. 12.1) (Dami, Odihi, & Ayuba, 2014; Uyigue & Agho, 2007).

12.4 PETROLEUM PRODUCTION IN THE NIGER DELTA

Prior to the discovery of and commencement of exploration for crude oil in the Niger Delta region of Nigeria, the vast and biodiversity-rich area had remained mainly an undisturbed, pristine, and vibrant ecosystem rendering varieties of ES to the inhabitants. Early in the 19th century, c. 1908, a German company started oil and gas exploration in the Niger Delta region, and in 1936, Shell D'Arcy Exploration Company from the Netherlands secured exclusive rights to oil and gas exploration for all of Nigeria (Steyn, 2009). The outbreak of the Second World War temporarily brought the search for oil to a halt and Shell/D'Arcy suspended its oil drilling in the Delta region after reaching agreement with the colonial government in 1941 (Pearson, 1970). Shell, in partnership with British Petroleum (BP), intensified its search for oil in the eastern Delta of Nigeria in 1947 after the end of the war by conducting an extensive gravity survey of southern Nigeria from 1948 to the early 1950s (Frynas, 2000). Full-scale oil exploration of crude oil in the area started in 1956/58 (Onuoha, 2008; Pearson, 1970) with the discovery in commercial quantity at Oloibiri (in present day Bayelsa state). Other multinational corporations, including Italian oil giant ENI, which owns the Nigerian Agip Oil Company, Chevron/Texaco, ExxonMobil, and Total were involved in the search for oil both onshore and offshore since independence in 1960. The Nigerian government in 1977 replaced the National Oil Company established in 1971 with Nigerian National Petroleum Corporation. There are also many local and fully indigenous-owned companies involved in oil and gas exploration in the Niger Delta. According to the BP Statistical Energy Survey (Raji & Abejide, 2013), Nigeria is presently the tenth largest oil producer in the world, the third largest in Africa, and the fifth largest supplier of crude oil to the United States, mostly in four of the Niger Delta states: Delta, Bayelsa, Rivers, and Akwa Ibom. Furthermore, the country also has significant natural gas reserves (5.29 trillion m^3, or ~3% of the world total) that are subexplored. The Niger Delta accounts for more than 95% of Nigeria's total export annual earnings and approximately 65% of government revenues (CIA World Fact Book, 2008). The two states of Rivers and Delta alone produced approximately 75% of Nigeria's petroleum, which represents more than 50% of the national government's revenues (World Bank, 2002).

12.5 STATUS OF THE ECOSYSTEM SERVICES IN THE NIGER DELTA

The Niger Delta region is one of the well-endowed ecosystems in the world with one of the highest concentrations of biodiversity on the planet. It supports an array of distinctive and abundant flora and fauna, arable terrain that can sustain a wide variety of crops, lumber or agricultural trees, and more species of fish than any ecosystem in West Africa (Omofonmwa & Odia, 2009; Wikipedia, 2006, 2014). Despite the growing awareness of the importance of these ES, environmental degradation, especially due to petroleum production in the area, has continued unabated, causing various land use and land cover changes. Arial views of the Niger Delta reveal a lush green and healthy environment. However, on the ground, the level of destruction of the ecosystem is enormous as many aquatic animals, including fish, have died, many agricultural lands have become infertile, and drinking water has become polluted. Human activities, especially crude oil exploration, which is the main driver of the changes or disruption of the supply of the ES, continue unabated. In terms of degradation, major oil spills have occurred that have devastated rivers, destroyed mangroves and coastal life, and affected the health and livelihoods of millions of inhabitants (WRM, 2003). Consequently, there has been enormous financial loss, extensive habitat degradation, and poverty leading to the continuous crises, including rising communal conflicts, kidnaping of oil workers, and vandalization of oil installations (Adedeji, Ibeh, & Oyebanji, 2011). With the decline in the supply of ES, there is a greater risk of ecosystem collapse and exacerbation of poverty, particularly among the resource-dependent poor communities. Much of the Niger Delta environment has been adversely affected by petroleum exploration and the incessant environmental, socioeconomic, and physical disasters that have accumulated over the years due to limited scrutiny and lack of assessment (Achi, 2003). Destruction of the ecosystems has led to reduced crop yield, polluted fishing systems, and decreased land productivity, which have reduced both income and standard of living. Indirect impacts of oil exploration in the Niger Delta region result from construction of infrastructures such as pipelines. More than 7000 km of pipelines were laid to transport oil and gas across all types of terrain in the delta, and extraction of sand or gravel during the construction of pipeline and roads causes siltation and erosion. Saltwater intrusions into previously freshwater areas are common due to the dredging of the waterways.

Pollutants from the oil industry, which are released into the environment, include hydrocarbons, drilling muds, cuttings, oil and greases, sulfides, suspended solids, heavy metals, phenol, cyanide, and toxic additives (Ukoli, 2005). Groundwater pollution by liquid from surface impoundments or spills from storage tanks, pipelines, improperly closed or abandoned oil wells, and poorly constructed injection wells is a major concern in many Niger Delta communities (Oteh & Eze, 2012). Petroleum hydrocarbon from oil spills due to their toxicity affects aquatic organisms such as birds and mammals, which are vulnerable to oil spills when their habitats become contaminated, and this may reduce reproductive rates and survival and could also cause physiological impairment (Briggs, Yoshida, & Gershwin, 1996). Oil spillages on the water bodies may also prevents natural aeration and leads to death of freshwater or marine life and on land, leads to retardation of vegetation growth and causes soil infertility for a long period of time (Ukoli, 2005). Another major environmental problem that has occurred because of petro-activities is gas flaring. For several years, the communities in the Niger Delta, health professionals, concerned activists, and nongovernmental organizations have been raising concerns about the impacts of gas flaring on human health in the area. Despite the obvious implications, Nigeria remains one of the major gas-flaring countries in the world.

12.6 IMPORTANCE OF ECOSYSTEM SERVICES IN THE NIGER DELTA REGION

The Niger Delta ecosystem is unique and very important to the people in the region, as the majority of the people in the area depends largely on the services (agriculture, industry, fishing, food, drinking water, wood, shelter, medicine, employment, and esthetics) provided by the ecosystem.

The Niger Delta ecosystem is a critical habitat for many species of fish and wildlife, serves as a coastal fish and shellfish nursery habitat, and produces large quantities of leaf material that becomes the basis for a detritus food web (James, Adegoke, Saba, Nwilo, & Akinyede, 2007; James, 2008). It is home to three endemic families represented by five plant species and the introduced family of exotic species (James et al., 2007; James, 2008). Studies from different parts of the world have shown that ES provided by nature are declining rapidly, including their capacity to buffer local communities from disasters (FAO, 2004; Millennium Ecosystem Assessment MA et al., 2005b; Ranganathan et al., 2008). These findings suggest that ES are often overlooked or assumed to be available as development decisions are made; the attainment of development goals is consequently often in jeopardy.

12.7 MAPPING AND MODELING APPLICATIONS

Efforts at mapping and assessing the ecosystems and their services are a major policy drive of many developed countries of the world. For instance in Europe, Action 5 of the EU Biodiversity Strategy to 2020 foresees that member states will, with the assistance of the commission, map and assess the state of ecosystems and their services in their national territory by 2014 (Maes et al., 2012). ES assessment examines changes in the ecosystem over the long term as well as the short term because dramatic decline from which it is difficult to recover may occur as an ecosystem reaches a tipping point, or threshold, at which rapid change occurs (Scheffer, Carpenter, Foley, Folke, & Walker, 2001). Consequently, to better represent the impacts related to these drivers, it is necessary to map not only the capacity for ES provision (e.g., according to land cover type) but also the actual ES provision and the remaining soil erosion (Nelson et al., 2009; Martínez-Harms & Balvanera, 2012). Mapping and modeling are important for the understanding of the drivers of changes in ecosystem health, including risks and trends over time and new tools and methods for assessing ecosystem health across diverse landscapes (Hauck, Winkler, & Priess, 2015; Winowiecki, Vågen, & Huising, 2016). The driving forces of environmental degradation in the Niger Delta are poorly understood due to lack of information on environmental and ES. There is a need to assess the breadth of assumptions concerning drivers of ecosystem change on global and European levels as used in various scenario approaches. Environmental modeling combined with scenarios provides insights into drivers of change, potential implications of different trajectories, and options for action. Assessment of ecosystems and their services in the Niger Delta region would depend on spatially explicit mapping to address the key drivers, including land use/land cover changes, environmental degradation, climate change, pollution (water, air, and land) flooding, coastal erosion et cetera, and their different gradients and variations in space and time. According to the Millennium Ecosystem Assessment (MA, 2005c), there are different methods of assessing ES. They include the use of remote sensing and geographical information systems, ecological models, and participatory approaches and expert opinion.

The MAES conceptual model builds on the premise that the delivery of certain ES upon which we rely for our socioeconomic development and long-term human well-being is strongly dependent on both the spatial accessibility of ecosystems as well as on ecosystem condition. Ecological models such as Ecosim, Ecopath, IMPACT, PODIUM, WaterGAP, and EcoServ-GIS (Durham Wildlife Trust, 2013) are simplified mathematical expressions of the complex interactions between the physical, biological, and socioeconomic elements of the ecosystems. These models have been successfully used in developed countries to assess, e.g., agroecological zones. In western China, the Agroecological Zoning Model was used to estimate the carrying capacity of land, whereas PODIUM was used in Southern

Africa to assess the trade-offs between food and water provisioning services. ES can also be assessed using ecological models that would help in quantifying the effects of management decisions on the condition of ES. It would also help to project long-term effects of changes in ES such as the loss of the ability of the ecosystem to regulate or provide vital services to the people. Mapping and assessing ES is an important biodiversity strategy in the European Union (EU), and member nations were required to map and assess the state of ecosystems and their services in their national territory by 2014 with the assistance of the commission. EU member countries were expected to assess condition and biodiversity as well as ES by conducting biodiversity and status assessment and also in-depth assessments of the ecosystems, including the forests, agroecosystems, freshwater, and marine (Millennium Ecosystem Assessment MA et al. 2005a). Mapping using different mapping approaches is an important part of the program, which will produce ecosystem maps and other thematic maps of the ecosystem and ES.

Remote sensing can be used for assessment of large areas such as the Niger Delta to examine the impacts of human activities such as petroleum exploration on land use/land cover, including the very important and vulnerable wetland biodiversity. This can be tackled by the acquisition of medium- to high-resolution satellite data from sensors such as LANDSAT, MODIS, and IKONOS whereas GIS software such as ArcMap, IDIRIS, Quantum GIS, and others provide the analytical power for spatial analysis of spatial and temporal changes in the wetland ecosystem. It can help to obtain trends of changes in the ES and the possible impacts of such changes. Hyperspectral remote sensing is now used in crude oil exploration either by adopting direct- and indirect-evidence detection (Bharti & Ramakrishnan, 2014; Van der Meer, van Dijk, & Kroonenberg, 2000). Direct detection involves mapping/identification of oil pools and alteration of minerals in soils and rocks due to seepages. Indirect detection, on the other hand, aims at recognition of secondary effects of volatile hydrocarbons on plants/crops (Ramakrishnan & Bharti, 2015; Van der Meer et al., 2000). Collection of spatio-temporal data will particularly contribute to mapping of the ecosystem and the services it supplies, even at local scales (Spalding, Kainuma, & Collins, 2010). Information on the current condition and trends of ES and identifying the drivers that affect human well-being result in an understanding of current changes to ES (Ranganathan et al., 2008). In Nigeria, several efforts to assess the state of the environment and the environmental stewardship of economic development along the Niger Delta ecosystem of Nigeria, as in many other developing countries, are often hindered by the lack of access to a comprehensive regional environmental information system (Adedeji et al., 2011). Nigeria as a country currently suffers from the lack or inadequacy of major geospatial scientific work that would offer an impartial evaluation of the destruction caused by oil and gas activities in the delta area of the South (Human Rights Watch, 1999). Mapping ecosystem "goods" in terms of describing their value relies on giving each area of land either an actual monetary value, an explicit quantitative value, or an explicit qualitative value (Egoh, Drakou, Dunbar, Maes, & Willemen, 2012; Medcalf, Small, Finch, & Parker, 2012; Haines-Young, Potschin, Medcalf, Small, & Parker, 2013a; MAES, 2013). Such valuation provides an indication of what ecosystem costs (e.g., risk of environmental degradation) and its benefits and may also reveal unexpected benefits and costs (Medcalf et al., 2012; Jackson et al., 2013; Haines-Young, Potschin, Medcalf, Small, & Parker, 2013b, 2013c). Supporting techniques such as overlay mapping, multicriteria analysis, and participatory mapping are often used for the "benefit mapping" exercise (Medcalf et al., 2012; BESS, 2013).

12.8 APPLICATION OF MODELING AND MAPPING OF ECOSYSTEM SERVICES IN NIGER DELTA

There are various applications and methods of modeling and mapping of ES, which are used to assess the key components of ES. Mapping and modeling of ES of the Niger Delta are crucial to identifying the types of services and its application. Spatial analysis as a form of mapping, and qualitative and quantitative modeling, are used to recognize the possible and appropriate trade-off that can be harnessed, including socioeconomic activities and ecological consideration (Duarte, Ribeiro, & Paglia, 2016). To apply modeling and mapping on ES in the Niger Delta, some criteria are required, such as the particular type of ES, availability and type of data sources, scale, and the type of methodological approaches (Martínez-Harms & Balvanera, 2012).

The type of ES, as mentioned in previous sections, and understanding the appropriate ES such as cultural, provisioning, regulating, and supporting, would enable one to use the appropriate modeling and mapping method. Availability and type of data are also important. The availability of primary and secondary data are important, including the validity and reliability of information considered. The type of data falls within biophysical, socioeconomic, and institutional sources of information, and integration of all these data sources are the criteria that is considered. The scale of these approaches differ, such as patch ($10-102$ km^2), local ($102-103$ km^2), regional ($103-105$ km^2), national ($105-106$ km^2), and global (>106 km^2). There are different approaches, such as look up structures, regression models, causal relationships, and extrapolation of data. Look up structures uses secondary information of ES, regression models allows experts to prioritize and rank land cover types, causal relationships uses existing information on the different type of layers of information to create proxy structures and layers of ES. Extrapolation of data uses adjustment of field data, which is weighed by cartography source of data. Regression models uses data of ES as response variables and proxies, especially data obtained (Martínez-Harms & Balvanera, 2012).

According to Bagstad et al. (2013), there are some criteria that can be considered for adopting of this mapping and modeling software, which include quantification and uncertainty, temporal factor, ability to adopt the tool independently, development level and documentation, the generality and scalability of the tool, social and nonsocial factors, and ability to be integrated with another tool.

Some of the applications for mapping and modeling ES that can be adopted for the Niger Delta include the following:

1. Automated Geospatial Watershed Assessment (AGWA) is a GIS-based and a water-related modeling tools for ES, used mostly to model hydrologically related flows. They use the desired parameters for indicators that are aquatic related. The AGWA can be used for any hydrological ecosystem−related studies in the Niger Delta because it gives a simple and replicable modeling approach and can accommodate basic GIS data,

which allows it to conform with GIS-based applications, particularly that the pattern of the analysis can be used to handle scenarios and multifaceted simulation scale. The usage of the AGWA is in four levels: delineate and discrete watershed, fixing parameters for soil and land covers, inputting and systematic modeling, and presentation of the visual result (Nedkov & Burkhard, 2012).

2. Integrated Valuation of Ecosystem Services and Tradeoffs (INVEST) is a GIS tool created by Stanford and Minnesota Universities, World Wildlife Fund (WWF), and the Nature Conservancy under the Natural Capital Project. This tool is used to map and quantifies ES, including the ecosystem quality (Duarte et al., 2016; Guerry et al., 2012). These models are adaptable to both terrestrial and aquatic ecosystems. It can be used to assess changes and flow of aquatic-related ES, including caron storage, ecosystem and water quality, plant pollination and production, erosion management, fishery production, and recreation. It is incorporated in ArcGIS as a toolbox placed on spatial and nonspatial biophysical, socioeconomic, and other relevant data. It is a tool that can be used in the Niger Delta, particularly for aquatic-based ES, climate adaption, and restoration of the stressed environment (Crossman et al., 2013; Prado et al., 2016).

3. Toolkit for Ecosystem Service Site-based Assessment (TESSA) is created by BirdLife International. It uses flow charts to explain the nature by which ES affect the society according to certain variables and parameters, including the setting of some scenarios. TESSA can be used on a site-scale level to consider stakeholders' involvement and process, including preliminary and background work, appraisal of drivers and services and its beneficiary, selecting an alternative state, selection of appropriate methods, data acquisition analysis, and communication. This can be adopted in the Niger Delta (Christin, Bagstad, & Verdone, 2016; Peh et al., 2013).

4. Artificial Intelligence for Ecosystem Services (ARIES) is a modeling tool that adopts multiple modeling components in spatial outcome. It adopts artificial intelligence—based data through semantic modeling to analyze ES to beneficiaries. It adopts machine learning, pattern recognition, and Bayesian and probability methods, and it is well-suited for spatial mapping and analysis of ecosystem service trade-offs. ARIES can assess small-scale fisheries and coastal management in the Niger Delta. It has two key assessment steps: setting the context and locating the concept that represents the particular purpose (Christin et al., 2016).

5. Costing Nature was developed by AmbioTEK and King's College of London. It is used for mapping and modeling multiple ES on a global data sets. It adopts ES as opportunity costs and is used to show spatial relationship between ES and conservation key areas (Bagstad et al., 2013).

6. Multiscale Integrated Models of Ecosystem Services (MIMES) is created at Louisiana State University, USA. They incorporate stakeholders' input to analyze tradeoffs within various ES, as well as simulation within the facet of sustainability such as ecological, social, economic, and estimate ES for intended scenarios. It is used to evaluate land use changes and impacts of ES on the local, regional, and global scale and can be considered for both long-term and short-term usage. It also investigates how ecosystem benefits are gained and lost. It shows how resources are changed between social, human, natural, and built capital. It is also used to forecast ES and anthropogenic changes based on various scenarios (Boumans, Roman, Altman, & Kaufman, 2015).

Other tools for mapping and modeling include Social Values for Ecosystem Services (SOVES), Natural Assets Information System (NAIS), Utilization and Capability Indicator (LUCI), Ecosystem Valuation Toolkit (EVT), Envision, EcoMetrix, EnSym—Environmental Systems Modelling Platform, SPASMO, and MOSES (Christin et al., 2016; Maes et al., 2012; Crossman et al., 2013).

There have been agitations from stakeholders to adopt a standardized application to measure and analyze ecosystems services using mapping and modeling, which will be applicable to all regions and localities, including the Niger Delta. Mapping and modeling of ES will properly evaluate tradeoff, linking the benefit of ES and the resources (Crossman et al., 2013).

12.9 RELEVANCE OF MAPPING AND MODELING FOR THE NIGER DELTA

There are increasing interest and attention in adopting mapping and modeling tools to solve problems relating to ES. The Niger Delta region is known to have abundant resources, and there is a need for policies from decision makers that will protect the environment (Maes et al., 2012). Mapping and modeling have been used to solve various ecosystem issues at local, regional, and global scales. Using them enhances the knowledge of the spatial relationship with biodiversity and other components of nature. It also allows creating relationships between different microcomponents of the services in the region. It is an initative that enables the evaluation of the costs and benefits of various ES and the analyses of the demand and supply of the various ES in the region. It allows recognition of areas and regions where there is need for intervention in sections of the environment that are degraded, to enable relationships between anthropogenic pressures and the available resources (Maes et al., 2012).

Mapping and modeling provide useful information for decision makers, assist to transform ambiguous information to simple image concept that is useful for sustainability of the resources and services (De Groot, Alkemade, Braat, Hein, & Willemen, 2010). Mapping and modeling consider spatial information to reveal informative knowledge. They create a relationship between indicators to analyze and supply the services to be revealed in a spatial image map to capture supplies for a particular region (Fig. 12.2).

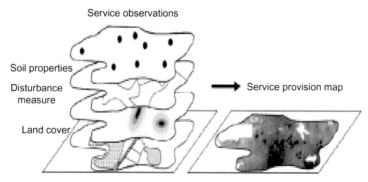

FIGURE 12.2   A sample of mapping of ecosystem services using data to relate services investigation and landscape features (De Groot et al., 2010).

It is important for land management, where policy makers adopt the result of the map to know the importance of land use and associated resources and strategies for providing important ES (de Groot et al., 2010).

12.10 INTEGRATION OF MAPPING AND MODELING OF THE ECOSYSTEM IN POLICY MAKING FOR THE NIGER DELTA

Mapping and modeling are a vital part of decision making to enable proper facilitation of ES and planning, and its benefits for humans. The information of mapping is vital for planning and policy making. It enables creating ES and demand stationed in areas of interest for planning. The requirement for enabling mapping and modeling into decision making depends on the availability of data, including spatial information. They need to be reliable sources of information across the region and scale of interest. Policy making requires associative incorporation of public understanding and participation to include mapping and modeling into action. This integration appreciates the use of standardized benefits that turn into sustainable results. It also helps to enable transparency in the trade-offs of the ES (Maes et al., 2012; Duarte et al., 2016).

According to NESP (2016), credibility, consistency, and unambiguous methods are some required tool to be considered when including mapping and modeling in ES (Christin et al., 2016). They help to consider the benefits of ES for planning and preparation of institutional aspects.

12.11 RECOMMENDATIONS

There is a need for a better integrated approach for mapping and modeling ES. It is observed that natural and manmade effects affect spatial and temporal aspects of mapping and modeling of ES. These should be considered in adopting the best methods and application for mapping and modeling. Also, the decision makers need to be well informed and carried along when considering indicators and other components (Birkhofer et al., 2015). To develop complex ES that will incorporate policy, biophysical, and management aspects, there is need to have a consistent method for quantification and mapping of these services. In addition, there is need to develop a framework that will be solely adaptable to the Niger Delta region and will consider all the environmental, social, economic, and specific policies that will be unique to the region. Best scale should be used appropriately and participatory approach should be considered while mapping and modeling ES.

12.12 CONCLUSION

This section shows the value and importance of mapping and modeling for ES without excluding the integration for decision making as shown in the structure of the paper. Consideration of the Niger Delta ecosystem is not complete if the spatial, computational, and modeling aspects are not appreciated. They give value to resources and are important for policy makers. The various mapping and modeling applications enumerated in this chapter are very vital for valuing ES in the Niger Delta. This concept of mapping and modeling is also important in the Niger Delta because it allows visualization of all the important locations with resources and areas that are degraded, which will be useful for decision making.

References

Aaron, S. (1996). Dying for oil. *World Watch*, May/June, 120.

Achi, C. (2003). Hydrocarbon exploitation, environmental degradation and poverty: The Niger Delta experience. *In Proceedings of the Diffuse Pollution Conference*. Dublin, Available at: http://www.ucd.ie/dipcon/docs/theme02/theme02_07.

Adedeji, O.H., Ibeh, L., & Oyebanji, F.F. (2011, September, 12–15). Sustainable management of mangrove coastal environments in the Niger Delta Region of Nigeria: The role of remote sensing and geographic information systems. In O. Martins, E.A. Meshida, T.A. Arowolo, O.A. Idowu, & G.O. Oluwasanya (Eds.), *Proceedings of the environmental management conference* (vol. 2, pp. 308–324). Abeokuta, Nigeria: Federal University of Agriculture. Available at: http://www.unaab.edu.ng.

Amnesty International, Nigeria (2009). *Petroleum, pollution and poverty in the Niger Delta* (pp. 65–79). London: Amnesty International Publications.

Amnesty International (2011). *The true tragedy: Delays and failures in tackling oil spills in the Niger Delta.* London: Amnesty International Publications.

Ash, N., Lucas, N., Bubb, P., Iceland, C., Irwin, F., Ranganathan, J., & Raudsepp-Hearne, C. (2008). Framing the link between development and ecosystem services. 2008. In J. Ranganathan, C. Raudsepp-Hearne, N. Lucas, F. Irwin, M. Zurek, K. Bennett, N. Ash, & P. West (Eds.), *Ecosystem services: A guide for decision makers* (pp. 15–21). World Resources Institute.

Bagstad, K. J., Semmens, D. J., Waage, S., & Winthrop, R. (2013). A comparative assessment of decision-support tools for ecosystem services quantification and valuation. *Ecosystem Services, 5*, 27–39. https://doi.org/10.1016/j.ecoser.2013.07.004.

Bharti, R., & Ramakrishnan, D. (2014). Uraniferous calcrete mapping using hyperspectral remote sensing. *IEEE International Geoscience and Remote Sensing Symposium*, 2902–2905.

Biodiversity and Ecosystem Services (BESS). (2013). Ecosystem service mapping gateway [online]. Available at: http://www.nercbess. net/ne-ess/ [Accessed 15th May 2016].

Birkhofer, K., Diehl, E., Andersson, J., Ekroos, J., Früh-Müller, A., Machnikowski, F., ... Smith, H. G. (2015). Ecosystem services: Current challenges and opportunities for ecological research. *Frontiers in Ecology and Evolution, 2*(87), 1–12. https://doi.org/10.3389/fevo.2014.00087.

Boumans, R., Roman, J., Altman, I., & Kaufman, L. (2015). The Multiscale integrated model of Ecosystem Services (MIMES): Simulating the interactions of coupled human and natural systems. *Ecosystem Services, 12*, 30–41 . https://doi.org/10.1016/j.ecoser.2015.01.004.

Boyd, J. W., & Banzhaf, H. S. (2007). What are ecosystem services? *Ecological Economics, 63*, 616–626.

Briggs, K. T., Yoshida, S. H., & Gershwin, M. E. (1996). The influence of petrochemicals and stress on the immune system of seabirds. *Regulatory Toxicology and Pharmacology, 23*, 145–155.

Christin, Z. L., Bagstad, K. J., & Verdone, M. A. (2016). A decision framework for identifying models to estimate forest ecosystem services gains from restoration. *Forest Ecosystems, 3*(3), 1–12. https://doi.org/10.1186/s40663-016-0062-y.

CIA. (2008). CIA World fact Book: United Kingdom. Available at: http://www.cia.gov/library/publications/the-world-factbook/print/uk.litml. [Accessed 15th May 2016].

Civil Liberties Organization (CLO). (2002). *Blood trail: Repression and resistance in the Niger Delta.* Ikeja: CLO.

Crossman, N. D., Burkhard, B., Nedkov, S., Willemen, L., Petz, K., Palomo, I., ... Maes, J. (2013). A blueprint for mapping and modelling ecosystem services. *Ecosystem Services, 4*, 4–14.

Daily, G., Polasky, S., Goldstein, J. Kareiva, M. P., Mooney, A. H., Pejchar, L., ... Shallenberger, R. (2009). Ecosystem services in decision making: Time to deliver. *Frontiers in Ecology and the Environment, 7*(1), 21–28.

Dami, A., Odihi, J. O., & Ayuba, H. K. (2014). Assessment of land use and land cover change in Kwale, Ndokwa-East Local Government Area, Delta State, Nigeria. *Global Journal of Human-Social Science: B Geography, Geo-Sciences, Environmental Disaster Management, 14*(6), 17−23.

De Groot, R. S., Alkemade, R., Braat, L., Hein, L., & Willemen, L. (2010). Challenges in integrating the concept of ecosystem services and values in landscape planning, management and decision making. *Ecological Complexity, 7*(3), 260−272. https://doi.org/10.1016/j.ecocom.2009.10.006.

Duarte, G. T., Ribeiro, M. C., & Paglia, A. P. (2016). Ecosystem services modelling as a tool for defining priority areas for conservation. *PLoS One, 11*(5), e0154573. https://doi.org/10.1371/journal.pone.0154573.

Durham Wildlife Trust. (2013). EcoServ-GIS Version 1 (England only): A wildlife trust toolkit for mapping multiple ecosystem services [online]. Available at: http://www.durhamwt.co.uk/wp-content/uploads/2012/06/EcoServ-GIS-Executive Summary-Only-WildNET-Jan-2013-9-pages.pdf. [Accessed 15th May 2016].

Egoh, B., Drakou, E.G., Dunbar, M.B., Maes, J., & Willemen, L. (2012). Indicators for mapping ecosystem services: A review. Report EUR 25456 EN. Publications Office of the European Union, Luxembourg.

Federal Ministry of Environment Abuja, Nigerian Conservation Foundation Lagos, WWF UK and CEESP-IUCN Commission on Environmental, Economic, and Social Policy, May 31. (2006). Niger Delta Resource Damage Assessment and Restoration Project.

Fisher, B., & Turner, R. (2008). Ecosystem services: Classification for valuation. *Biological Conservation, 141,* 1167−1169.

Food and Agriculture Organization of the United Nations (FAO). (2004). *Status and Trends in Mangrove Area Extent Worldwide*. Rome: FAO. Available at: http://www.fao.org/docrep/007/j1533e/J1533E01.htm#P181_4100 [Accessed 15th May 2016].

Frynas, J. G. (2000). *Oil in Nigeria: Conflict and litigation between oil companies and village communities* (p. 27). Hamburg: Lit Verlag.

Gbadegesin, A. (2008). *The impact of oil exploration and petroleum activities on the environment: Implications on peasant agriculture.* Ibadan: Macmillan Press.

Glavan, M., Pintar, M., & Urbanc, J. (2015). Spatial variation of crop rotations and their impacts on provisioning ecosystem services on the River Drava alluvial plain. *Sustain. Water Qual. Ecol, 5,* 31−48.

Guerra, C. A., Maes, J., Geijzendorffer, I., & Metzger, M. J. (2016). An assessment of soil erosion prevention by vegetation in Mediterranean Europe: Current trends of ecosystem service provision. *Ecological Indicators, 60,* 213−222.

Guerry, A. D., Ruckelshaus, M. H., Arkema, K. K., Bernhardt, J. R., Guannel, G., Kim, C. K., & Wood, S. A. (2012). Modelling benefits from nature: Using ecosystem services to inform coastal and marine spatial planning. *International Journal of Biodiversity Science, Ecosystem Services and Management, 8*(1-2), 107−121.

Haines-Young, R., Potschin, M., Medcalf, K., Small, N., & Parker, J. (2013a). Briefing paper 1 Bayesian Belief Networks. *Report to JNCC.*

Haines-Young, R., Potschin, M., Medcalf, K., Small, N., & Parker, J. (2013b). Briefing paper 2 Mapping ecosystem service trade-offs. *Report to JNCC.*

Haines-Young, R., Potschin, M., Medcalf, K., Small, N., & Parker, J. (2013c). Briefing paper 3 Mapping ecosystem service valuations. *Report to JNCC.*

Hauck, J., Winkler, K. J., & Priess, J. A. (2015). Reviewing drivers of ecosystem change as input for environmental and ecosystem services modelling. *Sustainability of Water Quality and Ecology, 5,* 9−30.

Human Rights Watch (HRW). (1999). *The price of oil corporate responsibility and human rights violations in Nigeria's oil producing communities (Environment)* [Online]. New York. Available at: http://www.hrw.org/reports/1999/nigeria/[Accessed 30th April 2016].

Ikelegbe, A. (2005). The economics of conflict in the oil rich Niger Delta Region of Nigeria. *Journal of Third World Studies, 43*(2), 24−50.

Jackson, B., Pagella, T., Sinclair, F., Orellana, B., Henshaw, A., Reynolds, B., . . . Eycott, A. (2013). Polyscape: A GIS mapping framework providing efficient and spatially explicit landscape-scale valuation of multiple ecosystem services. *Landscape and Urban Planning, 112,* 74−88.

James, G.K. (2008). *Assessment of environmental change and its socio-economic impacts in the mangrove ecological zone of the Niger Delta, Nigeria.* Doctor of Philosophy Degree, University of Missouri, USA.

James, G. K., Adegoke, J. O., Saba, E., Nwilo, P., & Akinyede, J. (2007). Satellite-based assessment of the extent and changes in the mangrove ecosystem of the Niger Delta. *Marine Geodesy, 30,* 249−267.

II. THE EFFECTS OF CRUDE OIL EXPLORATION ON THE SOCIO-CULTURAL AND ECO-ECONOMICS OF NIGERIAN ENVIRONMENT

Kadafa, A. A. (2012). Environmental impacts of oil exploration and exploitation in the Niger Delta of Nigeria. *Global Journal of Science Frontier Research Environment and Earth Sciences, 12*(3), 10—28.

Kia, M. (2009). The struggles for the survival of 12 million people in the Niger Delta. Environmental experts from UK, US and Nigeria affirmed that the Delta is the most polluted area in the world. Available at: http://niger-deltasolidarity.wordpress.com [Accessed 15th May 2016].

Lele, S., Springate-Baginski, O., Lakerveld, R., Deb, D., & Dash, P. (2013). Ecosystem services: Origins, contributions, pitfalls, and alternatives. *Conservation and Society, 11*(4), 343.

Maes, J., Egoh, B., Willemen, L., Liquete, C., Vihervaara, P., Schägner, J. P., . . . Drakou, E. G. (2012). Mapping ecosystem services for policy support and decision making in the European Union. *Ecosystem Services, 1*, 31—39. Available at: http://dx.doi.org/10.1016/j.ecoser.2012.06.004 [Accessed 15th May 2016].

Mapping and Assessment of Ecosystems and their services (MAES). (2013). An analytical framework for ecosystem assessments under Action 5 of the EU Biodiversity Strategy to 020. Discussion paper —version 9.6. Available at: http://biodiversity.europa.eu/ecosystemassessments/ [Accessed 15th May 2016].

Martínez-Harms, M. J., & Balvanera, P. (2012). Methods for mapping ecosystem service supply: A review. *International Journal of Biodiversity Science, Ecosystem Services and Management, 8*(1—2), 17—25, Available at: http://dx.doi.org/10.1080/21513732.2012.663792 [Accessed 15th May 2016].

Medcalf, K.A., Small, N., Finch, C., & Parker, J. (2012). Spatial framework for assessing evidence needs for operational ecosystem approaches. *JNCC Report No 469*.

Millennium Ecosystem Assessment (MA) (2005a). In S. Carpenter, P. Pingali, E. M. Bennett, & M. Zurek (Eds.), *Ecosystems and human well-being: Scenarios, volume 2*. Washington, DC: Island Press.

Millennium Ecosystem Assessment (MA) (2005b). In R. Hassan, R. Scholes, & N. Ash (Eds.), *Ecosystems and human well-being: Current state and trends, volume 1*. Washington, DC: Island Press.

Millennium Ecosystem Assessment (MA) (2005c). *Ecosystems and human well-being: Synthesis*. Washington, DC: Island Press, Available at: http://www.maweb.org [Accessed 15th May 2016].

National Ecosystem Services Partnership (NESP). (2016). *Federal resource management and ecosystem services guidebook* (2nd ed.). National Ecosystem Services Partnership (NESP), Duke University.

Nedkov, S., & Burkhard, B. (2012). Flood regulating ecosystem services—Mapping supply and demand in the Etropole municipality, Bulgaria. *Ecological Indicators, 21*, 67—79. https://doi.org/10.1016/j.ecolind.2011.06.022.

Nelson, E., Mendoza, G., Regetz, J., Polasky, S., Tallis, H., Cameron, Dr, . . . Shaw, M. R. (2009). Modeling multiple ecosystem services, biodiversity conservation, commodity production, and tradeoffs at landscape scales. *Frontiers in Ecology and the Environment, 7*, 4—11. Available at: http://dx.doi.org/10.1890/080023.

Nwilo, P.C., & Badejo, O.T. (2005). *Impacts and management of oil spill pollution along the Nigerian coastal areas*. Department of Survey and Geoinformatics, University of Lagos, Lagos, Nigeria. Available at: www.fig.net/pub.figpub36/chapters/chapter_8.pdf [Accessed 15th May 2016].

Obi, C. (2000). Globalised images of environmental security in Africa. *Review of African Political Economy, 27*(83), 47—62.

Omofonwan, S. I., & Odia, L. O. (2009). Oil exploitation and conflict in the Niger-Delta region of Nigeria. *Journal of Human Ecology, 26*(1), 25—30.

Onojeghuo, A. O., & Blackburn, G. A. (2011). Forest transition in an ecologically important region: Patterns and causes for landscape dynamics in the Niger Delta. *Ecological Indicators, 11*(5), 1437—1446. https://doi.org/10.1016/j.ecolind.2011.03.017.

Onuoha, F. C. (2008). Oil pipeline sabotage in Nigeria: Dimensions, actors and implications for national security L/C. *African Security Review, 17*(3), 99—115.

Oteh, C. O., & Eze, R. C. (2012). Vandalization of oil pipelies in the Niger Delta region of Nigeria and poverty: An overview. *Studies in Sociology of Science, 3*(2), 13—21.

Pearson, S. R. (1970). *Petroleum and the Nigerian Economy* (p. 15). Stanford: Stanford University Press.

Peh, K. S. H., Balmford, A., Bradbury, R. B., Brown, C., Butchart, S. H. M., Hughes, F. M. R., & Birch, J. C. (2013). TESSA: A toolkit for rapid assessment of ecosystem services at sites of biodiversity conservation importance. *Ecosystem Services, 5*, 51—57. https://doi.org/10.1016/j.ecoser.2013.06.003.

Potschin, M., & Haines-Young, R. (2011). Ecosystem services: Exploring a geographical perspective. *Progress in Physical Geography, 35*(5), 575—594.

Powell, C.B., White, S.A., Ibiebele, D.O., Bara, M., Dut Kwicz, B., Isoun, M., & Oteogbu, F.U. (1985, November 11—13). Oshika oil spill environmental impact; effect on aquatic biology. In *Paper presented at NNPC/FMHE international seminar on petroleum industry and the Nigerian environment* (pp. 168—178). Kaduna, Nigeria.

II. THE EFFECTS OF CRUDE OIL EXPLORATION ON THE SOCIO-CULTURAL AND ECO-ECONOMICS OF NIGERIAN ENVIRONMENT

Prado, R. B., Fidalgo, E. C. C., Monteiro, J. M. G., Schuler, A. E., Vezzani, F. M., Garcia, J. R., & Simões, M. (2016). Current overview and potential applications of the soil ecosystem services approach in Brazil. *Pesquisa Agropecuária Brasileira, 51*(9), 1021–1038. https://doi.org/10.1590/s0100-204x2016000900002.

Raji, A. Y., & Abejide, T. S. (2013). Shell D'Arcy exploration and the discovery of oil as important foreign exchange earnings in Ijawland of Niger Delta, C. 1940s-1970. Arabian. *Journal of Business and Management Review (OMAN Chapter), 2*(11), 22–33.

Ramakrishnan, D., & Bharti, R. (2015). Hyperspectral remote sensing and geological Applications. *Current Science, 108*(5), 879–891.

Ranganathan, J., Raudsepp-Hearne, C., Lucas, N., Irwin, F., Zurek, M., Bennett, K. . . . West, P. (2008). *Ecosystem Services: A guide for decision makers.* Washington, DC: World Resources Institute.

Salau, A.J. (1993). *Environmental crisis and development in Nigeria.* Inaugural Lecture, No.13 University of Port Harcourt, Choba, Nigeria.

Schägner, J. P., Brander, L., Maes, J., & Hartje, V. (2013). Mapping ecosystem services' values: Current practice and future prospects. *Ecosystem Services, 4,* 33–46.

Scheffer, M. S., Carpenter, S. R., Foley, J. A., Folke, C., & Walker, B. (2001). Catastrophic shifts in ecosystems. *Nature, 413,* 591–596.

Schneiders, A., Van Daele, T., Van Landuyt, W., & Van Reeth, W. (2012). Biodiversity and ecosystem services: Complementary approaches for ecosystem management? *Ecological Indicators, 21,* 123–133.

Seppelt, R., Dormann, C. F., Eppink, F. V., Lautenbach, S., & Schmidt, S. (2011). A quantitative review of ecosystem service studies: Approaches, shortcomings and the road ahead. *Journal of Applied Ecology, 48,* 630–636.

Small, N., Medcalf, K., Parker, J., Haines-Young, R., & Potschin, M. (2013). Further development of a spatial framework for mapping ecosystem services: User Manual. *Report to JNCC.*

Spalding, M., Kainuma, M., & Collins, L. (Eds.), (2010). *World Atlas of mangroves* London: Earthscan.

Steyn, M. S. (2009). Oil Exploration in the Colonia Nigeria, c. 1903-58. *Journal of Imperial and Commonwealth History, 37*(2), 249–274.

TEEB. (2010). The economics of ecosystems and biodiversity: Mainstreaming the economics of nature: A synthesis of the approach, conclusions and recommendations of TEEB.

Tolulope, A. O. (2004). Oil exploration and environmental degradation: The Nigerian experience. International society for environmental information science. *International Information Archives, 2,* 387–393.

Twumasi, Y. T., & Merem, E. C. (2006). GIS and remote sensing applications in theassessment of change within a coastal environment in the Niger Delta Region of Nigeria. *International Journal of Environmental Research and Public Health, 3*(1), 98–106.

Ukoli, M.K. (2005). Environmental factors in the management of the oil and gas industry in Nigeria. Available at: www.cenbank.org[Accessed 15th May 2016].

United Nations Environment Programme (UNEP) (2005). *Ecosystems and Human Well-being. Millennium Ecosystem Assessment (Program).* Washington, DC: Island Press.

United Nations Environment Programme (UNEP). (2011). Environmental assessment of Ogoni Land. ISBN:978-92-801-3130-9. Available at: www.unep.org/nigeria [Accessed 15th May 2016].

Uyigue, E., & Agho, M. (2007). *Coping with climate change and environmental degradation in the Niger Delta of Southern Nigeria.* Community Research and Development Centre Nigeria (CREDC).

Van der Meer, F.D., van Dijk, P.M., & Kroonenberg, S.B. (2000). Hyperspectral hydrocarbon microseepage detection and monitoring: Potentials and limitations. In *Proceedings of the 2nd EARSeL workshop on imaging spectroscopy.* Enschede. 9.

Viglizzo, E. F., Paruelo, J. M., Laterra, P., & Jobbágy, E. G. (2012). Ecosystem service evaluation to support land-use policy. *Agriculture, Ecosystems and Environment, 154,* 78–84. http://dx.doi.org/10.1016/j.agee.2011.07.007.

Volk, M. (2015). Modelling ecosystem services: Current approaches, challenges and perspectives. *Sustainability of Water Quality and Ecology, 5,* 1–2.

Volk, M. (2013). Modelling ecosystem services – Challenges and promising future directions. *Sustainability of Water Quality and Ecology, 1–2,* 3–9.

Wikipedia. (2006). Petroleum in Nigeria. Available at: http://en.wikipedia.org/wiki/petroleum. [Accessed 15th May 2016].

Wikipedia. (2014). The Niger Delta. Available at: http://en.wikipedia.org/wiki/Niger_Delta [Accessed 18th May 2016].

Winowiecki, L., Vågen, T. G., & Huising, J. (2016). Effects of land cover on ecosystem services in Tanzania: A spatial assessment of soil organic carbon. *Geoderma*, *263*, 274–283.

World Bank. (2002). *Memorandum of the President of the IDA and the IFC to the executive directors on an interim strategy update for the Federal Republic of Nigeria*. New York.

World Rainforest Movement (WRM). (2003). Nigeria: Gas Corporation NLNG Destroys Mangrove forest in the Niger Delta. Available at: http://wrm.org.uy/oldsite/bulletin/68/nigeria.html [Accessed 16th May 2016].

World Resource Institute (WRI) (1992). *World Resource 1992-93*. New York: Oxford University Press.

Zabbey, N. (2004). Impacts of extractive industries on the biodiversity of the Niger Delta. In *National workshop on coastal and marine biodiversity management*. NNPC, Nigerian National Petroleum Corporation, Monthly Petroleum Information. September, Lagos, Nigeria, 1984, 53 pp.

II. THE EFFECTS OF CRUDE OIL EXPLORATION ON THE SOCIO-CULTURAL AND ECO-ECONOMICS OF NIGERIAN ENVIRONMENT

The Socio-Cultural Implications of Crude Oil Exploration in Nigeria

Ike Nwachukwu[1] and Ikechukwu C. Mbachu[2]

[1]Michael Okpara University of Agriculture, Umudike, Abia State, Nigeria [2]Brandenburg University of Technology, Cottbus-Senftenberg, Brandenburg, Germany

13.1 INTRODUCTION

Nigeria holds the largest natural gas reserves in Africa since natural gas is associated with oil production (EIA, 2011). Until the recent drop in prices of crude oil, Nigeria's economy was largely oil-based. As of the year 2000, oil and gas exports accounted for more than 98% of export earnings and approximately 83% of the federal government's revenue, as well as generating more than 14% of its GDP. It also provides 95% of foreign exchange earnings and approximately 65% of government budgetary revenues. In 2015, Nigeria earned $76 billion from crude oil export and had a daily production of 1.8 million barrels of oil (OPEC, 2015).

Crude oil resource gave the Nigerian government approximately USD20 million a day. Nigeria boasts of more than 37 billion barrels of proven oil reserves (EIA, 2011; OPEC, 2015) and more than 176 trillion cubic feet of gas reserves, which make Nigeria Africa's largest crude oil producer. The Niger Delta region is considered the mainstay of the Nigerian economy for its significantly high level of oil reserves and has become prominent in the global oil market as Nigeria's largest oil region and one of the highly productive oil-exporting regions in the world (Osuntokun, 2000).

Nigeria's huge oil reserves make it the 10th most petroleum-rich nation, and by far, the most affluent in Africa. In mid-2001, her crude oil production was averaging around 2.2 million barrels (350,000 m^3) per day. It is expected that the industry will continue to be profitable based on an average benchmark oil price of $85–$90 per barrel. By the beginning of 2016, the daily production was approximately 2.4 million barrels per day. However, the oil price fell to approximately $38 per barrel in the same year, and oil production was reduced to approximately 1.4 million barrels per day, due to the disruption of production by militant youths demanding environmental restoration. The Niger Delta

The Political Ecology of Oil and Gas Activities in the Nigerian Aquatic Ecosystem
DOI: http://dx.doi.org/10.1016/B978-0-12-809399-3.00013-6

region is where the oil reserves are concentrated, but off-shore rigs are also prominent in the well-endowed coastal region. Nigeria is one of the few major oil-producing nations still capable of increasing its oil output.

Unfortunately, oil exploration for over five decades has left the environment of the region greatly degraded. This has negatively impacted on the economic, social, and cultural lives of the people. Although the country makes millions of dollars daily from crude oil extracted from the Niger Delta, the lives of the people are impoverished, and they live in abject poverty.

13.2 BEFORE THE ADVENT OF OIL EXPLORATION

According to reports, there was abundant evidence to show that the Niger Delta people were not a homogenous entity, although they have common interests and problems; there are different ethnic groups within the region. The Niger Delta region is extremely heterogeneous with respect to culture and ethnicity. The very rich culture and heritage of the region is based on the presence of approximately 40 different ethnic groups speaking 250 languages and dialects. The numerous ethnic groups include Ijaws, Ogonis, Ikwerres, Etches, Ekpeyes, Ogbas, Engennes, Obolos, Isoko, Nembes, Okrikans, Kalabaris, Urhobos, Itsekiris, Igbos, Ika-Igbos, Ndoni, Oron, Ibeno, and Yorubas. Other groups include Ibibios, Anang, Efiks, Bekwarras, Binis, and others. The heritage of the people is reflected in modes of dressing, marriages, traditional culture, and festivals (NDDC Report, 2006).

Before the advent of oil exploration, the Niger Delta was a peaceful traditional environment. The people had a well-organized social life that accentuated their community values, which protected the rights of every member in the land. Communities were administered under strict rules that defined interpersonal relationships. The knowledge of these rules was transferred from one generation to another, with modifications where necessary. Whoever violated the rules was appropriately sanctioned. Respect for elders and traditional authority, chastity, and good conduct were the cherished values in the communities.

The communities were headed by traditional rulers who wielded a great deal of authority. They appeared in awesome royal regalia. In some of these communities, the word of the king was law. They were well revered, and that gave them the authority to rule the community. The traditional ruler, with his Council of Chiefs, administered the affairs of the people and adjudicated where there were conflicts. The decision of the king on any matter was final.

The communities were agrarian and so their livelihoods were derived from natural resources. Agriculture was the dominant occupation, made up of crop and fish farming. In 1995, a World Bank study revealed that "the Niger Delta has been blessed with an abundance of physical and human resources, good agricultural land, extensive forest, excellent fisheries, as well as developed industrial base, and a vibrant private sector" (World Bank Report, 2011).

Here, agriculture was their way of life. Their sociocultural life was tied to their agricultural life. Their religious practices, festivals, and lifestyles were all derived from their agricultural practices and farming calendar. It meant that anything that happened to their agriculture affected their total wellbeing. The foregoing was the situation in the oil-producing communities before exploration started in 1958, which today has distorted the sociocultural environment of the people.

13.3 OIL EXPLORATION IN NIGERIA—THE REQUIREMENTS (PHYSICAL IMPACTS)

Commercial oil exploration in Nigeria started in 1958 at a place called Oloibiri in Bayelsa State. Oil exploration comes with necessary required actions that ultimately impact on the environment and influence the sociocultural lives of the people. The following activities are carried out in oil exploration.

Land take. The oil companies must of necessity acquire expanses of land for their operations. The land take will include the space for the oil well and operational installation of equipment. Land for pipelines, which in some instances run into kilometers, at a width not less than 30 m, from the oil well to the loading station. The UNDP (2006) states that since the discovery of oil in the region, more than 1481 oil wells have sprung up, producing from approximately 159 oil fields (the current estimate of oil wells is put at approximately 5000). There are more than 7000 km of pipelines and flow lines and 275 flow stations operated by more than 13 oil companies. Such installations invariably culminate in the loss of land and natural resources upon which indigenous communities have traditionally relied. This can jeopardize the survival of indigenous groups, as distinct cultures are inextricably connected to the lands they have traditionally inhabited. The cultural values of the Niger Delta are interlinked with their relationship with land; "no land means no culture"; thus, damage to land is direct damage to the people and their culture.

Influx of skilled and unskilled foreigners. Oil exploration requires many specialized skills that are usually obtained from abroad and other parts of the country. At any oil well operation, the communities suddenly witness the inflow of strangers who come with their own culture, which usually pollutes the existing culture of the indigenous people. The foreigners live with the people and produce a lifestyle that is peculiar to them but strange to the community.

On-farm labor migration. The exploration activities also require unskilled labor, which is abundant in the local communities. The oil companies offer very high pay for labor, far higher than what a subsistent farmer can ever earn. The result is that farmers (usually males) migrate from the farms to the oil companies to take up jobs.

Oil spills. During drilling and loading of crude oil, there would be some spills into the environment. Sometimes, the spill will be as a result of operational accidents that may lead to extensive spills. Most importantly, recently, extensive oil spills have been caused by vandalism by the youths from that region. They have been agitating against the neglect of the region by the oil companies and the federal government, which have extracted oil resources from their place and left them in poverty. There were 63 operational spills in 2011, 32 in 2010, but the total volume spilled decreased to 3595 barrels from 5270 in 2010 due to improved oil spill response time (SPDC, 2016).

When oil spills, it leads to soil contamination with hydrocarbon. This in some instances reaches a depth of approximately 5 m (UNEP Report, 2011). The spill also affects the vegetation, leaving the leaves and stems denuded, with the roots coated in a bitumen-like substance. The same oil spill triggers fire outbreaks that destroy the vegetation. When oil spills, it necessarily gets into the rivers, destroying the aquatic life of that environment. Fishes would migrate away from polluted waters into the deeper seas. UNDP (2006) estimates that in 30 years, more than 400,000 tons of oil had been spilled into the creeks and soils of southern Nigeria and about 70% of the oil has not been recovered. The vast majority of the spills are a consequence of aging facilities and human errors. In January 2008, the Nigerian National Oil Spill Detection and Response Agency (NOSDRA) declared that it had so far located more than 6727 of 10,340 spills sites abandoned by various oil companies within the Niger Delta (NOSDRA, 2016).

Gas flaring. Gas is produced in the process of oil drilling. Although this gas is a source of revenue to the government, the vast majority of it is flared into the atmosphere. This causes a great deal of air pollution, which is injurious to the health of the people in the community. It contributes to environmental problems, such as acid rain with it attendant impact on agriculture, rural livelihood, gender relations, health, forests, and other physical infrastructure. Gas flaring raises the temperature of the local environment to values beyond normal of 13,000–14,000°C and causing noise and light pollution around the vicinity.

13.4 EFFECTS OF OIL EXPLORATION ON THE SOCIAL ENVIRONMENT

The consequences of oil exploration usually produce negative effects on the environment. It is expected that before oil exploration commences in any community, an environmental impact assessment should be conducted and mitigation plans outlined for identified negative impacts. At the beginning of oil exploration in Nigeria, this was not done. Even in recent times where these studies were carried out, no serious attention is paid to any mitigation plans.

Outlined here are the effects of oil exploration on the social environment:

13.4.1 Inadequate Land for Farming

Land take by the oil companies greatly reduces the available land for farming. The land area within which the network of pipelines are located is about 31,000 km^2 (NDDC, 2006). This land take is made worse by the fact that the Delta Region is a wetland surrounded by water and creeks. The vegetation is mostly mangrove and swamp forest. The result is that limited land is available for farming. Also, the greater population of male farmers has been drawn to the oil companies as casual labor. This means that farming, which had hitherto been the mainstay of the economy, is no longer so. The farmhand is now largely made up of women. The result is that the traditional division of farm labor by gender is no longer there. When the men are not available to do the hard labor on the farm, such as bush clearing and ridge making, women have been forced to take up the task.

13.4.2 Land and Water Contamination

Oil spill contaminates the farmland. This implies that the available farmland is further degraded by oil pollution, resulting in crop failure. Studies have shown that about 100% loss in yield of all crops cultivated occurred 200 m away from the Izombe flow station; 45% loss in yield of crops planted 600 m away, and 10% loss for those cultivated 1 km away from the flare point (Anyanwu & Tanee, 2008; Okezie & Ekeke, 1987). Also, a reduction in cassava yield parameters (growth, fresh weight of shoot and tubers, total fresh weight, etc.) as a result of oil pollution has been observed.

The same thing happens to the river where oil spill pollutes the water. The result is that fishing is made more difficult because fishes will avoid polluted water and move further into deeper waters where fishers would not be able to access them due to the crude fishing equipment. What this means is that farming, which has been the way of life of the people, is distorted, thereby affecting their sociocultural life. It results in drastic reduction in local food production and dependence on imported food that is not indigenous to the people. This affects the health and general outlook of the people.

13.4.3 Acculturation

Foreigners who come into the oil-producing communities come with their own culture. Many of them come from the Western world with cultures that are totally different from the local ones. Since they are whites from the developed world and with better expertise, the local people are likely to see the foreign culture as better than theirs and then get acculturated to the strange foreign culture. This becomes detrimental to the indigenous people. This is usually the beginning of defiant behavior by the youths who are most likely to be acculturated. The elders would always resist strange culture, but the youths will always insist on their new ways of doing things. The result is often a breakdown of law and order in the community.

13.4.4 Weakened Traditional Authority

Three things have led to the weakening of traditional authority in the oil-producing communities. One is when the youths defy the authority of the king as a result of the newly acquired lifestyle. This is usually influenced by the increased income of the youths from their earnings from the oil companies and the newly acquired culture of the foreigners. Until very recently, youths unquestionably followed the dictates of the elders. The oil companies have often been accused of deliberately funding the conflict between the youths and the leaders in order to divert attention from the inadequacies of the companies.

Second is when a parallel authority structure emerges in the community. It is the practice of the oil companies to bring development projects to the community as a corporate social responsibility. To manage these projects effectively, a Community Development Association is usually established, whose executive usually manages the funds from the oil companies. This management committee usually sees itself as independent of the traditional control. This whittles down the authority of the king. The third one is when the benefits accruing to the king from the oil company begin to receive the jealousy of other men.

II. THE EFFECTS OF CRUDE OIL EXPLORATION ON THE SOCIO-CULTURAL AND ECO-ECONOMICS OF NIGERIAN ENVIRONMENT

Suddenly, the throne becomes more attractive to people, and there is a strong fight by different contending parties who were not in the ruling houses to ascend the throne. This leadership friction means that the traditional arrangements for resource use and management have virtually broken down.

13.5 SOCIO-CULTURAL IMPACTS OF OIL EXPLORATION

Impact is the outcome of effect. Oil exploration, over the years in the Niger Delta, has come with great consequences on the environment, especially on the human beings. It has become a necessary evil. Whereas the nation derives huge revenue from crude oil, the environment from where the oil is drilled is left in abject poverty and criminal negligence by both the State and the oil companies. The impacts of this are discussed below.

13.5.1 Breakdown of Cherished Cultural Heritage

Most communities of the Niger Delta are very traditional in their ways of life. Their cultures are central to their lives and guide their everyday activities and interactions (Odoemena, 2011). The massive influx of outsiders has, however, greatly diminished the cultural well-being of the people of the region. The most apparent and striking result of the influx of foreigners into the community (and subsequent acculturation) and a huge amount of money in the hands of the youths from oil companies and leadership tussle, is the breakdown of law and order. There are so many intra- and intercommunal conflicts in the oil-producing communities, which many times have led to violent destruction of lives and properties. There is hardly any such community that is at peace. The youths are perpetually in conflict with the elders, and the Management Committee Chairmen are in a power tussle with the king. The cherished respect for elders and the unchallenged authority of the king have all been eroded. This is what has been described as the "curse of oil."

13.5.2 Youth Restiveness

A corollary to the above is youth restiveness. A dominant feature of the population of the Niger Delta region is its significant level of young people, with more than 62% of the population below the age of 30 years (NDDC, 2006). In the past two decades, there has been a growing agitation by the youths against the perceived injustice meted out to their environment and the lack of adequate compensation by the federal government and the oil companies. They are also insisting that the resources of the region must be controlled by the people and pay adequate tax to the government.

It is not as if the government and the oil companies have not done anything. In 1992, the government established the Oil Mineral Producing Areas Development Commission (OMPADEC) to directly bring development to the area. However, the Commission was not successful in doing so. The agitation by the youths continued. This agitation for control of the resources attracted the attention of the federal government due to the level of poverty in the Niger Delta. This prompted the government into reviewing the law that established the Derivation Fund. This allocated more funds to the States where oil is drilled. The derivation percentage was raised from 3% to 6% in 1995 and then to 13% in 1999. However, the impacts of the derivation fund have not been able to ameliorate the poverty in the region. The government then established the Niger Delta Development Commission (NDDC) in 2000 to replace the moribund OMPADEC, as an agency that will manage the derivation fund.

In 2006, the youth restiveness in the region became militant. They acquired sophisticated weapons and destroyed oil pipelines, which reduced oil production to less than half of the quantity being produced. This period saw the emergence of different militant groups such as the Movement for the Emancipation of the Niger Delta (MEND), Niger Delta People's Volunteer Force (NDPVF), the Egbesu Boys, Martyrs Brigade, Niger Delta People's Salvation Front, Niger Delta Vigilante (NDV), and the Joint Revolutionary Council.

These militant groups blew up oil installations and kidnaped many expatriate oil workers. To stop the destructive militancy, the government declared a state of amnesty for the militants in 2009. The federal government made this declaration with the aim of restoring normalcy to the oil-producing areas. A number of activities were planned for the many repentant militants, and these included training in various skills both within and outside the country. In most cases, large sums of money were paid to the leaders of the various groups. A few were given contracts running into millions of dollars to protect the oil pipelines from vandalization. In the same year (2009), the government also established the Ministry of the Niger Delta to ensure adequate development for the region. The oil companies have equally been developing the region through many projects. The latest is the Global Memorandum of Understanding (GMoU). This is a joint venture between the oil companies and the government. Funds are pulled together from the partners and allocated to the oil-producing communities, which are grouped in clusters. The fund is administered by Shell Petroleum Development Company, and each cluster has a management committee that executes the different projects decided by each community.

In spite of all these, the youths are insisting that these ventures are not commensurate with the funds derived from the oil and the devastation of the environment. To them, there was no plan in place to mitigate the degraded environment. With the granting of amnesty and the fulfillment of the amnesty agreement, militancy in the Niger Delta abetted. However, with the change in government in 2015, the agitation in the region has resumed. In 2016, the militancy has escalated with the emergence of new militant groups such as the "Niger Delta Avengers" and once again, Nigeria's oil production has greatly reduced, triggering increased oil prices at the international market.

13.5.3 Antisocial Behavior

One of the major impacts of oil exploration in the region is the antisocial behavior of the youths. As stated earlier, the influx of foreigners has negatively influenced the behavior of the youths, both directly and indirectly.

Drug abuse: There have been increased incidences of drug abuse among the youths in recent times. Smoking and use of hard drugs have been on the increase. The reason for this is not farfetched. Many of the youths are unemployed and lacked any skills. The money they are able to make from casual labor or criminal activities is lavished on drugs.

Kidnaping and robbery: In 2006, the Niger Delta militancy snowballed into another phase of kidnaping and hostage taking for huge ransoms. The concentration then was on expatriate oil workers. To date, this trend has continued and has mushroomed into an everyday security challenge in most parts of the nation. Cases of armed banditry have increased as criminals have pitched their tents in the area as they expect higher bounty because of the comparative high cash flow in the area. Another dimension to this is the emergence of rival gangs, known locally as "cults," for selfish political and clandestine purposes. These cultists have brought so much violence and have unleashed mayhem on the citizenry. As was stated earlier, the militant activities exposed many of the youths to sophisticated weapons. By 2009 when the amnesty program took effect, there was a lull in the agitation for resource control. So those who refused to buy into the amnesty program and could not use a gun again to fight the government, started using it for kidnaping and robbery.

Prostitution and teenage pregnancy: Teenage and unwanted pregnancies have become the order of the day in these communities as a result of temporal participants (haulage drivers, suppliers, job seekers, researchers, etc.) who flood the area. The result of multiple sex partners in this area has led to high spread of HIV/AIDS. Parental care has broken down in most cases. The fathers are not home to cater to their children, especially the girls. Most times, these girls are raised by single mothers. The girls drop out of school and they are lured into prostitution. Most of the oil workers usually do not come to the worksite with their spouses, so they look out for girls for their sexual gratification.

Bunkering and pipeline vandalization: A corollary to the upsurge in militancy is the increased case of illegal "bunkerers" who steal crude oil by intentionally vandalizing pipelines. According to various estimates, Nigeria loses approximately 20% of total oil production to illegal bunkering annually. This represents a serious threat to the economic well-being of the nation. Illegal bunkering has been a key source of funds for the militant warlords who are either publicly or privately involved in such practices. Most of those involved in the practice consider it to be the only avenue for providing income for aggrieved and impoverished residents of oil-producing communities.

13.5.4 Lack of Education/Low Literacy Level

The lure of quick money from the oil companies, kidnaping, and militancy activities have made the youths drop out of school. Reports have shown that there has been a downward trend in educational attainment by the male youths in the region. The result is that the youths lack skills and are unemployable.

13.5.5 Lack of Food Security

The Niger Delta is one of the most food-insecure regions in the country. The reasons are obvious. One, the land and water have been largely polluted and consequently, are unproductive. This has made farmers abandon the farms for other vocations. Two, the required labor is not available for farming; the men have largely abandoned the farms.

Three, the different State governments in that region are not even interested in improving agricultural production because they get derivation funds, which are huge amounts of money. The result is that they are not self-sufficient in supplying any food consumed in the region. One of the implications is that if there is any food blockage from outside the region, the people will suffer starvation.

13.5.6 Poverty

According to the Copenhagen declaration, poverty is defined as "a condition character-ized by severe deprivation of basic human needs, including food, safe drinking water, san-itation facilities, health, shelter, education, and information. It depends not only on income but also on access to social services" (UNDP, 1995). In October 2015, the World Bank updated the international poverty line to USD1.90 a day. Going by this index, it is evident that the Niger Delta region is one of the poorest parts of the developing world. Per capita income is very low—66% of the population earns less than N10,000 (approx. USD75) per month and 76.6% earn less than N20,000 (NDDC, 2006). The incidence of poverty is very high, with more than 70% living at subsistence level in rural areas. Life expectancy in the region is low, 46.8 years, and is even lower in some of the more remote wetland areas where access to health care is difficult. Infant and child mortality is particularly high (20% die by the age of 5). This pernicious cycle includes a high degree of adult morbidity ema-nating from a wide variety of diseases that undermine individual employment and initia-tive. The quality of life is further affected by social unrest and threats to peaceful coexistence among ethnic groups, particularly in the core Niger Delta (NDDC, 2006).

Economically, petroleum does not just fail to offer long-term sustainable employment alternatives at the local level, but it can seriously disrupt preexisting patterns of produc-tion, hence leading to the destruction of traditional means of livelihood. As a result, most of the populace have become jobless since their local economic support system of fishing and farming is no longer sustainable. Since the discovery and exploitation of crude oil, excruciating poverty has been characteristic of the socioeconomic milieu of the Niger Delta. The poverty of the local people and its contrast with the wealth generated by oil has become one of the world's starkest and most disturbing examples of the "resource curse" (Amnesty International, 2009).

13.5.7 Corruption

Unbridled corruption on the part of government officials has prevented a large chunk of the earnings derived from the sale of oil from the Niger Delta from returning back to the region to provide the much needed development to the area. Some 70% of the oil mon-ies is accrued by 1% of the population, with 70% of the wealth held in private hands abroad (Lubeck, Watts, & Lipschutz, 2007).

The Nigerian oil industry has created a classic rentier state and economy in which the leaders have cultivated and imbibed the spirit of prebendalism. The result of this is the "corruption culture" prevalent in the Nigerian oil industry. Odoemena (2011) noted that crude oil has become the perfect conduit for corrupt government officials of successive administrations to enrich themselves. As a result of the deluge of agitations, derivation to the Niger Delta states was increased to 13%. This increase in derivation-based revenues has not produced any significant increase in the standard of living in the oil-rich region. Rather, the political elite has been engaged in large-scale financial embezzlement and corruption scandals. The implementation of the 13% derivation fund appears not to have led to radical transformation or development (Obi, 2005). Analysis of the revenue allocation data shows that the Niger Delta states have received far more revenue from the Federation Account than all the other states on account of the 13% derivation funds. Oluwaleye (2013) argued that if government officials in the region have judiciously spent their monthly allocations to improve the lots of the ordinary people through the creation of jobs, or embarked on the infrastructural development of the region, the situation would have been better than this current sorry state. There have been some attempts to tackle the corruption and lack of transparency in the oil sector. Nigeria's Federal Government has made important—albeit limited—strides toward becoming more transparent and responsible in its own use of public resources. For instance, Nigeria was the first nation to sign up to the voluntary Extractive Industry Transparency Initiative (EITI) aimed at eliminating corruption in the extractive industry.

13.5.8 Rise of Environmental and Economic Migration

The advent of oil has brought about different kinds and waves of migration within the Niger delta.

* Economic migrants: Rural–Urban and Urban–Urban
   In many parts of the developing world, large-scale oil fields act as magnets of economic opportunities for individuals and groups from a wide catchment area. The Niger Delta region has attracted a huge number of migrants, both foreign and local. They come seeking economic opportunities from oil production (UNDP, 2006). The emergence of petroleum as a major resource led to oil industry–induced migrations involving migrants moving in search of greener pasture in the major urban areas of the Niger Delta.
* Environmental migrants: Rural–Rural, Rural–Urban
   Aghalino and Eyinla (2009) noted that migration in the Niger Delta is informed by the wish of affected people to move elsewhere because of the degradation of aquatic bodies and land. Although many of them end up in the cities, a large proportion end up in the rural areas, thereby increasing the already existing pressure on the land and rural resources. Farmers and fishermen who are unable to migrate to other rural areas usually move to nearby urban centers such as Warri, Eket, Port-Harcourt, and Uyo in search of "economic redemption." The magnitude of the rural–urban exodus has severe consequences in terms of overcrowding, overstretching available resources and facilities, increasing the spread of communicable diseases, unemployment, and helping to provide a fertile training ground for class-consciousness and a stack awareness of social deprivation (Aghalino, 2011). Furthermore, the oil industries have created intra- and intercommunal conflicts that have displaced thousands of people from their communities.

13.5.9 Increased Social and Income Inequality

The rapid influx of skilled and the highly paid oil project workers introduced major distortions into the social and economic fabric of the local societies. Apart from the destruction of local economic activities, oil companies have created systems and levels of social and economic stratification by creating "islands of comfort" (i.e., areas where oil executives live quite lavishly in comparison to the miserable living conditions of the locals). This further exacerbates delinquent attitude among the poor. It breeds hatred for the affluent and the desire to rob them of their affluence.

The high cost of living in the Niger Delta can be partly traced to the presence of the multinational oil companies whose workers earn a comparatively higher sum of money; hence, they can easily afford high-priced goods and services. The effect is a significant increase in the cost of living, even for the people who do not share in the benefits of oil projects. With their low-income level and low purchasing power, the locals cannot compete with the migrants who are high-end income earners. This further entrenches and deepens the level of social and economic inequality in the Niger Delta.

In addition, compensation and royalty payments by the oil companies are seldom equally distributed between the different social groups. Kings, community leaders, or "big-men" often receive the lion share of the largesse, leaving the crumbs for the lowly. This translates to substantial income inequalities within the groups in the communities.

13.5.10 Human Rights Violation and Militarization of Social Space

The parlous environmental and economic state of the people has led to a great deal of disenchantment and subsequently not-so-peaceful agitation and clamor for a variety of causes. Such protests by the Niger Delta people have led to brutal repression by the central government and oil companies' security forces (Babatunde, 2009). In almost every oil-producing community in the Niger delta, there have been incidents in which security forces have beaten, detained, or even killed those involved in agitations.

In the 1990s, under the military junta, several communities were subjected to massive human rights violations orchestrated by the police and by special state security forces. Voices of dissent were either imprisoned or outrightly eliminated through extra-judicial means. A poignant case is that of the Ogoni writer and environmental activist, Ken Saro Wiwa, who was sentenced to death by hanging by the military junta. With the advent of the civilian rule, the crackdown and highhandedness were not abated. Another notable instance was in 2000; Odi town was invaded and occupied by military forces. Hundreds of people were killed and properties were deliberately destroyed and set ablaze by the military. It was and still is the policy of the government to adopt crude force in handling tense situations in the oil-producing communities.

The militaristic approach adopted by successive governments has contributed to the spate of militancy in the Niger Delta with the cumulative effect that the social space in the region has become heavily militarized. The militarization has worsened armed conflict in the region, for such violent response worsens the very conditions that gave rise to the violence in the first place, thereby creating a "conflict trap" (Okumagba, 2011).

13.5.11 Gender Issues

Although in most parts of the Niger Delta traditionally, females neither own nor inherit land, they are primarily responsible for the sustenance of the household. Women of the Niger Delta depend mainly on land and rivers for sustaining their household. However, decades of oil exploitation with the resultant oil spillages and gas flaring have led to the destruction of farmlands and aquatic flora and fauna. Furthermore, the constant land-take for oil production activities has left women in the region with little or no means to feed or support their families; hence, they have to go further away from home to find unpolluted land and water to sustain themselves and their dependents. As the livelihood of these women depends totally on the viability of their environment, a degraded environment is a challenge on their socioeconomic status. So the women have to over-exploit the slim natural resources available to them, in a bid to squeeze out a living for their families (Odoemena, 2011). This has placed an extra burden on women in the Niger Delta, as they have to strive even harder to meet their daily needs.

13.5.12 Intractable Communal Conflicts

Exploitation in the Niger Delta has fueled and aggravated conflicts within and among communities in the region. While territorial disputes in the delta predate the discovery of oil and while they continue in other parts of the nation, it is undoubtedly the case that many of the conflicts between neighboring communities in the delta are fueled by the presence of oil. The conflict that has emerged in the Niger Delta as a result of the extraction of oil has its roots in the violation of the rights of local community people as a result of the promulgation of obnoxious legislations (Achi, 2003). Such legislation includes the land use decree, which stipulates that ownership of all land in the country is vested in the Governor of the State who would hold such land in trust for the people. This decree effectively denies the communities the right to claim royalties from oil derived from their own land and only get paltry sums as compensation for loss of surface rights. This has inevitably led to greater poverty and landless groups of people whose basic source of sustenance as peasant farmers has been denied them. Such a situation leads to violent communal clashes. Inter- and intraethnic clashes are caused by the struggle for the ownership of resources, usually land.

States and the oil companies operating in the area are complicit in a majority of the conflicts in this region as they use financial inducement as a "divide and rule" tactic (Iyayi, 2004). Most of these conflicts occur among groups with a common ancestry and no previous history of antagonism. The discovery and subsequent exploitation of oil in the area have, however, dislocated this bond as history and heritage count for little when it comes to fighting for the spoils of hosting oil companies and the sharing of largesse and the spoils of oil from either the State or the oil companies. Several communal clashes have resulted in loss of lives and properties worth billions of naira.

13.6 CONCLUSION

From the foregoing, the devastating impacts of oil exploration on the sociocultural lives of the people are self-evident. The oil resource and its revenue have been poorly managed by the government and the oil companies. These negative impacts have been neglected for a long time by the regulatory agencies of the government. What one sees from the region is a degraded environment that has distorted the cultural lives of the people. Now the people have become wiser and are now demanding the restoration of their environment in a very militant way. What is clear about oil exploration in Nigeria is that the people matter, irrespective of the amount of money to be made from oil exploration. There have been violent disruptions of exploration activities, which is now forcing the government to consider the agitations of the people.

The government and local as well as international agencies (NDDC, UNEP, TI) have conducted investigations on the social impacts of oil exploration in the Niger Delta. Far-reaching recommendations have been made. It is here recommended that the government and the oil companies should implement those recommendations forthwith. This will restore the environment and reduce violent agitation by the youths.

References

Achi, C. (2003). Hydrocarbon exploitation, environmental degradation and poverty: The Niger Delta experience. *Diffuse Pollution Conference, Dublin 2003.*

Aghalino, S. O. (2011). Oil and cultural crisis: The case of the Niger Delta, Nigeria. *Africana: The Niger Delta, 5*(1), 1–21 . Available on <http://unilorin.edu.ng/publications/aghalinoso/Aghalino%20Oil%20and%20Cultural%20Crisis.pdf>

Aghalino, S. O., & Eyinla, B. (2009). Oil exploration and the marine pollution: Evidence from the Niger Delta, Nigeria. *Journal of Human Ecology, 28*(3), 178–185.

Amnesty International (2009). *Petroleum, pollution and poverty in the Niger Delta.* London: Amnesty International Publications.

Anyanwu, D. I., & Tanee, F. B. G. (2008). Tolerance of Cassava (Var. TMS 30572) to different concentrations of post-planting crude oil pollution. *Nigeria Journal of Botany, 21*(1), 203–207.

Babatunde, A. (2009). Oil exploitation and conflict in the Nigeria's Niger Delta. A study of the Ilaje, Ondo State, Nigeria. *Journal of Sustainable Development in Africa, 11*(4), 134–159.

EIA. (2011). Energy Information Administration. *Crude oil inventory report.*

Iyayi, F. (2004). An integrated approach to development in the Niger Delta. In *A paper prepared for the Centre for Democracy and Development (CDD).*

Lubeck, P., Watts, M., & Lipschutz, R. (2007). *Convergent interests: US energy security and "Se-curing" of Nigeria's democracy.* Washington, DC: Center for International Policy.

National Oil Spill Detection and Response Agency (NOSDRA). (2016). *Oil spill statistics.* Retrieved January 7, 2017 from <https://oilspillmonitor.ng/>.

Niger Delta Development Commission (NDDC). (2006). *Niger Delta regional development master plan.* Abuja: NDDC.

Obi, C. (2005). Oil and federalism in Nigeria. In E. Onwudiwe, & R. Suberu (Eds.), *Nigerian federalism in crisis.* Ibadan: PEFS.

Odoemena, A. (2011). Social consequences of environmental change in the Niger Delta of Nigeria. *Journal of Sustainable Development in Africa, 11*(2), 123–135.

Okezie, D.N., & Okeke, A.O. (1987). Flaring of associated gas in oil industry: Impact on growth, productivity and yield of selected farms crops, Izombe flow station experience. In *Paper presented at NNPC Workshop, Port Harcourt, Rivers State, Nigeria.*

Okumagba, P. O. (2011). Militancy and Human Rights Violation in the Niger Delta. *International Review of Social Sciences and Humanities, 3*(2), 28–37.

Oluwaleye, J. M. (2013). Militancy and the dilemma of sustainable development: A case of Niger Delta in Nigeria. *IOSR Journal of Humanities and Social Science, 15*(6), 96–100.

Organisation of Petroleum Exporting Countries (OPEC). (2015). *Organisation of Petroleum Exporting Countries annual statistical bulletin.* Geneva.

Osuntokun, A. (2000). *Environmental problems of the Niger Delta.* Lagos: Friedrich Ebert Foundation.

Shell Petroleum Development Company (SPDC). (2016). *Oil spill data.* Shellnigeria.org. Retrieved on February 8, 2017.

United Nation Environmental Program (UNEP). (2011). *Environmental assessment of Ogoni Land.* ISBN: 978-92-801-3130-9 Retrieved from <www.unep.org/nigeria>.

United Nations Development Programme (UNDP) (2006). *Niger Delta Human Development Report.* Abuja: UNDP.

United Nations Development Programme (UNDP). 1995. *Report of the world summit for social development: Copenhagen, 6–12 March 1995.*

World Bank. (2011). *Nigeria: Associated gas usage.* World Bank Group Press Release. Global Gas Flaring Reduction Public-Private Partnership. Washington, DC.

The Eco-Economics of Crude Oil Exploration in Nigeria

Anthony Kola-Olusanya[1] and Gabriel Olarinde Mekuleyi[2]

[1]Osun State University, Osogbo, Nigeria [2]Lagos State University, Lagos, Nigeria

14.1 INTRODUCTION

The significant role of crude oil in human civilizations since the industrial revolution in the 1700s cannot be ignored. It has provided the much-needed finances for infrastructural development across nations and has contributed to corruption and civil strife in many parts of the world. Nigeria is blessed with a huge deposit of oil resources, and this resource has provided the seed capital for development in other sectors. For instance, crude oil exploration has contributed significantly to the gross domestic product (GDP) of the country. The GDP of Nigeria has risen over the years from a meager 1.6% in 1960 to 40% presently. This fact is adequately articulated in Otiotio (2014), Okorie (2005), and Kwaghe (2015) that oil is the main source of Nigeria's GDP with a gross output that includes accruals from oil exports, local sale of refined products, and natural gas with export revenue accounting for approximately 70% of the oil revenue (Larsen & Butler, 2013; Odulari, 2008).

The accrued revenues and incomes from the oil industry have contributed immensely to the economic well-being of Nigeria. However, its exploration has also resulted in various socio-ecological problems. The critical nature of these environmental problems can be seen in oil spills, which have had a major impact on the ecosystem of the oil-producing areas. Oil spills in the Niger Delta region of Nigeria have caused severe depletion of natural capital. They have also caused hunger, poverty, and social insecurity in oil-producing communities. Specifically, the negative impacts of the introduction of undesirable by-products (flared gas and spilled oil) into the environment during petroleum prospecting and production cannot be ignored (Audu, Jimoh, Abdulkareem, & Lawrence, 2016; Okoli, 2013). From extraction to combustion, oil is hazardous to health, environment, and security. In relation to Nigeria's Niger Delta area, oil exploration is responsible for the grave environmental degradation and conflict and its underdevelopment since the discovery of

The Political Ecology of Oil and Gas Activities in the Nigerian Aquatic Ecosystem
DOI: http://dx.doi.org/10.1016/B978-0-12-809399-3.00014-8

oil in 1937 and the production in commercial quantities in 1958 in Oloibiri in present-day Bayelsa State. But the crisis of underdevelopment, environmental degradation, and conflict in Nigeria's Niger Delta area cannot adequately be understood without contextualizing the problem.

The Niger Delta has estimated crude oil reserves of approximately 22.5 billion barrels. The region produces approximately 2.5–3 million barrels of petroleum per day, or approximately 1 billion barrels per year with a large amount of natural gas deposit, thereby making Nigeria the ninth-largest producer of crude oil in the world and a major African petroleum producer south of the Sahara. Most of Nigeria's crude oil is extracted from the oil and gas-rich Niger Delta area of the country, which also serves as the economic fulcrum of the nation, with petroleum accounting for 90% of the country's revenue.

Being the goose that laid the golden egg, the expectation, therefore, is that the Niger Delta area and its indigenous people should enjoy the good things of life or at the very least have access to basic infrastructures such as health, education, roads, and pipe-borne water. Rather, the area, which in its 30 years of oil exploration has generated $280 billion for the Nigerian economy, is worse than other average non-oil-producing communities in Nigeria. The region has many challenges; development is difficult because of the region's terrain, whereas almost six decades of crude oil exploitation by multinational oil companies in conjunction with local oil corporations have continued to compound the natural problems. As a result, these two major challenges have been worsened in the last 40 years due to lack of political will by successive governments to develop the region (Ojefia, 2004).

As noted by Bisina (2005), the discovery and production of petroleum in commercial quantities since the 1950s as well as the absence and lack of enforcement of appropriate environmental laws and regulations, have resulted in significant damages to communities, streams, and farm and sacred lands, which are either acquired for crude oil and gas prospecting and development or polluted in the course of production. In addition, the lack of adoption of best practices by some of the multinational oil companies operating in the Niger Delta region, coupled with their inability to comply with environmental regulations, has further contributed to the ecological devastation within the region. The chapter will attempt a discourse on eco-economics by attempting an understanding into the concept. The chapter will also examine the conceptual understanding: ecological economics of oil production, promoting eco-economics/eco-sustainability in crude oil exploration, oil exploration, and ecological destruction. The chapter will conclude by summarizing and drawing conclusions on the issues discussed.

14.2 WHAT IS ECO-ECONOMICS?

Eco-economics, or ecological economics, is a combination of two words—ecology and economics. While ecology deals with the study of living organisms and their interactions with the environment, economics studies human behavior in relation to ends and scare means, which have alternative uses. Ecological economics is a new and growing multidisciplinary field of endeavor that seeks the expansion and improvement of existing economic theory to include natural systems of the earth, human health, values, and their well-being (Czech, 2009; Martinez-Alier, 1998). The field examines economic issues from a broader perspective by considering things such as human (health, education), social (family, friends), and natural (earth's natural resources) capitals, which are not factored into conventional economic theories. Eco-economics attempts to integrate scientific understanding of the link between economics of human and natural systems and how to utilize the concept to develop effective policies that could lead to even distribution of resources and sustainable ecological systems. The ecological economics paradigm offers a refreshing and apt perspective on the relationship between economy and ecology for progressives seeking sustainable alternatives to current patterns of economic growth and environmental degradation (Sheeran, 2006).

According to Costanza (2010),

> Ecological economics is a trans-disciplinary field bridging across not only ecology and economics but also psychology, anthropology, archaeology, and history. That's what's necessary to get a more integrated picture of how humans have interacted with their environment in the past and how they might interact in the future. It's an attempt to look at humans embedded in their ecological life-support system, not separate from the environment. It also has some design elements, in the sense of how do we design a sustainable future? It's not just analysis of the past but applies that analysis to create something new and better.

Ecological economics breeds economic efficiency and good decision-making by taking full cognizance of all the cost-effects of economic activities on natural system and resources. Ecological economics also take into account the impacts of humans and their economic activities on ecological systems and services and vice versa. In a nutshell, the central concept of ecological economics is sustainability, and this could be approached both qualitatively and empirically.

14.3 CONCEPTUAL UNDERSTANDING: ECOLOGICAL ECONOMICS OF OIL PRODUCTION

Oil and gas resources are the mainstay of the government's revenue in Nigeria, accounting for approximately 90% of export earnings and 70% of total revenue generated by the government (Bayode, Adewunmi, & Odunwole, 2011; Larsen & Butler, 2013; Odulari, 2008). However, extraction of the oil is associated with the release of contaminants or pollutants into the environment, which has caused ecological degradation and socio-economic hardship to the indigenes of the Niger Delta. These problems have culminated into militancy, kidnapping, oil theft, and other social vices because of, for example, inadequate scrutiny, corruption, poverty, and lack of political will by the government. The conceptual issue in ecological economics is anchored on the understanding of humans about the notion of development and productivity. This suggests that humans see natural or environmental resources from the value derived therein. Mabogunje (1981) and Bayode et al. (2011) viewed development as a concept that integrates the following factors: economic growth, justice in resource distribution, modernization, and socio-economic reform.

This negative realization has been responsible for the type of development witnessed since the industrial revolution that caused monumental environmental devastation. The devastation occasioned by oil exploration is captured in environmental pollution, which pervades the areas. Pollution has different definitions, depending on discipline, perception, interest, and so on. However, in the Nigerian constitution, *Section 41 of the Federal Environmental Protection Agency (FEPA) Act Cap.F10 Laws of the Federation 2002* defines pollution as "man-made or man aided alterations of chemical, physical or biological quality of the environment to the extent that is detrimental to that environment or beyond acceptable limits."

Recently, oil has become a very decisive factor in the politics, rhetoric, and diplomacy of states. Ever since the discovery of oil in Nigeria in the 1950s, the Niger Delta area has been bearing the brunt of the environmental decadence caused by oil spill. Several properties have been destroyed, and health as well as livelihood have been hampered. Yet, very little has been done to fully assess the impacts of pollution on the Niger Delta region.

14.4 CRUDE OIL EXPLORATION AND ECOLOGICAL DESTRUCTION

The Niger Delta, which was once a globally renowned ecosystem due to her unique wetland, mangrove, and biodiversity, is now in a state of dilapidation. Reports have shown that the Niger Delta is the most degraded environment in Africa (Sumenitari, 2003). This has been caused by decades of crude oil spill and gas flaring, which have resulted in atmospheric and aquatic pollution, biodiversity loss, deforestation, destruction of farmland and fisheries, loss of livelihood, and so on, and all these problems are associated with petroleum exploration in the Niger Delta region of Nigeria (Okoli, 2013). Another major problem caused by oil spill in the Niger Delta is the destruction of mangroves, which are the breeding grounds of many aquatic organisms. Mangroves are also important in climate change mitigation because of their high carbon sequestration potential. According to Kwaghe (2015) and Baird (2010), an estimated 10% of the indigenous mangrove vegetation of the Niger Delta has been lost due to continuous oil spills and environmental degradation resulting from careless activities of the oil companies. Kwaghe (2015) also reported that:

> Based on quantities reported by the operating companies, the NNPC estimated that approximately 2,300 cubic meters of oil are spilled in 300 separate incidents every year. Due to under-reporting, the aggregate spillage is believed to be at least 10 times higher.

Another issue that has exacerbated the pollution crisis in the Niger Delta area is flaring of gas. Gas flaring is the burning of natural gas, which is associated with oil extraction processes. Flaring of gas is noted to release dangerous and poisonous compounds, including hydrogen sulfide, xylene, nitrogen dioxide, benzene, sulfur dioxide, as well as greenhouse gases such as carbon dioxide and methane. Global statistics show that the natural gas released into the atmosphere by Nigeria between 2007 and 2011 was up to 11 billion cubic meters (bcm) (Fig. 14.1), the second highest behind Russia, and this translates into annual loss of USD1.8 billion.

Although gas faring is prohibited under law in Nigeria (Associated Gas Reinjection Act of 1979 Sections 2 and 3), gas flaring has continued unabated (Fig. 14.2) as a result of weak enforcement, government protocol, and abysmally low flaring fines to be paid by the

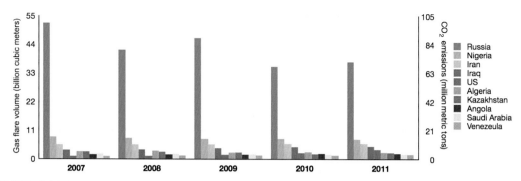

FIGURE 14.1 Volume of gas flared by major crude oil producers (left axis) and the equivalent carbon dioxide emissions (right axis). *Source: World Bank data.*

FIGURE 14.2 Satellite image of gas-flaring activities in the Niger Delta. *Source: National Oceanography and Atmospheric Administration (NOAA).*

polluting oil companies, which stands at #10 per million cubic feet (mcf) in 2003. According to Ebrahim and Friedrichs (2013), gas flaring makes economic sense to the multinational oil companies, and this has thrived because of Nigeria's dysfunctional petroleum sector (Hassan & Kouhy, 2013).

The flaring of 11 mcf of natural gas daily also gives an idea of the level of environmental hazard experienced by the people of the area. In addition, flaring of gas also "raises the temperature of the surrounding environment to levels beyond the normal value of 13°C to 14,000°C" (Pyagbara, 2007).

According to Aboriobo (2007):

> Oil exploration affects people living in oil producing areas negatively. Most farmers face the problem of displacement without resettlement during oil spills. Apart from loss of farms, oil spills have led to extensive deforestation with no adequate replanting practices. This in effect has shortened fallow periods, compounded land degradation and led to loss of soil fertility and consequently erosion of the top soil.

Pollution has also caused deterioration in the physical attributes and biological activities of the soils of the Niger Delta environment (Kola-Olusanya, 2014). The combined effect of this is reduction in farm output, which often leads to displacement of farmers and their dependents in what is described as "rural flight." The displaced farmers with their families are compelled to relocate to nearby urban centers, which do not have adequate infrastructure to accommodate the increasing population (Achebe, 2004; Kola-Olusanya, 2014).

Of all the costs of oil exploration on the peoples of the Niger Delta, environmental degradation and damage to human health are by far the highest.

According to Mariano and Rovere (2002):

> The oil industry holds a major potential of hazards for the environment, and may impact it at different levels: air, water, soil, and consequently all living beings on our planet. Within this context, the most widespread and dangerous consequence of oil and gas industry activities is pollution. Pollution is associated with virtually all activities throughout all stages of oil and gas production, from exploratory activities to refining.

Undesirable environment consequences in the Niger Delta have become aggravated due to absence of plans to envisage and ameliorate the impacts of oil spill. The effects have also been worsened by climate change.

14.5 PROMOTING ECO-ECONOMICS/ECO-SUSTAINABILITY IN CRUDE OIL EXPLORATION

There is a visible trend of degradation in ecological health and viability of the Niger Delta region as a result of oil exploration. These are seen from the conflicts and insecurity over natural resources, which has lingered over the last two decades.

> The challenge for our generation is to reverse these trends before environmental deterioration leads to long term economic decline, as it did for so many earlier civilizations. These increasingly visible trends indicate that if the operation of the sub-system, the economy, is not compatible with the behaviour of the larger system -the earth's ecosystem - both will eventually suffer (Brown, 2001a, 2001b).

Anwana (2011) has suggested the adoption of new methods of oil exploration that are ecologically safe and that will put an end to the environmental devastation in the region.

The question that naturally follows is: Can there be environmentally responsible forms of oil production? Siegel (2013) argued that:

> The adoption of Octopus technique could reduce environmental impact of oil on the land surface by about 75%. This would result in drillers accessing more oil with less "on-the-ground" damage to the land. It also allows drillers to disturb less underground water sources, certainly a major benefit these days.

What it suggests is that there are emerging technologies that abhor throwaways and externalities currently prevalent in the oil industry, which has been linked to decades of wastage as well as degradation of natural environment/resources within the oil-producing areas. This type of technologies point to the very message of ecological economists who see a more sustainable way of drilling crude oil with minimal loss and wastages. One of the major problems with the Nigerian government is that they are good dreamers but bad interpreters. For instance, green economy was envisioned to be attained in 2020 but as of the time of this writing, there was no proper preparation for fulfilment of the vision.

Therefore, if eco-sustainability in the oil-producing areas of Nigeria would be attained, then urgent steps must be taken to resuscitate the rich ecosystem and unique biodiversity of the Niger Delta. Also, necessary measures should be adopted to reduce oil spill, stop gas flaring, clean polluted air and water, as well as implement a positive change in the mentality and attitude of the people in order to achieve a green economy.

14.6 CONCLUSION

The role of oil in our society is significant. However, it pollutes and creates several problems when it is not well managed or processed. The oil region (the Niger Delta) in Nigeria is fast losing its natural biodiversity, health, and environment to the aftermath of oil exploration. The products and services provided by the ecosystem are declining while human-made capital is increasing. The reformation of the devastated environment begins with a shift in economic mindset by recognizing that the economy is part of the earth's ecosystem. Eco-economics is needed to create new political contexts that will shift the focus of economic activity from just wealth generation to those that also consider sustainability of the ecosystem.

References

Aboribo, R. I. (2007). The conflict of globalization and globalization of conflictin the Niger Delta area of Nigeria. In *7th annual conference of Nigerian Sociological Society*. Sharon Ultimate Hotel, Area 3, Garki, Abuja.

Achebe, C. C. (2004). *Oil: Prize or curse?* Retrieved 29 March 2005, from http://www.nigeriavillagesquare1.com/Articles/CCAchebe/2004/07/oil-prize-or-curse_04.html.

Anwana, E. O. (2011). Impact of oil exploitation on sustainable development and green economy in Nigeria: the Niger delta case. *International Journal of Economic Development Research and Investment, 2*(2), 155−167.

Audu, A., Jimoh, A., Abdulkareem, S. A., & Lawrence, O. (2016). Economics and environmental impacts of oil exploration and exploitation in Nigeria Energy Sources. *Energy Sources, Part B: Economics, Planning, and Policy, 11*(3), 251−257.

Baird, J. (2010). *Oil's shame*, from www.connection.ebscohost.com.

Bayode, O. J. A., Adewunmi, E. A., & Odunwole, S. (2011). Environmental implications of oil exploration and exploitation in the coastal region of Ondo State, Nigeria: A regional planning appraisal. *Journal of Geography and Regional Planning, 4*(3), 110–121.

Bisina, J. (2005). *Oil and corporate recklessness in Nigeria's Niger Delta Region*. Retrieved 17 March 2005, from http://www.afrika.no/noop/page.php?p=Detailed/6906.html&d=1.

Brown, L. (2001a). *Eco-economy—Building an economy for the earth*. London, UK: Earthscan: Earth Policy Institute.

Brown, L. R. (2001b). *Eco-economy: Building an economy for the earth*. New York, NY: W. W. Norton & Co.

Costanza, R. (2010). In T. O'Callahan (Ed.), *What is ecological economics?* New Haven, CT: Yale Insights.

Czech, B. (2009). Ecologcal economics. In R. J. Hudson (Ed.), *Animal and plant productivity*. Oxford, UK: Eolss Publishers.

Ebrahim, Z., & Friedrichs, J.. (2013). *Gas flaring: The burning issue*. Retrieved 13 March 2017, from http://www.resilience.org/stories/2013-09-03/gas-flaring-the-burning-issue/.

Hassan, A., & Kouhy, R. (2013). Gas flaring in Nigeria: Analysis of changes in its consequent carbon emission and reporting. *Accounting Forum, 37*(2), 124–134.

Kola-Olusanya, A. (2014). A narrative analysis of environmental-related conflicts interventions in Niger Delta. *Annals of Social Science, 1*(2&3), 205–219.

Kwaghe, Z. E. (2015). Black gold and the Nigerian State (1956–2014): A critical review. *Chinese Business Review, 14*(2), 72–86.

Larsen, L., & Butler, L. (2013). The governance of the global oil and gas industry: A foundation for developing a comprehensive framework. Retrieved from http://www.opecweb/en/aboutus/167htm.

Mabogunje, A. L. (1981). *The development process*. New York, NY: Holmes.

Mariano, J. B., & Rovere, E. L. L. (2002). *Environmental impacts of the oil industry*. Paris, France: UNESCO.

Martinez-Alier, J. (1998). *Ecological economics as human ecology*. Lanzarote: Fundacion Cesar Manrique.

Odulari, G. O. (2008). *Crude oil and the Nigerian economic performance*. Retrieved 2 April 2017, from http://www.ogbus.ru/eng.

Ojefia, I. A. (2004). *The Nigerian state and the Niger Delta question*. Retrieved 20 March 2005, from http://www.deltastate.gov.ng/oyefia.htm.

Okoli, A. C. (2013). The political ecology of the Niger Delta crisis and the prospects of lasting peace in the post-amnesty period. *Global Journal of Human Social Science, 13*(3), 36–47.

Okorie, A. (2005). *Nigerian oil: The role of multinational oil companies*. Retrieved 12 February 2017, from http://www.web.stanford.edu/...Nigerian%20oil%20.

Otiotio, D. (2014). *An overview of the oil and gas industry in Nigeria*. Retrieved 4 March 2017, from http://www.academia.edu2654835.

Pyagbara, L. S. (2007). The adverse impacts of oil pollution on the environment and wellbeing of a local indigenous community: The experience of the Ogoni people of Nigeria. International Expert Group Meeting on Indigenous Peoples and Protection of the Environment. Khabarousk, Russian Federation.

Sheeran, K. A. (2006). Ecological economics: A progressive paradigm. *Berkeley La Raza Law Journal, 17*(1), 26–28.

Siegel, J. (2013). *Eco-friendly oil drilling*, from http://www.energyandcapital.com/articles/eco-friendly-oil-drilling/3170.

Sumenitari, I. (2003). Of victims and villan. Nigerian Weekly Independent, no. 4.

II. THE EFFECTS OF CRUDE OIL EXPLORATION ON THE SOCIO-CULTURAL AND ECO-ECONOMICS OF NIGERIAN ENVIRONMENT

Managing Nigeria's Aquatic Resources

Monday Ilegimokuma Godwin-Egein, Udensi Ekea Udensi and Sunday Omovbude

University of Port Harcourt, Port Harcourt, Nigeria

15.1 INTRODUCTION

Nature is complete in itself. It is self-sustaining. It is capable of perpetuating itself indefinitely. It applies "biological control" of some sort in perpetuating, maintaining, and conserving itself in all ways and aspects. The fact is evident in the various services (called ecosystem services) it renders to itself, which man encroaches on and takes advantage of, most of the time to plunder. The ultimate beneficiary of these services is man. Ecosystem services are ecological goods and services that are benefits derivable from the ecological functioning of a healthy ecosystem. These benefits are shared by all living organisms and the environment, and not man alone.

Ecosystem services are categorized into four broad groups, namely: (1) provisioning, which has to do with the production of water and food; (2) regulating, which deals with the control of climate and disease; (3) supporting, which deals with nutrient cycles and crop pollination; and (4) cultural, which deals with spiritual and recreational services. Ecosystem services are essential to human survival on earth, but human activities are impairing the flow of the services on a large scale due to population growth. They are utilized, at present, without any plan for the future.

For perpetuation and conservation of these services, payments are sometimes made to humans at a particular catchment (e.g., landowners, farmers, etc.). These payments are called "payment for ecosystem services." Payment for ecosystem services "is a system for the additional provision of environmental services through conditional payments to voluntary providers." Markets can be created to trade ecosystem services following the use of policy tools, which could be regulation, cross compliance programs, environmental marketing scheme, voluntary participation programs, market-based approaches, one-time direct payments, or ongoing direct payments.

There are questions regarding the economic values of ecosystem services. To address such questions, a transparent system of valuation is pertinent. Valuation of ecosystem services in monetary terms is guided by avoided cost, replacement cost, factor income, travel

The Political Ecology of Oil and Gas Activities in the Nigerian Aquatic Ecosystem
DOI: http://dx.doi.org/10.1016/B978-0-12-809399-3.00015-X

cost, hedonic pricing, and contingent valuation. Although monetary pricing continues with respect to the valuation of ecosystem services, the challenges in policy implementation and management are significant and multitudinous.

The main focus of this chapter is the provisioning category, which overlaps with the three other categories. The specific concern here is "Aquatic resources of Nigeria," specifically, resources availability, utilization, payment systems, management implementation, and valuation.

Ninety-five percent of the water on earth is saltwater, and only three percent is freshwater and is found mainly as groundwater, with only a small fraction present above ground or in the air. Freshwater is a renewable resource. It is water in the river, lake, or freshwater wetland. Surface water is naturally replenished by precipitation and naturally lost through discharge to the oceans, evaporation, evapotranspiration, and groundwater recharge. Human activities can have a large and sometimes devastating impact on these resources. Water is used for agricultural, industrial, domestic, and recreation purposes. Water stress occurs and it applies to situations where there is not enough water for all uses, whether agricultural, industrial, or domestic. Climate change has significant impacts on water resources because of the close connections between the climate and the hydrological cycle. Water pollution is one of the main contributors to water stress.

Nigerian aquatic resources have limited available data. However, some assessment of water reserves indicates that there are large supplies of aquatic resources, and their exploration and exploitation are at an early stage (Oteze, 1981).

Nigeria's water resources are unevenly distributed, with intense drought in the north and high precipitation and flooding in the south. Attempts are being made to implement a water resource management strategy that seeks to address these challenges. Most of Nigeria's agriculture is still based on rain-fed peasant farming and other natural sources. This makes agriculture and food security climate-sensitive and fragile.

Approximately 850 km of Nigeria's Atlantic coastline is characterized by sensitive ecosystems that are mainly marine ecosystems. Nigeria has a fair share of freshwater, which runs through lakes, rivers, and streams, or as soil moisture or buried deep underground. It is not only important for the survival of humans, but also for the survival of all the existing species of animals and plants. Aquatic resources provide the raw materials needed for the manufacturing of clothing, building materials (lime extracted from coral reefs), ornamental items, and personal-use items. In addition, processes within aquatic environments can be used for the production of renewable energy.

Nigeria has given high priority to her freshwater resources due to the growing concern of the increasing stress on water supplies caused by poor use patterns. To mitigate this, the government has prepared the Water Resources Decree 101 of 1993 to give the legal framework for the development of the water resources of the country, and there is a Water Resource Assessment Sector, which is a repository of water resources facilities. Hydrogeological mapping of the country is currently being conducted.

For the protection of water resources, water quality, and aquatic ecosystem resources, the government has carried out five activities: Nigeria Register of Dams, rehabilitation of dams, rehabilitation of soil erosion sites, flood control, and establishment of Water Quality Laboratories. Nigeria is endowed with abundant surface and groundwater resources whose availability varies with rainfall, location, and geological formations. The highest annual precipitation of about 3000 mm occurs in the Niger Delta and mangrove swamp belt of the southeast where rainfall lasts for almost 10 months of the year. Although various human activities draw on the surface water resources, the groundwater resource potential in the country has not been fully tapped.

Nigeria has recently instituted an Integrated Coastal Area Management (ICAM) approach to address the various environmental problems and to promote sustainable utilization of coastal resources, and this is handled by the Federal Ministry of Works and Housing and the Federal Ministry of water resources.

Over the years, various legislation, regulations, and policy instruments have been developed, and they have been modified from time to time as new scientific evidence emerges. The national strategy or policy that will cover the relevant aspects of integrated coastal zone management and sustainable development, including environmental impacts of activities affecting the coastal and marine areas, are also provided for. Policies and plans have been developed to preserve and sustain specific aquatic resource bases. For example, policies and plans that have been developed specifically to address the preservation and sustainable use of the fragile ecosystem (mangroves) of the Niger Delta include: creation of mangrove forest reserves, planting of mangroves in areas of mangrove deforestation, the Akassa declaration, development of the Niger Delta Action Plan, and the Niger Delta Environmental Survey, among others.

Three major sectors, artisanal, industrial, and aquiculture activities, exist in the Nigerian coastal zone. Percentage of the economy contributed to by fishing is estimated to be 1.4%.

The national information available from the aquatic resources sector that assists both decision makers and planners working in the country is dismal.

Ecosystem services seem not to be organized in Nigeria in terms of their systematic exploration, valuation, measurement, exploitation, management policies, payment systems, conservation, and growth. This chapter gives an insight into aquatic resources of Nigeria, their availability, utilization, payment system(s), management policies, and their implementation and conservation efforts.

Nigeria lies between longitudes 2°49′E and 14°37′E and latitudes 4°16′N and 13°52′N of the equator, and occupy land area totaling 924,000 km^2, with a population of approximately 170 million (Anon, 2016). The total annual rainfall varies between the north and the south, with the north ranging between 500 and 1000 mm, and increases southward with an annual rainfall ranging between 1500 and 4000 mm. The country has two major rivers, the Niger River and the Benue River, with a confluence in Lokoja and discharge into the Gulf of Guinea through several creeks and tributaries that form the popular Niger Delta. However, there are a few other tributary rivers that drain into the Niger-Benue trough and Lake Chad, and these are the Sokoto-Rima, Kaduna, Anambra, Gongola, Hadejia, Jama'are and Yobe rivers. Other major rivers that empty directly into the Atlantic Ocean are Cross, Imo, Ogun, Oshun, Benin and Qua Iboe.

15.2 AQUATIC RESOURCES AND AVAILABILITY

The main category of aquatic resources identifiable in Nigeria is marine and freshwater. The marine aquatic ecosystem, which is endowed with diverse resource zones, is made up of the oceanic, benthic, intertidal, and neritic zones (estuaries, salt marsh, coral reefs, lagoons, and mangrove) with numerous aquatic flora and fauna, their genetic pool, the abiotic elements, and their interactions with the biotic elements and each other defines the services they offer that are of value to mankind. The freshwater aquatic ecosystem is also rich in diverse resources, and these resources include wetlands, streams, lakes, rivers, springs, seeps, ponds, groundwater, and reservoirs (Table 15.1). It also includes the biotic element of the aquatic ecosystem, the genetic resources of the organisms or parts thereof and populations, among others, with actual or potential use or value for humanity. The country is blessed with abundant inland water resources that emanate from both surface and underground waters (Ayoade, 1981), as well as rainfall. According to Ita, Sado, Balogun, Pandogari, and Ibitoye (1985), Nigeria inland waterways covers approximately 149,919 km^2 made up of lakes, rivers, ponds, floodplains, and mining and stagnant pools. However, Satia (1990) put the statistics at 5000 fish ponds, 839 floodplains, and 347 reservoirs and lakes. Besides this, there are several cattle drinking ports, many earth wells, mine pits, and excavation ponds of quarries that hold a significant amount of water all year round as well as boreholes. The determination to solve problems of drought and water shortages resulted in the development of dams, lakes, and reservoirs that are now abundant in the northern parts of the country such as Kano, Jigawa, and Katsina states that are mostly affected by drought. The southern half of Nigeria, towards the coast, is covered with sedimentary rocks and is therefore richer in underground water. It is estimated that Nigeria can utilize approximately >50% of its underground water resources without depleting surface reserves sources (Mitchelle-Thome, 1961). The inland aquatic resources of Nigeria offer a diverse natural habitat for various aquatic flora and fauna.

TABLE 15.1   Major Rivers and Coverage

Major rivers	Location in Nigeria	Length (km)
Anambra River	Anambra State	210
Benue River	Benue, Adamawa, Kebbi, Taraba States	1400
Cross River	Cross-River State, Akwa Ibom and Rivers States	80
Imo River	Imo State	241
Kaduna	Kaduna and Plateau State	550
Kwa Iboe River (Qua Ibo)	Umuahia in Abia State and Ibeno in Akwa Ibom State	3
Niger (with Kainji and Jebba lakes) River	Anambra and Kogi States	4180
Ogun River	Ogun State	–
Oshun River	Osun State	276
Sokoto-Rima River	Funtua in Katsina State, Sokoto and Zamfara States	275

Source: From Ita, E.O., Sado, E.K., Balogun, J.K., Pandogari, A., & Ibitoye, B. (1985). Inventory survey of Nigerian inland waters and their fishery resources I. A preliminary checklist of inland water bodies in Nigeria with special reference to ponds, lakes, reservoirs and major rivers. Kainji Lake Research Institute Technical Report Series, No. 14, KLRI, New Bussa, 51.

II. THE EFFECTS OF CRUDE OIL EXPLORATION ON THE SOCIO-CULTURAL AND ECO-ECONOMICS OF NIGERIAN ENVIRONMENT

15.3 WATERSHEDS, MAJOR RIVERS, AND RIVER BASINS

There are three main watersheds in Nigeria, and they are the Western Highlands, the Udi Plateau, and the Jos Plateau. The main rivers draining from these watersheds are the River Niger and its tributaries, both seasonal and perennial, the River Benue and its tributaries, and other large river systems, which include the Hadejia-Jama'are-Yobe Rivers, the Cross River and its tributaries, the Kaduna River, the Gongola River and its tributaries, the Sokoto-Rima River complexes, and the Oshun, Ogun, and Imo Rivers (Table 15.1). Osun, Ogun, Imo, and Cross rivers have double-peak flow and flood regimes, corresponding to the two peaks of rainfall each year associated with the coastal areas. To the second category belong rivers such as the Sokoto-Rima, the Gongola, Kaduna, and Hadejia. These are rivers mainly of the plateau and north, which usually have one main peak flood and a flow pattern corresponding to the single maximum rainfall season common in the northern part of the country. The third hydrological pattern is found in the catchments of the River Niger and the River Benue. These very long rivers, each with several tributaries, have a complex flow pattern. There are usually two floods (the white and black flood), which depend on rainfall outside Nigeria.

15.3.1 Lakes and Reservoirs

The lakes and reservoirs have a surface area of between 4000 and 550,000 ha, with a total surface area of approximately 853,600 ha, and this represents about one percent of the total area of Nigeria (Table 15.2) (Ita et al., 1985).

15.3.2 Wetlands

Wetland has been defined based on the perception of values, services, and need of these aquatic water resources. According to McCartney, Rebelo, SenaratnaSellamuttu, and de Silva (2010), wetland is a sink for surface or groundwater flows from a surrounding catchment. They maintained that wetlands are natural harvesters of rainwater within landscapes and by definition, sites where water occurs at or close to the ground surface. On the other hand, wetlands are land area covered with water at or near the soil surface at varying periods of the year (USEPA, 2009). These areas support the prevalence and abundance of aquatic flora and fauna that are typically adapted to life in water-saturated conditions. Ramsar Convention (1971) defined wetlands as "Areas of marsh, fen, peat land or water whether natural or artificial, permanent or temporary, with water that is static or flowing, fresh, brackish or salt, including areas of marine water, the depth of which at low tide does not exceed 6 meters." There are, of course, other definitions of wetlands driven by research purpose and an organization's mandate. Due to the various definitions and the differences in mapping and inventory of wetlands, it has been difficult to ascertain the extent and number of wetlands globally (McCartney et al., 2010). A recent global estimate puts it at 917 Mha (Lehner & Doll, 2014). Nigeria as a country is richly endowed with abundant wetlands ecosystem (Table 15.3), the majority of which are found in the Niger, Benue, and Chad basins, and are estimated to cover 3% (28,000 km^2) of the land surface

TABLE 15.2   Some Major Lakes and Reservoirs

Major lakes and reservoirs	Location	Approximate surface area coverage (km²)
NATURAL LAKES		
Lake Chad	Borno State, Kainji	1350
Oguta	Imo State (Oguta)	2.48
Ndakolowo (Tatabu)	Downstream of Jebba reservoir	3
MAN-MADE LAKE		
Kainji Lake	This resource is situated on the Niger River	1250–1280 km², with a volume 15 km³
Jebba	Jebba, North western Nigeria	350 (Ita, 1984)
Shiroro Lake	Niger State (Kaduna river, a tributary of upper Niger)	312
Bakolori Lake Goronyo	Talata-Mafara area of Sokoto State within the Sokoto River basin	80
Tiga Lake	Tiga on the Kano River (about 80 km from Kano town) in Kano State	178
IITA-Reservoir	Idi-Ose, Ibadan Oyo State	0.78

Source: From Ita, E.O., Sado, E.K., Balogun, J.K., Pandogari, A., & Ibitoye, B. (1985). Inventory survey of Nigerian inland waters and their fishery resources I. A preliminary checklist of inland water bodies in Nigeria with special reference to ponds, lakes, reservoirs and major rivers. Kainji Lake Research Institute Technical Report Series, No. 14, KLRI, New Bussa, 51.

TABLE 15.3   Wetlands and Their Distributions in Nigeria

Types of wetlands and distribution	Approximate total surface area coverage (ha)
1. DELTAS AND ESTUARIES	
Niger Delta	617,000
Cross River estuary	95,000
Imo and Kwa Iboe (Qua Ibo) estuary	36,000
Others	110,000
2. FRESHWATERS	
Niger Delta freshwater	362,000
Apex of delta to Lokoja	635,000
Niger/Sokoto Basin	470,000
Niger Kaduna Basin	150,000
Lower Niger: Jebba to Lokoja	385,500
Benue River floodplain	312,000
Hadejia-Komadugu Yobe	624,000
Ogun/Oshun floodplains	
Cross River floodplains	250,000
Imo River floodplains	26,000
Kwa Iboe (Qua Ibo)	7000

Source: From Ita, E.O., Sado, E.K., Balogun, J.K., Pandogari, A., & Ibitoye, B. (1985). Inventory survey of Nigerian inland waters and their fishery resources I. A preliminary checklist of inland water bodies in Nigeria with special reference to ponds, lakes, reservoirs and major rivers. Kainji Lake Research Institute Technical Report Series, No. 14, KLRI, New Bussa, 51.

area of Nigeria (Uluocha & Okeke, 2004). The Niger Delta is one of the most important wetlands in Nigeria, the largest in Africa, and third largest in the world (Asibor, 2009; Oyebande, Obot, & Bdiliya, 2003).

15.3.2.1 *Coastal Wetlands*

These are the most extensive wetlands in Nigeria, found within the southern region bordering the Atlantic Ocean (Nwankwoala, 2012). This group of wetlands includes:

1. Lagos Lagoon and Lekki peninsula
2. The Niger Delta wetlands
3. Cross River wetlands
4. Badagary and Yewa Creeks
5. Ologe Lagoon

15.3.2.2 *Inland Water Wetlands*

Moving inland are the riverine wetlands scattered across the country, and these are:

1. Floodplains of Niger and Benue rivers
2. Ogun/Oshun wetlands
3. Anambra/Imo
4. Sokoto-Rima
5. Komadugu-Yobe
6. Ngada
7. Yedseram
8. El beid River

15.3.2.3 *Lake Chad Wetlands*

These are made up of Lake Chad wetlands that are very important because of proximity to the edge of the Sahara Desert and provision of water to 20 million people from the four countries of Nigeria, Chad, Niger, and Cameroon living within the Lake Chad region (Gophen, 2008) and also its importance for fisheries (Bene et al., 2003).

15.3.2.4 *Interior Wetlands*

These are interior wetlands that are not associated with any major river system. They are seasonal but support a wide variety of livelihood activities, including material collection, fishing, and farming.

15.3.2.5 *Artificially Impounded Wetlands*

These are artificially impounded wetlands, which include the Kainji Lake wetlands noted for their importance in fisheries and irrigation.

However, all the wetland belts in Nigeria, according to Oyebande et al. (2003), can come under three broad categories, namely: freshwater, man-made, and saltwater wetlands. Other preferred classes of wetland include:

- Marine wetlands (saltwater intertidal areas associated with the ocean)
- Estuarine wetlands (brackish water areas where fresh water streams enter the sea)

- Riverine wetlands (wetlands associated with rivers and streams)
- Palustrine (marsh and swamps)
- Lacustrine (wetlands associated with lakes)

15.4 AQUATIC FLORAL AND FAUNAL RESOURCES

15.4.1 Aquatic Floral Resources

Aquatic plants, besides their ecological role, contribute greatly to the economic, scientific, and recreational importance of Nigerian water resources. However, if present in large quantities, navigation may be impaired, resulting in water losses, and this might then become an opportunistic habitat for vectors of water-borne diseases. Plants occurring in most Nigerian water resources are similar to those in other tropical regions. Approximately 29 families of aquatic plant resources covering approximately 76 species have been reported across freshwater, inland, and wetland resources (Table 15.4). Of all the species and families of aquatic plant resources studied in Nigerian freshwater ecosystem, approximately 13 species had ≥50% occurrence across the freshwater systems studied (Table 15.5). Species such as *Ludwigia* spp., *Ipomoea* spp., *Nymphaea* spp., and *Pistia* spp. are ubiquitous across the freshwater resources (Table 15.5).

These plants are usually zoned according to their proximity to land and water as follows: the fringe, seasonally flooded zone, mud zone, and permanently flooded (open water) zone (Obot, 1987). These zones are usually abundant in diverse flora.

15.4.1.1 The Fringe Zone

The floras of this zone are similar to those of terrestrial and savannah ecosystems. Species here are mainly herbaceous and woody, such as, *Syzygium guineense*, *Cassia mimosoides*, *Digitaria horizontalis*, *Andropogon* spp., *Daniellu* spp., *Khaya senegalensis*, *Isoberlinia* spp., *Terminalia* spp., *Vitellaria paradoxa*, *Crotalaria microcarpa*, *Ficus* spp., *Parkia clappertoniana*, and *Alternanthera sessilis*.

15.4.1.2 The Seasonally Flooded Zone

This zone is generally covered for a couple of months of the year. A few woody and herbaceous species generally seen in this zone include: *Mimosa pigra* and *Mitragyna* spp., and *Vossia cuspidata*, *Leersia hexandra*, *Oryza* spp., and *Ludwigia erecta*.

15.4.1.3 The Mud Zone

This zone is constantly inundated by water but may be free for a few weeks. The regular species in this zone are *Echinochloa* spp., *Polygonum senegalense*, and *V. cuspidata*.

15.4.1.4 Permanently Water-Logged (Open Water) Zone

This zone is usually free of vegetation except at the edge where aquatic plants such as *Pistia stratiotes* and *Nymphaea* spp. encroach into this zone when floods connects the two areas. In plant ecology, zonation in lentic habitat is similar to that in rivers. and four major vegetation zones may be identified. Obot (1986) grouped the vegetation in this zone into

TABLE 15.4 Some Families of Aquatic Plants of Nigeria

S/No.	Family	No. of species reported
1	Alismataceae	7
2	Araceae	1
3	Amaryllidaceae	1
4	Aponogetonaceae	2
5	Azollaceae	
6	Ceratophyllaceae	1
7	Convolvulaceae	2
8	Cyperaceae	5
9	Gramineae (poaceae)	12
10	Hydrocharitaceae	2
11	Lemnaceae	3
12	Lentibulariaceae	8
13	Marantaceae	1
14	Menyanthaceae	1
15	Mimosacea	2
16	Najadaceae	1
17	Nymphaeaceae	4
18	Onagraceae	7
19	Parkeriaceae	1
20	Podostemonaceae	2
21	Polygonaceae	3
22	Pontederiaceae	
23	Potamogetonaceae	2
24	Rubiaceae	1
25	Salviniaceae	1
26	Scrophulariaceae	1
27	Sphenocleaceae	1
28	Trapaceae	1
29	Typhaceae	1
	Total species	**76**

Source: From White, E. (1965). The first scientific report of the Kainji Biological Research Team. England: Liverpool (White, 1965); Imevbore, A.M.A. (1971). Floating vegetation of Lake Kainji. Nature, 230, 599–600 (Imevbore, 1971); Obot, E.A. (1987). Echinochloa stagnina, a potential dry season livestock fodder for arid regions. Journal of Arid Environments, 12, 175–177 (Obot, 1987).

II. THE EFFECTS OF CRUDE OIL EXPLORATION ON THE SOCIO-CULTURAL AND ECO-ECONOMICS OF NIGERIAN ENVIRONMENT

TABLE 15.5　Distribution of Some Freshwater Aquatic Plants in Selected Aquatic Resources of Nigeria

S/No.	Species	Lake Kainji	Jebba Lake	Bagauda Lake	Tiga Lake	Ruwa Kenyah	Asejire Dam	Ebub-Ochani River (Port Harcourt)	Cross River Floodplains
1	Ludwigia decurrens	+	+	+	+	+	+	+	+
2	Ludwigia erecta	+	+	+	+	+	+	+	+
3	Ludwigia suffruticosa	+	+	+	+	+	+	+	+
4	Ipomoea aquatica	+	+	+	+	+	+	+	+
5	Ipomoea asarifolia	+	+	+	+	+	+	+	+
6	Echinochloa pyramidalis	+	+	+	+	−	−	+	+
7	Sorghum arundinaceum	+	+	+	+	+	+	+	+
8	Typha australis	+	+	+	+	+	−	−	−
9	Pistia stratiotes	+	+	+	+	+	+	+	+
10	Lemna aequinoctialis	+	+	+	+	−	+	+	+
11	Nymphaea lotus	+	+	+	+	+	+	+	+
12	Vossia cuspidata	+	+	+	+	−	+	+	+
13	Salvinia nymphellula	+	+	−	−	−	+	+	+

Source: Adapted from Obot, E.A., Abolaji, J., & Daddy, F. (1991). Contributions to the biology and utilization of wild guineacorn (Sorghum arundinaceum Stapf). Discovery and Innovation, 3(4),107 (Obot, Abolaji, & Daddy, 1991); Obot, E.A. (1984). The Kainji Lake basin of Northern Nigeria (Ph.D. thesis). Ile-Ife: University of Ife (Obot, 1984); FAO. (1979). Handbook of utilization of aquatic plants. FAO, Fisheries Technical Paper, No. 187. Rome: FAO (FAO, 1979).

four broad categories based on the water-holding duration of the area into: flooded annually (December−January), terrestrial species dominates here; flooded only for a period of two months (January−February) is the second zone dominated by a few woody species and grasses such as *M. pigra* and *Sorghum arundinaceum*; the third zone is mostly flooded for approximately 8 months and is usually dominated by species such as *Echinochloa stagnina*, *Echinochloa crus-galli*, and other members of the poaceae family; and the fourth zone is continuously flooded except in extremely dry years, and is dominated by floating plants. The floral composition of the communities that can develop in these zones is

II. THE EFFECTS OF CRUDE OIL EXPLORATION ON THE SOCIO-CULTURAL AND ECO-ECONOMICS OF NIGERIAN ENVIRONMENT

dependent upon the size, properties, distribution, and circulation of water in this area. Other highly productive aquatic and wetland macrophytes are *Azolla africana* (which fixes atmospheric nitrogen) and *Neptunia oleracea* (a free-floating nitrogen-fixing legume).

15.4.2 Aquatic Faunal Resources (Fishery)

Ita et al. (1985) reported that Nigeria has more than 12 million ha of freshwaters with the capacity to yield over 500,000 tonnes of fish. These waters, in addition, have many freshwater-dependent vertebrates, some of which rely on fish as their main source of food. According to the author, more than 200 species of fish, 14 species of reptiles, 7 species of mammals, 59 species of amphibians, and 72 species of water-associated birds are abound in Nigerian inland waters. In terms of total number of vertebrate species associated with inland waters, fish alone constitute approximately 57%, and other vertebrate species make approximately 49% (Ita et al., 1985).

15.4.2.1 *Fish Species Diversity in Rivers, Lakes, and Reservoirs of Nigeria*

Researchers have identified more than 200 species of fish from main river systems of Nigeria, including natural and artificial lakes, reservoirs, marine, and estuarine (Banks, Holden, & McConnell, 1965; Reed, Burchard, Hopson, Jenness, & Ibrahim, 1967; Welman, 1948). The river systems of Nigeria are richer in species diversity than all natural lakes, wetlands, and reservoirs. Hence, they are fed with fish from their inflowing rivers. The fish stocks in the rivers are in turn replenished from their adjacent floodplains after each flood season during which the fish breed. Therefore, any natural or artificial phenomenon (such as drought or dam construction) that disrupts the natural cycle of flooding is definitely going to affect fish population and species diversity both in lakes (natural or artificial) and wetlands.

15.4.2.1.1 FISH SPECIES DIVERSITY IN MAJOR RIVER SYSTEMS OF NIGERIA

Fish families and species found in major rivers in Nigeria are presented in Table 15.6. It is estimated that approximately 46 fish families covering approximately 239 species abound in Nigerian major rivers. Of these estimated fish families, 7 families were prominent and were found across the major rivers (Banks et al., 1965; Reed et al., 1967; Welman, 1948). These families are Bagridae, Characidae, Cichlidae, Clariidae, Cyprinidae, Mochoidae, and Schilbeidae. Approximately 39 out of the 46 fish families were ubiquitous across the major rivers, with approximately 181 species (Table 15.6).

15.4.2.1.2 FISH SPECIES DIVERSITY IN THE MAJOR NATURAL LAKES AND WETLANDS

Fish species diversity in the following lakes and wetlands of Nigeria has been reported.

1. *Oguta Lake*: Is estimated to have approximately 40 fish species that have been reported.
2. *Lake Ndakolowo*: This is a floodplain lake of the Niger River on the downstream of Jebba Reservoir. It is estimated to have between 9 and 25 fish species according to a survey conducted in 1978 by the National Institute for Freshwater Fisheries Research survey.

TABLE 15.6   Fish Family in the Major Rivers of Nigeria

S/No.	Fish family	Cross River	Ogun River	Oshun River	Anambra River	Sokoto/Rima River	Niger-Benue	Kaduna River
1	Amphiliidae	–	–	–	–		1	–
2	Anabantidae	1					2	
3	Ariidae						3	
4	Bagridae	2	3	3	3	2	11	4
5	Carangidae						1	
6	Centropomidae		1	1	1		1	1
7	Channidae	1	1	1	1		1	
8	Characidae	3	2	2	2	3	14	2
9	Cichlidae	4	3	3	3	4	15	3
10	Citharinidae	2	1	1	1		9	
11	Clariidae	3	2	2	2	2	6	2
12	Clupeidae	1	–	–	–	–	2	
13	Cromeriidae						1	–
14	Cynoglossidae						1	–
15	Cyprinidae	2	2	2	2	5	12	6
16	Cyprinodontidae					4		–
17	Distichodontidae	1				3		1
18	Eleotridae					2		–
19	Gymnarchidae		1	1	1		1	–
20	Hepsetidae	1	1	1	1		1	–
21	Ichthyoboridae	2					2	–
22	Kneriidae						1	–
23	Lepidosirenidae						1	–
24	Lutjanidae	1					1	–
25	Malapteruridae					1	1	–
26	Mastacembelidae						1	
27	Mochokidae	1	1	1	1	2	20	2
28	Mormyridae	8	2	2	2		25	5
29	Mugilidae	1					1	–

(*Continued*)

II. THE EFFECTS OF CRUDE OIL EXPLORATION ON THE SOCIO-CULTURAL AND ECO-ECONOMICS OF NIGERIAN ENVIRONMENT

TABLE 15.6   (Continued)

S/No.	Fish family	Cross River	Ogun River	Oshun River	Anambra River	Sokoto/Rima River	Niger-Benue	Kaduna River
30	Nandidae							−
31	Notopteridae						2	−
32	Osteoglossidae		1	1	1		1	−
33	Pantodontidae						1	−
34	Phractolaemidae							−
35	Polynemidae						1	−
36	Polypteridae	1					5	−
37	Pomadasyidae						1	−
38	Schilbeidae	3	2	2	2	2	4	2
39	Synbranchidae					−		−
40	Tetraodontidae	1				−	1	−
41	Trigonalidae					−	1	−
42	Pristidae					−		−
43	Sphyraenidae					−		−
44	Sciaenidae					−		−
45	Monodactylidae					−		−
46	Pristipomatidae							−
	Total family (Species = 239)	**19 (39)**[a]	**14 (23)**	**14 (23)**	**14 (23)**	**11 (22)**	**35 (161)**	**10 (28)**

[a]*Total figures in parentheses are values for species.*
Source: Adapted from Welman, J.B. (1948). Preliminary survey of the freshwater fisheries of Nigeria. *Lagos: Government Printer; White, E. (1965).* The first scientific report of the Kainji Biological Research Team. *England: Liverpool; Reed, W., Burchard, J., Hopson, A.J., Jenness, J., & Ibrahim, Y. (1967).* Fish and fisheries of Northern Nigeria. Ministry of Agriculture, Northern Nigeria (p. 226). *Zaria: Gaskiya.*

3. *Lake Chad*: As one of the most intensively studied natural lakes in Nigeria, it is estimated to have more than 80 species of fish (Reed et al., 1967). Another survey conducted in 1985 estimated a total of approximately 19 species (Bukar & Gubio, 1985), which was regarded as an improvement over the previous reports.
4. *Hadejia/Nguru Wetlands*: This is the most extensive wetland area in the northern part of the country and was estimated to have approximately 40 fish species (Matthes, 1990). The majority of the species identified are of little or no economic significance on account of their sizes. A severe decline in the population of the larger fish species such as *Lates, Gymnarchus, Heterotis,* and *Heterobranchus* have been observed; however, not to the point of extinction.

II. THE EFFECTS OF CRUDE OIL EXPLORATION ON THE SOCIO-CULTURAL AND ECO-ECONOMICS OF NIGERIAN ENVIRONMENT

15.4.2.1.3 DIVERSITY OF FISH SPECIES IN NIGERIAN MAJOR RESERVOIRS

The major reservoirs of interest in this discussion are:

- Kainji and Jebba reservoirs
- Shiroro Reservoir
- Goronyo and Bakolori reservoirs
- Tiga Reservoir

Kainji Lake leads with an estimate of approximately 104 species followed by approximately 50 species in Jebba (Table 15.7).

15.4.3 Nonfish Aquatic Faunal Resources (Other Vertebrates)

Besides fish, there are other aquatic vertebrates that feed and breed on aquatic biota and these include: snails, aquatic insects, and plants, as well as other invertebrates. This group is frequently related to the varying aquatic habitats. Within these varying habitat types, these animals may exhibit or have certain roles or occupy different positions relevant to their activities in the system. Hence it is rational to assume that the presence or absence of nonfish aquatic vertebrates within a given water body would affect the dynamics of the habitat and fisheries.

Studies directed at fauna and flora of water bodies in Nigeria have concentrated mostly on fishery potential, the fish fauna, management difficulties, strategies for improving

TABLE 15.7 Diversity of Fish Species in Nigerian Reservoirs (1978−86)

Lakes/reservoirs	Total number of species
Kainji	104
Jebba	50
Shiroro	27
Kiri	21
Tiga	18
Goronyo	22
Bakolori	17

Source: Ita, E.O. (1978). A preliminary report on the fish stock assessment and management proposal for the IITA (Ibadan) irrigation and domestic water supply reservoir. Kainji Lake Research Institute Technical Report Series No.1. New Bussa (Ita, 1978); Ita, E.O., & Mohammed, A. (1979). A preliminary report of expeditions to inlet and outlet channels of Lake Ndakolowu (Tatabu) and a proposal for refilling the Lake: Report submitted to Kainji Lake Research Institute (p. 13) (Ita & Mohammed, 1979); Ita, E.O. (1984). Lake Kainji, Nigeria: Fishery synthesis. In J.M. Kapetsky, & T. Petr (Eds.), Status of African reservoir fisheries (pp. 43−103). CIFA (FAO) Tech. Paper 10 (Ita, 1984); Ita, E.O., & Balogun, J.K. (1982). Planning for fisheries in new reservoirs: a case study of Bakolori and the proposed Goronyo Reservoirs in Sokoto State, Nigeria. Kainji Lake Research Institute Newsletter, 8, 1−8 (Ita & Balogun, 1982); Ita, E.O., Mohammed, A., Omorinkoba, W.S., Bankole, N.O., & Awojoodu, S. (1986). A preliminary report on the immediate post-impoundment fishery survey of Shiroro reservoir, Niger State, Nigeria. Kainji Lake Research Institute 1985 Annual Report, pp. 25−29 (Ita et al., 1986).

inland fisheries, zooplankton, phytoplankton, and on the biology and utilization of aquatic macrophytes (Okaeme, Haliru, & Wari, 1989). Scanty information exists on the contribution of wetland wildlife and biodiversity to the aquatic system. This group of aquatic faunal resources is facing decline in terms of their abundance and habitat status, and this is related to the threat of increasing human population pressure and the need for sectoral development (Adams, Kimmel, & Ploskey, 1983). The aquatic resources of the wetland wildlife encompass amphibians, mammals, reptiles, birds, and others that are adapted to life in aquatic ecosystems.

In Nigeria, the popular reptiles reported are the crocodiles and monitor lizards. Three crocodile species, the Nile crocodile (*Crocodylus niloticus*), long-snouted (*Crocodylus cataphractus*), and the West African dwarf crocodile (*Osteolaemus tetraspis*) have been reported (Dore, 1983). The Nile monitor lizard (*Varanus niloticus*) is the most common type found in rivers with stony terrain (Ayeni, Afolayan, & Ajayi, 1983). Other reptiles found in this system are tortoises, terrapins, and snakes associated with aquatic habitats in Nigeria. Frogs and toads are the most abundant amphibians in the aquatic systems of Nigeria, and more than 67 species of frogs have been reported from Nigeria (Schiotz, 1963). Snakes, including the cobras, night and puff adder, and water snakes, are also in abundance in the aquatic system because they feed on the frogs and toads (Adebayo, 1988). The popular mammals within these habitats are the marsh mongoose, civets, genets, and otters.

Also found in Nigerian aquatic systems, especially in Parks and Game Reserves, are two species of hippopotamus; the African giant *Hippopotami amphibious* mostly found in the savannah areas and confined to big rivers and lakes, and the pigmy hippopotamus, *Choeropsis liberiensis*, which is confined to the Niger River deltas in the forest zone.

The manatee is another mammal whose presence has been reported in Nigeria. They are confined to large water bodies of the Sahel and are also spreading along the coastal region of Nigeria. However, the population of this mammal has been threatened by human pressure and development, and they are now confined to large water bodies in the coastal and delta regions of Cross River and Akwa Ibom States. Only the species *Trichechus senegalensis* is indigenous to Nigeria and is abundant in West Africa along the estuaries of Senegal, the Niger River, and the Congo River (Anon, 1976). In terms of birds, the most common species include herons, water fowl, fish eagle, and darters, but approximately 67 of the bird species that are purely aquatic are reported in Nigeria (Elgood, 1982).

15.5 ECOSYSTEM SERVICES AND UTILIZATION OF AQUATIC RESOURCES IN NIGERIA

Approximately 15% of global water need is for household activities that include drinking, cooking, bathing, gardening, and sanitation (Gleick, 2000). Nigeria as a country is blessed with more than 14 million hectares of aquatic resources with the potentials of providing close to one metric tonne of fishes annually (FDF, 2003). Various aquatic flora and fauna are potential sources of food for man and other aquatic organisms, spawning and breeding places for fishes and birds, habitat for nonfish vertebrates and invertebrate aquatic resources. Aquatic resources of water provide many services that include agricultural uses (irrigation and aquaculture) and generation of electricity through damming and navigation. Aquatic resources also provide basic and transitional materials and water resources for the industries in Nigeria.

15.5.1 Aquatic Ecosystem Services and Utilization in Nigeria (Major Rivers, Lakes, Reservoirs, and Wetlands)

15.5.1.1 Major Rivers, Streams, Lakes, and Reservoirs

Rivers and perennial streams are key features in the lowlands of Nigeria and important sources of fisheries and other aquatic produce. They sustain a range of aquatic organisms throughout the year, although they are also subject to large annual fluctuations in volume and flow between the rainy and dry seasons. As permanent water bodies, they serve as critical habitats for a number of strictly aquatic and riverine species.

15.5.1.1.1 HYDROPOWER AND INDUSTRIAL

Globally, the location of many industrial areas is largely dependent upon the availability and accessibility of suitable water supply. Most of these industries use water for cooling or as power source (i.e., hydroelectricity), e.g., crude oil refineries, chemical processes, and manufacturing plants. For instance, Kanji dam supplies energy (i.e., electricity) at low cost to most industries in Nigeria for their various operations, although, the electricity supply still needs to be improved on. The hydroelectric dams also generate energy for household consumption and employment for their workers. Also, in the case of Kanji dam in Nigeria, the hydroelectric power supply has strengthened the international relationship between Nigeria and her neighboring countries.

15.5.1.1.2 NAVIGATIONAL IMPORTANCE

Inland water systems are relevant in transportation of goods and humans from one destination to another. Water transport is important for importation and exportation of goods and for international relationships. Substantial investments are being directed to waterways to expand the natural navigable systems, thus enabling sea transport right into inland industrial areas. Lagos State government in Nigeria inaugurated a water transport service that moves from Mariner to Apapa, thereby easing the pressure on road transport in the state. In the southeast and south-south part of the country, there are several ferry and boat transport services that use the major rivers and waterways. These have improved economic activities and growth of the communities in those areas.

15.5.1.1.3 RECREATIONAL/TOURISM IMPORTANCE

The increasing human population and their affection for leisure and pleasure have placed an enormous demand on natural resources for recreational purpose. This is especially true for water bodies that are often the focus for a variety of recreational activities such as sailing, power-boating, water skiing, general picnicking, relaxation, and angling. However, among these recreational activities, angling (i.e., sport fishing) is highly economical in terms of domestic and foreign earnings. For example, the Argugun fishing festival of Kebbi state attracts tourists from all over the world to Nigeria. Some water bodies are used as tourist centers to generate income. For instance, there is Ikogosi warm spring in Ekiti state and Wikki spring in Bauchi, among others. Natural tourist centers such as these are accepted as having a wide variety of uses, including nature study, photography, fishing, camping, and picnicking, as well as economic pursuits.

15.5.1.1.4 AGRICULTURAL IMPORTANCE

Aquatic resources have contributed significantly to the development of agriculture and Nigerian economy. The importance of aquatic resources to agriculture depends on the types of agricultural operation (cropping, livestock, etc.) in the area. The major areas of contribution of water resources are in irrigation for fodder and food crop production as well as aquaculture. In the extreme savannah arid zone of Nigeria, irrigation supplements rainfall for cropping to boost agricultural yield and output that will cater to the teeming population of the country. Farmers are now being encouraged to supplement water supply from rainfall by providing artificial irrigation for their crops. This will boost the availability of food for household consumption and income generation.

15.5.1.2 Wetlands

Wetlands before now were regarded as waste aquatic resources. However, in recent times, proponents of ecosystem services and valuation have come to recognize that wetland resources and features are beneficial to humans, fishes, and wildlife (USEPA, 2006). Wetlands have both marketable and nonmarketable functions and values. Fish is one of the most important and cheap sources of protein in the diet of most Nigerians, and approximately 45% of animal protein consumed in Nigeria comes from fish. Wetlands, according to Olomukoro and Ezemonye (2007), have played a considerable role in the development of Nigerian economy; and their economic valuation can enhance the placement of wetlands on the agenda of decision makers for conservation and development (USEPA, 2006).

15.5.2 Aquatic Floral Resources (Plant)

There are several types of aquatic plant resources, and they provide several aquatic ecosystem services and functions through supporting and regulating these functions. They provide spawning and breeding grounds and shelter for a variety of fish species in Nigeria (Imevbore & Bakare, 1974). Certain aquatic macrophytes have been reported as feedstuffs for carnivorous fish fingerlings (Agbogidi, Bamidele, Ekokotu, & Olele, 2000). Duckweed (*Lemna* spp.) has been reported as a supplementary low-cost feed for Nile tilapia (*Oreochromis niloticus*) monoculture (Chowdhury, Shahjahan, Rahman, & Islam, 2008). Duckweed also serves as a phytoremediation plant for sewage, animal waste, nitrogen, and phosphate-polluted water, as well as feedstuff for cows, pigs, and chickens.

15.5.2.1 Food, Livestock Feed, and Medicine

Aquatic macrophytes have also been reported as food, as well as providing some medical and health-related services for humans (Mbagwu & Adeniji, 1988). They have been reported as food for the humans in the Kainji shores and within the Delta of River Niger and food for aquatic fauna such as turtle and fishes (Ayeni, Obot, & Daddy, 1999; Kio &

Ola-Adams, 1987). In the mangrove swamp of the Niger Delta, mangrove palm (*Nypha fruticans*) is harvested for alcohol, sugar, and vinegar (Maltby, 1986). Two wild rice varieties (*Oryza longistaminata* (northern states) and *Oryza punctata* (floodplains of the Cross River)) are important species collected at the local level for food. However, the importance and potential of these varieties have not yet been determined for Nigeria. Rhizome, floral receptacle, and fruits of *Nymphaea lotus* (water lily) are either eaten raw or cooked for food (Kio & Ola-Adams, 1987), and *Ludwigia stolonifera* is used as an ingredient for soup in the Yelwa area of Kebbi State (Obot & Ayeni, 1987).

Aquatic plants provide important food for many animals such as songbirds, ducks, and geese. Songbirds eat the seeds of many emergent plants. Duck and geese eat the seeds, leafy parts, and tubers of plants such as pondweeds (*Potamogeton* spp.), watershield (*Brasenia schreberi*), arrowhead (*Sagittaria latifolia*), water pepper (*Polygonum* sp.), and duckweed (*Lemna* sp.) (Anon, 2016). Aquatic macrophytes used as fodder include *V. cuspidata*, *L. hexandra*, *Brachiaria mutica*, *Echinochloa pyramidalis*, *S. arundinaceum*, *Paspalum vaginatum*, and *E. stagnina*. Approximately 7%–9% of Lake Kainji has been reported to be covered at various times by plants.

15.5.2.2 Habitat

Other ecosystem services provided by aquatic plant resources include living space for small aquatic invertebrates (insects, snails, and crustaceans), which in turn supply food for fish. In addition, vegetated area of the aquatic ecosystem provides shelter for spawning, substrate, and nursery or breeding sites for fishes, and amphibians use the vegetated areas as cover from predatory fish (Agbogidi et al., 2000; Anon, 2016; Mitchell, 1974). Several species of game use aquatic plants as a nesting place and food source.

15.5.2.3 Phytoremediation of Polluted Aquatic Ecosystem

Aquatic macrophytes also play an important regulating role physiologically by removing mineral nutrients and heavy metal from water bodies (Uka, Mohammed, & Birnin-Yauri, 2009), as well as serve as indicators for water quality (Ghavzan, Gunale, Mahajan, & Shirke, 2006). Some aquatic plants have in recent time been reported with the potentials of remedying polluted and contaminated water bodies. Water hyacinth (*Eichhornia crassipes*) is valuable in the treatment of agro-industrial waste pollutant (Dar, Kumawat, Singh, & Wani, 2011). Phytoremediation potentials of water hyacinth, water lettuce, and several aquatic macrophytes, especially for a heavy metal–polluted aquatic environment, have been widely documented (Boyd, 1970; Chantiratikul, Atiwetin, & Chantiratikul, 2008; Ndimele & Jimoh, 2011; Uka et al., 2009). This regulating role of aquatic macrophytes suggests their global importance in the treatment of municipal and industrial wastewater. They have also been reported as efficient indicators of water quality and biomonitoring agents (Brix & Schieriup, 1989; Petre, 1990).

15.5.2.4 Erosion Control

Emergent and submersed aquatic plant species shield shorelines from erosion caused by water currents and wave actions. These plants also stabilize sediment, which enhances water clarity.

15.5.2.5 Potential Soil Amendment

Aquatic plants have also been reported to accumulate large quantities of nutrient elements such as nitrogen and phosphorus in their tissues that can enhance and improve soil productivity when added as organic soil amendment. *Azolla* sp. is a free-floating fern, which is known to fix nitrogen. This particular ecosystem service is widely exploited as bio-fertilizer for rice production. In Nigeria, the local species *A. africana* is yet to be exploited in this manner (Maltby, 1986). The use of water hyacinth (*E. crassipes*) for biogas, pulp, organic fertilizer, and sewage treatment is documented. However, there is the need to assess the profitability of utilizing water hyacinth for the same purpose in Nigeria.

15.5.2.6 Aquatic Plants as a Potential Source of Energy and Industrial Raw Material

Aquatic plants also have the potentials for use as fuel in cooking and fish smoking, especially in the coastal and delta regions of Nigeria (Kio & Ola-Adams, 1987). Some of the species used include *Aeschynomene crassicaulis*, *Echinochloa* spp., *Cyperus papyrus*, and *Rhizophora* spp. The potential and the economics of using aquatic plants as biogas resources are yet to be assessed in Nigeria. They also offer a variety of resources that could be utilized by innovative industry for a wide range of products, such as for construction, matting, bedding, and pulp/paper. *V. cuspidata*, *C. papyrus*, and *E. crassipes* have economic potential for use as pulp, paper, and fiber. According to Kio and Ola-Adams (1990), strips of the young fronds and mid-ribs of *Phoenix reclinata* are used for weaving sleeping mats, sieves, and bags, and *Laguncularia racemosa* yields timber, tannin, and dyeing materials.

15.5.3 Aquatic Resources—Fisheries and Aquatic Wildlife/Nonfish Resources

15.5.3.1 Fisheries and Aquaculture

The freshwater ecosystems of Nigeria are the richest in fish species diversity in West Africa, with an estimate of approximately 200 fish species (Meye & Ikomi, 2008; Todor, 1992). This plays an important role in the provision of protein to Nigerians, with a population close to 170 million people (FDF, 2008). Fisheries and aquaculture have been of great importance to the national economy in the areas of fish production, raw materials to industries, employments, household empowerment, and income generation. Fisheries and the aquaculture sector of a nation also make use of some finished products from small and large industries. These industries provide employment opportunities to citizens, which contribute to both domestic and foreign earnings, thereby improving the per capita income and standard of living of the country. Fisheries and aquaculture resources have also provided both direct and indirect employment opportunities to those into fishing gear fabrication, canoe building, sale of frozen aquatic products, etc. Through all these and more, exploitation of aquatic resources have contributed greatly to reduction of unemployment and improvement of income earnings.

15.5.3.2 Nonfish and Aquatic Wildlife

Close to or more than 80% of the rural dwellers in Nigeria depend on the aquatic ecosystem for food and protein supplement. The livelihoods of the riverine and coastal line communities depend predominantly on fish and nonfish aquatic vertebrates such as crocodiles, frogs, toads, snakes, terrapins, hippopotamus, and manatee as their main source of animal protein (Nicol, 1953). These animals are also hunted for other products such as their skins for leather, their cartilaginous parts, feces, and skin and hoofs as aphrodisiacs for additional income and for traditional medicine. Crocodiles and monitor lizards are hunted extensively for their meat and skins, which are important foreign exchange earners. The high demand for crocodile skins, meat, and body parts for traditional medicine certainly has contributed to the observed decline in their populations in Nigeria. These animals also play an important role in the population dynamics and biological function of the aquatic community, especially in the different trophic levels of the aquatic system's food chain.

15.6 VALUATION OF AQUATIC RESOURCES: SCOPE AND SOURCES

Total economic value (TEV) is a concept in cost—benefit analysis that refers to the value derived by people from a natural resource, a man-made heritage resource, or an infrastructure. It is represented by the maximum amount a consumer is willing to pay for an item in a free market economy, or the amount of time an individual will sacrifice waiting to obtain a government-rationed good in a socialist economy. Economic valuation provides a means for measuring and comparing the various benefits derived from aquatic resources and is a powerful tool to aid and improve their judicious use. It attempts to assign quantitative values to the goods and services provided by aquatic resources, whether or not market prices are available. In conducting an economic valuation exercise, it is essential to know the use and nonuse values. Use value involves some interaction with the resource, either directly or indirectly, which could be direct use values and indirect use values. Direct use values involve human interaction with the ecosystem itself rather than via the services it provides, and it may be consumptive, nonconsumptive, or extractive use. Indirect use value is derived from the services provided by the ecosystem. Nonuse value is associated with benefits derived simply from the knowledge that the ecosystem maintained, which could be existence value (derived from the satisfaction of knowing that ecosystems continue to exist); bequest value, associated with the knowledge that ecosystems and their services will be passed on to descendants to maintain the opportunity for the services to be enjoyed in the future; altruistic value, derived from knowledge that contemporaries can enjoy the goods and services the ecosystems provide; option value, where an individual derives benefit from ensuring that ecosystem services will be available for his or her own use in the future; quasioption value, derived from the potential benefits of waiting for improved information prior to giving up the option to preserve a resource for the future; and philanthropic value, the satisfaction gained from ensuring that resources are available to contemporaries in the current generation. The classification of TEV of water resources is presented in Table 15.8.

15.6.1 Economic Valuation of Wetland Resources

The general economic value of wetlands and the functions of various wetlands in Nigeria are summarized in Tables 15.9 and 15.10, respectively.

TABLE 15.8 Classification of Total Economic Value of Water Resources

Use			Nonuse
Direct use value	**Indirect use value**	**Option and quasioption value**	
Consumptive used, such as use of water for irrigation or the harvesting of fish	Flood protection provided by wetlands or the removal of pollutants by aquifer recharge	Potential future uses (as per direct and indirect uses)	Existence value
Nonconsumptive such as recreational swimming, or the aesthetic value of enjoying a view		Future value of information	Bequest
Flood protection provided by wetlands or the removal of pollutants by aquifer recharge			Philanthropic value

Source: *Adapted from Turner, R.K., Bateman, I., & Adger, N. (Eds.). (2001).* Economics of coastal and water resources: Valuing environmental functions. *Dordrecht: Klumer (Turner, Bateman, & Adger, 2001); WRS. (1997). Approach for comprehensive water resources management in Sri Lanka.* Colombo: WRS (WRS, 1997).

TABLE 15.9 Classification of Total Economic Value for Wetlands

Used values			Nonused values
			Existence value
Direct use value	**Indirect use value**	**Option and quasioption value**	
Fish	Nutrient retention	Potential future uses (as per direct and indirect uses)	Biodiversity
Agriculture	Flood control	Future value of information	Culture, heritage
Fuel wood	Storm protection		
Recreation	Groundwater recharge		
Transport	External ecosystem support		
Wildlife harvesting	Microclimatic stabilization		
Peat/energy	Shoreline stabilization		

Source: *Adapted from Barbier, E.B. (1989). The economic value of ecosystems: 1—Tropical wetlands.* LEEC gatekeeper series 89-02. *London: London Environmental Economics Centre (Barbier, 1989); Barbier, E.B. (1993). Valuing tropical wetland benefits: Eco-methodologies and applications.* Geographical Journal. *Part 1, 59, 22—32 (Barbier, 1993); Scodari, P.F. (1990). Wetlands protection: The role economics environmental law institute monograph.* Washington, DC (Scodari, 1990); Scott, D.A. (1989). Design of wetland data sheet for database on Ramsar sites. *Gland: Mimeographed report to Ramsar convention Bureau (Scott, 1989).*

II. THE EFFECTS OF CRUDE OIL EXPLORATION ON THE SOCIO-CULTURAL AND ECO-ECONOMICS OF NIGERIAN ENVIRONMENT

TABLE 15.10 Values and Functions of Wetland Types in Nigeria

	Estuaries (without mangrove)	Mangroves	Open coasts	Floodplains	Freshwater marshes	Lakes (natural and man-made)	Peat lands	Swamp forest
FUNCTIONS								
Groundwater recharge	Absent	Absent	Absent	Common and important	Common and important	Common and important	Present	Present
Groundwater discharge	Present	Present	Present	Present	Common and important	Present	Present	Common and important
Flood control/regulation	Present	Common and important value	Absent	Common and important	Common and important	Common and important	Present	Common and important
Shoreline stabilization/erosion control	Present	Common and important value	Present	Present	Common and important	Absent	Absent	Common and important
Sediment/toxicant retention	Present	Common and important value	Present	Common and important value	Common and important	Common and important	Common and important	Common and important
Nutrient retention	Present	Common and important value	Present	Common and important value	Common and important	Present	Common and important	Present
Biomass export	Present	Common and important value	Present	Common and important value	Present	Present	Absent	Present
Storm protection/wind break	Present	Common and important value	Present	Absent	Absent	Absent	Absent	Present
Microclimate stabilization	Absent	Present	Absent	Present	Present	Present	Absent	Present
Water transport	Present	Present	Absent	Present	Absent	Present	Absent	Absent
Recreation/tourism	Present	Present	Common and important value	Present	Present	Present	Present	Present

Products	Estuaries (without mangrove)	Mangroves	Open coasts	Floodplains	Freshwater marshes	Lakes (natural & man-made)	Peat lands	Swamp forest
Forest resources	Absent	Common and important	Absent	Present	Absent	Absent	Absent	Common and important
Wildlife resources (including fowls)	Common and important	Present	Present	Common and important	Common and important	Present	Present	Present
Fisheries	Common and important	Common and important	Common and important	Common and important	Common and important	Common and important	Absent	Present
Forage resources	Present	Present	Absent	Common and important	Common and important	Absent	Absent	Absent
Agricultural resources	Absent	Absent	Absent	Common and important	Present	Common and important	Present	Absent
Water supply	Absent	Absent	Absent	Present	Present	Present	Present	Present
ATTRIBUTES								
Biological diversity	Present	Present	Present	Common and important	Present	Common and important	Present	Present
Uniqueness to culture/heritage	Present	Present	Present	Present	Present	Present	Present	Present

Source: Modified from Nwankwoala, H.O. (2012). Case studies on coastal wetlands and water resources in Nigeria. European Journal of Sustainable Development, 1(2), 113–126.

15.7 CONSERVATION OF AQUATIC RESOURCES IN NIGERIA

Conservation is the planned, controlled exploitation or judicious use of natural resources to ensure their continuous availability and to preserve the quality or original nature of the environment.

15.7.1 Methods of Conserving Water Resources

- Rainwater harvesting with the help of some infrastructures such as storage tanks and dams
- Prevention of water pollution by sewage and chemicals from industries and homes
- Use of gray water
- Saving of water by prompt repair of burst pipes or turning off taps immediately after use
- Tree planting

15.7.2 Methods of Conserving Wetland

Wetlands act like the kidneys of the earth, cleaning the water that flows into them. They trap sediment and soils, filter out nutrients, and remove contaminants. These can reduce flooding and protect coastal land from storm surge and are important for maintaining water tables. They also return nitrogen to the atmosphere. However, human activity seems to be one of the major threats to Nigerian wetlands. Some of these threats include the following:

- Sand and gravel extraction that causes changes in water levels, damages existing vegetation, and provides access for weeds
- Reclamation of lake and river margins, lagoons, and estuaries, and draining of farm swamps reduces wetland areas
- Pollution by excess runoff of sediment and nutrients from farmlands
- Plant and animal pest invasion
- Stock grazing in wetlands and surrounding catchments damages vegetation, decreases soil stability, and contributes to pollution
- Careless recreation practices, including misuse of jet-skiing, hunting, kayaking, power-boating, and white baiting disturbs plant and animal life and may destroy parts of the physical wetland environment
- Forest harvesting near wetlands
- Loss of vegetation in surrounding catchments
- The drawing of water away from groundwater systems by pine forests, leaving depleted supplies
- Wetland drainage for urban or rural development

Some of the approaches that can be used to conserve Nigerian wetlands include:

Sanitary Measures: The aquatic resources arm of Nigerian quarantine services should employ sanitary measures to prevent the entry of foreign exotic pests, pathogens, and diseases of aquatic resources. These measures could employ the following:

- Declaration of all import/export or trans-shipment of aquatic resources and their products to the Nigerian Agricultural Quarantine Service (NAQS) at the point of entry and exit from the country
- Regulation and control of all import and export of ornamental/live aquatic resources through licensing of importers and exporters; issuance of health certificates
- Verification of import and export permits for aquatic resources
- Control and prevention of illegal trading in endangered or prohibited aquatic species and products to safeguard human and aquatic resource health
- Prevention of the introduction of highly invasive aquatic resource species, aquatic resource diseases/pests, and diseases associated with various types of seafood-fish, bivalves, mollusks, shellfish, and crustaceans
- Assurance of regular monitoring and inspection of approved premises with facilities for holding, quarantining, and packaging of ornamental/live aquatic resources
- Conducting disease and pest surveillance of all aquatic resources to detect and monitor the presence and spread of such pathogens

Establishment of Agencies for Conservation: Examples of these agencies are: the Nigerian conservation foundation, forest departments, the Federal Environmental Protection Agency (FEPA), the River Basin Development Authorities (RBDA), the Ministries of Agriculture, and the Department of Wildlife Conservation.

Establishment of Game Reserves or National Parks: Some game reserves in Nigeria are: Yankari Games Reserve in Bauchi State, Borgu Game Reserves in Niger State, Shasha River Forest in Ogun State, Olomu Forest Reserve in Kwara State, Mamu River Forest Reserve in Anambara State, and Zamfara Forest Reserve in Zamfara State.

15.7.2.1 *Enacting Conservation Laws*

Conservation Education: Conservation education serves to inform the populace about the need to conserve natural resources.

Setting Standard for Pollution Control: These standards help to protect land, water, and air resources.

Manage the Riparian Margin: Managing the riparian margin responsibly will help protect wetlands.

Plant Around Wetlands: Planting appropriate flora around wetlands will help stabilize soil and stop nutrient runoff flowing directly into water, especially when coupled with fencing to exclude stock.

Control Over Forest Fire: Destruction or loss of forest by fire is fairly common in a forested wetland because trees are highly exposed to fire, and once started it becomes difficult to control.

II. THE EFFECTS OF CRUDE OIL EXPLORATION ON THE SOCIO-CULTURAL AND ECO-ECONOMICS OF NIGERIAN ENVIRONMENT

Minimization of Storm Water: Runoff is an important component of a wetland's hydrologic budget.

Buffers and Greenbelts: A greenbelt is a strip of upland surrounding the wetland that is maintained in a natural vegetated state.

Fencing: In areas where livestock grazing in wetlands or excessive human use is degrading wetlands, fencing is one of the simplest ways to protect such wetland.

Land Reclamation: Wetlands are reclaimed by filling them with rocks and clay.

15.8 CONCLUSION

- Ecosystem services exist in Nigeria like anywhere else.
- Efforts are being made for their systematic exploration, valuation, measurement, exploitation, management, payment systems, conservation, and growth, but at the moment, aquatic resources in Nigeria have limited available data.
- The above assertion is truer for the aquatic resources, as their potentials are barely harnessed.
- Nigeria is endowed with abundant surface and groundwater resources whose availability varies with rainfall, location, and geological formations.
- Over the years, various legislation, regulations and policy instruments have been developed, modified from time to time as new scientific evidence emerges.
- Aquatic resources have contributed to the development of agriculture.
- Most of Nigeria's agriculture is still based on rain-fed peasant farming and other natural sources. This makes agriculture and food security climate-sensitive and fragile.
- Aquatic resources also provide basic and transitional materials for industries, recreational and spiritual sites for the many people in Nigeria.
- Ecosystem services seem not to be organized in Nigeria in terms of their systematic exploration, valuation, measurement, exploitation, management, payment systems, and conservation.

References

Adams, S. M., Kimmel, B. L., & Ploskey, G. R. (1983). Source of organic matter for reservoir fish production. A trophic-dynamics analysis. *Canadian Journal of Fisheries and Aquatic Sciences, 40*, 1480–1485.

Adedayo, O.F. (1988). *A taxonomic study of some Nigerian Anurans (Amphibians) in Lagos and it's environ* (M.Sc. thesis). University of Lagos.

Agbogidi, O. M., Bamidele, J. E., Ekokotu, P. A., & Olele, N. F. (2000). The role and management of aquatic macrophytes in fisheries and aquaculture. *Issues on Animal Science, 10*, 221–235.

Anon (1976). *Larouse encyclopedia of animal life*. London: The Hamlyn Publishing Group Ltd.

Anon. (2016). *The uses and benefits of aquatic plants*. Available online at <http://www.ecy.wa.gov/programs/wq/plants/native/uses.html>. Retrieved 31.05.16.

Asibor, G. (2009). Wetlands: Values, uses and challenges. In *A paper presented to the Nigerian Environmental Society at the Petroleum Training Institute, Effurun, 21ˢᵗ November, 2009*.

Ayeni, J. S. O., Afolayan, T. A., & Ajayi, S. S. (1983). *Introductory handbook on Nigerian wildlife*. Ilorin: Saclog Printing Production, 80 pp.

Ayeni, J. S. O., Obot, E. A., & Daddy, F. (1999). Aspect of biology, conservation and management of aquatic vascular plant resources of Nigeria wetland based on the Kainji lake experience. *Proceedings of a workshop on* sustainable management and conservation of fisheries and other aquatic resources of Lake Chad and the Arid Zone of Nigeria, Jan. 16–17, Maiduguiri, , 64–73.

Ayoade, J. O. (1981). On water availability and demand in Nigeria. *Water Supply and Management*, 5, 361–372.

Banks, J. W., Holden, M. J., & McConnell, R. M. (1965). Fishery report. In E. White (Ed.), *The first scientific report of the Kainji biological Research Team*. Ife-Ife: University of Ife.

Barbier, E.B. (1989). *The economic value of ecosystems: 1–tropical wetlands*. LEEC gatekeeper series 89-02. London: London Environmental Economics Centre.

Barbier, E. B. (1993). Valuing tropical wetland benefits: Eco-methodologies and applications. *Geographical Journal*. *Part 1*, 59, 22–32.

Bene, C., Neiland, A., Jolley, T., Ovie, S., Sule, Q., Ladu, B., ... Quensiere, J. (2003). Inland fisheries, poverty and rural livelihood in the Lake Chad basin. *Journal of Asian and African Studies*, 38, 17–51.

Boyd, C. E. (1970). Vascular aquatic plants for mineral nutrient removal from polluted waters. *Economic Botany*, 24, 95–103.

Bris, H., & Schierup, H. H. (1989). The use of aquatic macrophytes in water pollution control. *Ambio*, 18, 100–107.

Bukar, T. A., & Gubio, A. K. (1985). The decline in commercially important species of fish and predominance of *Clarias lazera* in Lake Chad. In E. O. Ita, et al. (Eds.), *Proceedings of Fisheries Society of Nigeria (FISON)* (pp. 35–41).

Chantiratikul, A., Atiwetin, P., & Chantiratikul, P. (2008). Feasibility of producing selenium-enriched water lettuce (*Pistia stratiotes* L.). *Journal of Biological Science*, 8, 644–648.

Chowdhury, M. M. R., Shahjahan, M., Rahman, M. S., & Islam, M. S. (2008). Duckweed (*Lemna minor*) as a Supplementary feed in Monoculture off Nile Tilapia, Oreochromis niloticus. *Journal of Fish Aquatic Science*, 3, 54–59.

Dar, S. H., Kumawat, D. M., Singh, N., & Wani, K. A. (2011). Sewage treatment potential of water hyacinth (*Eichhornia crassipes*). *Research Journal of Environmental. Science*, 5, 377–385.

Dore, M.P.O. (1983). Crocodile conservation in Nigeria. In *13th annual conference of the Forest Association of Nigeria, Benin-City, Bendel State*.

Elgood, J. H. (1982). The birds of Nigeria. An annotated checklist. *British Ornithologist*, 4, 246.

FAO. (1979). *Handbook of utilization of aquatic plants*. FAO, Fisheries Technical Paper, No. 187. Rome: FAO.

Federal Department of Fisheries (FDF). (2003). Presidential forum on fisheries and aquaculture (status and opportunities) (49 pp.). Abuja: Federal Department of Fisheries Report.

Federal Department of Fisheries (FDF). (2008). *Fisheries statistic of Nigeria projected human population; fish demand and supply in Nigeria, 2000–2015* (56 pp.).

Ghavzan, N. J., Gunale, V. R., Mahajan, D. M., & Shirke, D. R. (2006). Effects of environmental factors on ecology and distribution of aquatic macrophytes. *Asian Journal of Plant Science*, 5, 871–880.

Gleick, P.H. (2000). *The world's water, 2000–2001: The biennial report on freshwater, Island press, 33, ISBN 1-55963-792-7*; online at Google Books.

Gophen, M. (2008). Lake management perspectives in arid, semi-arid, sub-tropic and tropical dry climates. In M. Sengupta, & R. Dalwani (Eds.), *12 World lake conference* (pp. 1338–1348). Jaipur: *International Lake Environment Committee*.

Imevbore, A. M. A. (1971). Floating vegetation of Lake Kainji. *Nature*, 230, 599–600.

Imevbore, A. M. A., & Bakare, O. (1974). Pre-impoundment studies of the swamps in Lake Kainji Basin. *The African Journal of Tropical Hydrology and Fish*, 3, 79–93.

Ita, E.O. (1978). *A preliminary report on the fish stock assessment and management proposal for the IITA (Ibadan) irrigation and domestic water supply reservoir*. Kainji Lake Research Institute Technical Report Series No.1. New Bussa.

Ita, E. O. (1984). Lake Kainji, Nigeria: Fishery synthesis. In J. M. Kapetsky, & T. Petr (Eds.), *Status of African reservoir fisheries* (pp. 43–103). CIFA (FAO) Tech. Paper 10.

Ita, E. O., & Balogun, J. K. (1982). Planning for fisheries in new reservoirs: a case study of Bakolori and the proposed Goronyo Reservoirs in Sokoto State, Nigeria. *Kainji Lake Research Institute Newsletter*, 8, 1–8.

Ita, E.O., & Mohammed, A. (1979). *A preliminary report of expeditions to inlet and outlet channels of Lake Ndakolowu (Tatabu) and a proposal for refilling the Lake: Report submitted to Kainji Lake Research Institute* (p. 13).

Ita, E. O., Mohammed, A., Omorinkoba, W. S., Bankole, N. O., & Awojoodu, S. (1986). *A preliminary report on the immediate post-impoundment fishery survey of Shiroro reservoir, Niger State, Nigeria. Kainji Lake Research Institute 1985 Annual Report* (pp. 25–29).

II. THE EFFECTS OF CRUDE OIL EXPLORATION ON THE SOCIO-CULTURAL AND ECO-ECONOMICS OF NIGERIAN ENVIRONMENT

Ita, E.O., Sado, E.K., Balogun, J.K., Pandogari, A., & Ibitoye, B. (1985). *Inventory survey of Nigerian inland waters and their fishery resources I. A preliminary checklist of inland water bodies in Nigeria with special reference to ponds, lakes, reservoirs and major rivers.* Kainji Lake Research Institute Technical Report Series, No. 14, KLRI, New Bussa, 51.

Kio, P. R. O., & Ola-Adams, B. A. (1990). Utilization and development of wetlands. In T. V. A. Akpata, & D. U. U. Okali (Eds.), *Nigerian wetlands* (pp. 48–54). Port Harcourt: UNESCO/MAB.

Kio, P. R. O., & Ola-Adams, B. A. (1987). Economic importance of aquatic macrophytes. In C. Iloba (Ed.), *Ecological implications in the development of water bodies in Nigeria.* New Bussa: National Institute for Freshwater Fisheries Research Institute.

Lehner, B., & Döll, P. (2014). Development and validation of a global database of lakes, reservoirs and wetlands. *Journal of Hydrology, 296,* 1–22.

Maltby, E. (1986). *Waterlogged wealth: Why waste the world's wet places?* London & Washington, DC: Institute for Environmental Development; Earthscan Publ.

Matthes, H. (1990). *Report on the fisheries related aspects of the Hadejia-Nguru wetlands conservation project.* Report to the Hadejia-Nguru wetlands conservation project, Nguru, Nigeria.

Mbagwu, I. G., & Adeniji, A. H. (1988). The nutritional content of Duckweed (Lemna paucicostata Hegelm) in the Kainji Lake Area, Nigeria. *Aquatic Botany, 29*(4), 357–365.

McCartney, M. P., Rebelo, L. M., SenaratnaSellamuttu, S., & de Silva, S. (2010). *Wetlands, agriculture and poverty reduction* (p. 39). Colombo: International Water Management Institute (IWMI), (IWMI Research Report 137).

Meye, J. A., & Ikomi, R. B. (2008). A study of the fish fauna of Urie Creek at Igbide, Niger Delta. *Zoologist, 6,* 69–80.

Mitchell, D. S. (1974). *Aquatic vegetation and its use and control.* Paris: UNESCO.

Mitchelle-Thome, R.C. (1961). Preliminary estimate of average annual available groundwater supply in Nigeria. In *Paper presented at the conference of the Science Association of Nigeria in Enugu, December, 1961.*

Ndimele, P. E., & Jimoh, A. A. (2011). Water hyacinth *Eichhornia crassipes* (Mart.) Solm. in phytoremediation of heavy metal polluted water of Ologe Lagoon, Lagos, Nigeria. *Research Journal of Environmental Sciences, 5,* 424–433.

Nicol, B. M. (1953). Protein in the diet of the Isoko tribe in Niger Delta. *Proceedings of the Nutrition Society, 12,* 66–69.

Nwankwoala, H. O. (2012). Case studies on coastal wetlands and water resources in Nigeria. *European Journal of Sustainable Development, 1*(2), 113–126.

Obot, E.A. (1984). *The Kainji Lake basin of Northern Nigeria (Ph.D. thesis).* Ile-Ife: University of Ife.

Obot, E. A. (1986). Ecological comparison of the pre and post-impoundment macrophytes flora of the River Niger and Lake Kainji. *Vegetation, 68,* 67–70.

Obot, E. A. (1987). Echinochloa stagnina, a potential dry season livestock fodder for arid regions. *Journal of Arid Environments, 12,* 175–177.

Obot, E. A., Abolaji, J., & Daddy, F. (1991). Contributions to the biology and utilization of wild guineacorn (*Sorghum arundinaceum* Stapf). *Discovery and Innovation, 3*(4), 107.

Obot, E. A., & Ayeni, J. S. O. (1987). *A handbook of common aquatic plants of the Kainji Lake Basin, Nigeria.* Ilorin: Kainji Lake Research Institute/Saolog Printing Production.

Okaeme, A. N., Haliru, M., & Wari, M. (1989). Utilization of aquatic birds fishing communities of Lake Kainji. *NIFFR Annual Report* (pp. 40–45).

Olomukoro, J. O., & Ezemonye, L. I. N. (2007). Assessment of macro vertebrate fauna of rivers in Southern Nigeria. *African Zoology, 42,* 1–11.

Oteze, G. (1981). Water resources of Nigeria. *Environmental Geology, 3,* 171–184.

Oyebande, L., Obot, E.O., & Bdiliya, H.H. (2003). *An inventory of wetlands in Nigeria.* Report prepared for World Conservation Union – IUCN, West African Regional Office, Quagadougou, Burkina Faso.

Petre, T. (1990). Fish, fisheries aquatic macrophytes and water quality in inland waters. *Water Quality Bulletin, 12,* 103–106.

RAMSAR Convention. (1971) *Proceedings of the international conference on conservation of wetlands and waterfowl, Ramsar, Iran, 30th January–3rd February 1971.* Slimbridge: International Wildfowl Research Bureau.

Reed, W., Burchard, J., Hopson, A. J., Jenness, J., & Ibrahim, Y. (1967). *Fish and fisheries of Northern Nigeria. Ministry of Agriculture, Northern Nigeria* (p. 226). Zaria: Gaskiya.

II. THE EFFECTS OF CRUDE OIL EXPLORATION ON THE SOCIO-CULTURAL AND ECO-ECONOMICS OF NIGERIAN ENVIRONMENT

Satia, B.P. (1990). *National reviews of aquaculture development in Africa* (p. 29). Nigeria: FAO, Fisheries Circular, No. 770,192.

Schiotz, A. (1963). Amphibians of Nigeria. *Videnskabelige Meddelesler fra Dansk Naturhistorisk Forening, 125*, 1–92.

Scodari, P.F. (1990). *Wetlands protection: The role economics environmental law institute monograph*. Washington, DC.

Scott, D. A. (1989). *Design of wetland data sheet for database on ramsar sites*. Gland: Mimeographed Report to Ramsar Convention Bureau.

Todor, J. G. (1992). Fish and shellfish conservation interest in Nigeria. Nig. Institute. *Oceanography and Marine Research Technical Paper No, 79*, 30 pp.

Turner, R.K. Bateman, I., & Adger, N. (Eds.). (2001). *Economics of coastal and water resources: Valuing environmental functions*. Dordrecht: Klumer.

Uka, U. N., Mohammed, H. A., & Birnin-Yauri, Y. A. (2009). In O. A. Fagbenro, O. A. Bello-Olusoji, E. O. Adeparusi, L. C. Nwanna, O. T. Adebayo, A. A. Dada, & M. O. Oluayo (Eds.), *Aquatic macrophytes role in nutrient and heavy metal regulation* (pp. 39–42). Nigeria: Proceedings of Fisheries Society of Nigeria.

Uluocha, N. O., & Okeke, I. C. (2004). Implications of wetlands degradation for water resources management: Lessons from Nigeria. *Geojournal, 16*, 151–154.

USEPA. (April 2006). Application of elements of state water monitoring and assessment program for wetland. Wetland division, office of wetlands, oceans and watersheds, US environmental protection agency, Washington DC. 2006.

USEPA.(2009). *What are wetlands? Wetlands 2009.* <http://www.epa.gov/owow/etlands/vital/what>. Retrieved November 17, 2013 @ 1645 hours.

Welman, J. B. (1948). *Preliminary survey of the freshwater fisheries of Nigeria*. Lagos: Government Printer.

White, E. (1965). *The first scientific report of the Kainji Biological Research Team*. England: Liverpool.

WRS (1997). *Approach for comprehensive water resources management in Sri Lanka*. Colombo: WRS.

Further Reading

Tobor, J. G. (1973). A survey of the fisheries of the lower Yobe in the North Western Basin of Lake Chad. *Annual Report of the Fed. Dept. of Fisheries* (pp. 82–89).

II. THE EFFECTS OF CRUDE OIL EXPLORATION ON THE SOCIO-CULTURAL AND ECO-ECONOMICS OF NIGERIAN ENVIRONMENT

Trade-off Analyses of Ecosystem Services in Nigerian Waters

Prince Emeka Ndimele[1], Adeniran Akanni[2], Jamiu Adebayo Shittu[1], Lois Oyindamola Ewenla[1] and Oluwanifemi Esther Ige[1]

[1]Lagos State University, Ojo, Lagos State, Nigeria [2]Lagos State Ministry of Environment, Ikeja, Lagos, Nigeria

16.1 INTRODUCTION

Aquatic resources are one of the natural assets Nigeria is bestowed with, noticeable in her abundant annual rainfall, numerous streams and lakes, large rivers, extensive underground network, as well as marine resources. The sizes and potentials of these resources have not been fully assessed and exploited. Even the little that has been harnessed has been managed ineffectively, posing problems and hindering the nation's development. This management problem characterizes most aquatic resources in Africa and the world at large. For instance, exploitation of the world's freshwater resources is still at a low level of approximately 0.08% (Fry, 2008).

Water resource management is essential for optimum use of aquatic resources, and it involves proper planning and management skills. The end point of these management options is expected to be the provision and allocation of water and its resources on an equitable basis. Inadequate and ineffective management options will lead to a state of scarcity of water and its resources, which in turn impinge human survival presently and in the future, since sustainability and conservation are threatened.

The term "ecosystem services," which can simply be thought of as services provided by natural ecosystems, is an evolving concept popularized by the Millennium Ecosystem Assessment (MA) in the early 2000s. This has led to the valuation of natural resources, incorporating this valuation into decision-making and sensitizing the world about their importance, since the depletion in the services supplied by these ecosystems will have disastrous short- and long-term effects on man. Aquatic ecosystem services, such as hydropower generation and provision of clean drinking water, vary in space and time.

Since all services delivered by a particular ecosystem cannot all be at maximum, multiple ecosystem services supplied within an area involve some forms of trade-off, whether intentional or nonintentional. While some services are given priority and subsequently developed in order to meet human needs, some services are neglected and underdeveloped (Deng, Zhao, & Wu, 2011; Seppelt, Lautenbach, & Volk, 2013). However, these services exhibit some forms of complexity in their dependence.

Management schemes and projects in Nigeria are tailored in one direction as they only intend to solve immediate human needs by enhancing services needed to solve it without considering how the provisioning of other services is affected, which will have negative feedbacks over the long term. Trade-off analysis of ecosystem services can aid in making optimal decision to balance the costs and benefits of multiple human usage of ecosystems.

This chapter aims to create adequate awareness on why trade-off analysis can be a solution to problems associated with the management of aquatic ecosystems in Nigeria. These water bodies have the potential of supplying multiple benefits. The decisions made on ecological and socioeconomic functions of ecosystems will become easier if adequate information is conveyed to decision-makers on where specific trade-offs are identified among multiple ecosystem services at different times and scales (Ruhl, Kraft, & Lant, 2007; Tallis, Kareiva, & Marvier, 2008).

16.2 GENERAL IDENTITY OF ECOSYSTEM SERVICES IN NIGERIA

Ecosystem services can generally be defined according to three citations; the conditions and processes in which human life is satisfied and sustained by natural ecosystems and the organisms inhabiting them (Daily, 1997a, 1997b), the direct or indirect benefits humans gain from ecosystem functions (Costanza, 2006), and simply put, the benefits human derive from natural ecosystems (MA, 2005). Despite the general ideas and agreements demonstrated by these definitions, there are still some obvious differences. The benefits were extended to also cover the actual life-support functions within the ecosystem as well as the processes within it according to Daily (1997a, 1997b). Millennium Ecosystem Assessment (2005) explicitly described the services derived as benefits. Costanza et al. (1997) described it as a form of linkage between functions and the provision of goods and services and from the latter to utility enjoyed by human population.

An alternative definition different from those stated earlier was offered by Boyd and Banzhaf (2007). They defined ecosystem services as the ecological segment enjoyed to improve human welfare and not the benefits they obtain from the ecosystem. They concluded that services and benefits are not the same.

It can be deduced from all these definitions that there is a little disparity between services and benefits. Services are the directly consumed, used, or enjoyed components of nature by man. A benefit is regarded as something that directly impacts on the welfare of people; i.e., something that changes the level of well-being (welfare). For instance, service such as clean water may lead to a benefit, such as drinking.

Knowledge of the identity or the characteristics of ecosystem services will make management, maintenance, and evaluation of ecosystems and their services easier and effective. Understanding the ways these characteristics are related will make classifying them easy and help decision-makers arrive at proper decisions on the usage of natural ecosystems by analyzing the ecology—society link of these ecosystem services.

16.2.1 Public—Private Good Aspect

Market goods generally have two characteristics. The terms "rival" and "excludable" are used by economists to describe these characteristics. *Rival* is used to describe the fact that a particular good is not meant for a user, so all potential users are rival to themselves. *Excludable* is the total prevention of a good from being used by other users, such as consuming it or keeping it.

These goods and services generally range from rival to nonrival and from excludable to nonexcludable (Table 16.1).

Basically, the foods provided by our ecosystems, such as tomato or fish, are goods that are regarded as both rival and excludable. These goods are sold in markets where all potential buyers are rival to themselves, and purchasing a particular part of that good prevents another buyer from using it, which makes it also excludable.

Other services are grouped under a category known as toll or club goods. Although, this type of good is nonrival, but it is excludable. An example is information gotten from nature. My usage of this information does not reduce the information you can also get, but I can actually prevent you from using it by, let's say, controlling it. For instance, everyone has the right to visit a game reserve to learn about nature, making it nonrival, but a parent can prevent his or her child from visiting, which makes it excludable. Some goods are also classified as rival and nonexcludable. For instance, in open-access fisheries, all potential users are rivals since usage of such a fishery leaves less for another user, but it is not possible for a user to prevent another user from exploring such fishery.

Finally, we have the set called public goods. These are goods available for all since they are nonrival and nonexcludable, for instance, rainfall. My use of this service does not mean less to you, and neither can I prevent you from using it.

However, these four sets of goods have other grades of goods and services between them. For example, deep-sea fisheries may be excludable at some point if some form of monitoring is put in place during management of the fisheries. For instance, during a closed season, such fisheries become excludable as people are prevented from using them. Also, some goods may be nonrival at low levels and, as its usage increases to a certain level, it becomes competitive and therefore rival. For instance, sustainable inshore fisheries at low level.

TABLE 16.1 The Nature of Market Goods and Services

	Excludable	Nonexcludable
Rival	Private or market good; ecosystem service benefit, e.g., tomatoes, seafoods	Open access; some fisheries
Nonrival	Club or toll goods; information from nature	Pure public good; UV protection, rainfall

There is an existing overlap between the ecosystem dynamics and social system, which will help us understand how goods and services fit into the public—private space. Different societies have different ways, which is affected by governance, market, on utilizing ecosystem services.

16.2.2 Spatial and Temporal Dynamism

The services provided by an ecosystem vary spatially (space) and temporally (time). Periodic change in time affects ecosystem services. For instance, provision of clean water will be reduced during hot seasons. These services are not homogeneous across landscapes but are heterogeneous as they evolve through time and space. Some services are also provided in a place at a particular time, and their benefits are reaped in another place in another time, for instance, climate regulation.

16.2.3 Joint Production

Ecosystem services are characterized with the derivation of multiple benefits as a result of joint products. For instance, stream flow gives services such as water for irrigation, water for drinking, as well as hydroelectric power. Joint production must be properly understood by decision-makers because provision of all these multiple benefits cannot be maximized. Therefore, trade-off can be an efficient management tool. Also, accounting and classification schemes might also be derived from analysis of joint production.

16.2.4 Complexity

Ecosystem services are complex systems with nonlinear relationships between them, having feedbacks, time lags, and nested phenomena (Limburg, O'Neill, & Costanza, 2002). Valuation of ecosystem services has been difficult due to this complexity. Valuing the waste absorption service from a place will be difficult with a certain level of productivity. Some ecosystems may abruptly change beyond a certain threshold after being initially insensitive to species loss, while other ecosystems may become destabilized if as low as a species is lost, for instance, sea otter decline may be caused by the collapse of the kelp system and subsequent release of sea urchin populations (Estes, Tinker, Williams, & Doak, 1998).

16.2.5 Benefit Dependence

Services provided by an ecosystem are often determined by the benefit(s) a beneficiary is interested in. Services can switch from being an intermediate one to final service (Table 16.2). Clean water provision may be a final service when we want a good water quality benefit. However, in the provision of fish, clean water provision moves to an intermediate service. The beneficiaries of a service determine the position of such service, and this will have an effect on the valuation and maintenance of such service. Therefore, for

TABLE 16.2 Interdependence between Intermediate and Final Services

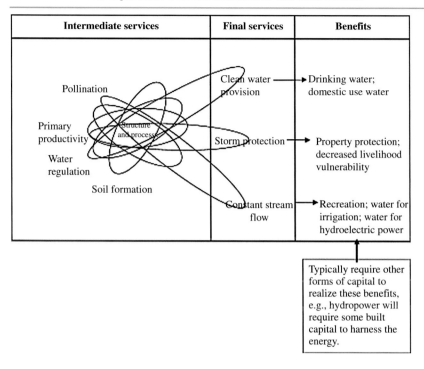

Adapted from Fisher, B., Turner, R.K., & Morling, P. (2008). Defining and classifying ecosystem services for decision making. Ecological Economics, 68, 643–653. doi:10.1016/j.ecolecon.2008.09.014; On JRC's side, we encourage the reuse of the material. It is in fact already accessible online: http://capacity4dev.ec.europa.eu/eu-au-evidence-and-policy/documents?gterm[0] = 38122. However, credit is due to colleagues who developed the case: JoAnne Bayer and Piotr Magnuszewski from IIASA, as well as Isayvani Naicker from Department of Science and Technology in South Africa.

people who depend solely on an ecosystem for fish production, the benefit from clean water provisioning may not be fully utilized, so it may not be valued. It is therefore important for decision-makers to take into consideration these intermediate services as final services that stakeholders are interested in depend on them.

16.3 CLASSIFICATION OF ECOSYSTEM SERVICES

MA classification of ecosystem services has been generally accepted and widely used. MA classified them into four classes, which are as follows:

1. Provisioning services
2. Cultural services
3. Regulating services
4. Supporting services.

II. THE EFFECTS OF CRUDE OIL EXPLORATION ON THE SOCIO-CULTURAL AND ECO-ECONOMICS OF NIGERIAN ENVIRONMENT

16.3.1 Provisioning Services

Outputs such as energy or materials gotten or provided by an ecosystem are regarded as provisioning services. This can be generally understood as provision of food and raw materials. However, provision of water and also medicinal resources are also included. Food is needed for survival, and it comes from any well-managed ecosystem—terrestrial or aquatic, fresh or marine. Aquatic ecosystems provide foods that are mainly proteinous in nature such as fish and shrimp. Freshwater provision serves as a linkage to benefits such as drinking, hydropower, and irrigation. Energy outputs are gotten from fuels such as wood. Genetic resources that lead to developments in biotechnology is another provisioning service. Others include biochemicals, food additives, ornamental resources, and so on.

16.3.2 Regulating Services

One of the functions of the natural ecosystem is the ability to maintain and control their qualities, which have effects on human survival. Examples of regulating services are air and water quality regulations. Carbon sequestration is the ability of an ecosystem to store excess carbon (blue carbon) removed from the atmosphere as carbon dioxide (CO_2)—a chief greenhouse gas. Several oceans and some coastal habitats, such as salt marshes and mangroves, have been discovered to act as potential carbon sinks as they remove a significant amount of carbon dioxide from the atmosphere. There is a potential reduction in the susceptibility of man to natural disasters such as floods, storm surges, tsunamis, and disease outbreaks due to the buffering actions of these ecosystems and the organisms inhabiting them against these events, or instance, soaking up of flood water by wetlands. Treatment of wastes by aquatic habitats is a regulating service that is now valued all over the world. These habitats serve as filters for human and animal wastes, eliminating pathogens and reducing pollution. Pollination and biological control are other important regulating services (Ndimele & Ndimele, 2013; Ndimele, Jenyo-Oni, Chukwuka, Ndimele, & Ayodele, 2015).

16.3.3 Cultural Services

Some services provided by natural ecosystems come in nonmaterial forms. People obtain certain cultural benefits, which include spiritual enrichment, aesthetic values, and also recreational activities. Lagos State, for instance, has imbibed planting of trees as a trending culture. LASPARK (Lagos State Parks and Garden Agency) is an agency in Lagos State that sensitizes the people on the need for planting trees in order to preserve their aesthetic values. Cultural services have a close relationship with human life—social, economical, and political aspects, among others. These services also include cultural diversity, spiritual and recreational values, educational values, cultural heritage, and sense of place. An example of an ecosystem in Nigeria where the cultural services are enjoyed is the Osun-Oshogbo River. The people of Osun believe that this water body has so many cultural benefits, which led to the prohibition of fishing in this particular water body. There is a festival annually to celebrate this cultural heritage. Recreation and ecotourism are benefits people derive from cultural services. Again, cultural services may differ among individuals of different groups since perception of culture is dynamic. People of different cultures derive different cultural values from ecosystems around them.

16.3.4 Supporting Services

These services are indispensable as provision of other services depend on them. An optimum provision of supporting services will mean no problem to the provision of other services. For instance, a supporting service, such as provision of a sustainable habitat for species, will have a positive effect on provisioning services such as provision of food, raw materials, ornamentals, and genetic resources. The impacts of supporting services on people's life is not an immediate or direct one, unlike other services such as provisioning, cultural, and regulating. However, some services can assume both supporting and regulating forms, depending on timescale and how long their impacts are felt. Water cycling is classified as a supporting service, since changes to this will indirectly affect humans through a clean water provisioning service. Other examples of supporting services include primary production, soil formation, production of oxygen, and nutrient cycling.

16.4 SIGNIFICANCE OF AQUATIC ECOSYSTEM SERVICES IN NIGERIA

Human societies derive many essential goods from natural ecosystems, especially aquatic ecosystems. Good such as seafoods represent an important and familiar part of the Nigerian economy. Until recently, the fact that aquatic ecosystems perform fundamental life-support services, which include water purification, regulation of climate, and production and maintenance of biodiversity, was less appreciated. All these are essential for human civilization, and their deterioration will seriously threaten human survival and sustainability.

Solar energy seems to be the driving force of a productive aquatic ecosystem. For instance, the process of photosynthesis by algae, as well as water purification, is worth many trillion dollars annually. Yet, because most of these benefits are not traded in economic markets, they carry no price tag that could alert the society to changes in their supply or deterioration of underlying ecological systems that generate them.

Based on available scientific facts, we are certain that aquatic ecosystem services are essential to human survival, aquatic ecosystem services operate on such a grand scale and in such intricate and little-explored ways that are irreplaceable by technology, and Nigerian activities such as oil spillage and persistent waste disposal, if not properly checked, will dramatically alter all aquatic natural capital in Nigeria within a few decades. The biodiversity of the Nigerian aquatic ecosystems is increasingly being destroyed and spoiled by a persistent threat of aquatic pollution resulting from intense human activities such as indiscriminate use of fertilizers and pesticides in agriculture; industrialization; mal-utilization and mismanagement of natural aquatic resources during dam, road, and bridge construction; irrigation; and the draining and filling of wetlands. The negative environmental and societal impacts of these projects are beginning to manifest and must be checked if we are to avoid the sorts of problem that they have brought in other parts of the world (NEST, 1991).

Several aquatic ecosystems in Nigeria, such as Ikpoba River (Benin city), tropical man-made lake (Moro dam) (Kwara state), Baguma Creek (River state), Lower Cross River, and Badagry Creek (Lagos State), have their biodiversities being threatened by human activities such as dredging, road and bridges construction, leaching of fertilizers from farmlands, obnoxious fishing practice, continuous cutting of mangrove, anthropogenic inputs, and so on. All these reduce the quality of services supplied by these ecosystems.

In summary, aquatic ecosystem services are important to us because of the following:

1. *Boundless benefits*: Valuing nature in a way that can speak to decision-makers may help promote conservation efforts in the future. It brings nature back into the cost—benefit discussion in a way that can be easily understood.
2. *Foundation for sustainable development*: Companies have started to use ecosystem services in conservation offset planning, where they can buy and sell credits to offset a development or set aside land to meet a specific offset.
3. *Essential for our survival*: We depend on healthy aquatic ecosystems for purified water, carbon sequestration, nutrient recycling, and most importantly provision of food so we do not go hungry.

16.4.1 Beneficiaries of Ecosystem Services in Nigeria

Ecosystem services beneficiaries benefit from ecosystem goods or services either through active or passive consumption, or through appreciation resulting from awareness of these services (Nahlik, Kentula, Fennessy, & Landers, 2012). Identification of benefits and beneficiaries from ecosystem services is paramount in order to identify enhanced ecosystem management options (Kettunen et al., 2009). The particular set of group benefiting from a particular aquatic ecosystem service will determine the extent of implementation and utilization of several management options. These benefits vary, depending on their individual characteristics, spatial scale, and distance between production area and location of beneficiaries (Fisher et al., 2008).

Generally, the benefits from aquatic ecosystem services in Nigeria are enjoyed by the following sets of people:

1. Private (large companies, small medium enterprises, small holder with hired labor)
2. Public (government agencies at various levels)
3. Household entities.

Services such as water supply, recreation, transportation path, carbon sequestration and storage, microclimate regulation, sediments and nutrients regulation, aesthetic value, other habitat-derived ecosystem services (including migration support), and nonuse value such as existence, hydroelectric generation, and flood control are supplied by Nigeria's aquatic ecosystem.

The beneficiaries also include artisanal and industrial farmers, farm workers, the local, national and international biodiversity fanatics, tourists, rural dwellers, and so on.

A case study is the hydroelectric power service provision enjoyed by most Nigerians. The hydropotential is high, as it accounts for 32% of the total installed commercial electric power capacity. These hydropotential sources are exemplified by large rivers, small rivers, and streams, and various basins being developed.

There are more than 278 unexploited sites with total potential of 134.3 MW. Availability of electricity has positively affected the lives of women, children, and the aged more than men because more of their time and work are carried out at home. The inland water mass was estimated to be approximately 12.5 million hectares of inland waters capable of producing 512,000 metric tons of fish annually (Shimang, 2005). The Osun-Oshogbo festival, which largely depends on the Osun River, is an example of cultural benefits derived from our aquatic ecosystem. This festival has been in existence for more than 400 years. The value of this festival to the Osun people goes beyond monetary benefits; it is more of preserving the cultural heritage.

The artisanal fisheries in Nigeria depend solely on the provisioning services of these aquatic ecosystems. As of 2007, the yield from artisanal fisheries was approximately 240,000 metric tons annually (FDF, 2008). The artisanal fishermen that depend on these ecosystems for livelihood and income is approximately 1.6 million, which is a significant percentage of Nigeria's population.

16.4.2 Distribution of Ecosystem Services in Nigeria

Nigeria is a West African maritime state with a coastline that is approximately 853 km long, stretching from the western border with the Benin Republic to the eastern border with the Cameroun Republic. The Nigerian coastal zone can be defined as the area that extends from the shelf break, inland to the limit of tidal influence. This coastline is interrupted by a series of estuaries, which open into an extensive lagoon system in Lagos and Ondo States (Fig. 16.1). In Lagos State, the creeks, floodplains, lagoons, and rivers account for approximately 22% of the 790 km^2 land mass. There are at least 22 estuaries from Benin River in Delta State. The Nigerian coastal zone is generally low lying, resulting in extensive wetlands and mangrove swamps. Nigeria has the largest area of mangrove forest in Africa (Ajao, Oyewo, & Unyimadu, 1996). The Nigerian land mass is enclosed between latitudes 4° 16′–13° 52′ N and longitudes 2° 49′–14° 32′ E and being 1100 km on a North/West axis. The finite natural resource base enjoyed by 158 millions humans and an unknown number of plants and other animals is estimated at 923,700 km^2 of land.

The Ogun River, which rises in Oyo State near Shaki, flows through Ogun into Lagos State. It has a reservoir capacity of 690 million cubic meters (Berga, 2006). It provides recreational facilities to tourists. Oyan River dam, which crosses Oyan River, a tributary to the Ogun River, is used for bathing, washing, and drinking and also as a drain for most organic wastes from abattoirs located along the river's course (Ayoade, Sowunmi, & Nwachukwu, 2004). The four main lagoons in Lagos State are the Lagos, Epe, Ologe, and Lekki lagoons (Ndimele & Jimoh, 2011; Ndimele & Kumolu-Johnson, 2011). These lagoons are used for fishing as well as transportation. The size of the Lekki Lagoon is being reduced as a result of filling of the lagoon for land use. There are also many sand-mining activities in these lagoons, which has affected the biodiversity, ecology, and aesthetic value

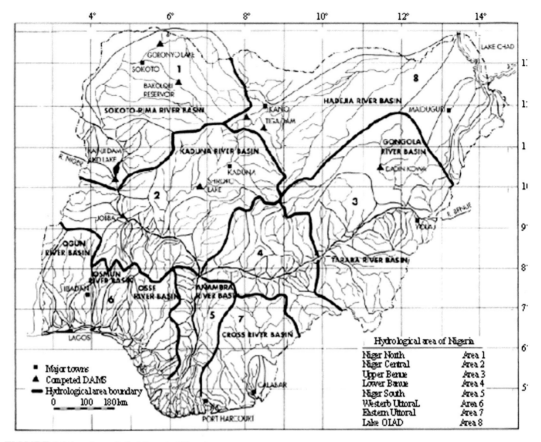

FIGURE 16.1   Aquatic habitats in Nigeria.

of these aquatic ecosystems. The Bonny Estuary is an estuary on the coast of Rivers State dominated with mangrove swamp, a good site for carbon sequestration.

The basin of the Niger River is partially regulated through dams. The Kainji Dam and the Jebba Dam are used mainly to generate hydropower. The aquatic ecosystem services are under pressure due to increased water abstraction for irrigation and due to the impacts of climate change. The FAO estimated the irrigation potential as 1.68 Mha of the rivers in Nigeria. Dredging activities also take place, which is intended for easier transportation of goods.

Lake Chad, which has a primary flow from the Chad River, is a lake that has varied in size over the centuries. The surface area was 1350 km^2 as of 2005 (Odada, Oyebande, & Oguntola, 2005). Its ecosystem service is mainly provision of purified water for drinking, to more than 68 million people living not only in Nigeria but also in Chad, Cameroun, and the Niger Republic (Hassan, 2012). Oguta Lake in Imo State is the largest natural lake in Imo State (Nfor & Akaegbobi, 2012). The lake provides services such as fish, water, tourism, and an outlet for sewage.

The Sokoto River, another tributary of the Niger River, has its plains around the river; these plains are widely cultivated and the river used as a source of irrigation and also a means of transport (Akané & Jürgen, 2005). Another tributary of the Niger River is the Benue River, which is also an important transportation route. The Taraba River, which is a tributary of the Benue River, is used for farming of rice, yam, and groundnut, as well as for fishing. The Great Kwa River flows through Cross River State. Small-scale farming, aquaculture, and artisanal fisheries, mainly shrimp, are the human activities that depend on this river.

The Forcados River is an important channel in the Niger Delta for small ship transportation. People have also been fishing on this river for years for sale and personal consumption. The Kaduna River, which is another tributary of the Niger River, provides a habitat for crocodiles that live in the river. Kaduna in the native Hausa dialect is the word for crocodile. The Ikogosi warm spring is a tourist attraction located at Ikogosi in Ekiti State.

16.5 ECOSYSTEM CHARACTERISTICS OF NIGERIAN WATERS

In analyzing the ecosystem characteristics of Nigeria water, a case of Lagos lagoon is described to give a semblance of what ecosystem characteristics of Nigeria waters looks like.

The Lagos Lagoon is the largest lagoon along the West African coastline. The lagoon, due to its one estuary mouth, is best described as an incomplete delta. This estuary is located at Onikan and near Bar Beach, and it is known as the Commodore Channel, which releases its Lagoon content into the Atlantic Ocean through three sites—Badagry Creek, Lekki Creek, and Ologe Lagoon (Ogunbambi, 2009). The lagoon's geological features are also similar to those that govern the existence of estuarine ecosystems all over the world (Ogunbambi, 2009). The diameter and size in area of this lagoon are 285 km and 6354 km², respectively. The lagoon supplies both regulating and supporting services. This lagoon's lower water temperature is high and fairly constant at all times in a year (Ogunbambi, 2009).

However, a gradual temperature decrease is noticed during the rainy period (from May to November) when the water experiences great influx of river water and heavy cloud cover. The salinity gradient of this lagoon is a dynamic one both seasonally and semi-diannually due to the effect of the Atlantic Ocean. The lagoon is located in the southwest of the country and under the GOGLME (Gulf of Guinea Large Marine Ecosystem)—a project developed to monitor coastal waters (Okusipe, 2008). Additionally, the lagoon is a cultural landscape for indigenous people (the Ilajes) from Ondo State who live in the riverine ecosystem.

Lagos, through the lagoon, is richly blessed with both coastal and inland aquatic resources, which have ecological, economic, sociocultural, scientific, and recreational significance. These are the functions to be considered during sustainable development.

16.5.1 Characteristics of the Lagoon

16.5.1.1 *The Land and River Factor*

Lagos is a region that is close to the Atlantic Ocean and that has a flat topography. Its lagoons and creeks are channels through which water and sediments from rivers are discharged into the Atlantic Ocean. The inland shoreline arrangements as well as the behavior of its coastline are both influenced by the high sediment loads discharged from these rivers (Ogunbambi, 2009).

16.5.1.2 The Atlantic Ocean Factor

The amount of sediment retained in the estuary depends on oceanic wave, tide, and currents as well as the wind. These features also affect inland regression (Ogunbambi, 2009). The ocean will also affect the salinity gradient of this estuary as well as the species diversity.

16.5.1.3 The Biotic Factor

These are the living components—the flora and fauna—of the lagoon. These organisms have an effect on the survival of Lagos foreshores and wetlands (Ogunbambi, 2009).

16.6 TRADE-OFF IN NIGERIAN WATERS

16.6.1 Conceptual Framework

Water resource management over the decade has received both scientific and social concern. Even when credible water supply is not guaranteed globally, human activities that have negative impacts on this low water resource supply have been on the increase. This has also resulted in the frequent occurrence of natural disasters such as flood and drought. The effects of all these are increasing scarcity of water, food insecurity, and outbreak of water-borne diseases.

The Nigerian waters seem to be vulnerable to these impacts, hence the need for national attention over water-related risks. As a response to these challenges, there is a need to address the issues in a coordinated approach, putting into consideration associated trade-offs involved in water management. However, is an attempt to bundle key Ecosystem Services provided by Nigeria territorial waters, services offered for the benefit of the nation include those identified by Millennium Ecosystem Services (MA, 2005):

* Provision of drinking water in the country
* Linking water management with food security in country through increased fisheries and other seafoods
* Linking water management and habitat protection with biodiversity protection
* Linking water management with carbon sequestration and climate change.

The main reason why the concept of ecosystem services was introduced was to develop solutions to problems relating to human survival, keeping an eye on sustainability (Nelson et al., 2009). People's survival has always depended on the supply of these ecosystem services, whether intermediate or final (Constanza, 2008). In managing water-related ecosystem services, the case is not different.

Spatial and temporal trade-offs are the two basic types of ecosystem service trade-offs. When the provision of one ecosystem service in a particular place results in a decrease in another service at that place or its surrounding area, spatial trade-off is said to have occurred. Quality of water available for other use is affected when production of services that supply food is enhanced. For instance, excess use of fertilizer will lead to degraded water quality, reduction in biodiversity, and also release of greenhouse gases. Spatial trade-off also covers a scenario where a particular set of people residing in an area benefit from services generated in another distant area. For instance, agricultural produce from a rural area benefits those living in urban areas.

Temporal trade-off, on the other hand, is controlled by immediate societal needs. It depicts a situation where the future production of a service is affected due to excessive exploitation of such service or another dependent service at the present.

Trade-off is at different levels, and it is expected from different stakeholders in relation to water use, which can enhance their overall well-being. And since, in most cases, the multiple ecosystem services provided by water cannot be bundled together at the same time, trade-off often exists by sacrificing one service for the other. This is why trade-off is needed with regard to water governance at the country level. The concept of trade-off is succinctly demonstrated in the example/case study presented in Section 16.9.

16.7 LAWS, REGULATIONS, AND POLICIES ON TRADE-OFF

The Federal Ministry of Water Resources and the Federal Ministry of Agriculture are the two organs of government responsible for the formation of policy and coordination of issues that affect development and management of water resources in the country.

Due to the overdependence on these water resources and the technicality in their management, there are other government institutions that assist the two apex organizations. The Federal Ministry of Environment, the ministries at state levels in charge of these resources, and the Departments of Agriculture at various local governments are bodies that are created to manage water resources in the country. This analysis shows comanagement of water resources by the central government, the states, and Local Government Authorities (LGA) that are close to the water basin, playing important roles with regard to water management. However, there has to be a clear direction with policy directives from the central government to other tiers of governments at the state and local government levels. This is possible if water issue is addressed from the Ministry of Water Resources, as the constitution of the country gives clear direction on how this should be coordinated. The constitution gives the federal government prerogative for managing inland and transboundary waters. This will be difficult, for example, if management of water is coordinated from the Ministry of Environment because the constitution of the country places environmental issues under a concurrent list. By this, the state government has a right to evolve their own policies.

The governmental institutions, namely the Federal Ministry of Agriculture, Water Resources, and the Federal Ministry of Environment through the Coastal Zone Department are saddled with utilization of water resources in the country in a sustainable way. Other institutions are concerned only with meeting their water requirements, without considering how their activities affect the ecosystem. This also seems to be a good division of tasks and responsibilities with regard to water management.

Another means of shaping water management in the country is the use of legal instruments. The "Water Use Decree Number 101 of August 1993" and the "Environmental Impact Assessment Decree Number 86 of 1992" are the two legal frameworks that if rightly enforced, can control the coordination of water resource management. Both decrees are already in use, but there are modalities affecting their effective implementation.

The execution of these water strategies has been characterized with weak coordination. These policies have also suffered harmonization; for instance, allocation of land for agriculture and establishment of hydro-agricultural schemes have suffered from improper or lack of strategy (UNDP, 2006).

There is no proper functioning mechanism for integrated water resource management despite the functional catchment management plans. Again, national sectorial policies that affect the management of land and water resources are often uncoordinated or not stream-lined. For instance, the National Agricultural Policy reduces dependence on rain-fed farming through creation of dams and large-scale irrigation schemes without proper EIA (Environmental Impact Assessment). The increased water demand places environmental sustainability at risk (UNDP, 2006).

However, for effective and sound management of water resources in the country, Nigeria needs to also domesticate the United Nations Conventions on the Law of the Sea (UNCLOS) and the International Maritime Organization (IMO) Instruments for which Nigeria is a party, which are equally substantial policy instruments for the needs of the country, especially for transboundary river basins to which the country is a member.

16.8 HOW AQUATIC ECOSYSTEM SERVICE TRADE-OFF ARE GOVERNED IN NIGERIA

Several efforts have been made universally to maximize the benefits and profits derived from natural ecosystems with regard to sustainability and conservation. These efforts are frequently reviewed, which prompts the development of new concepts that can better shift our natural ecosystems toward more sustainable and conservable ones.

Trade-off analysis of ecosystem services is an evolving concept that has been applied to the management of our natural ecosystems since choices have to be made on the qualities these ecosystems deliver. Trade-off simply means finding a compromise between two incompatible qualities. Aquatic ecosystem services, which include hydropower generation, irrigation, provision of clean drinking water, fisheries, and so on, are services supplied by aquatic ecosystems in Nigeria. However, the provisioning of these services varies spatially (space) and temporally (time) (Rodríguez et al., 2006), as all these services cannot be constantly supplied in a particular habitat at every point in time and space. These services may have a nonlinear relationship but are dependent on one another at some point, and being ignorant about them leads to making aimless and unintentional trade-offs among them (Table 16.5) (Rodriguez et al., 2006). Currently, trade-off analysis of Nigerian waters is characterized with two words: aimless and unintentional.

The type and magnitude of services provided by aquatic ecosystems can be altered as a result of management choices made by humans during trade-off. The Osun-Oshogbo River is an aquatic ecosystem known for its cultural services. Provisioning service, such as supply of food, is impossible, as fishing activity is banned. Trade-off has simply occurred here as the people have given up these provisioning services to protect their cultural belief. Although, this trade-off is controlled by traditional and mythical axioms rather than by optimizing the benefits derived from this ecosystem.

Water resource development across the world is characterized by the inability to maximize the benefits of all aquatic ecosystem services supplied. One has to be reduced for the other to be maximized. Generally, the agricultural sector in Nigeria takes the greatest share of water abstraction, which is approximately 5510 million m^3 (44% of the total water withdrawn) for livestock and aquaculture (FAO, 2016). This statistic depicts that aquatic ecosystem services such as provision of clean drinking water, hydropower, and water transport are under pressure. It can also be deduced that hydropower generation, which is one of the major sources of electricity production, can be improved if proper trade-off is done where there will be a balance between services that depend on water abstraction and those that do not. There is a looming gap between the supply of clean drinking water and the rise in human population, especially in the northern part of the country. Maximization of this aquatic ecosystem service requires proper trade-off analysis.

It is apparent that our water policies in Nigeria that develop different schemes used to exploit and manage aquatic resources are unidirectional, as they only try to improve a particular aquatic ecosystem service mainly needed by people at a particular time without putting into consideration how this development will affect the provision of other services from this source, which are also paramount to people's survival presently and also in the future. For instance, the irrigation scheme will not consider how clean water provisioning is affected (Fig. 16.2).

We dwell more on the enhancement of provisioning services in order to favor poverty reduction. Also, constant supply of provisioning services over the long term leads to ecosystem degradation. For instance, constant fisheries services may lead to overfishing, depletion of stock, and extinction of species. Nevertheless, as a result of the dynamics and relationship between provisioning, cultural, regulating, and supporting services, the reduction in poverty will go on until feedback loops from ecosystem degradation will cascade back through these services, obviously reducing overall well-being and thus, the poor suffer again.

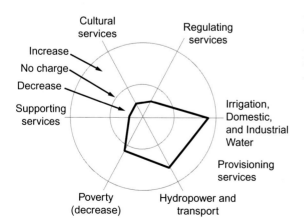

FIGURE 16.2 The current trend in trade-off between ecosystem services. *Adapted from Millennium Ecosystem Assessment (MA) (2005).* Ecosystems and human well-being: Synthesis. *Washington, DC: Island Press.*

II. THE EFFECTS OF CRUDE OIL EXPLORATION ON THE SOCIO-CULTURAL AND ECO-ECONOMICS OF NIGERIAN ENVIRONMENT

Trade-off depicts two outcomes, which are (1) a situation in which there is a decrease in the quality or quantity of a service utilized by a stakeholder owing to the utilization of such service or other ecosystem services and (2) a condition whereby there is a reduction in the well-being of a user or stakeholder as a result of another stakeholder utilizing the ecosystem service.

Again, most trade-offs that arise due to management options happen between the financial gains of the private sector and the social loss experienced by the public sector (Zhang, Ricketts, Kremen, Carney, & Swinton, 2007). The Hadeija-Nguru wetland is currently unable to supply the regulating (climate change) and supporting (species endangered) services. This is because these services were ignorantly traded-off long ago when provisioning services such as provision of water for agricultural purposes and drinking were highly rated in order to meet people's immediate need, and this resulted in the susceptibility of other services to decrease.

In some cases, there has also been trade-off between the present provision of a service and its provision in the future. In this case, sustainability is threatened. Lake Chad, which is the main source of water for both agriculture and people in the Chad region, has undergone a drastic decrease from 25,000 km^2 in 1960 to 4800 km^2 in 2014. Also, between 2005 and 2016, the population has moved from 17 million to 38 million (Oyedele, 2017). When those provisioning services were supplied, they came at a cost, which was water abstraction, and it is definitely obvious that over the long term such is expected. Balance between provisioning service, supporting service, and regulating service must be ensured, otherwise the unintentional future supply of ecosystem services would be traded-off (Fig. 16.3).

The Cross River basin has services that are underexploited. Less than 5% of the water is presently used for drinking, irrigation, industry, commerce, farming, and fishing. Proper trade-off analysis can help maximize the provision of one or two services, whereas other services' supply will be maintained at a level where sustainability is ensured, as all services cannot be supplied at climax. Trade-off analysis helps decision-makers to come to a conclusion on where trade-offs among services are inevitable and where win—win is possible.

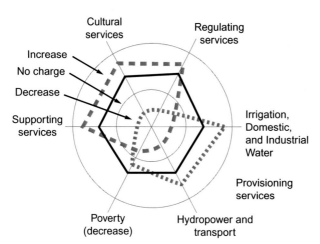

FIGURE 16.3 How trade-off can be balanced between ecosystem services. *Adapted from Millennium Ecosystem Assessment (MA) (2005). Ecosystems and human well-being: Synthesis. Washington, DC: Island Press.*

16.8.1 The Trade-off Analysis Process

For efficient trade-off analysis between two (or more) ecosystem services of interest, valuation may not be necessary to formally evaluate trade-offs among management options (Kristin, White, Gaines, Costello, & Anderson, 2013). Different forms of data can be used, such as empirical, quantitative, or conceptual models for this analysis. The society's influence on the choice of how natural resources are allocated makes the feature of ecosystem service trade-off analysis important whether the economic value is known or not (Ruhl et al., 2007).

Fig. 16.4 shows trade-off between two provisioning services (irrigation and drinking). The different points on the curve represent the level of supply of both services at a particular fixed budget, legal framework, and regulatory policy. However, for simplicity purpose, let's assume that the cost of implementation is the same and decision-makers are only concerned about the two ecosystem services. The curve shape represents how many units of ES1 must be sacrificed in order to gain one additional unit of ES2 and vice versa.

The outer boundary represents the efficient frontier, and any chosen point will be a good trade-off option. The efficiency frontier is the outcome where improving any particular objective without decreasing performance of another attribute is possible. Accuracy of the plotted points determines the usefulness of the trade-off analysis as a policy-making tool. Moving from point A to point B increases ES1, and ES2 decreases on a fair scale. Points A, B, and D are efficient points. Points that fall inside the curve, such as C and other internal points, are inefficient. It would never be best to choose a point such as C, as efficient management options can still move point C vertically to B or horizontally to D. At B, more of ES2 while ES1 constant and at D, more of ES1 and ES2 is constant.

In addition to the fact that trade-off analysis determines inefficient outputs, trade-off analysis can also determine how a policy change can cause a change in the efficiency frontier, increasing it outward or decreasing it inward. Thus, while ecosystem service trade-off has great potential to help inform natural resource decision-makers on the best output an ecosystem can deliver, models must be carefully designed for their intended purpose.

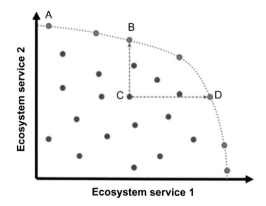

FIGURE 16.4 Trade-off analysis between two ecosystem services supplied by a particular water body (irrigation and clean water provisioning). All points on the graph represent the possible outcomes of combining the two ecosystem services. Points inside the curve such as C are inefficient. The efficient frontier, which contains points A, B, and D, is the outer curve showing the most efficient use of the two services at a given fixed budget and policy. *Adapted from Kristin, C., White, C., Gaines, S.D., Costello, C., & Anderson, S. (2013). Ecosystem service trade-off analysis: Quantifying the cost of a legal regime.* Originally published in the Arizona Journal of Environmental Law and Policy, Vol. 4, No. 1 (Fall 2013).

16.9 A CASE STUDY OF TRADE-OFF ANALYSIS INVOLVING THE LOWER NIGER RIVER IN NIGERIA AND A PART OF THE NIGER RIVER BASIN

Nigeria waters have many inland water basins and also some are transboundary in nature. Management of these basins posits some sustainability issues. Political stability is important in achieving high economic growth, high agriculture productivity, and reduction of poverty due to livelihood linkage as well as reducing the risk of drought and flood for human security.

In the case of the lower Niger River, farmers and residents of the river basin are facing high and increasing risks from drought and floods. The main aim of the Basin's water policy is to strengthen the role of the water resources sector in sustaining economic growth and reducing poverty.

Of particular importance for the Niger River Basin is to expand irrigation to increase yields from crops, especially sugarcane as the major crop in the region, and this is essential for jobs and development. Food security is also threatened by floods, resulting in many fatalities and huge damage to crops, implying that available public resources should ensure the greatest protection for droughts and floods as much as possible.

Hydrologists are of the view that the greatest protection would result from completing the expansion of the Kanji Dam, which would expand the reservoir and nearly double the area of irrigated agriculture, mostly sugarcane, which will be exported for sugar and ethanol production. The completed dam expansion will also regulate river levels and flooding downstream by temporarily storing the flood volume and releasing it later, through a government-supported integrated water management plan. Besides drought and flood control, the expanded Kanji Dam will provide water to the water-scarce city in the neighborhood of the river.

Water efficiency measures, and even water pricing, have been suggested to reduce water demand for agriculture and urban centers, but it is the job of the government to make sure every farmer and city dweller has sufficient water. Market may not be needed for this. However, it is important, of course, to ensure that the dam and resulting reservoir meet all requirements for ecosystem protection and that those persons who must be relocated are appropriately compensated.

Even with the heightened dam, there will be residual risks, especially extreme droughts and floods. Natural flood measures for increasing water retention, such as wetlands and forests, are far too limited in their scope to be of much use for drought and flood control. The government should provide access to long-term forecasts for droughts, and an early warning system, combined with emergency plans and rescue operations for floods. Public investments in education can also improve resilience against floods and droughts.

With regard to security and economic growth, the preferred options are:

- The completion of Kainji Dam expansion + + o one-stage investment (not incremental) o Cost: 90 budget units.
- Expand irrigation system + + o two-stage investment.

- Note, second stage of irrigation system will require the completion of Kanji Dam expansion o Cost: 30 budget units/stage.
- Long-term weather forecast combined with early warning system + o one-stage investment o Cost: 80 budget units.
- All farmers can use long-term information to reduce flood and drought losses.
- Education: Entrepreneurship and flood resilience + o three-stage investment o Cost: 10 budget units/stage o Addressing new income and subsistence options as well as flood risk reduction.

Further instructions:

The aim here is to reduce flood/drought risk and improve food production. However, we are trying to fund solutions with "+ +,"which are significantly more effective for achieving our goals than solution with only " + ."

The Niger basin has vast potential to eventually become a major food producer in the region. Only 16% of land suitable for farming is currently cultivated. Improving agricultural productivity and ensuring access to food are now top priorities. Agriculture contributes more than a quarter of the basin's GDP (gross domestic product) and employs 80% of its labor force.

The Kanji Dam was built to harness the heavy precipitation in the rainy seasons and provide protection from floods and drought. Flow volumes of the river are highly variable, with some years characterized by very low flows, while other years display runoff volumes equivalent to six times the mean value.

If shortages are to be avoided in the future, then additional sources of water are immediately required to serve the rapidly expanding new areas and meet a rapidly increasing demand that is predicted to increase fourfold. In the dry season, the reservoir dries out, creating a gap between supply and demand.

Further expansion of the Kanji Dam has been identified as the only option capable of providing sufficient water to augment supply before the predicted shortage will severely limit economic growth. Even in the dry season, the river water flow is sustained with the water stored during the rainy season. The expanded reservoir will enable the doubling of irrigated sugarcane acreage. It will also help solve the water deficit in Minna (the state capital of Niger State) by increasing water supply to the city by 25%. Finally, the dam will protect against floods. In 2015, a major flood in the Lower Niger basin resulted in 71 deaths, including 15 schoolchildren. Over the past 20 years, more than 500 people have lost their lives in floods. The estimated reduction of downstream flow by 5%, resulting from the new construction, is not significant enough to affect biota negatively downstream. Although family graves will be inundated, and persons will be relocated, this will be dealt with under the Resettlement Action Plan.

16.9.1 Careful Stewardship of the Lower Niger River

The Lower Niger River is the lifeline of the inhabitants of Minna in Niger State. Not only does the river support a rich biodiversity, but by depositing sediment and nutrients, it is essential for agriculture along the river's entire reach. The freshwater reaching the estuary is vital for maintaining healthy mangroves that support fisheries, which are a major source of livelihoods for the local communities. High water abstractions driven by water-intensive agriculture, especially sugarcane, are jeopardizing these ecosystem functions. Overconsumption of water is further worsened by plans to divert the river's water to the major urban center.

Completing the Kanji Dam expansion is not necessarily the answer; rather, it is the problem. By reducing the sediment deposited downstream and eventually into the delta, the fish habitat is endangered and fisheries decline; the mono-cropping of sugarcane has led to eroded soils, and the application of fertilizers could result in serious water pollution. More worrisome is the fact that the profits from sugarcane accrue mainly to the large commercial farmers. In addition, the greatly expanded reservoir will force whole villages to relocate with huge social and economic cost—the government cannot be relied upon to adequately compensate these misplaced persons. Finally, sacred burial grounds will be inundated.

What is needed instead of expanding the Kanji Reservoir is more holistic and sustainable development of the basin. The region has an unacceptably high number of stunted children under 5 years of age. Instead of sugarcane, it is important to foster investments in oilseed, cashew, and fruit value chains because of the nutritional benefits they offer to the poor, as well as their resilience to flood events. Small farmers planting less water-intensive crops should be supported with smart subsidies for seeds and other inputs, and with long-range weather forecasts.

Floods are increasing in their intensity, mainly as a result of deforestation and destruction of wetlands, as well as climate change, which is leading to increased precipitation in this basin. The small farmers along the fertile Lower Niger Basin have learned to live with floods by building temporary and transportable homes that can be moved to higher ground in periods of heavy rains. This adaptive strategy could be supported with early warning systems. Moreover, investments in forestation and wetland restoration, by retaining moisture in the soil, will go a long way in reducing both drought and flood risk.

16.9.2 Preferred Options

- Invest in wetland restoration (+ +), one-stage investment.
- In order to have significant effect, wetlands should be combined with forestation. Cost: 50 budget units.
- Invest in forestation (stage 1 and stage 2) + +. This is a two-stage investment. In order to get significant effect, it has to be done together with the wetland restoration. Cost: 30 budget units/stage.
- Subsidize small-scale agriculture (+) one-stage investment. Cost: 10 budget units/stage.
- Flood warning system and improved systems for communicating long-term weather forecasts (+) one-stage investment o Cost: 80 budget units.
- Education: entrepreneurship and flood resilience (+) three-stage investment. Cost: 10 budget units/stage.

The aim is to reduce flood/drought risk and improve food production.

II. THE EFFECTS OF CRUDE OIL EXPLORATION ON THE SOCIO-CULTURAL AND ECO-ECONOMICS OF NIGERIAN ENVIRONMENT

Additional goals may include health, ecosystems, and education, which are consistent with the narratives given earlier.

We are trying to fund solutions with "++." They are significantly more effective for achieving our goals than solution with only one "+."

16.9.3 Research-Supported Facts

The following critique of the Kanji Dam expansion project is based on an expert report from an international NGO, which champions the global struggle to protect rivers and the rights of communities that depend on them. According to this report:

* The Kanji Dam project will benefit mainly the large sugarcane interests, partly owned by foreign companies. Indeed, the analysis reported that the increased irrigated acreage belongs to only one person.
* While the annual flow downstream of the dam has been estimated at 5%, this is an average, and during dry periods, the flow can be so low that it negatively impacts agriculture, fisheries, and biota downstream.
* The dam will reduce flooding downstream for the 1 in 10- to 1 in 50-year floods; however, the dam can be overtopped for the more extreme floods occurring less frequently than 1 in 50 years. The dam thus creates a false sense of security.
* Estimates show that the increase of the size of the reservoir will lead to an increased risk (+20%) of intestinal and urinary bilharzia and malaria as observed in many other reservoirs on the continent.
* In addition, family graves are considered sacred places, and 100 will be inundated. Moreover, at least 1000 households will need to be relocated. Although the government claims that this will be dealt with under the Resettlement Action Plan, past experiences show that persons will be relocated to desolate areas with little outlook for continuing their livelihoods.

16.9.4 Alternatives to the Dam

To conserve the available water, the government should consider funding water efficiency investments. A study by the African Association of Water Engineers shows that by reducing conveyance losses in irrigation systems, it is possible to reduce the annual water use rate of the irrigated areas by 25%. Moreover, Minna (the state capital of Niger State) is facing serious water shortages, but this need not be solved by the ecologically questionable option of diverting water from another river basin since this same study showed that by preventing water leakages, urban water usage could be reduced by 50%.

Finally, according to a recent report by the African Development Bank (ADB), poverty and food insecurity are the main underlying causes of chronic undernutrition, which currently affects 44% of children under 5 years old. Food availability is limited by low yields and inadequate access to markets for many citizens. In the poorest areas of the river basin, more than 50% of children under 5 years of age suffer from stunted growth. The ADB recommends that agriculture should focus strongly on vitamin-intensive and water-saving crops.

16.9.5 Rational Choice

Households and farmers in the Lower Niger River Basin are at risk of floods and droughts; yet, these are not the only concerns of the people, and probably not the main ones. Diseases, including HIV/AIDS, pneumonia, malaria, and diarrhea, are the highest causes of death in the basin, and floods and droughts contribute to malnutrition and disease. It is very important to allocate scarce public resources, taking into account all the basin priorities, i.e., the costs and benefits based on empirical risk analyses.

The costs and benefits of investments in drought and flood protection should thus determine how the public invests, whether in structural measures (e.g., heightening the Kanji Dam) or more ecological measures (e.g., forestation, support for small farmers). Above all, the government should consider encouraging investments in health and education, which, by building human capital, enable the population to not only increase their resilience to flood and drought risk, but also to reduce their higher risks from disease. Education not only equalizes opportunities, but helps pull the poor out of rural poverty.

The problem with subsidized agricultural schemes is that they encourage farmers to produce crops that may be economically unsustainable. It is important that prices reflect true costs. For instance, offering free water to the large sugarcane producers results in more sugarcane production, crowding out crops that use less water. Moreover, free water will result in its overuse and waste. This is not a socially optimal use of water resources, nor is it equitable if the publicly financed dam and reservoir are disproportionally benefitting the large growers of sugarcane. Water pricing would enable farmers to make their own choices with regard to their crop choices, irrigation, and even seeking alternative livelihoods. Of course, it is important that water is affordable to the poor through a needs-based pricing strategy or income supports.

16.9.6 Preferred Options

- Consideration of the costs and benefits of investments in the Kanji Dam and irrigation scheme, forestation, and wetlands
- Support for water pricing through public investment in monitoring systems (+ +), one-stage investment
- Support for education (entrepreneurship and flood resilience) to empower individuals to choose alternative livelihoods and reduce risk (+) three-stage investment

The aim is to reduce flood/drought risk and improve food production.

Additional goals such as health, ecosystems, and education may be included. These should be consistent with the narrative given earlier.

Now we are trying to fund solutions with ("+ +"). They are significantly more effective for achieving our goals than solution with only one (" + ").

Table 16.3 provides rational choices in trade-off analysis of Lower Niger River in Nigeria while Table 16.4 provides the cost effectiveness of risk reduction.

TABLE 16.3 Rational Choices in Trade-off Analysis of Lower Niger River in Nigeria

Solution	Preference	Cost	Stages	Flood Risk Reduction (/Stage)	Source or Loss of Income (or Subsistence)	Dependencies	Other Benefits	Other Costs or Barriers
Unit	++, +, −, −−	$	1–3	0–10	++, +, −, −−			
1 Completion of the upstream dam	++	90	1	10	++ Income for the sugarcane farmers; − Reduced income for small farmers − Reduced income from fisheries downstream		Water supply to Kanji; total cost much higher but supported by the World Bank subsidy	Reduction of downstream flow because of evaporation; social and psychological costs of displaced persons
2 Irrigation scheme for sugarcane		30	2	0	++ Income for sugarcane farmers	Only one stage can work without the dam completion		
3 Wetland restoration		50	1	3	+ Income from fisheries	In order to get significant effect, it has to be done together with forestation (at least one stage)	Contribution to biodiversity (including migration corridors)	
4 Forestation		30	2	1	+ Subsistence (food, firewood)	In order to get significant effect, it has to be done together with the wetland restoration	Sequestration of CO_2; erosion reduction; contribution to biodiversity (including pollination); increased yield from remaining crops	Reduced crop acreage

(Continued)

TABLE 16.3 (Continued)

| Solution | Preference | Cost | Stages | Flood Risk Reduction (/Stage) | Source or Loss of Income (or Subsistence) | Dependencies | Other Benefits | Other Costs or Barriers |
Unit	++, +, -, --	$	1–3	0–10	++, +, -, --			
5 Support for small farmers for high nutrition crops	+	10	3	0	+ Subsistence food			
6 Long-term forecast combined with early warning system	+	80	1	5	+ All farmers can use long-term information to reduce flood and drought losses			
7 Education: entrepreneurship and flood resilience	+	20	3	1	+ Supporting new income and subsistence options		Evidence shows that educated households have fewer deaths from drought and flood	
8 Water pricing	++	80	1	0	- Reduces farmers' income		Increases water use efficiency; domestic water use for sanitation remains free	

On JRC's side, we encourage the reuse of the material. It is in fact already accessible online: http://capacity4dev.ec.europa.eu/eu-au-evidence-and-policy/documents?gterm[0] = 38122. However, credit is due to colleagues who developed the case: JoAnne Bayer and Piotr Magnuszewski from IIASA, as well as Isayvani Naicker from Department of Science and Technology in South Africa.

TABLE 16.4 Cost Effectiveness of Risk Reduction (Cost in Budget Units of Reducing the Risk-One Level; Lower Values Mean Better Efficiency)

Completion of the upstream dam	9
Wetland restoration	17
Forestation	30
Long-term forecast combined with early warning system	16
Education: entrepreneurship and flood resilience	20

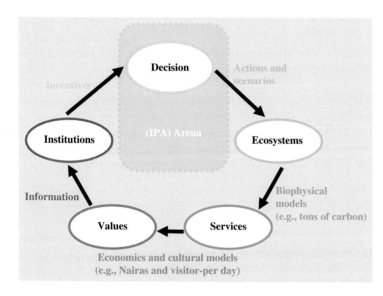

FIGURE 16.5 A framework showing how the IPA decision-making model can be infused into ecosystem services. One could link any two ovals, in any direction; the simplest version is presented here. Institutions feed into the IPA, and the outcomes are on the ecosystem (Daily & Matson, 2008).

16.10 MANAGEMENT OPTION AND DECISION

The need to embrace trade-off with regard to water management in developing countries with a case study of Nigeria becomes imperative given the fact that we are living in a world of scarcity. However, we can minimize risk associated with trade-off by applying some management measures. In achieving the goal of a well-managed water-related ecosystem services trade-off, we have come up with the Integrated Planning Approach (IPA) (Fig. 16.5), a decision-making framework with cross-cutting strategies and cross-scale approaches to managing water ecosystem services trade-offs.

The concept of IPA boils down on the need among others to:

1. Protect the water ecosystem holistically; an important problem limiting the development of water ecosystem in most countries is that agencies handling the different aspects of ecosystems such as wildlife, fish, forest, and agriculture are separate and most especially their decisions are not harmonized. Different sectors such as

energy, transportation, and trade affect all the mentioned agencies. There is need for integrated decision-making among these agencies to develop national sustainable development strategies.

2. Achieve a transparent and accountable government and private sector; when all stakeholders are involved and well enlightened during decision-making, an overall increase in ecosystem welfare is expected.

3. Incorporate ecosystem protection in decision-making; policies on ecosystem managements should be extended to cover trade aspect. Also, people in this sector should be relevant during decision-making because they will also be affected if the ecosystem is not in good shape. For effective management of aquatic ecosystem services trade-off, a level playing field must be established for the different players involved, especially now that we are faced with the negative impacts of climate change such as sea level rise, drought, flood, and so on.

If rightly enforced and practiced, IPA could provide adequate information on the different winners and losers during ecosystem usage and how decision-making considers both parties.

In analyzing the trade-off of ecosystem services of Nigerian waters, there is the need to further understand the linkages between land use conversion, decline, or extinction of ecosystems services in water and the effects on the people that depend on these for survival. Future study may encompass analysis of the water resource users to ascertain the number of people dependent on the water ecosystem services, the resources obtained, and its value.

Acknowledgment

The authors acknowledge the European Commission Joint Research Centre (EC-JRC) for using the narratives developed by JoAnne Bayer and Piotr Magnuszewski from IIASA, as well as Isayvani Naicker from the Department of Science and Technology in South Africa for the capacity building in water—food—energy nexus.

References

Ajao, E.A., Oyewo, E.O., & Unyimadu, J.P. (1996). A review of the pollution of coastal waters in Nigeria. *Nigerian Institute for Oceanography and Marine Research (NIOMR) Technical Paper No. 107 of March, 1996* (pp 20.).

Akané, H., & Jürgen, S. (2005). *Bakolori Dam and Bakolori Irrigation Project—Sokoto River, Nigeria.* Eawag Aquatic Research Institute, Retrieved January 10, 2010.

Ayoade, A., Sowunmi, A., & Nwachukwu, H. I. (2004). Gill asymmetry in Labeo ogunensis from Ogun River, Southwest Nigeria. *Revista de Biología Tropical, 52*(1), 171—175.

Berga, L. (2006). *Dams and reservoirs, societies and environment in the 21st century: Proceedings of the international symposium on dams in the societies of the 21st century, 22nd international congress on large dams (ICOLD)* (p. 314). Barcelona, Spain: Taylor & Francis, June 18, 2006. ISBN 0-415-40423-1.

Boyd, J., & Banzhaf, S. (2007). What are ecosystem services? The need for standardized environmental accounting units. *Ecological Economics, 63*(2—3), 616—626.

Costanza, R., d'Arge, R., de Groot, R., Farber, S., Grasso, M., Hannon, B., ... van den Belt, M. (1997). The value of the world's ecosystem services and natural capital. *Nature, 387*, 253—260.

Costanza, R. (2006). Nature: Ecosystems without commodifying them. *Nature, 443*, 749.

II. THE EFFECTS OF CRUDE OIL EXPLORATION ON THE SOCIO-CULTURAL AND ECO-ECONOMICS OF NIGERIAN ENVIRONMENT

Costanza, R. (2008). Ecosystem services: Multiple classification systems are needed. *Biological Conservation, 141,* 350–352.

Daily, G. C. (1997a). Introduction: What are ecosystem services? In G. C. Daily (Ed.), *Nature's services: Societal dependence on natural ecosystems* (pp. 1–10). Washington, DC: Island Press.

Daily, G. C. (1997b). Valuing and safeguarding Earth's life support systems. In G. C. Daily (Ed.), Nature's services: Societal dependence on natural *ecosystems* (pp. 365–374). Washington, DC: Island Press.

Daily, G. C., & Matson, P. A. (2008). Ecosystem services: From theory to implementation. *Proceedings of the National Academy of Sciences, 105*(28), 9455–9456.

Deng, X. Z., Zhao, Y. H., & Wu, F. (2011). Analysis of the trade-off between economic growth and the reduction of nitrogen and phosphorus emissions in the Poyang Lake Watershed. *China Ecological Modelling, 222*(2), 330–336. Available from http://dx.doi.org/10.1016/j.ecolmodel.2010.08.032.

Estes, J. A., Tinker, M. T., Williams, T. M., & Doak, D. F. (1998). Killer whale predation on sea otters linking oceanic and near shore ecosystems. *Science, 282*(5388), 473–476.

Federal Department of Fisheries (FDF) (2008). *Fisheries statistics of Nigeria* (4th ed.). Garki, Abuja, Nigeria: Federal Department of Fisheries.

Fisher, B., Turner, R. K., & Morling, P. (2008). Defining and classifying ecosystem services for decision making. *Ecological Economics, 68,* 643–653. Available from http://dx.doi.org/10.1016/j.ecolecon.2008.09.014.

Food and Agriculture Organization (FAO) (2016). *AQUASTAT website.* Food and Agriculture Organization of the United Nations (FAO), Accessed 16.03.17.

Fry, C. (2008). *The impact of climate change: The world's greatest challenge in the twenty-first century 2008.* New Holland Publishers Ltd.

Hassan, T.A. (2012). Nigeria: Helping to save Lake Chad. Daily Trust Newspaper of May 24, 2012.

Kettunen, M., Genovesi, P., Gollasch, S., Pagad, S., Starfinger, U., ten Brink, P., & Shine, C. (2009). *Technical support to EU strategy on invasive species (IAS)—Assessment of the impacts of IAS in Europe and the EU (Final draft report for the European Commission).* Brussels, Belgium: Institute for European Environmental Policy (IEEP).

Kristin, C., White, C., Gaines, S. D., Costello, C., & Anderson, S. (2013). Ecosystem service trade-off analysis: Quantifying the cost of a legal regime. *Arizona Journal of Environmental Law and Policy, 4,* 39–87.

Limburg, K. E., O'Neill, R. V., & Costanza, R. (2002). Complex systems and valuation. *Ecological Economics, 41*(3), 409–420.

Millennium Ecosystem Assessment (MA) (2005). *Ecosystems and human well-being: Synthesis.* Washington, DC: Island Press.

Nahlik, A. M., Kentula, M. E., Fennessy, M. S., & Landers, D. H. (2012). Where is the consensus? A proposed foundation for moving ecosystem service concepts into practice. *Ecological Economics, 77,* 27–35. Available from http://dx.doi.org/10.1016/j.ecolecon.2012.01.001.

Ndimele, P. E., & Jimoh, A. A. (2011). Water hyacinth (*Eichhornia crassipes* [Marts.] Solms) in phytoremediation of heavy metal polluted water of Ologe Lagoon, Lagos, Nigeria. *Research Journal of Environmental Sciences, 5*(5), 424–433.

Ndimele, P. E., & Kumolu-Johnson, C. A. (2011). Preliminary study on physico-chemistry and comparative morphometric characterisation of *Cynothrissa mento* (Regan, 1917) from Ologe, Badagry and Epe Lagoons, Lagos, Nigeria. *International Journal of Agricultural Research, 6*(10), 736–746.

Ndimele, P. E., & Ndimele, C. C. (2013). Comparative effects of biostimulation and phytoremediation on crude oil degradation and absorption by water hyacinth (*Eichhornia crassipes* [mart.] Solms. *International Journal of Environmental Studies, 70*(2), 241–258.

Ndimele, P. E., Jenyo-Oni, A., Chukwuka, K. S., Ndimele, C. C., & Ayodele, I. A. (2015). Does fertilizer ($N_{15}P_{15}K_{15}$) amendment enhance phytoremediation of petroleum-polluted aquatic ecosystem in the presence of water hyacinth (*Eichhornia crassipes* [Mart.] Solms)?. *Environmental Technology, 36,* 2502–2514.

Nelson, E., Mendoza, G., Regetz, J., Polasky, S., Tallis, H., Cameron, D., ... Shaw, M. (2009). Modeling multiple ecosystem services, biodiversity conservation, commodity production, and trade-offs at landscape *scales. Frontiers in Ecology and the Environment, 7,* 4–11.

NEST (1991). *Nigeria's threatened environment: A national profile* (pp. 124–131). Ibadan: Intec Printers Ltd.

Nfor, B. N., & Akaegbobi, I. M. (2012). Inventory of the quaternary geology and the evolution of the Oguta Lake, in Southeastern Nigeria. *World Journal of Engineering and Pure and Applied Science., 2,* 2.

II. THE EFFECTS OF CRUDE OIL EXPLORATION ON THE SOCIO-CULTURAL AND ECO-ECONOMICS OF NIGERIAN ENVIRONMENT

Odada, E. O., Oyebande, L., & Oguntola, J. A. (2005). *Lake Chad: Experiences and lessons learned brief. Managing lakes and their basins for sustainable use.* International Lake Environment Committee (ILEC) Foundation, Retrieved February 15, 2016.

Ogunbambi, H.A. (2009). Lagos State Ministry of Environment monthly environmental seminar. June 2009 (Unpublished).

Okusipe, O.M. (2008). Lagos Lagoon Coastal profile: Information database for planning theory. <http://proceedings.esri.com/library/userconf/proc04/dsc/pap1579.pdf>.

Oyedele, D. (2017). The dwindling lake. D + C, development and cooperation. Retrieved June 14, 2017.

Rodríguez, J. P., Beard, T. D., Jr., Bennett, E. M., Cumming, G. S., Cork, S., Agard, J., . . . Peterson, G. D. (2006). Trade-offs across space, time, and ecosystem services. *Ecology and Society, 11*(1), 28. <http://www.ecologyandsociety.org/vol11/iss1/art28/>.

Ruhl, J. B., Kraft, S. E., & Lant, C. L. (2007). *The law and policy of ecosystem services.* Cambridge University Press.

Seppelt, R., Lautenbach, S., & Volk, M. (2013). Identifying trade-offs between ecosystem services, land use, and biodiversity: A plea for combining scenario analysis and optimization on different spatial scales. *Current Opinion in Environmental Sustainability, 5*(5), 458−463. Available from http://dx.doi.org/10.1016/j.cosust.2013.05.002.

Shimang, G. N. (2005). *Fisheries development in Nigeria, problems and prospects.* Abuja: The Federal Director of Fisheries, The Federal Ministry of Agriculture and Rural Development.

Tallis, H., Kareiva, P., & Marvier, M. (2008). An ecosystem services framework to support both practical conservation and economic development. *Proceedings of the National Academy of Sciences, 105*(28), 9457−9464.

United Nations Environment Programme (UNDP) (2006). *Challenges to international water: Regional assessments in a global perspective.* Nairobi: UNEP. <www.unep.org/Documents.Multilingual/Default.asp?Document>.

Zhang, W., Ricketts, T. H., Kremen, C., Carney, K., & Swinton, S. M. (2007). Ecosystem services and dis-services to agriculture. *Ecological Economics, 64,* 253−260.

II. THE EFFECTS OF CRUDE OIL EXPLORATION ON THE SOCIO-CULTURAL AND ECO-ECONOMICS OF NIGERIAN ENVIRONMENT

Land Use/Land Cover Change in Petroleum-Producing Regions of Nigeria

Saheed Matemilola[1], Oludare Hakeem Adedeji[2] and Evidence Chinedu Enoguanbhor[1]

[1]Brandenburg University of Technology Cottbus — Senftenberg, Cottbus, Germany
[2]Federal University of Agriculture, Abeokuta, Ogun State, Nigeria

17.1 BACKGROUND ON THE NIGER DELTA REGION

The Niger Delta region is located in the southern part of Nigeria where it formed one of the world's largest acute fan-shaped river deltas. The Niger Delta region lies between 4.01°N and 7.90°N and between 4.50°E and 10.56°E, extending over approximately 70,000 km², which makes up approximately 7.5% of Nigeria's total land mass. It is found in the tropical rainforest belt (Kadafa, 2012a) and has a humid tropical climate characterized by wet and dry seasons. The wet season occurs between March and September, while the dry season is between October and February with the mean annual rainfall that ranges from 2000 mm to 4000 m (Akpokodje, 2000). Approximately 85% of this rainfall occurs within the months of August and October. Geologically, the area is made up of sedimentary rocks that include the Benin Formation, the Agbada Formation, and the Akata Formation. The Niger delta ecosystem is quite distinctive and renowned to be one of the largest wetlands in the world, with a very high biodiversity (CLO, 2002; Powell et al., 1985; Twumasi & Merem, 2006). The ecosystems are comprised of diverse species of flora and fauna, both aquatic and terrestrial species, and consists of four main ecological zones; coastal inland zone, freshwater zone, lowland rainforest zone, and mangrove swamp zone (WWF UK, CEESP-IUCN, 2006). The rich mangrove swamp that covers more than 20,000 km² of the vast wetland was formed primarily by sediment deposition from intricate networks of rivers and creeks that transverse the region. Several locally and globally endangered species of plants and animals can be found in the area, which harbors approximately

60%–80% of all plant and animal species found in Nigeria (WRI, 1992; Zabbey, 2009). The Niger Delta has one of the highest population densities in the world, with approximately 265 inhabitants per square kilometers. The area covers such states as Abia, Akwa Ibom, Bayelsa, Cross River, Delta, Edo, Imo, Ondo, and Rivers, whose large expanse of land and huge resources has attracted a high human population for several decades. According to the Nigerian 2006 census, approximately 33 million people lived in the Niger Delta region (>265 people per km²), making it one of the most densely populated regions in Africa and the world (Dami, Odihi & Ayuba, 2014; Uyigue & Agho, 2007). The Niger Delta has a huge deposit of crude oil, which is one of the largest in the world.

Petroleum exploration in the vast expanse of wetlands that make up the Niger Delta region of Nigeria began in 1908 by a German company and in 1936, Shell D'Arcy Exploration Company from the Netherlands secured exclusive rights to oil and gas exploration for all of Nigeria (Steyn, 2009). The discovery in commercial quantity at Oloibiri (in present day Bayelsa state) by 1956/58 led to full-scale exploration of crude oil in the area (Onuoha, 2008). The huge deposits of crude oil and natural gas deposits within the Niger Delta region of Nigeria have invariably brought about tremendous changes in the natural ecosystem and the land use/land cover pattern in the area. Petroleum-producing activities have altered the once undisturbed, pristine, and vibrant ecosystem, causing changes in biodiversity and the supply of ecosystem services to the inhabitants. The Niger Delta region, with its unique ecological system that supports an array of distinctive and abundant flora and fauna, arable terrain that can sustain a wide variety of crops, lumber or agricultural trees, and species of fresh fish (Omofonmwa & Odia, 2009), has witnessed various land use and land cover changes. In the last 60 years of petroleum production in the Niger Delta, the impacts of the activities on land use-land cover changes (LULCC) in the area have assumed dimensions of significant proportions (Abbas & Fasona, 2012). The main driver of these changes is the exploration of crude oil and production of petroleum in the area. In terms of degradation, major oil spills have occurred that have devastated rivers, killed mangroves and coastal life, and affected the health and livelihoods of millions of inhabitants (WRM, 2003). The UNEP's Millennium Ecosystem Assessment (2005) summed up the impacts of oil spill in the Niger Delta to include high mortality of aquatic animals, impairment of human health, loss of biodiversity in breeding grounds, vegetation hazards, and loss of portable and industrial water resources, poverty, rural underdevelopment, and bitterness. Quite a number of the communities in the Niger Delta region have been negatively affected by petroleum production in the area. Destruction of the ecosystems has led to reduced crop yield, polluted fishing systems, and decreased land productivity that have reduced both income and standard of living. Furthermore, petroleum hydrocarbon contamination of soils and sediment is a global concern because of the toxicity (Ite & Semple, 2012).

For several decades, the Niger Delta region had undergone serious environmental changes because of oil exploitation in the area, much of which was carried out without adequate spatial information. Furthermore, increasing population and industrial activities coupled with continuous oil spills and other environment degradation in the Niger Delta would require the establishment of a land use and land cover baseline. This can be achieved by a holistic approach to land use and land cover inventory of the area, with a focus on establishing the geospatial infrastructure for policy makers as well as for proper planning and management of the environmental conditions of the region (Dami et al., 2014; Fasona & Omojola, 2005). Geographic Information Systems (GIS), Global Positioning Systems (GPS), and Remote Sensing (RS) have become indispensable tools in almost all environmental endeavors (UN, 1986). These concepts have been employed in various studies, including atmospheric studies (Fagbeja, 2008), lithospheric (Maruo, Hiwot, Lulu & Gorfu, 2002), hydrologic (Nwilo & Badejo, 1995), biodiversity (Salami & Balogun, 2006), assessment of developmental change over time (Twumasi & Merem, 2006), land use and land cover categories (Ehlers, Jadkowski, Howard & Brostuen, 1990), as well as groundwater studies (Maruo et al., 2002).

17.2 LAND USE /LAND COVER CHANGE IN THE NIGER DELTA

Petroleum and other derived products, including several types of fuels, lubricants, and other oil materials for countless manufactured products from crude oil are essential for the functioning of the society worldwide (Rosales, Martínez-Pagán, Faz & Bech, 2014). However, petroleum production and other ancillary activities in the Niger Delta have had a great impact on the land use and land cover pattern in the area. Land use/cover change impacts cause local environmental changes, which may eventually cumulatively lead to global issues (Meyer & Turner, 1996; Salami & Balogun, 2006). In the Niger Delta, activities of petroleum exploration companies have had drastic and profound impacts on land use/land cover. LULCC in the Niger Delta and the impacts of such changes have assumed a critical dimension in view of the proximate and underlying factors that influence biodiversity, livelihoods, and a wide range of socio-economic and ecological processes (Fasona & Omojola, 2005). These activities directly affect the hydrology of the area, water quality, agriculture, fishing, and biodiversity. The indirect impacts include the altering of the local and global climate (Weng, 2001). The impacts are felt in both the short and long terms. In the short term, food security, human vulnerability to hazards, health, and safety are adversely affected. In the longer term, the viability of the earth as a whole is being threatened (Briassoullis, 1999). Land use and land cover have become very important parameters in highlighting such environmental changes that have taken place over time within the earth's surface (Ademiluyi, Okude & Akanni, 2008; Matiko, Mtalo & Mwanuzi, 2012). Changes in land use can be categorized by the complex interaction of structural and behavioral factors associated with technological capacity, demand, and social relations that affect both environmental capacity and the demand, along with the nature of the environment of interest (Ahmad, 2012; Verburg, van Eck, de Hijs, Dijst & Schot, 2004). It has become one of the major parameters for environmental change monitoring and natural resource management (Seif & Mokarram, 2012; Zhang, Peterson, Zhu & Wright, 2008). However, there are fewer studies on the land use/land cover changes in the petroleum-producing region of Nigeria despite the reported damage to the ecosystem by the activities of the oil companies. Land-cover patterns reflect the underlying natural and social processes, which, thus, help to provide essential information for modeling and understanding many phenomena on Earth (Liang, 2008). Understanding the complex interaction between human activities and global change requires the analysis of land cover data (Gong et al., 2013; Okoro, Schickhoff, Böhner & Uwe, 2016).

17.3 LAND USE CONFLICTS

Despite the decades of oil exploration in the Niger Delta, the region got very little developmental attention from the Nigerian State, with less than 5% of revenue from oil spent directly on development projects in oil-producing areas at some point. This caused widespread resentment and agitation among the inhabitants whose land and resources were affected by oil exploration activities, which contaminated the water table and destroyed many farming and fishing grounds. There is evidence of acute and chronic toxicity, which demonstrates the potential toxic and negative impacts of petroleum-derived wastes on the Niger Delta environment (Holdway, 2002). Many people were displaced from the areas of their livelihood, which affected their economic activities and social lives. Consequently, there has been enormous financial loss, extensive habitat degradation, and poverty leading to continuous crises, including rising communal conflicts, kidnaping of oil workers, and vandalization of oil installations (Adedeji, Ibeh & Oyebanji, 2011). Although there are other sources of pollution in the Niger Delta, the oil industry is a major contributor. According to Suberu (1996), the difficulties and deprivations of the oil-producing communities in the Niger Delta have invariably brought them into direct confrontation with not only the oil-prospecting companies in the area but even government agencies. Conflicts often occur near oil facilities, in pipelines, or near oil spills (Figs. 17.1–17.3). When oil spills occur, the oil spreads over a wide area, affecting terrestrial and marine resources and in some instances, oil spills have necessitated the complete relocation of some communities and loss of ancestral homes. Conflict theorists have for a long time argued that disparity in access to and share of resources as well as a high level of inequality are parts of a structural condition, which increases the threat of community conflicts and frequent disruptions of business activities (Williams & Timberlake, 1984). In fact, conflict within and between communities is also common (often related to access to the benefits of oil operations).

17.4 THE NATURE AND CHARACTERISTICS OF THE NIGER DELTA LAND USE/LAND COVER

According to UNDP's Niger Delta Human Development Report (UNDP, 2006), the Niger Delta has an enormously rich natural endowment in the form of land, water, forests, and fauna. However, these assets have been subjected to extreme degradation due to oil prospecting. Activities associated with oil extraction, including pipe laying, building infrastructure, and making the area accessible by road and water have done considerable damage to the Niger Delta environment. These activities have changed the land use/land cover pattern in the region where vast amounts of land have been degraded to the extent that it would take several years before it can be restored to its original status. Oil spill is the major impact on land use in the area, with approximately 9 to 13 million barrels of crude oil reported to have been spilled into the Niger Delta ecosystem over the past 50 years (IUCN/CEESP, 2006). Steiner (2008) stated that approximately 31,000 km^2 of networks of pipeline transverse much of the Niger Delta area, including land and water, with approximately 5284 wells drilled in more than 1500 different locations.

FIGURE 17.1   Cases of violence in the Niger Delta.

II. THE EFFECTS OF CRUDE OIL EXPLORATION ON THE SOCIO-CULTURAL AND ECO-ECONOMICS OF
NIGERIAN ENVIRONMENT

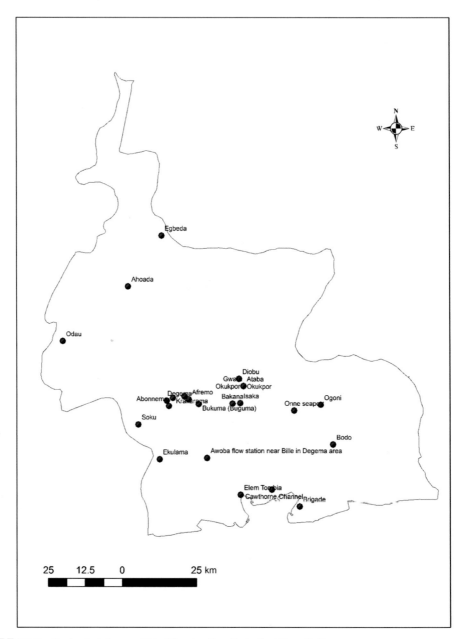

FIGURE 17.2   Spots of violence within 5 km to oil spills in the Niger Delta.

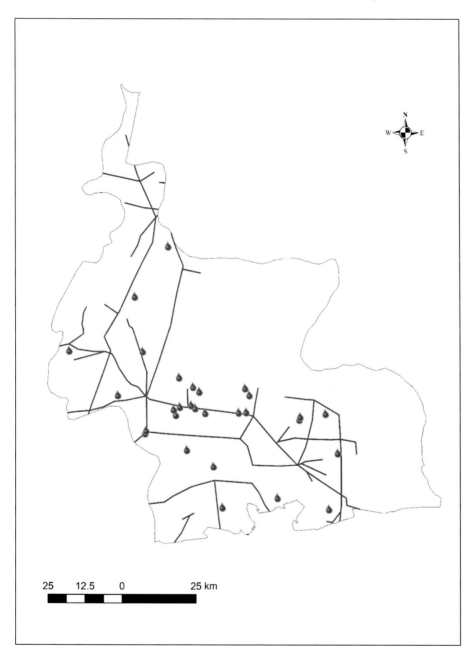

FIGURE 17.3   Spots of violence within 5 km to pipelines in the Niger Delta.

II. THE EFFECTS OF CRUDE OIL EXPLORATION ON THE SOCIO-CULTURAL AND ECO-ECONOMICS OF
NIGERIAN ENVIRONMENT

In another report by the World Bank (1995), the SPDC alone has seismic line covering 66,000 km^2 of land in Bayelsa and Rivers states, with 22 flow stations and 349 drilling sites. This shows that sizeable portions of the Niger Delta land are occupied by oil-related activities that have relegated other land use to the background and degraded the quality of the land. Activities associated with petroleum exploration, development, and production operations have local detrimental and significant impacts on the atmosphere, soils and sediments, surface and groundwater, marine environment, and terrestrial ecosystems in the Niger Delta (Anochie & Mgbemena, 2015).

17.5 PETROLEUM RESOURCE AND THE ASSOCIATED PROBLEMS IN THE NIGER DELTA REGION

Nigeria is ranked the number one oil producing African country and sixth amongst the Organization of Petroleum Exporting Countries (OPEC). The economic landscape of the Nigerian state has witnessed a tremendous level of transformation with the discovery of oil (Ugochukwu & Ertel, 2008). Oil was first discovered in Nigeria in the year 1956, while commercial production and first exportation was recorded in 1958. As of 1981, crude oil contributes 90% of Nigeria's exports by value and 80% of the federal revenue (Achi, 2003; Kadafa, 2012a).

However, the impacts of the Nigerian oil industry come in different forms. For instance, the impacts of petroleum resource exploration and exploitation on the physical environment of the Niger Delta region of Nigeria have become very emphatic in recent years. According to Achi (2003), despite the inmense benefits the Nigerian state has accrued from the oil wealth, the oil-bearing Niger Delta region has endured a disastrous socio-physical environment within the first four decades of oil discovery in the region, a situation that has threated the economy, livelihood, and survival of the local community. In concurrence, Kadafa (2012b) opined that the Niger Delta region is ranked amongst the 10 most important wetlands and marine ecosystems globally, but persistent unsustainable exploration practices have condemned the region amongst the five most severely devastated/impacted ecosystems in the world from petroleum activities. In fact, studies have shown that within 50 years of the commencement of exploration activities in the Niger Delta, not less than 9 to 13 million barrels of the oil resource has been spilled Kadafa (2012a). In his study, Nwankwo (1984) showed that 1360 spill incidences were recorded from 1976 to 1983, resulting in 1,425,794 barrels of oil spilled into the environment of the Niger Delta (Bayode, Adewunmi & Odunwole, 2011). Based on the foregoing, there is the need to review the costs and associated problems of oil exploration activities to the Niger Delta community.

17.5.1 Environmental Issues Arising From Petroleum Activities

Oil exploration operations requires vast land area, a situation that has accentuated land scarcity, as virtually the whole Niger Delta region (land and waters) sits on top of the oil resource. Thus, Oil Mining Leases (OMLs) or Oil Prospecting Leases (OPLs), which entitle the multinational oil companies to encroach on peasant lands and fish ponds have left the locals competing for smaller portions of land. Shell Petroleum Development Corporation (SPDC), for example, acquired land to the tune of 1723.6 acres in the Isoko region of Delta State for oil exploration. A protest statement delivered by the Uzere community to the SPDC clearly underscores the threat that continuous land alienation poses to the community. It reads, "39 Oil wells and pipelines have taken all over our land, where shall we farm?" (Aghalino, 2001).

Also, periodic operational or accidental spills are inevitable during crude oil exploration due to human fallibility and equipment malfunction regardless of precautionary measures and process sophistication. During the various phases of oil exploration, fluids are discharged into the environment. Such discharge may result from activities that include drill cuttings, drilling mud and production stimulating fluids, and produced fluids such as oil and water as well as other chemicals injected into them to enhance the separation of oil from water or to control corrosion (Bayode et al., 2011; Ite, Ibok, Ite & Petters, 2013).

The result of a study conducted by Kadafa (2012b) to evaluate the trend in spillage resulting from exploration activities in the Niger Delta region showed that although the annual volume of spillage from oil exploration activities declined significantly from the years 1981 to 2000, the number of spill incidences was on a steady rise from 1976 to 2000 (Figs. 17.4 and 17.5). Beyond the period represented in Figs. 17.4 and 17.5, Kadafa (2012b) further stated that numerous incidences of oil spills have been witnessed from 2001 to 2010 for which estimates could not be obtained for the study. For instance, there was the spillage caused by a broken pipeline at Yorla oilfield of Royal/Dutch Shell in Ogoniland on April 29, 2002, which flowed for several days and polluted vast amounts of vegetation and farm lands.

In a different study by Emuedo and Abam (2015), it was estimated that approximately an average of 546 million gallons of oil is spilled per annum into the Niger Delta ecosystem since the inception of exploration and 6817 incidences of oil spills were documented between the years 1976 and 2001, although approximately 70% of the spilled oil was

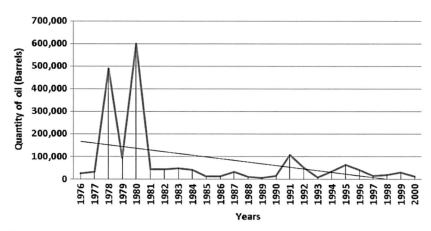

FIGURE 17.4   Trend in annual volume of oil spillage (1976–2000). *From Kadafa, A.A. (2012b). Oil exploration and spillage in the Niger Delta of Nigeria. Civil and Environmental Research, 2(2), 38–52.*

II. THE EFFECTS OF CRUDE OIL EXPLORATION ON THE SOCIO-CULTURAL AND ECO-ECONOMICS OF NIGERIAN ENVIRONMENT

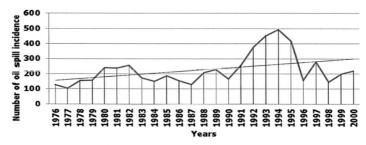

FIGURE 17.5   Trend in the annual incidence of oil spillage (1976–2000). *From Kadafa, A.A. (2012b). Oil exploration and spillage in the Niger Delta of Nigeria.* Civil and Environmental Research, 2(2), 38–52.

recovered. Furthermore, the National Oil Spill Detection and Response Agency (NOSDRA) documented 2405 oil spillage incidences between 2006 and 2010 with an upward trend in the number of recorded incidences as years progress: 252 incidences were recorded in 2006, 598 were documented in 2007, 927 in 2008, and 628 in 2009 (Emuedo & Abam, 2015). Table 17.1 highlights some critically impacted regions in the Niger Delta. However, it is important to mention here that the vast majority of the oil spill incidences in the Niger Delta are not documented because they are often regarded as "minor." Some of the incidences considered "major" and documented are the Bomu II blowout in 1970, the Forcados terminal spillage in 1980, the Funiwa 5 oil well blowout in 1980, the Oyakana pipeline spillage in 1980, the Okoma pipeline spillage in 1985, the Oshika pipeline spillage in 1993, and the Goi Trans Niger pipeline oil spill in 2004 (Ugochukwu & Ertel, 2008).

Petroleum exploration in the Niger Delta also involves massive gas flaring. According to Emoyan, Akoborie, and Akporhonor (2008), Nigeria flares more gas than any other country in the world, with approximately 46% of Africa's total, while also flaring the highest volume of gas per ton of oil produced. Further analysis showed that Nigeria flared an average of 76% of the total quantity of gas produced every year starting from 1970. Natural gas from oil wells is directly flared at a rate of approximately 70 million m^3 per day. This figure equals approximately 40% of total natural gas consumption in Africa (Emoyan et al., 2008). It was estimated that approximately 123 natural gas flaring station exist in the Niger Delta region as of 2004, releasing approximately 45.8 billion kilowatt of heat into the environment from 1.8 billion cubic feet of gas everyday and rendering a vast region of the host community inhabitable (Kadafa, 2012b). Based on the report of Sagay, I. at the National Political Conference in 2005, the natural gas flared away daily in the Niger Delta is worth approximately 86 million US dollars and if utilized could generate sufficient electricity for the whole of West Africa Ugboma (2015).

17.5.2 Implications of Petroleum Resource Exploration on the Niger Delta Communities

The Niger Delta people understandably have a generic belief that the region, despite being the economic mainstay of Nigeria, has nothing but pain, anguish, and destruction to show for it. This is mainly because of the negligent attitude of the successive government

TABLE 17.1   Some Severely Oil-Impacted Sites in the Niger Delta Region

Location	Environment	Impacted Area (ha)	Nature of incidence
BAYELSA STATE			
Biseni	Freshwater Swamp forest	20	Oil spillage
Etiama/Nembe	Freshwater Swamp forest	20	Oil spillage and fire outbreak
Etelebu	Freshwater Swamp forest	30	Oil spill incident
Peremabiri	Freshwater Swamp forest	30	Oil spill incident
Adebawa	Freshwater Swamp forest	10	Oil spill incident
Diebu	Freshwater Swamp forest	20	Oil spill incident
Tebidaba	Freshwater Swamp forest mangrove	30	Oil spill incident
Nembe Creek	Mangrove forest	10	Oil spill incident
Azuzuama	Mangrove	50	Oil Spill Incident
9 sites			
DELTA STATE			
Opuekebe	Barrier Forest Island	50	Salt water intrusion
Jones Creek	Mangrove forest	35	Spillage and burning
Ugbeji	Mangrove	2	Refinery waste
Ughelli	Freshwater Swamp forest	10	Oil spillage-well head leak
Jesse	Freshwater Swamp forest	8	Product leak/burning
Ajato	Mangrove		Oil spillage incident
Ajala	Freshwater Swamp forest		Oil spillage incident
Uzere	Freshwater Swamp forest		Oil spillage incident
Afiesere	Freshwater Swamp forest		Oil spillage incident
Kwale	Freshwater Swamp forest		Oil spillage incident
Olomoro	Freshwater Swamp forest		Oil spillage incident
Ughelli	Freshwater Swamp forest		Oil spillage incident
Ekakpare	Freshwater Swamp forest		Oil spillage incident
Ughuvwughe	Freshwater Swamp forest		Oil spillage incident
Ekerejegbe	Freshwater Swamp orest		Oil spillage incident
Ozoro	Freshwater Swamp forest		Oil spillage incident
Odimodi	Mangrove forest		Oil spillage incident
Ogulagha	Mangrove forest		Oil spillage incident
Otorogu	Mangrove forest		Oil spillage incident
Macraba	Mangrove forest		Oil spillage incident
20 sites			
RIVERS STATE			
Rumuokwurusi	Freshwater swamp	20	Oil spillage
Rukpoku	Freshwater swamp	10	Oil spillage

Source: Kadafa, A.A. (2012a). Environmental impacts of oil exploration and exploitation in the Niger Delta of Nigeria. Global Journal of Science Frontier Research, 12, 18–28.

TABLE 17.2   Incidence of Poverty in the Niger Delta Region—1980—2004

Country/states	1980	1985	1992	1996	2004
Nigeria	28.1	46.3	42.7	65.6	54.4
Edo/Delta	19.8	52.4	33.9	56.1	Delta: 45.35 Edo: 33.09
Cross River	10.2	41.9	45.5	66.9	41.61
Imo/Abia	14.4	33.1	49.9	56.2	Imo: 27.39 Abia: 22.27
Ondo	24.9	47.3	46.6	71.6	42.15
Rivers/Bayelsa	7.2	44.4	43.4	44.3	Rivers: 29.09 Bayelsa: 19.98

Source: From Oviasuyi, P.O., & Uwadiae, J. (2010). The dilemma of Niger-Delta region as oil producing states of Nigeria. Journal of Peace, Conflict and Development, 16, 110—126.

of Nigeria towards the welfare of the people of the region (Oviasuyi & Uwadiae, 2010). Inadequate administration of the oil resource has led to deteriorating socio-economic and socio-political conditions as well as complex interaction between people, economy, and environment (Ite et al., 2013). There is no nexus between the wealth generated from the oil resource, the multinational oil companies, and the human development of the Niger Delta region. Yet, exploration activities keep destroying the social, economic, and environmental structures of the oil-bearing communities through occupational dislocation, unemployment, rural-urban drift, and impaired human health (Iwejingi, 2013).

Analysis of data provided by the National Bureau of Statistics on poverty and human development indicators reflects the disheartening condition of the region (Table 17.2). The incidence of poverty in the region surged between the years 1980 and 2004 (Oviasuyi & Uwadiae, 2010). In effect, the Nigerian oil industry has been faced with rising incidence of oil theft and bunkering (Ite et al., 2013). This should be anticipated because the incidence of poverty has continued to take an upward trend.

The rising level of impoverishment is occasioned by contamination of agricultural land and water bodies by oil spillage. On land, oil spillage could cause retarded vegetation growth or even fire incidence, whereas in water, it could prevent natural aeration, a situation that could lead to suffocation of marine life (Bayode et al., 2011). The result is rendering of arable lands infertile and extermination of fishes in ponds and rivers, thereby threatening the sustenance of communities that might depend on them for livelihood. Also, there are records of disastrous oil spillage clean-up incidences where the designate chemicals for clean-up were recklessly applied, thereby creating additional contamination problems (Iwejingi, 2013).

Agriculture represents the most important and major economic activity in the Niger Delta region, with farming and fishing constituting 90% of the economic occupation (Achi, 2003). In a bid to establish the effects of oil exploration activities on the soil and agricultural productivity of the Niger Delta region and beyond, several empirical studies have been conducted including those of Bello, Aladesanwa, Akinlabi, and Mohammed (1999), Minai-Tehrani, Shahriari, and Savagbebi (2007), Abii and Nwosu (2009), Idodo- Umeh and Ogbeibu (2010), Ojimba (2011). For example, in a study of gas flaring effects on the growth and yield of maize on farmlands located near the points of gas flaring by Bello et al. (1999), it was shown that the average percentage of plant survival and grain yield diminished significantly in all the samples located close to gas flaring points. In fact, farmlands located within a 200-m radius to the flaring stations failed to yield (Oshwofasa, Anuta & O, 2012).

On the whole, the effects of oil spillage and gas flaring as well as other exploration-related activities on the biota and ecosystem come in manifolds (Ugochukwu & Ertel, 2008). Several other environmental impacts resulting from petroleum exploration activities have been variously identified, including site clearance, access roads construction, brine pits, tank farms, and pipelines, as well as other forms of land manipulations and modifications that could be required for the various phases of the exploration activities (Ite et al., 2013). Table 17.3 highlights some of the environmental, heath, and socio-economic problems associated with petroleum exploration activities in the Niger Delta.

17.6 IMPACTS OF PETROLEUM ACTIVITIES ON NIGER DELTA LAND USE/COVER

Petroleum activities such as oil and gas exploration and exploitation as well as gas flaring have been causing significant impacts on land use/cover change in the Niger Delta region of Nigeria. The spatial and temporal assessments of land use/cover of some selected petroleum-producing regions of the Niger Delta are presented in Dami et al. (2014).

In the oil and gas production community of Kwale in Delta State, land use/cover has changed over the years. From the study carried out by Dami et al. (2014), it was discovered that the bare suface that covered an area of 61,374 km^2 in 2001 has reduced drastically to 2296 km^2 in 2008. Furthermore, the forest vegetation also reduced from 46,873 km^2 to 31,309 km^2. Dami et al. (2014, p. 20) stated that "the global agitation for more environmentally friendly practices and subsequently the various mitigative tendencies of oil companies must have influenced the trend in 2001 while the further reduction in forest vegetation in 2008 could be due to increased exploration and exploitation activities of the area."

However, agriculture, setlement, and water body had increased land use within this period. Scattered cultivation, settlement, and water body increased from 154,104 km^2 to 193,725 km^2, 13,375 km^2 to 16,420 km^2, and 4556 km^2 to 41,050 km^2, respectively. The increase in land use by settlement was as a result of people from other communities coming to settle down in Kwale to seek and pick up employment in the oil and gas industries. As people migrated to Kwale, the demand for foodstuff increased, leading to more agricultural production and hence, the increase in agricultural (scatered cultivation) land use. The woodland increased from 217,064 km^2 to 225,293 km^2 within the same period, and Dami et al. (2014) concluded that despite the reduction in vegetation caused by oil and gas exploration and exploitation, woodland increased due to aforestation efforts as part of environmental remediation in the area.

In the oil- and gas-producing communities of Ilaje and Ose-Odo, local government areas of Ondo State, in the western Niger Delta, land use and cover have experienced some changes between 1986 and 2008 (Figs. 17.6 and 17.7 and Table 17.4).

TABLE 17.3 Environmental Impacts Associated With Upstream and Downstream Petroleum Operations

Activities	Potential/associated risks	Environmental, health, and safety issues
EXPLORATION OPERATIONS		
• Geological survey • Aerial survey • Seismic survey • Gravimetric and magnetic survey • Exploratory drilling • Appraisal	a. Noise pollution b. Habitat destruction and acoustic emission c. Drilling discharges (e.g., drilling fluids (water-based and oil-based muds) and drill cuttings) d. Atmospheric emission e. Accidental spills/ blowout f. Solid waste disposal	Ecosystem destruction and interference with land use to access onshore sites and marines resource areas; environmental pollution (air, soil, and controlled water) and safety problems associated with the use of explosives; land pollution, which affects plants and poses human health risks; groundwater contamination and adverse effects on ecological biodiversity
DEVELOPMENT AND PRODUCTION		
• Development drilling • Processing: separation and treatment • Initial storage	a. Discharges of effluents (solids, liquids, and gases) b. Operation discharges c. Atmospheric emission d. Accidental oil spills e. Deck drainage f. Sanitary waste disposal g. Noise pollution h. Transportation problems i. Socio-economic/cultural issues	Ecosystem destruction and interference; contamination of soils and sediments with petroleum-derived wastes; atmospheric emissions from fuel combustion and gas flaring/venting; environmental pollution (air, soil and sediments, controlled waters) and groundwater contamination; ecological problems in the host communities, adverse human health risks; safety-related risks and interference with socio-cultural systems
DECOMMISSIONING AND REHABILITATION		
• Well plugging • Removal of installations and equipment • Site restoration	a. Physical closure/removal b. Petroleum-contaminated waste disposal c. Leave in situ (partial or total) d. Dumping at sea	Environmental pollution and human safety; pollution related to onshore and offshore operations; hazard to other human activities such as fishing and navigation; marine pollution, fishing, and navigation hazards
REFINING OF PETROLEUM PRODUCTS		
	a. Atmospheric emissions and air pollution b. Discharges of petroleum-derived wastes	Atmospheric emissions and air pollution; oil spillages; water effluents and production discharges
TRANSPORTATION AND DISTRIBUTION		
• Pipelines • Barges, ships, tankers, and FPSOs • Road tankers and trucks	a. Emissions and accidental discharges b. Discharges from transporting vessels (e.g. ballast, bilge, and cleaning waters)	Air emissions (hydrocarbons from loading racks and oil spills); accidental discharges and operational failures; disposal of sanitary wastes; contamination of soils and sediments
MARKETING OPERATIONS		
• Product importation • Storage	a. Operational discharges b. Wastes disposal	Spillage; contamination of soils and sediments; emission of organic contaminants and environmental pollution

Source: Ite, A.E., Ibok, U.J., Ite, M.U., & Petters, S.W. (2013). Petroleum exploration and production: Past and present environmental issues in the Nigeria's Niger Delta. American Journal of Environmental Protection, 78—90.

II. THE EFFECTS OF CRUDE OIL EXPLORATION ON THE SOCIO-CULTURAL AND ECO-ECONOMICS OF
NIGERIAN ENVIRONMENT

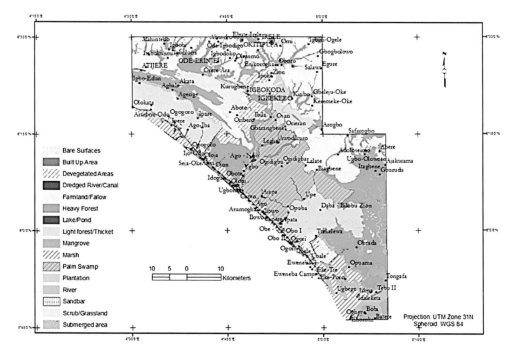

FIGURE 17.6   Land use map of Ilaje and Ose-Odo in 1986. *From Abbas, I.I. (2012). An assessment of land use/cover changes in a section of Niger Delta, Nigeria. Frontiers in Science, 2(6):137–143.*

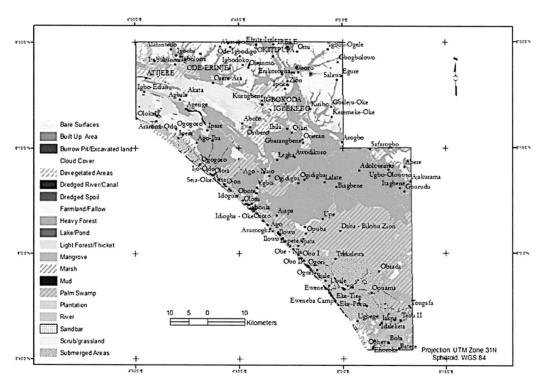

FIGURE 17.7   Land use map of Ilaje and Ose-Odo in 2008. *From Abbas, I.I. (2012). An assessment of land use/cover changes in a section of Niger Delta, Nigeria. Frontiers in Science, 2(6):137–143.*

TABLE 17.4   Land Use/Cover Change of Ilaje and Ose-Odo, Ondo State for 1986 and 2008

Land use class	Area (Ha) (1986)	Percentage (1986)	Area (Ha) (2008)	Percentage (2008)
Built-up areas	4976.13	1.5	7934.11	2.39
Degraded lands	21,633.74	6.6	94,202.48	28.46
Wetlands	33,559.67	10.2	14,562.16	4.4
Agricultural lands	81,315.57	24.6	57,053.44	17.24
Natural/semi-natural vegetation	179,509.76	54.3	144,249.91	43.81
Open area	484.68	0.01	1710.66	0.52
Cloud cover	–	–	329.64	0.10
Ground total	331,465.23	100	331,464.43	100

Source: Extracted from Abbas, I.I. (2012). An assessment of land use/cover changes in a section of Niger Delta, Nigeria. Frontiers in Science, 2(6):137–143.

According to the study conducted by Abbas (2012), it was discovered that the built-up area increased from 4976.13 Ha to 7934.11 Ha, representing an approximately 2.18% increase, which was attributed to the massive construction works of Ondo State Oil-Producing Areas Development Commission (OSOPADEC) and the Niger Delta Development Commission (NDDC). There was also an increase in degraded land, from 21,633.74 Ha in 1996 to 94,202.48 Ha in 2008, due to oil and gas pollution in the area. Comparing the situation of the water bodies in Ilaje and Ose-Odo to that of Kwale, it is clear that although the water bodies in Kwale have been increasing, those in the Ilaje and Ose-Odo communities have been decreasing. A similar situation occurred on agricultural land use (Table 17.4). Vegetation in both study areas decreased, and this was due to oil and gas activities. Abbas (2012) opined that since oil and gas exploration and exploitation commenced in the Niger Delta, environmental problems associated with oil spillage, gas flaring, canalization, forest resources depletion, costal erosion, and others, have been brought by the petroleum activities, resulting in changes in the land use/cover of the area.

In the southern part of Niger Delta with a wider scope of the spatial assessment of land use/cover change, Mmom and Fred-Nwagwu (2013) carried out a study using the application of image classification methods of supervised and maximum likelihood parameter.

The analysis shows that water bodies decreased by 18% from 1996 to 2007 and swamp by 16% within the same period, and it was concluded that it could be as a result of the various land reclamation activities around the city and communities during the period (Mmom & Fred-Nwagwu, 2013). Residential land use increased due to econmic pressure, including petroleum activities. In 1996, residential area covered 22.35% of the study area while in 2007, its percentage coverage rose to 23.94%. There was an increase in vegetation cover within the period, from 39.11% in 1996 to 47.01% in 2007. Farmland decreased drastically from 3641.42 km^2 (4.06%) in 1996 to 406.41 km^2 (0.45%) in 2007. It was concluded that the increase in economic activities, including petroleum production, has attracted people to Port Harcourt, which has led to an increase in residential land use and a decrease in farmland as well as changes in other land uses/cover in the study area (Mmom & Fred-Nwagwu, 2013).

17.7 REMEDIAL ACTIONS AND WAYS FORWARD

"Landuse and landcover change have become a central component in current strategies in managing natural resources and monitoring environmental changes" (Mmom & Fred-Nwagwu, 2013, p. 077). The Niger Delta region of Nigeria should not be left out of such strategies in order to protect the abundant natural resources deposited in this region. Solving the environmental problems caused mostly by petroleum activities in the Niger Delta would contribute to the achievement of sustainable land use development in particular and sustainable development in general.

However, scholars have suggested remedial actions as a way forward towards addressing petroleum-induced environmental problems and unsustainable use of land resources in the Niger Delta region of Nigeria. Some of the recommendations include:

- The legislations on petroleum issues should be revised and updated. The fines and license of petroleum companies should be reviewed and efforts should be made to ensure compliance. Technologies in the petroleum industry should be environment-friendly (Ayuba, 2012). It has also been suggested that the environmental restoration program should be jointly carried out by the oil companies and the government.
- Reclaimation of areas occupied by bare surface and marshlands and converting them to agricultural lands in order to reduce poverty and increase food production and food security in the region should be undertaken (Dami et al., 2014).
- Ugboma (2015) suggested that the 1958 Willinks Commission report and the Niger Delta Development Board Act of 1961 that encouraged carrying out a feasibility study of the Niger Delta for the promotion of physical, human, and infrastructural developments should be revisited. Ugboma (2015) further suggested that the petroleum companies in Nigeria should adopt the methods used by crude oil companies in the developed world when handling environmental issues.
- Onwuteak (2014) suggested the idea of implementing a decision support system for land use data and information gathering. This would help to keep land use/cover change update and would be useful for planning and economic development.
- In addition to the recommendations given by various scholars, this study recommends the efficient implementation of Land Use Planning in the Niger Delta. Furthermore, the Land Use Planning should be assessed environmentally, economically, and socially to achieve sustainable land use development. In relation to environmental assessment of Land Use Planning, there should be implementation of Strategic Environmental Assessment (SEA) in Nigeria, and this environmental management tool should be applicable on Land Use Planning in the Niger Delta. In addition, the Environmental Impact Assessment of petroleum projects should be monitored for effective implementation and to ensure environmental sustainability of the Niger Delta.

References

Abbas, I. I. (2012). An assessment of land use/cover changes in a section of Niger Delta, Nigeria. *Frontiers in Science*, 2(6), 137–143.

Abbas, I. I., & Fasona, M. J. (2012). Remote sensing and geographic information techniques: veritable tools for land degradation assessment. *American Journal of Geographic Information System, 1*(1), 1–6. Available from http://dx.doi.org/10.5923/j.ajgis.20120101.01.

Abii, T. A., & Nwosu, P. C. (2009). The effect of oil-spillage on the soil of Eleme in Rivers State of Niger-Delta area of Nigeria. *Resource Journal of Environmental Sciences, 3*, 316–320.

Achi, C. (2003). *Hydrocarbon exploitation, environmental degradation and poverty: The Niger Delta experience.* Dublin, s.n.

Adedeji, O.H., Ibeh, L., & Oyebanji, F. (2011). Sustainable management of mangrove coastal environments in the Niger Delta region of Nigeria: The role of remote sensing and geographic information systems. In O. Martins, E. A. Meshida, T. A. Arowolo, O. A. Idowu, & G. O. Oluwasanya (Eds.), *Proceedings of the environmental management conference, September 12-15, 2011, Federal University of Agriculture, Abeokuta, Nigeria.* Vol. 2:308–324. Available at http://www.unaab.edu.ng.

Ademiluyi, I., Okude, A., & Akanni, C. (2008). An appraisal of landuse and landcover mapping in Nigeria. *African Journal of Agricultural Research, 3*(9), 581–586.

Aghalino, S. O. (2001). Oil exploration and its impact on the Nigerian environment. *Kiabara Journal of Humanities, 7*(1), 103–111.

Ahmad, F. (2012). Detection of change in vegetation cover using muti-spectral and multi-temporal information for District Sargodha. *Pakistan. Sociedade Natureza, 24*, 557–572.

Akpokodje, J. (2000). *Oil pollution, distribution and land degradation in the Niger Delta of Nigeria* (unpublished M.Sc. thesis). United Kingdom: University of York.

Anochie, U. C., & Mgbemena, O. O. (2015). Evaluation of some oil companies in the Niger Delta region of Nigeria: An environmental impact approach. *International Journal of Environment and Pollution Research, 3*(2), 13–31.

Ayuba, K. A. (2012). Environmental impacts of oil exploration and exploitation in the Niger Delta of Nigeria. *Global Journal of Science Frontier Research Environment and Earth Scinces, 12*(3), 18–28.

Bayode, O. J. A., Adewunmi, E. A., & Odunwole, S. (2011). Environmental implications of oil exploration and exploitation in the coastal region of Ondo State, Nigeria: A regional planning appraisal. *Journal of Geography and Regional Planning, 4*(3), 110–121.

Bello, E. I., Aladesanwa, R. D., Akinlabi, S. A., & Mohammed, T. I. (1999). Effects of gas flaring on the growth and yield of maize (*Zea mays* L.) in South-Easter Nigeria. *Applied Tropical Agriculture, 4*, 42–47.

Briassoullis, H. (1999). *Analysis of land use change: Perspectives, theoretical and modelling approaches.* Earths- Can Publication Ltd, London www.idosi.org/aejsr/1 (1)06/7.

Civil Liberties Organization (CLO). (2002). *Blood trail: Repression and resistance in the Niger Delta.* Ikeja: CLO.

Dami, A., Odihi, J. O., & Ayuba, H. A. (2014). Assessment of land use and cover change in Kwale, Ndokwa East Local Government Area, Delta State, Nigeria. *Global Journal of Human Social Sciences: B Geography, Geo-Sciences and Environmental Disaster Management, 14*(6), 16–24.

Ehlers, M. M. A., Jadkowski, R. R., Howard, & Brostuen, D. E. (1990). Application of SPOT data for regional growth analysis and local planning. *Photogrammetric Enginering and Remote Sensing, 56*, 175–180.

Emoyan, O. O., Akoborie, I., & Akporhonor, E. E. (2008). The oil and gas industry and the Niger Delta: Implications for the environment. *Journal of Applied Science and Environmental Management, 12*(3), 29–37.

Emuedo, C., & Abam, M. (2015). Oil, land alienation and improverishment in the Niger Delta, Nigeria. *European Journal of Research in Social Sciences, 3*(2), 8–23.

Fagbeja, M. (2008). Applying remote sensing and GIS techniques to air quality and carbon management: A case study of gas flaring in the Niger Delta (A Ph.D. thesis), Britain: University of the West of England.

Fasona, M.J. and Omojola, A.S. (2005). Climate change, human security and communal clashes in Nigeria. *Proceedings of the international workshop on human security and climate change,* Holmen Fjord Hotel, Asker, near Oslo, June 22–23, 2005. Available at www.cicero.uio.no/humsec/papers/Fasona&Omojola.pdf.

Gong, P. J., Wang, L., Yu, Y. C., Zhao, Y. Y., Zhao, L., Liang, Z. G., ... Liu, Y. (2013). Finer resolution observation and monitoring of global land cover: First mapping results with landsat TM and ETM + data. *International Journal of Remote Sensing, 34*(7), 2607–2654.

Holdway, D. A. (March 2002). The acute and chronic effects of wastes associated with offshore oil and gas production on temperate and tropical marine ecological processes. *Marine Pollution Bulletin, 44*(3), 185–203.

Idodo-Umeh, G., & Ogbeibu, A. E. (2010). Bioaccumalation of the heavy metals in cassava tubes and plantain fruits grown in soils impacted with petroleum and non-petroleum activities. *Resource Journal of Environmental Sciences, 4*, 33–41.

Ite, A. E., & Semple, K. T. (2012). Biodegradation of petroleum hydrocarbons in contaminated soils. In R. Arora (Ed.), *Microbial biotechnology: Energy and environment* (pp. 250–278). Wallingford: Oxfordshire CAB International.

Ite, A. E., Ibok, U. J., Ite, M. U., & Petters, S. W. (2013). Petroleum exploration and production: Past and present environmental issues in the Nigeria's Niger Delta. *American Journal of Environmental Protection*, 78–90.

IUCN/CEESP. (2006). Niger Delta natural resources damage assessment and restoration project–phase I scoping report. May 31, 2006 report from the IUCN commission on environmental, economic, and social policy, 14.

Iwejingi, S. F. (2013). Socio-economic problems of oil exploration and exploitation in Nigeria's Niger Delta. *Journal of Energy Technologies and Policy*, 3(1), 76–81.

Kadafa, A. A. (2012a). Environmental impacts of oil exploration and exploitation in the Niger Delta of Nigeria. *Global Journal of Science Frontier Research*, 12, 18–28.

Kadafa, A. A. (2012b). Oil exploration and spillage in the Niger Delta of Nigeria. *Civil and Environmental Research*, 2(2), 38–52.

Liang, S. (2008). Advances in land remote sensing. System, modeling, inversion and application–Dordrecht.

Maruo, T., Hiwot, A.G., Lulu, S., & Gorfu, S. (2002). Application of GIS for ground water resource management: Practical experience from groundwater development and water supply.Training centre. UNECA. Available online at http://www.uneca.org/groundwater/Docs/Application%/20%20GIS-%20-JICA%20NO-2.pdf Accsessd 12.06.15.

Matiko, M., Mtalo, I.F., & Mwanuzi, F. (2012). *Land use and land cover changes in Kihansi river catchment and its impact on river flow*. Department of water resources engineering, University of Dar Es Salaam, Dar Es Salaam, Tanzania. 22.

Meyer, W. B., & Turner, B. L., II (1996). Change in landuse and land cover change: Challenges for geographers. *Geojournal*, 39(3), 237–240.

Millennium Ecosystem Assessment (2005). *Ecosystems and human well-being: Synthesis*. Washington, DC: Island Press.

Minai-Tehrani, D., Shahriari, M. H., & Savagbebi, G. (2007). Effect of light crude oil-contaminated soil on growth and germination of *Festuca Arundinacea*. *Journal of Applied Sciences*, 2, 2623–2628.

Mmom, P. C., & Fred-Nwagwu, F. W. (2013). Analysis of landuse and landcover change around the city of Port Harcourt, Nigeria. *Global Advanced Research Journal of Geography and Regional Planning*, 2(5), 076–086.

Nwankwo. J.N. (1984). Oil and environmental pollution. Paper presented at the *conference on strategies for the fifth national development plan: 1986–1990*. NISER, Ibadan, November 25–29.

Nwilo, P. C., & Badejo, O. T. (1995). Management of oil spill dispersal along the Nigerian coastal areas. *Journal of Environmental Mangement*, 4, 42–51.

Ojimba, T. P. (2011). Socio-economic variables associated with poverty in crude oil polluted crop farms in Rivers State, Nigeria. *Journal of Applied Sciences*, 11(3), 462–472.

Okoro, S. U., Schickhoff, U., Böhner, J., & Schneider, U. A. (2016). A novel approach in monitoring land-cover change in the tropics: oil palm cultivation in the Niger Delta, Nigeria. *DIE ERDE*, 147(1), 40–52.

Omofonmwan, S. I., & Odia, L. O. (2009). Oil exploitation and conflict in the Niger-Delta region of Nigeria. Kamla-Raj. *Journal of Human Ecology*, 26(1), 25–30.

Onuoha, F. C. (2008). Oil pipeline sabotage in Nigeria: Dimensions, actors and implications for National Security L/C. *African Security Review*, 17(3), 99–115.

Onwuteak, J. (2014). Predicting changes in landuse and landcover in Niger Delta using post classification analysis. *Civil and Environmental Research*, 6, 110–117.

Oshwofasa, B. O., Anuta, D. E., & O, A. J. (2012). Environmental degredation and oil industry activities in the Niger-Delta Region. *African Journal of Scientific Research*, 9(1), 443–460.

Oviasuyi, P. O., & Uwadiae, J. (2010). The dilemma of Niger-Delta region as oil producing states of Nigeria. *Journal of Peace, Conflict and Development*, 16, 110–126.

Powell, C.B., White, S.A., Ibiebele, D.O., Bara, M., Dut Kwicz, B., Isoun, M., & Oteogbu, F.U. (1985). Oshika oil spill environmental impact; effect on aquatic biology. Paper presented at *NNPC/FMHE international seminar on petroleum industry and the Nigerian environment, November 11–13, 1985*, Kaduna, Nigeria 168–178.

Rosales, R. M., Martínez-Pagán, P., Faz, A., & Bech, J. (2014). Study of subsoil in former petrol stations in SE of spain: Physicochemical characterization and hydrocarbon contamination assessment. *Journal of Geochemical Exploration*, 147, 306–320.

II. THE EFFECTS OF CRUDE OIL EXPLORATION ON THE SOCIO-CULTURAL AND ECO-ECONOMICS OF NIGERIAN ENVIRONMENT

Salami, A. T., & Balogun, E. E. (2006). *Utilization of Nigeria Sat-1 and other satellites for forest and biodiversity monitoring in Nigeria. National Space Research and Development Agency (NASRDA)* (p. 142). Nigeria: Abuja.

Seif, A., & Mokarram, M. (2012). Change detection of Gil Playa in the Northeast of Fars Province, Iran. *American Journal of Scientific Research, 86*, 122–130.

Steiner, R. (2008). *Double standards? International standards to prevent and control pipeline oil spills, compared with shell practices in Nigeria.* A report submitted to Friends of the Earth, Netherlands.

Steyn, M. S. (2009). Oil exploration in the colonia Nigeria C, 1903-58. *Journal of Imperial and Commonwealth History, 37*, 249–274.

Suberu, T. R. (1996). *Ethnic minority conflict and governance in Nigeria.* Ibadan: Spectrum Books Ltd.

Twumasi, Y. T., & Merem, E. C. (2006). GIS and remote sensing applications in the assessment of change within a coastal environment in the Niger Delta Region of Nigeria. *International Journal of Environmental Research and Public Health, 3*(1), 98–106.

Ugboma, P. P. (2015). Environmental degradation in oil producing areas of Niger Delta Region, Nigeria: The need for sustainable development. *International Journal of Science and Technology, 4*(2), 75–85.

Ugochukwu, C. N. C., & Ertel, J. (2008). Negative impacts of oil exploration on biodiversity management in the Niger Delta area of Nigeria. *Impact Assessment and Project Appraisal, 26*(2), 139–147.

United Nations Development Programme. (UNDP) (2006). *Niger Delta. Human development report, UNDP* (p. 48). Nigeria: Abuja.

United Nations. (UN). (1986). Principles relating to remote sensing of the earth from space. *95th plenary meeting of the general assembly, December 3, 1986.*

Uyigue, E., & Agho, M. (2007). Coping with climate change and environmental degradation in the Niger Delta of Southern Nigeria. Community Research and Development Centre Nigeria. (CREDC).

Verburg, P. H., van Eck, J. R., de Hijs, T. C., Dijst, M. J., & Schot, P. (2004). Determination of land use change patterns in the Netherlands. *Environment and Planning. B: Planning and Design, 31*, 125–150.

Weng, Q. (2001). *Modeling urban growth effects on surface runoff with the integration of remote sensing and GIS.* New York: Springer-Verlag.

Williams, K. R., & Timberlake, M. (1984). Structured inequality, conflict, and control: A cross-national test of the threat hypothesis. *Social Forces, 63*(2), 414 ,1984.

World Bank (1995). *Africa: A framework for integrated coastal zone management. Environment Department.* Washington DC: The World Bank.

World Resource Institute (WRI) (1992). *World resource 1992–93.* New York, NY: Oxford University Press.

WRM. (2003). *Nigeria: Gas corporation NLNG destroys forest in the Niger Delta.* 22nd November. Available at http://www.wrm.org.uy/bulletin/68/AF.html Accessed 16.05.16.

WWF UK, CEESP-IUCN. (2006). Commission on environmental, economic, and social policy and federal ministry of environment Abuja, Nigerian conservation foundation Lagos, May 31, (2006). Niger Delta resource damage assessment and restoration project.

Zabbey, N. (2009). Impacts of oil pollution on livelihoods in Nigeria. Paper presented at the conference on petroleum and pollution – how does that impact human rights? Co-organized by Amnesty International, Forum Syd and Friends of the Earth, Sweden. At Kulturhuset, Stockholm, Sweden, April 27, 2009.

Zhang, Z., Peterson, J., Zhu, X., & Wright, W. (2008). *Modelling land use and land cover change in the strzelecki ranges. Proceedings of international congress on modelling and simulation (MODSIM07)* (pp. 1328–1334). New Zealand: Christchurch.

Petroleum Industry Activities and Climate Change: Global to National Perspective

Michael Adetunji Ahove[1] and Sewanu Isaac Bankole[2]

[1]Lagos State University, Lagos, Nigeria [2]Ogun State Institute of Technology, Ogun, Nigeria

18.1 INTRODUCTION

The glow of our beautiful blue planet, our only known home to humanity, is gradually fading away, comparable to the luster of the morning stars, as the day dawns, always lost to the overwhelming light energy of the sun. The quest for human survival, improved quality of life, and especially affluence has made modern human the culprit of the Earth's degradation due to our uncontrolled consumption pattern of natural resources, especially energy and numerous unfavorable environmental behaviors.

The first energy revolution, powered by steam and coal, gave wings to the first industrial revolution with the era of the printing press and the locomotive rail, which may be described as the most fundamentally significant paradigm shifts for modern society. This revolution took place within a short dispensation and created the platform for an extremely rapid transformation of human life and society from barely surviving to improved quality of life. This dispensation may be rightfully described as one that gave birth to the death of abject poverty in most countries in Europe and America as millions became rich, different types of jobs emerged, and there were improvements made in science and technology. This increased prosperity soon spread to other countries around the world.

Historically, the first and second industrial and energy revolutions have intertwined to produce multiple inventions that significantly altered the global industrial landscape. The second industrial revolution was energized by the use of fossil fuels, with mass production of goods, individual mobility, and the division of labor. However, rapid economic growth over the last quarter of the century, especially energy production, consumption, and industrial advancement of agriculture and technology, have unmistakably altered some essential natural processes and systems on the earth (UNIDO, 2015). Thus the astronomical advancements in science and technology in our globalized world have seriously contributed to dramatic changes in human societies, economics, and consumption patterns (Singseewo, 2011).

The energy derived from fossil fuels gave rise to a rapid industrial growth with greenhouse gas (GHG) emissions as a by-product, inducing global warming, which over the centuries has resulted in what is now referred to as anthropogenic climate change. The current concern of people globally may appear to revolve around three fundamental issues: economy, climate, and peace. Before the industrial revolution, this tripartite relationship was apparently favorable to our beautiful blue planet, making it a sustainable home for humanity (Ahove, 2016). Currently however, a contrary tune is being sung globally.

We are currently consuming fossil fuel energy not only at an alarming rate but more importantly, it has become a major threat to the survival of mother Earth and as Omotayo (2016) describes it with a rhetorical question, "Are we living in a dying Earth?" Obviously we are living in a dying planet where a significant amount of nonrenewable natural resources have been consumed unsustainably as a result of greed. Yes, greed! This economic greed within this context may be described from a translated proverb in Yoruba (a major language in southwest of Nigeria) as "the hydra-headed wood causing disorder in the flaming fire underneath the cooking pot" (Ahove, 2017). If we have a "just enough" consumption attitude and consideration for natural resources, a sizeable quantity of natural capital will be most probably available for future generations as well as good air quality without significant climate change challenges. This theoretical position is based on observable evidence from the pattern of living among the people of the Ekuri community in the Cross River State of Nigeria where a significant quantity of natural resources in the community's environment is not altered significantly, and nature is very well respected (UNEP, 2008).

18.2 THE PETROLEUM INDUSTRY AND FOSSIL FUEL

As modern humanity took the bold step of the industrial revolution circa the year 1750, our activities through the burning of fossil fuel produced an approximately 40% increase in the atmospheric concentration of carbon dioxide from 280 ppm in 1750 (Blasing, 2013) to 315.97 by January 1, 1959 and currently to 404.21 ppm by January 1, 2016 (Dlugokencky & Tans, 2016). This gap has not shrunk by any significant figure but rather has persisted irrespective of the various carbon sinks in nature such as the oceans, forest trees, and other processes in the carbon cycle. The principal culprit behind this increase is fossil fuels (coal, oil, and natural gas) combustion and unsustainable forest consumption, soil erosion, and massive production of cattle. This proves that carbon dioxide released into the atmosphere outweighs what is being removed. Nature is now in the battle of balancing this equation, which has proved to be near-impossible except that humans are ready to change their behavior and actions.

Hydrogen and carbon are the basic constituents of fossil fuels formed from the remains of ancient organic matter such as plants and animals having gone through millions of years of exposure to high heat and pressure with practically no oxidation within the Earth's crust. This causes intense pressure on the organic matter for hundreds of millions of years. Fossil fuels exist as volatile materials with low carbon-hydrogen proportions as it is found in butane gas, to the dark liquid petroleum oil and the nonvolatile near-pure carbon such as anthracite coal existing as a solid. The consumption of fossil fuels results in the release of approximately 21.3 billion tonnes of carbon dioxide (CO_2) yearly (Wikipedia, 2017). However, data show that natural mechanisms have the propensity to sink approximately 50% of that amount within a year. Thus we have an excess of 10.65 billion tonnes of CO_2 in the atmosphere, which has the capacity to increase global temperature.

18.3 THE BIRTH OF FOSSIL FUEL: OFFSHORE SCHEMATIC APPROACH

The origin and build-up of fossil fuel are described with schematic illustrations below. Millions of years ago, small animals and plants died and fell to the bottom of the sea. Their remains are buried in the earth (Fig. 18.1).

The dead organisms are buried in the soil for millions of years, and as the soil builds up over time, they are transformed into rocks where the organic matters are preserved in the absence of oxygen (Fig. 18.2). This rock mounts a great deal of pressure on the dead organic matter, transforming it mostly into dark liquid oil. The crude oil is then extracted from the rock where they are preserved (Fig. 18.3).

The extracted fossil fuels, which are finite and nonrenewable energy resources, are eventually purified and used in homes and organizations for heating, electricity generation, and transportation, as well as in driving different types of engines for industrial use. Industrial consumption of fossil fuel has been found to be significant relative to other sectors of the society. The global industrial sector is currently on record as one of the largest emitters of GHG, representing almost 30% of global emissions. Undeniably, it is essential

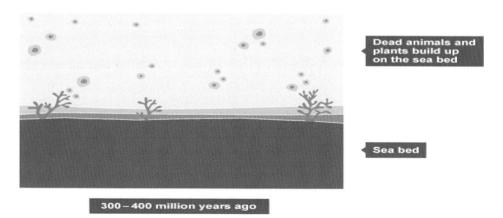

FIGURE 18.1   Deposition and build-up of organic matter on the sea bed. *From BBC Bitesize. (2017). National 4 chemistry: Fuels revision 1 Retrieved 18:01, March 3, 2017, from www.bbc.co,uk/education/guides/z8yj6sg/revision.*

FIGURE 18.2   Decomposed fossil in rock under high temperature and pressure. *From BBC Bitesize. (2017). National 4 chemistry: Fuels revision 1 Retrieved 18:01, March 3, 2017, from www.bbc.co,uk/education/guides/z8yj6sg/revision.*

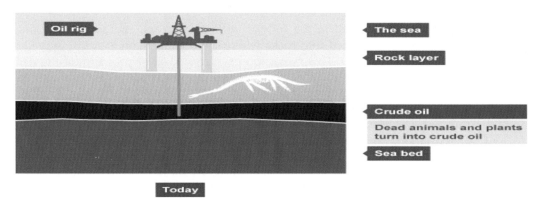

FIGURE 18.3   Extraction of fossil fuel. *From BBC Bitesize. (2017). National 4 chemistry: Fuels revision 1 Retrieved 18:01, March 3, 2017, from www.bbc.co,uk/education/guides/z8yj6sg/revision.*

that conventional industrial development patterns be transformed to become more climate resilient (UNIDO, 2015).

18.4 CONCEPTS OF GREENHOUSE EFFECT, GLOBAL WARMING, CLIMATE CHANGE AND OZONE LAYER DEPLETION

The atmosphere is made of mixtures of gases. This blanket of gases above the surface of the Earth is labeled as atmosphere. The atmosphere of the Earth is mostly composed of nitrogen (approximately 78%), oxygen (approximately 21%), and argon (Ar) (approximately 0.9%), with carbon dioxide and other gases in minute amounts. Water vapor accounts for approximately 0.25% of the atmosphere by mass. The concentration of water vapor may alter by volume in view of location; cold or hot regions of the Earth vary by the level of humidity. The air within the atmosphere has trace amounts of many other chemical compounds, depending on the location and influence of human activities. In the atmosphere, many substances would be present in an unfiltered air sample. This includes dust from organic material and mineral deposits. Aside from this, there are industrial emissions such as chlorine and its compounds, fluorine compounds, sulfur gas, and traces of mercury. Therefore, the atmosphere is made of GHGs and non-GHGs.

Greenhouse effects, global warming, and ozone layer depletion are three other terms related to the concept of climate change apparently requiring clarifications based on the opinions from previous studies of Grima, Filho, and Pace (2010) and Ahove (2015) that learners and readers alike often confuse and conflate. These terms are discussed below:

- Greenhouse effect and global warming
- Non-GHGs
- GHGs
- Climate change
- Ozone layer depletion

18.4.1 Greenhouse Effect and Global Warming

The concept of greenhouse effect is better explained with the illustration of a greenhouse. The concept of greenhouse effect emanated from the experience of warmth in a greenhouse utilized in some farms for an all-year place to nurture and grow plants for maximum output. The glass or polyethylene (plastic) structure where plants are grown is called a greenhouse. Usually the roof and walls are made of glass or polyethylene. A greenhouse creates an artificial environment by careful control of temperature, light, humidity, air quality, soil moisture, and heat levels (Ahove, 2001).

The transparent glass or translucent polyethylene roof and windows of a greenhouse allow in sunlight that warms up objects inside the greenhouse. These objects then give off heat. The glass or polyethylene of the greenhouse, however, does not let out the heat. Thus the greenhouse inhibits thermal ventilation and as such all the heat stays locked inside and the temperature rises, providing the essential warmth in all seasons. This increase in temperature as a result of trapped warmth is referred to as "greenhouse effect." The greenhouse effect also accounts for the increased temperature in a car of which its windows are wound up on a sunny day.

The earth and its atmosphere are analogous to a giant greenhouse. Like the glass or polyethylene roof and windows of a greenhouse, the atmosphere (basically the influence of GHGs, mostly carbon dioxide) is nearly transparent to shortwaves and visible solar radiation. Part of the energy absorbed by the Earth's surface from the sun is radiated to the atmosphere as long waves, much of which are absorbed by GHG before partial radiation back to the surface of the earth (Fig. 18.4). This causes the Earth and its atmosphere to warm up, which had remained nearly constant until the 20th century. If nature has not placed carbon dioxide in the atmosphere at an optimum level, the earth will not experience the appropriate and natural greenhouse effect for a normal impact sufficient to keep the earth comfortably warm on the average and not cold like most planets (Ahove, 2001).

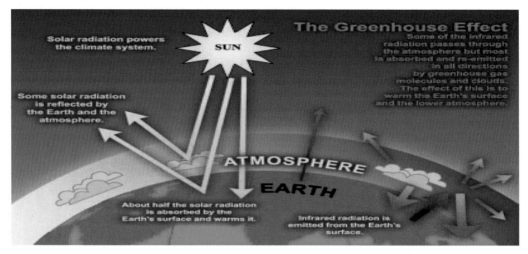

FIGURE 18.4   The earth and its atmosphere and the process of radiation and re-radiation of solar energy. *From Intergovernmental Panel on Climate Change (IPCC). (2014).* Climate Change 2014: Impact adaptation and vulnerabilisty, WG2AR5 summary for policymakers. *IPCC Secretariat, Geneva.*

The advent of the burning of fossil fuel was the beginning of the spurious release of large quantities of CO_2 into the atmosphere. The combustion of fossil fuel has brought about an ever- increasing rise in the CO_2 concentration in the atmosphere relative to its natural quantity. The rapid eradiation of forests, especially the tropical rainforests, is depleting the Earth's vegetation growth and diminishing its capability for absorbing carbon dioxide. The unabsorbed CO_2 rises to the upper atmosphere and inhibits the reradiation of solar energy back into space (Fig. 18.4). This trapped heat induces a global temperature increase resulting in global greenhouse effect, which most scientists have attributed to the release of GHGs, mainly carbon dioxide, from anthropogenic activities (IPCC, 2014).

This experience of heat within the greenhouse is what is used as an analogy to illustrate the phenomenon called global warming. The earth is a giant greenhouse where heat is trapped, leading to global warming. Global warming may be described as the gradual increase in global temperature (average increase in temperature of the earth surface, sea surface, and the troposphere) as a result of the effects of GHGs induced by human activities.

Based on estimates from the literature, if GHG emission continues at the present rate, the Earth's surface temperature could outstrip historical values as early as 2047, with harmful effects on the ecosystem, biodiversity, and the livelihood of humans worldwide. Recent estimates indicate that with the current emissions trajectory, the earth could pass the threshold of 2°C global warming, which the United Nations' IPCC designated as the upper limit to avoid dangerous global warming by 2036 (Wikipedia, 2017).

18.4.2 Non-Greenhouse Gases

Aside from the GHGs found in the atmosphere, there are other gases categorized as non-GHGs. They include nitrogen, oxygen, and Ar and are so called because molecules containing two atoms of the same element as found in O_2 and N_2 and monoatomic molecules such as Ar have no net change in the distribution of their electrical charges when they vibrate, and hence are almost totally unaffected by infrared radiation. Although molecules containing two atoms of different elements such as carbon monoxide (CO) or hydrogen chloride (HCl) absorb infrared radiation, these molecules are short-lived in the atmosphere, owing to their reactivity and solubility. Therefore, they do not contribute significantly to the greenhouse effect and usually are omitted when discussing GHGs.

18.4.3 Greenhouse Gases

Some of the GHGs emitted as a result of oil and gas activities are water vapor, CO_2, and methane, whereas nitrous oxide and fluorocarbons are emitted by other human activities. Table 18.1 shows the major anthropogenic sources of these gases, their average duration in the atmosphere, and their global warming potential as well as pre-industrial concentration. The expected concentrations in 2030 were also outlined.

It must be noted, however, that water vapor is the most abundant GHG but with the least greenhouse effect and thus insignificant impact on climate change.

These gases are projected to cause more increase in the average temperature of the troposphere. According to Miller (1991, 2012), the major ones are:

1. Carbon dioxide
2. Methane
3. Nitrous oxide
4. Chlorofluorocarbons (CFCs)

Carbon dioxide (CO_2): This gas is projected to be responsible for 49% of anthropogenic input of GHGs. However, approximately 80% of this contribution comes from fossil fuel burning, whereas deforestation appears to provide the balance of 20%. Industrial countries

TABLE 18.1   Major GHGs That Are Influencing Climate Change

Gas	Major anthropogenic sources	Amount released per year (millions of tons)	Average time in the atmosphere	Global warming potential* (more than 100 years)	Preindustrial concentration (approximately 1860) (ppb)	Current average concentration (ppb)	Expected concentration in 2030 (ppb)
CO_2	Burning of fossil fuels	5500	100 years	1	290,000	350,000	500,000
CH_4	Fossil fuel production, rice fields	500	10 years	21	850	1700	2300
N_2O	Fertilizers, deforestation, burning biomass	30	days	310	001–7	001–50	001–50
CFCs	Aerosol sprays, refrigerants	1	60–100 years	1500–8100	0	approximately 3	2.4–6

account for approximately 76% of annual emissions. The emitted CO_2 is thought to remain in the atmosphere for 50–200 years.

Methane (CH_4): This gas is responsible for approximately 18% of the human input of GHGs. It is produced by bacteria that decompose organic matter in oxygen-poor environments. Approximately 40% global methane emissions come from water-logged soils, bogs, marshes, and rice paddies. Warming may increase methane emissions from these sources by 20%–30% and thus amplify global warming. Other sources are the guts of termites and the digestive tracts of billions of cattle, sheep, pigs, goats, horses, and other livestock.

Some methane also leaks from coal seams, natural gas, wells, pipelines, storage tanks, furnaces, dryers, and stoves. Natural sources produce an estimated one-third of the methane in the atmosphere and human activities produce the rest. Methane remains in the troposphere for 7–15 years, and each molecule is approximately 25 times more effective in warming the atmosphere than a molecule of carbon dioxide.

Nitrous Oxide (N_2O): This gas is responsible for 6% of the global warming. It is released from the breakdown of nitrogen fertilizers in the soil, livestock wastes, nitrate-contaminated groundwater, and biomass burning. Its average stay in the troposphere is a few days. It also depletes ozone in the stratosphere. The global warming caused by each molecule of this gas is approximately 230 times that of a carbon dioxide molecule. These gases are referred to as GHGs, not because they are green in color but because they induce the greenhouse phenomenon on earth.

Chlorofluorocarbons (CFCs): These gases, irrespective of the variance, have been predicted to be responsible for approximately 14% of the anthropogenic input of GHGs and by 2020 will probably be responsible for approximately 25% of the input. CFCs are also guilty for the depletion of stratospheric ozone. Basic human sources of CFCs are leaking air conditioners and refrigerators, evaporation of industrial solvents, production of plastics, and foams and propellants in aerosol cans. CFCs are expected to remain in the atmosphere for between 65 and135 years, depending on the variant, but generally have 1500–7000 times the impact per molecule on global warming than each molecule of carbon dioxide. It takes between 10 and 20 years to reach the stratospheric ozone layer. In essence, when CFCs are released from the earth by humans, it takes between 10 and 20 years for the chemistry of the depletion to begin. Thereafter, the negative impacts will occur and be felt on earth. CFCs have been replaced by hydrofluorocarbons (HFCs), which have less GHG impact.

18.4.3.1 *How Greenhouse Gas Emissions can be Reduced*

To check the challenges of climate change requires simple actions from everyone irrespective of location on the globe. Governments, businesses, and individuals need to take significant steps to reduce emissions by using resources more efficiently and adopting new and cleaner technologies for energy use. The following are some strategic simple and practicable steps:

1. *Adopt energy-saving lifestyle*: Make it a habit to turn off the lights as you leave a room. Also replace standard light bulbs with energy-efficient fluorescent bulbs. Turn off all electronic appliances and unplug them when they are not in use.

2. *Think and use a better transportation method*: Walk or bike at every possible opportunity. This will surely reduce your carbon footprint and improve your health status. Take advantage of public transit or carpool as the opportunity opens up. Traveling long distances by train or bus is a better alternative to reduce your carbon footprint rather than flying or driving. If you must buy a car, go for environment-friendly ones.

3. *Use water wisely*: Be at alert over every drop of water. Fix drips and leaks, and install low-flow shower heads and toilets. Use sizeable water in a container while brushing teeth or shaving. Treating and transporting water requires energy, whereas water conservation results in reduced energy requirements and carbon emissions.

4. *Avoid a warm water wash and use a drying rack*: The new formula for saving energy while washing is to avoid a warm water wash. Washing clothing in cold water and hanging clothing to dry outside or indoors on a drying rack is an excellent option. These techniques will reduce your electricity bill and prolong the durability of your clothing by reducing wear on the fabric caused by electric dryers.

5. *Switch to high-efficiency appliances*: As you buy or replace appliances, high-efficiency equipment is your best option. Appliances with energy star ratings, an international standard for energy-efficient consumer products, are known to consume a minimum of 20% less energy.

6. *Embrace the "green energy revolution"*: Investigate the source of your power generation; wind, water, coal, or solar, and talk to your power provider to ensure that a greater percent is coming from renewable resources. Challenge power providers to switch to green energy. Better still, switch to a company offering power from renewable resources.

7. *Insulate your home*: Insulate yourself, your loved ones, and your home. This ensures heat conservation during cold or winter. This can be done by purchasing windows and window coverings that hinder heat from escaping and by sealing any existing cracks. In warm weather, use fans to ventilate your home and regulate the air conditioners to have a comfortable temperature at home. Regulating the temperature on your water heater to between 55°C and 60°C and insulating your pipes will also make a difference.

8. *Recycle*: Make recycling part of your daily lifestyle. Recycle every piece of packaging and consumer goods that are in your possession. One of your goals must be to purchase items with minimal and recyclable packaging. For electronics, facilities now exist that can dispose of electronics in an environmentally responsible manner.

9. *Second chance*: Avoid discarding or recycling clothing and household goods; rather, give them a second chance. Sparingly used clothing can be given to charity, friends, or family. Old clothing, which may have gone through second chance, may be given a third chance as rags for cleaning. Household goods can be donated to charity or sold at a garage sale. The concept of second chance will reduce the amount of waste being sent to our landfill sites. There is no need to use energy for recycling when others can benefit from your used items. Isn't that excellent?

10. *Plant trees*: As you work on your garden, select plants that are well adapted to your climate and require minimal watering and attention. Better still, plant a tree, and it will provide shade and sink carbon from the atmosphere into the soil. Everyone, including children, should plant and care for at least one tree every 6 months. This is an important form of Earth care but we must realize that tree planting is only a stop-gap

II. THE EFFECTS OF CRUDE OIL EXPLORATION ON THE SOCIO-CULTURAL AND ECO-ECONOMICS OF NIGERIAN ENVIRONMENT

measure for slowing CO_2 emissions. To absorb the CO_2 emitted into the atmosphere each year, we would have to plant an average of 1000 trees per person per year.

11. *Population control*: Reducing population growth is extremely important since humans are at the heart of energy consumption. As population increases so does the consumption of energy. If GHG emissions are reduced by half and human population doubles with consumption pattern, then we are back where we started or in a more terrible condition. Countries with high illiteracy rate and poverty may require more attention to population education.

12. *Green thinkers converts*: The simple techniques outlined in this subsection must not only be practiced but we must also seek to convert members of our family, friends, and acquaintances to do the same. The mandate of preaching green is in our hands for the survival of our big beautiful blue planet. Let us seek to convert others into thinking green and the world will soon be a better place for us and the coming generations.

18.4.4 Climate Change

Climate change may be simply described as a statistically significant difference in either the mean state of the climate or in its variability, persisting for an extended period (typically decades or longer). Climate change may be due to natural internal processes or external influences or persistent anthropogenic changes in the composition of the atmosphere or land use. The United Nations Framework Convention on Climate Change (UNFCCC) defines "climate change" as: "a change of climate which is attributed directly or indirectly to human activity that alters the composition of the global atmosphere and which is in addition to natural climate variability observed over comparable time periods."

It is essential to mention that climate change evolved as a result of decades of global warming. Thus, a gradual increase in global temperature for decades has led us into a climatic condition now referred to as climate change.

18.4.5 Ozone Layer Depletion

Nitrous oxide is the new and most important culprit damaging the ozone layer (Fig. 18.5). It is the largest cause of ozone layer depletion. This is because CFCs and many

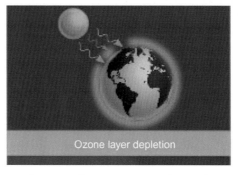

FIGURE 18.5   The earth showing the stratospheric ozone around its surface. *From BBC Bitesize. (2017). National 4 chemistry: Fuels revision 1 Retrieved 18:01, March 3, 2017, from www.bbc.co.uk/education/guides/z8yj6sg/revision.*

other gases that damage the ozone layer were banned by the Montreal Protocol (MP), and currently their atmospheric concentrations have reduced substantially. Nitrous oxide is not restricted by the MP, so while the level of other ozone layer depleting substances (ODS) are declining, nitrous oxide levels are increasing. These impacts are expected to become more severe, unless concerted efforts are made to reduce emissions.

18.5 FINAL DEADLY BLOWS ON CLIMATE CHANGE REALITY DEBATE

Mike Tyson goes down in the professional boxing history as the number one boxer in the heavyweight category with the deadliest blows. He won 26 of his 28 fights by knock-out/ technical knockout, and 16 were in the first round. That's a wow! These are undeniable and inexcusable defeats. One can easily predict that almost all of his first round−defeated opponents will never call for a second−time challenge. We can only metaphorically imagine that these deadly feats were unleashed against the ideas that oppose the reality of climate change. Ahove (2016) analyzed this victory for the future citizenry of our beautiful blue planet from three clusters. These are: the undisputable proof by the scientific community, consistent pressure from environmentalists, and perceived threats and actual experiences by citizens. A background discussion will serve as an appetizer before we engage in this three-course meal.

In the 19th century, what may be arguably considered as the first warnings that brought the debate on the reality of climate change to the fore was in 1827. Jean-Baptiste Fourier proposed an argument that an atmospheric effect kept the earth warmer than it would otherwise be, using the analogy of a greenhouse. In 1896, Svante Arrhenius (a Nobel Prize winner, obviously not in climate change) first proposed the idea of global warming. His line of argument was on CO_2 emissions from the burning of coal, which enhances the Earth's greenhouse effect and results in global warming. Baron (2006) further mentioned that this great scientist's prediction on global warming agrees fairly with the modern computer model. In the 1930s, global climate change was first mentioned by the media when a prolonged period of warm weather required explanation.

The lobbies and politics of climate change are among the most controversial in the international arena. This is regularly reflected in past conference debates emerging from the ratification of the Kyoto protocol. Climate change due to global warming is an issue with scientific, environmental, economic, development, and political dimensions. The largest percentage contribution to GHGs is by CO_2, emitted by fossil fuel during combustion for the release of energy for daily activities. Climate change includes issues ranging from burden-sharing among countries with widely differing economic vulnerabilities and adaptabilities to change, including choices made by individuals about everyday energy consumption, power use, and transportation.

18.5.1 Naturalistic Versus Anthropogenic Schools of Thought

There has been a protracted debate on the reality of climate change based on two schools of thought—naturalistic and anthropogenic (Ahove, 2016). The naturalistic school of thought holds the view that the basic reason for climate change is due to natural factors such as earthquake, volcanic activities, tsunami, and other seismic activities. Their perception also centers on the Earth's capacity to revert itself from the negative effects of climate change even if it will last as long as centuries or millennia. Therefore, the effects of climate change are only temporary and can only be reversed by natural means; human efforts will not result in any meaningful solution. The argument of the naturalistic school of thought hinges on past major global climate changes such as the global ice age, with devastating atmospheric challenges, which eventually reverted to a normal global climatic condition.

The anthropogenic school of thought holds the view that contemporary climate change is basically induced by human activities. Humans' quest for improved quality of life, affluence, and greed has resulted in unsustainable consumption of natural resources. These are the fundamental driving forces that have resulted in numerous human activities and environment-unfriendly behaviors that have continued to increase the amount of global GHGs, especially CO_2 in the Earth's atmosphere. The argument of this school of thought is principally based on several contemporary scientific studies, evidences from climate change models, and current realities as experienced globally. Global climate change is not an unusual occurrence of nature or a climatically engineered phenomenon by some secret scientists or aliens from outside our solar system (Ahove, 2016). Rather, it is a phenomenon that is exacerbated by human activities. The challenges of anthropogenic climate change can only be reversed by conscious human efforts. This is the view held by this school of thought and that it is better for our economies to be hurt than for us to be out of existence, leaving no economy to run. An important thought-provoking question to reflect on is: which school of thought is more appropriate in resolving the challenge posed by climate change—the naturalistic or the anthropogenic view? Whatever choice you make, ensure that it will leave the Earth in a sustainable state after your demise. Three fundamental issues are analyzed below as mentioned earlier that may help you make the right choice. This is the victory for the future citizenry of our beautiful blue planet.

18.5.2 Scientific Community With Undisputable Proofs

Consistent evidence accumulated for decades by the global scientific community indicates overwhelming evidence pointing to human activities as the major factor behind climate change. Recent data showed that the Earth's surface has warmed by more than 0.8°C over the past century and by approximately 0.6°C in the past three decades. These changes are largely caused by human activities, mainly the burning of fossil fuels releasing CO_2 that traps heat within the atmosphere. This CO_2 emission continues to rise, and climate models project the average surface temperature will rise between 1.1°C and 6.4°C by 2100 if we do not take appropriate actions to control global GHG emissions (IPCC, 2014). Sea level rise, coastal flooding, biodiversity loss, heat waves, droughts, desertification, oxygen depletion, forest fires, decreased crop yield, and negative health impacts are some of the negative consequences of climate change identified in literature (IPCC, 2014). Some of the implications of these consequences are increased malnutrition, increased deaths, disease and injury due to heat waves, and increased frequency of cardio-respiratory diseases.

The intellectual build-up of similar results from several scientists over the years indicating that evidence for anthropogenic climate change has been the reason for the numerous global, regional, and local conferences, workshops, seminars, technical, and experts meetings, including ministerial and heads of nations' gathering on this subject. These gatherings are fundamentally focused on the reality of climate change based on scientific evidence and real-life experiences from different communities. These meetings are organized by governments and organizations to develop strategies for climate change mitigation and adaptation as well as for raising funds to execute the programs. Some of these gatherings may appear to have failed; some may be rated as fair, whereas others have been very successful.

The conference of parties held in December 2015 (COP 21) in Paris has been rated as exceptionally successful by many of the participants as reported by many international news media across the world. This may not be unconnected with the long-awaited global climate change agreement, which was sealed after it has suffered several setbacks in the past. This agreement was designed to ensure that the global average temperature increase does not significantly exceed 1.5°C relative to global average temperature before the industrial revolution. What in the world would have been running through the minds of 185 heads of state or their representatives to have signed the COP 21 agreement in spite of a few dissenting voices and its economic implications? The very strong possibility that may be adduced to this is that the vast majority of the world now shares the view that climate change is real.

The challenges of climate change have been well-documented in notable scientific literature for decades by several scientists from different regions of the world. Thus the scientific community has a body of common evidence that is yet to be significantly disputed with empirical proofs. This scientific community has always being true and trusted with evidence from several theories, principles, laws, discoveries, and several other inventions. Why the skepticism about the reality of climate change? This signed agreement ended the skepticism substantially and proved that the vast majority of global citizens share the views of contemporary science that climate change is real.

18.5.3 Consistent Pressure From Environmentalists

Many of the ardent believers of the scientific evidences took up the responsibility to defend this noble and worthy cause for the global citizenry of today and for future generations. For decades, from one conference to the other, whether local, regional, or international, several climate change proponents have made their voices heard and stood their ground in favor of climate change, creating important landmarks on the road to a successful agreement so far reached.

These well-meaning environmentalists include individuals, corporate groups, nongovernmental organizations (NGOs), students, organized labor, the media (mainstream and social), government representatives, and numerous others from different backgrounds, have been tenacious in favor of policies and actions to slow down global GHGs and to especially develop mechanisms for mitigation and adaptation. For instance, Tella (2016) reported the position of the African group at COP 21 as they articulated how Africa bears a disproportionate burden of the adverse effects of climate change. Thus, Africa was said to have been short-changed by climate change. So, Africans should not allow themselves to be short-changed by climate finance. Consequently, more funding was demanded by the African group than the 14% initially allotted to the continent to tackle climate change. The attitude of the climate change faithfuls of consistent pressure on national governments and the international community is an indication that climate change is real.

18.5.4 Perceived and Actual Threat

The last straw that broke the backbone of climate change skeptics is predicated on the concept of perceived and actual threat of climate change. This concept of threat expresses the idea that individuals or groups will normally be willing to take a decision or action in favor of climate change if the threat of the consequences of climate change is perceived or actual (Ahove, Okebukola, Oshun, Okebukola & Bankole, 2016). The argument here is that there is an increase in perceived threat of climate change. Many people who are yet to have a personal devastating blow of the consequences of climate change may have seen, heard, or read about it in the media how several thousands of people and communities suffer. They are likely to share the view that one day "it may be my turn." Thus the fear of this possibility has accounted for the reason why nations of the world who are yet to have a major devastation agreed on COP 21.

Actual threat relates to individuals or group who have actually gone through the harrowing challenges of climate change effects. They know what it means to be a survivor, and the ordeal they have passed through has affected different aspects of their lives. These experiences will make anyone know how it feels to be devastated by climate change. Therefore, it may not be out of context to hold the view that a substantial number of individuals or groups at the local or national level may have actually passed through these experiences. Thus the signing of agreement to eliminate this threat may be the best option and indeed, it is the best alternative. Several studies have confirmed that various citizens around the globe exhibit perceived and actual threats of the consequences of climate change, which is an indication that climate change is real (Ahove, 2015; Gordon, Deinis & Havic, 2010; Grima et al., 2010; Zhao, Rolfe-Redding & Kotcher, 2014).

The position of this discussion is clearly illustrated with the analogy of a simple tripod (Fig. 18.6), indicating the three stands of undisputable proofs by the scientific community,

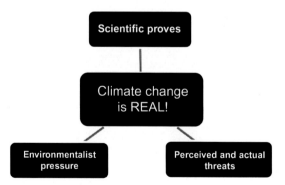

FIGURE 18.6   Reality of climate change: victory for future generations. *From Ahove, M. A. (2016). Research trends in climate change education: A vision for Nigeria. In P. Okebukola (Eds.),* Imperatives for Nigeria's development *(pp. 185–203). Lagos, Nigeria: Okebukola Science Foundation.*

consistent pressure from environmentalists, and perceived threat and actual experiences by citizens. Based on these arguments, is climate change real to you? The undeniable fact is that anthropogenic climate change is not only real but very clear and convincing to the vast majority of global citizens, now more than ever.

18.6 CONCLUSION

Human existence is unequivocally better and has been transformed with the discovery of fossil fuel. However, the activities of the oil and gas industry and unsustainable consumption of this energy have resulted in increases in the amount of GHGs trapped in our beautiful blue planet. Over the years, this has resulted in climate change with its hydra-headed challenges.

Climate change is currently labeled as the most pervasive of all global environmental challenges confronting humanity. It impacts all aspects of human life and livelihoods: health and well-being, food security, mobility, energy, water, vegetation, biodiversity, land, peace, and security. This is obviously a major threat to the ecosystem and the global economy. This is especially true with developing economies with rich ecosystems, which is often the major source of their income for survival. The impacts of climate change in these economies are exacerbated by poor political structures and governance. If this trend persists, the well-being and survival of mankind is at risk.

Scientists, governments, and all well-meaning individuals must rise and take actions that will reduce dependency on fossil oil as well as embrace other alternative sources of energy that are affordable to most people. Renewable energy is one of the best options that humanity must explore for sustainable energy consumption and a cleaner atmosphere for the sustenance of the Earth, which is the only known and best home for humanity.

References

Ahove, M.A. (2001). *Environmental management and education: An introduction.* Lagos golden-pen books.

Ahove, M. A. (2015). Will pedagogy influence urban and rural students' knowledge attitude and anxiety towards climate change? *Ife Journal of Behavioural Research, 7*(1&2), 44−59.

Ahove, M. A. (2016). Research trends in climate change education: A vision for Nigeria. In P. Okebukola (Ed.), *Imperatives for Nigeria's development* (pp. 185−203). Lagos Nigeria: Okebukola Science Foundation.

Ahove, M. A. (2017). Palliatives for environmental impact of recession in Nigeria. In P. Okebukola (Ed.), *lessons from a recession for national development* (pp. 107−118). Lagos Nigeria: Okebukola Science Foundation.

Ahove, M.A., Okebukola, P.A., Oshun, G.O., Okebukola, F.O., & Bankole, I.S. (2016). Influence of nollywood film in mother tongue on students' learning outcomes on contemporary environmental concepts. *Poster session presentation at the annual meeting of the National Association for Research in Science teaching Baltimore, Maryland, U.S.A. 14th − 17th April, 2017.*

Baron, J. (2006). Thinking about global warming. *Climate Change, 77,* 137−150.

BBC Bitesize. (2017). National 4 chemistry: Fuels revision 1 Retrieved 18:01, March 3, 2017, from www.bbc.co,uk/education/guides/z8yj6sg/revision.

Blasing, T.J. (2013). Current greenhouse gas concentrations, doi:10.3334/CDIAC/atg.032, on CDIAC 2013 downloaded 1st February 2017.

Dlugokencky, E., & Tans, P. (2016). Trends in atmospheric carbon dioxide. NOAA/ESCC.downloaded on 6th February 2017 from www.esrl.noaa.gov/gmd/ccgg/trends.

Gordon, J. C., Deinis, T., & Havic, H. (2010). Global warming coverage in the media: Trends in a Mexico City Newspaper. *Science Communication*, 32(2), 143–170.

Grima, J., Filho, W. L., & Pace, P. (2010). Perceived frameworks of young people on global warming and ozone depletion. *Journal of Baltic Science Education*, 9(1), 35–49.

Intergovernmental Panel on Climate Change (IPCC). (2014). *Climate Change 2014: Impact adaptation and vulnerability, WG2AR5 summary for policymakers*. IPCC Secretariat, Geneva.

Miller, G. T. (1991). *Environmental science sustaining the earth*. Califonia: Wadsworth Publishing.

Miller, G. T., & Spoolman, S. E. (2012). *Living in the environment* (17th ed.). Australia: Brooks/Cole.

Omotayo, A. (2016). Are we living in a dying Earth, 56th Inaugural lecture series of the Lagos State University delivered on Tuesday, 13th December.

Singseewo, A. (2011). Awareness of environmental conservation and critical thinking of the undergraduate students. *European Journal of Social Science*, 25(1), 136–144.

Tella, A. (2016). Implications of COP 21 agreements for global national environmental security. In P. Okebukola (Ed.), *Imperatives for Nigeria's development* (pp. 167–184). Okebukola Science Foundation.

United Nation Environment Programme (UNEP). (2008). Green breakthroughs solving environmental problems through innovative policies and laws. United Nations Environment Programme, Nairobi.

United Nations Industrial Development Organisation (UNIDO) (2015). *Promoting climate resilient industry*. Vienna: United Nations Industrial Development Organisation.

Zhao, X., Rolfe-Redding, J., & Kotcher, J. E. (2014). Partisan differences in the relationship between newspaper coverage and concern over global warming. *Public Understanding of Science*, 23(1), 1–17.

PETROLEUM INDUSTRY CHALLENGES AND THEIR SOLUTIONS

Politics of State/Oil Multinational Alliance and Security Response

Fidelis Allen

University of Port Harcourt, Port Harcourt, Nigeria

19.1 INTRODUCTION

Oil has political, social, economic, and environmental character—within and between exporting and importing countries—that cannot be contradicted (Painter, 2014). This alone suggests alliances that usually result from pursuit of interests among actors in any political system. Besides, every economic system or practice is, first and foremost, embedded in a political process that results in policy. Governments and multinational oil companies (MNCs) usually are the main formal actors in the oil business. Over time, local content policies of government have brought in local companies, offering mainly subsidiary services. In all, the industry is characterized by interests that inform alliances and opposition at the local, national, and global fronts. Local actors, such as host communities, in many instances occupy marginal and informal positions of influence in matters of policy in the face of poverty, environmental despoliation, and inadequate company compliance to national regulations and laws. Despite economic benefits, alliances around the oil business appear dangerous to humanity, environment, and interests in many ways (Raphael & Stokes, 2011). Scholars tend to see an unholy alliance of MNCs and governments of producing countries. On the other hand, consumer and powerful nations in the developed world exert a great deal of influence in decisions and behavior of producing countries. Interpretations of security by MNCs, governments, and communities affected by activities of these companies have shaped their behaviors.

MNCs are linked to a few realities, making it difficult to exonerate them from dangerous politics in host countries. They have extensive influence over local actors. This is easily noticed in the informal and formal relationships they forge with national and local political, military, and economic elites. A section of the literature suggests an outrageous alliance between the companies and host governments. In fact, scholars point to the reality of an alliance between governments and MNCs in the politics of oil at national and local fronts of countries endowed with the resource. But these companies deny that reality. Depending on oil for the survival of the economy easily makes exporting countries desire alliances that promote growth through revenue and foreign exchange earnings (Majinge, 2011; Stopford, 1999). As Stopford argued, "the relationship between governments and multinationals is characterized by a complex distribution of benefits." Adequate attention has not been given to the dynamics of political relations of MNCs and governments around oil matters at the domestic level, especially in the case of Nigeria, from where most of the examples are drawn in this chapter. What is the nature of oil politics, alliances, and implications for security response? This chapter addresses this question, relying on the content analysis of qualitative data derived from the literature.

19.2 OIL POLITICS/ALLIANCE(S)

To understand the political character of oil, a rapid mapping of interests, actors, nature of alliances, implications for security, and responses is essential. Energy security, profit, political stability, national security, interests, revenues, foreign exchange, and power are easily linked to actors, including governments, MNCs, political elites, non-governmental organizations, local communities, and so on. At the global front, oil continues to shape foreign policies and relationships between importing and exporting countries. In 1970, in the context of an emerging oil boom and crises, Libya threatened to alter supplies (energy insecurity) to powerful consumer countries of the developed world. It was a period of rising demand by consumer nations (Levy, 1971), one in which oil-producing countries had started to experience increased revenues, upward of $8.5 billion, from $7 billion in 1970 and 1975, respectively. It was an emerging age of extensive reliance of countries on energy from fossil fuel. Oil has remained a key economic and political commodity, driving commerce, aviation, transportation, manufacturing, and prosecution of war. This is to say that countries are concerned about where and how to meet their energy needs. The United States is probably the number one country to look at when it comes to efforts by governments, or powerful countries, at ensuring energy security, in a world where those without the luck of having the oil have to depend on those that have. Even when a country has the commodity, it may have a need for importing products in order to meet local needs or check unhealthy behavior of countries that use the advantage of the oil against the interest of others. Some outrightly use the oil as an instrument of bargain and extraction of concession in international relations.

The case of the United States is well-illustrated in the title of John Coleman's book, *We Fight For Oil, A History of U.S. Petroleum Wars*. The relationship between that country and those in the Middle East has mainly been shaped by oil. The United States pointed to other reasons, though, in the case of Iraq some years ago. Preventing Iraq from developing weapons of mass destruction has been dismissed in many quarters as smokescreen. In reality, energy security and preventing Iraq from becoming a threat to regional security underlined the attack by the United States. Oil, at the global level, at the moment, is embedded in the nature of demand for energy security across climes with members of the G7 as centers of power. This motivates particular politics, with governments of importing and exporting countries, and transnational corporations playing influential roles. That is also to say that alliances or alliance structures exist at the global and national fronts. The G7, for example, is a powerful platform for exerting collective influence over global issues of interest. Member countries exert the same influence at the International Monetary Fund (IMF), the World Bank, and the World Trade Organization (WTO). These are powerful multilateral institutions, dominated by developed countries, with implications for international economic relations with the developing world.

Oil-endowed countries in the developing world are vulnerable to decisions taken in these institutions. Often, the dynamics of economic relations through decisions at these institutions have made oil-exporting countries in the developing world somewhat helpless, in terms of control. The rule of market forces at the global economic system and its spread to national economies, as a matter of policy, depicts dangerous political economic trajectories that offer little to the poor, other than mere growth, at the local fronts. Efforts at securing aid (loan) from the IMF or the World Bank, or bilateral agreements with richer countries by oil-exporting countries in Africa, are linked to interests. Receiving countries make commitments of importing needed skills and construction of facilities in the oil industry by donor nations. The cost and economic implications are usually against the national interest of the receiving nation. Nigeria is currently negotiating a loan facility from China to build its rail system. The labor and skill for constructing it have to come from China. The money expected to service the loan will come from the oil industry. Servicing loans often is a huge burden on citizens who were never part of the decisions for such loans in the first place.

At the global front, multilateral, inter-governmental, and bilateral relationships end up as alliance structures that exert enormous influence on the politics of natural resource production. In the first place, technology plays a crucial role. Developed countries wield immense power in this direction. The oil industry in Nigeria has depended mainly on western technology for its development. The refineries and maintenance are usually carried out by foreigners, who often are unwilling to transfer skills. Many employment and service-providing openings in the industry that require modest technological knowledge, for which local content laws would discourage foreigners from hijacking, are often taken by them (foreigners).

Instability in the price of oil at the global market puts exporting countries in the developing world in bad shape. But a solution to the problem is both a matter of governance at the national front and resistance of globalizing forces to policies that undermine the interest of corporations and governments of importing countries. At a time, when even the Organization of Petroleum Exporting Countries (OPEC) seems helpless about what to do in order to secure good prices and keep revenues flowing, the real question is how this contributes to insecurity and shapes response by relevant actors. In this case, a drastic fall from $100 plus to somewhere between $40 and $50 per barrel of crude oil at the global market in recent times, means a great deal in terms of national security, well-being, and financial inflows. That is, oil is conceptualized mainly from an economic growth perspective with national elites and transnational business groups as partners.

In an age of emerging new trends and skepticism around the ability of neoliberal capitalism to close widening gaps between poor and rich countries, corporations do actually continue to earn more disrepute. The oil industry has seen many risks with corruption and bribery cases associated with multinational corporations, desperate for licenses to operate. The ease with which some of them collaborate with willing national political elites to circumvent law in the course of pursuing their interests is an issue that the global energy system lacks proper legal instrument to check. It is interesting to note the clear absence of oil regimes with enforceable instruments at the global level. Powerful countries, where corporations originate and repatriate profits to, do not neglect their role of ensuring inter-governmental and multilateral settings conducive for the energy business. National governments are making efforts at checking the excesses of corporations, when it comes to violation of human rights of host communities, through formulation of relevant laws, but the bulk of those laws remain unimplemented, as the political will to confront the companies with punitive measures in an atmosphere of partnership remains a key hindrance.

There are approximately 20 countries in the world, with leading proven large oil reserves. Nigeria comes next with 37 billion barrels, after Libya's 48 billion barrels in the case of Africa. In other countries: Venezuela (298,4000,000), Saudi Arabia (268,300,000), Canada (171,000,000), Iran (157,000,000), Iraq (144,200,000), Kuwait (104,000,000), Russia (103,000,000), and United Arab Emirates (97,000,000) fall within the uppermost class of countries with large reserves globally (World Atlas, 2017). It is significant that the United States has 36,520,000 billion barrels in reserve and consumes approximately 25% of the total global consumption of oil. Many of the leading producers belong to the OPEC. Government policies and multinational corporation/government/host community relations are among the key areas in discussing oil politics at the global and domestic fronts. It is important to note that there are yet no enforceable global regimes to take care of oil-related environmental problems at domestic levels. This is sometimes attributed to the nature of global energy security needs with powerful countries such as the United States, France, the United Kingdom, Italy, and Russia always covertly collectively or individually checking multilateral and inter-governmental policies against risks likely to work against their interests in bilateral relations and at major settings such as the United Nations' Security Council, IMF, World Bank, WTO, and so on. Africa occupies a strategic position when it comes to investment by MNCs. Specifically, Nigeria has been a key destination in the case of West Africa. Nigeria's oil is rated high because of its low sulfur content, and it is admired by buyers in Europe and North America. In many parts of Africa, oil has been associated with dangerous politics. Many countries in North Africa and the Middle East are faced with conflicts and disturbing international energy interests that have painted these countries badly within the energy world.

Discovery of large oil reserves in North Africa in the 1950s galvanized European antics at ensuring control and benefit. France, in particular, saw a great opportunity for oil politics, as it started to explore areas of influence of countries outside conventional geopolitical control of the Soviet Union and the United States. The new oil states of North Africa became easy targets. The United States through its seven sisters—registered oil companies—dominated other oil territories (Muso, 2017).

Varied social, economic, and political alliances underhandedly exist, formally and informally, in a world in which oil has become a source of power and bargain (Muso, 2017). France promoted the idea of *EuroAfrica* as a strategy to maintain influence on North Africa in context of the emerging oil economy. In the case of Algeria, struggle for decolonization carried a nationalistic campaign against foreign dominance of the country's newly found oil resource. The commodity soon turned into a bargaining tool, against the colonialists.

Remarkably, MNCs operating in Algeria played a role. They influenced then-existing governments in the region and earned condemnation from those involved in decolonization struggles.

Energy security is of essence, but its insecurity in the 1950s and 1960s meant different political response for importing and exporting countries. Rising rate of consumption attracted disturbing concern on the part of Europeans whose governments needed to take decisive actions to protect their interests. One of the manifestations of that era was the formation of the Organization of European Economic Cooperation (OEEC). It sought to plug gaps by initiating relationships with North African countries with proven oil reserves. The United States, at that time, occupied an enviable position of having the privilege of being home country to 36% of global oil companies. Venezuela and Russia easily followed as next. The Middle East was running fast to emerge as a key producing region (Muso, 2017).

Prior to the 1950s, the global oil industry was controlled by an oligopoly of five American (Socony Mobil; Gulf Oil; Standard Oil, California; Standard Oil, New Jersey; and Texaco) and two European oil companies (Royal Dutch Shell and British Petroleum). These companies regulated production volume and price in the 1920s. Oil is the most globalized industry. This alone, to date, means that multinational corporations and home governments exert significant influence on policy and governance processes at the global and national fronts.

The local or national front provides more evidence of alliances or alliance structures that tend to shape actors' responses to insecurity. Issues have defined such alliances. They vary according to local conditions characterized by the *resource-curse* environmental, social, economic, and political disorder. This trend itself portends insecurity.

The relationship between these companies and government is mainly collaborative in the case of Nigeria. Convergence of interest is at the core of this type of relationship. The initial involvement of Royal Dutch/British Petroleum in the upstream sector (prospecting and production) in Nigeria during the colonial era is seen as part of a colonial project under British imperial policy. It is no surprise that no other company had license to prospect for oil in the entire country for a long time, before other companies emerged on the scene. Noteworthy is the mode of entrance of multinational corporations into the oil industry.

The joint venture approach with a national government, which by law claims total ownership of natural resources and serves both as investor and regulator, evokes particular forms of alliances, influenced by economic interests. This has implications for security decision-making. In the Nigerian case, this relationship is, more often than not, collaborative. The nature of government rules, policies, and regulations guiding the relationship with MNCs in the upstream and downstream sectors tend to create diverse forms of alliances. Between the government and the companies is securing license to prospect and produce. The state's oil company, NNPC, comes directly into the picture as representative of government in the joint venture relationship. Nigeria has 55% equity holding in this relationship. This is more or less a business relationship that many believe has discouraged the oil majors from fully complying with many government oil-related regulations. It is another way to say that this pattern of relationship weakens government regulatory agencies from responding adequately to security threats, especially the aspect of environmental security. This scenario is partly a result of government avoidance of its obligations or liabilities as a joint venture partner. For example, financial contributions towards acquiring technology for ensuring environment-friendly prospecting, production, and marketing is required by joint venture partners for which both the government and MNCs have failed to consider in a meaningful way.

It would appear that these companies believe government environmental regulations hurt production and profits (Sampson, 2011). Nothing explains well why government exhibits signs of weakness when it comes to enforcing relevant environmental laws other than interest. Ensuring uninterrupted pumping of product is crucial for sustenance of those in government and their plans for society.

This economic relationship is enslaving. It suggests a likelihood of undue influence of the companies on government processes. This was seen in the hanging of founder and leader of the Movement for the Survival of Ogoni People (MOSOP), Kenulu Saro-Wiwa in 1995 by Nigeria's then—military government under General Sani Abacha. Saro-Wiwa faced trial on murder charges after the deaths of four prominent Ogoni indigenes who died at the hand of angry youth members of MOSOP on accusations of unpatriotic position in the struggle of MOSOP against environmental injustice in Ogoniland. Saro-Wiwa was perceived to have challenged the collusion of MNCs and political leaders. Despite popular denunciation locally and internationally, the regime survived because of uninterrupted flow of oil money to the government. In other words, it was oil revenues that kept the regime going for a long time despite popular resistance and a self-inflicted pariah status. As argued by Bucheli and Aguilera (2010), the "relationship between a government and MNC depends on the power of the home government over the company's host country. The constraint this imposes on government action poses a threat to national, environmental and individual security."

When it comes to destructive conflicts such as wars, MNCs have been accused of playing negative roles in Africa (Patey, 2007). Throughout the years of the civil war in Angola, MNCs in that country continued to produce oil. Revenues were critical for the government as it depended mainly on it for the financial resources needed to keep the war on against rebel soldiers. Political survival of regimes depends on the oil companies that provide the money needed by the government through uninterrupted production and sales at the international market. This equally applies to Sudan, where Chevron (United States) and Canada's Talisman were forced out of the country for their role in that country's civil war that lasted 20 years. Oil companies contributed extensively to the revenues that supported the federal government in the prosecution of that war. Their ouster, however, paved way for state-owned oil companies from China, India, and Malaysia.

19.3 LOCAL- AND NATIONAL-LEVEL POLITICS OF OIL

As early as 1996, Jean Damu and David Bacon had written an article with the title, *Oil Rules Nigeria*. This suggests the power of oil in the politics of the country. MNCs give special attention to the politics of any country where they operate. In Nigeria's case, mutual sensitivity of government and the companies exists. Both think closely about the meaning of security. This alone put the companies in a political position. Historically, declaration of a state of Biafra in 1967 remains a moment to remember in context of the oil industry. Oil companies were glued to the federal government against the secessionist government in Enugu under the Biafran state, which had expected these companies to seize from further payment of taxes and other financial obligations to the Nigerian government. With an emerging booming oil industry and fair price at the international market, the Nigerian government had the benefit of financial inflow that helped her in tackling the rebels to the point of surrendering. But the rebellion itself was in part, a reflection of the character of growing ethnic and regional oil politics in the post-colonial state.

The growing importance of the oil industry to government activities has brought the companies closely into the economic, political, and security issues in government. The energy needs of powerful countries such as the United States, United Kingdom, France, and so on, influence public decisions about investments. This is controversially the case, when issues of oil policies and related environmental regimes are considered. As earlier noted, being a highly globalized industry attracts more of its investments from share-holders in the developed countries. Home governments of these companies constantly advise the companies on the security situation in their countries of operation.

The overall contribution of these companies to meeting the energy needs of the world is crucial. Nigerian laws such as those intended to ensure sustainable oil development and protection of the environment are hardly fully implemented. Nigeria has, at least, no less than 25 environmental laws yet to be fully implemented. It would appear that economic growth considerations dominate the behavior of relevant government agencies and their political masters. Certainly, economic growth aspirations tend to weaken government's commitment to ensuring full compliance with many of her laws.

As already alluded, Nigeria's Niger Delta, where the country's oil is presently being drilled, is also the site of intense local, state, and national politics, with local actors relating in different manners with MNCs and government. The relationship depends on their individual perception of security. This will be discussed more extensively in the next section of this chapter. In any case, the local political situation with local community people, hooped as youth associations, armed militia groups, traditional rulers, women, and community development committees easily carry messages of conflict and collaboration. Demands for economic opportunities, environmental protection, and basic social amenities such as roads, schools, health care, and scholarships by host communities bring these companies and the communities into conflict−prone relationships. These demands were initially non-violent, via visits and letters to the management of these companies.

Violence became an option when the internal politics of these companies, which took more of a security approach with the backing of government, failed to convince the communities of a positive disposition of both government and the companies towards their demands. The security approach involved the use of force. This became severe in the 1990s in the Ogoni case, which saw the hanging of Saro-Wiwa. The company was accused of supporting the hanging by doing little to convince then military government under Sani Abacha from carrying out the act.

The hanging of Saro-Wiwa was intended to put an end to the agitations by host communities in the Niger Delta, which will guarantee uninterrupted pumping of product. It was also meant to give the companies a better sense of security in terms of threats to their anticipated profit in the context of a joint venture relationship with government. But a legal perspective to his death sees retribution for the murder of four prominent leaders of Ogoni who died in the hands of angry youths of MOSOP. They accused their victims of sabotaging their struggle for environmental justice.

Resource control struggle is at the heart of many political conflicts in the Niger Delta. This is compounded by ethnic- and religious-driven dispositions of political actors who are involved in oil politics. Efforts are rife covertly by political actors at the federal government level to secure control of the oil resource. This became clear in the response of the government to threats posed to the industry in 1976 when it promulgated the Land Use Decree. This law made the federal government owner of all land and natural resources in Nigeria and tried to put an end to likely conflicting formal claims to land by different groups. It gave MNCs a better sense of positioning for seasons of exploration and production with limited interruption by the local community and groups. But communities in the Niger Delta have continued to argue that they are the real owners of the oil.

Oil politics is severe. In other words, the industry has a political character that, directly or indirectly, makes alliances inevitable. Anger of the local communities against damage to their environment and loss of livelihood due to the activities of oil companies remains a key motivator. Venting this anger and the response to it requires a relationship with government and the companies. This means regular cooperation between the duo to ensure swift response. MNCs have provided equipment and operational vehicles to the Nigerian police and kept close contact with the Navy and Army. They support the Joint Military Task Force (JTF) and have played a significant role in diverse ways to support the government at local, state, and federal fronts to enable them to tackle the anger of locals from a national security perspective.

The question of fair distribution of profits from production has remained unresolved. Local and state governments and the federal government receive varied portions of oil money from the Federation Account on a monthly basis, with the federal government having the largest portion. The power to determine how much each level of government should have has been a subject of much debate. Oil politics in the context of the relationship between state and MNCs and their response to security issues has a great deal to do with the rentier character of the state itself. Depending on oil by the government means that rent is her main source of income (Losman, 2010). This is rentier politics. Revenue sharing and perception of its injustice have generated animosity between groups in the Niger Delta and the Nigerian state. The sense of exclusion felt by the Niger Delta communities is one of the reasons for the agitations. The government under the leadership of Goodluck Jonathan engaged armed militia groups to provide security for oil facilities through various contracts worth millions of Naira. This arrangement has collapse since Muhammad Buhari got elected into office in 2015. Prior to this dispensation, the government led by Umaru Musa Yar' Adua initiated an amnesty program, which was successful to a large extent. Oil has contributed to massive corruption in Nigeria, and the present government led by Muhammad Buhari is making an effort to stamp out corruption from the country.

No one doubts the power of oil when it comes to shaping the political behavior of those who occupy political offices. One area in which this has to be checked against alliance between MNCs and the political elite is the nature and content of the oil contract. In many African countries including Nigeria, a contract signed between MNCs and the national government regarding different stages of exploration, development, and marketing of product involves a great deal of politics. It involves government, national oil companies, MNCs, local oil companies, banks, transportation and marketing companies, rig operators, and so on. Securing a license to operate comes with challenges for some of the companies due to politics of interest with political figures in charge of the oil business. Contracts between MNCs and national governments are important for understanding possible alliances in the politics of oil. The point being made is that a contract between the duo defines particular economic and political relationships. Contracts provide insights into the nature of the political economy of oil. Many contracts signed with governments during the colonial era and shortly after did not have the benefits of citizens' contributions.

Until now, efforts at formulating a comprehensive petroleum law to govern the industry have not yielded the desired result. A recent report of the Senate Joint Committee on the Petroleum Industry Governance Bill (PIGB, 2017) is devoid of voices of communities who are seeking a percentage of profits from the oil business. Previous versions of the draft bill by the National Assembly contained the provision of 10% of profits going to host communities. It is hoped that this provision will be retained to give the oil-producing communities a sense of belonging in the Nigerian oil business.

19.4 (IN)SECURITY RESPONSE

Perception of (in)security by actors in the oil industry in Nigeria, as with many countries in Africa endowed with the resource, is crucial for understanding the responses to threats and the appreciation of related socio-economic, political, and environmental issues. National and local governmental leaders who depend on the oil for allocation of public goods easily see national security in terms of protecting the oil industry for the purposes of uninterrupted flow of the commodity (Ibeanu & Luckham, 2006). Threats to facilities posed by activities of armed militias, which started in the 1990s, have continued to trouble government at all levels in Nigeria's Niger Delta. There was a time when the country and its political institutions depended on agriculture both for export and domestic earnings. As earlier alluded, the oil economy has made all the difference; making other sectors take a backseat when it comes to sources of revenue for the government. The three arms of the government—local, state, and federal-depend nearly entirely on oil revenues, except a handful.

National security calculations mean responses by government based on the idea of preventing damage to an industry responsible for supplying the bulk of money needed by government to provide services to the governed. Barrels of oil produced daily in Nigeria fell drastically from 2.5 million to less than 1 million due to the threats posed by armed militia. Different armed militia groups, including the Niger Delta Volunteer Force (NDVF) and the Movement for the Emancipation of the Niger Delta (MEND) became prominent globally for being responsible for a series of kidnappings of oil company workers and attacks on many oil installations. It was easy to understand their arguments of pollution and loss of livelihoods, as well as failure of government to hold oil companies to account for these issues.

MNCs logically see security in terms of protection of their investments and dealing with those who challenge their business relationship with the state. They maintain private police and help equip the Nigerian police, as well as support the JTF—Nigeria's main security ad hoc outfit for the protection of oil facilities in the Niger Delta—to do its job.

These companies have been accused of dividing communities with financial payments to selected people in order to make them unable to reach consensus in matters of community interest with the companies. Security of staff and facilities has time and again come under threat of violent aggression of armed militia groups in the region. Regular kidnapping of staff and blow up of facilities, which characterized the region, beginning from the 1990s, subsided shortly after a presidential amnesty was pronounced by the government under Umaru Musa Yar' Adua in 2009. Nearly 7 years after that amnesty, which was intended to address threats to production and secure facilities, events of the period beginning from 2015 to date have shown that security issues around the aforementioned remain mainly unresolved. New armed groups have emerged, threatening the oil industry on the grounds of environmental pollution and what they perceive as gaps in the distribution of the oil wealth.

National security perspectives have dominated response by the federal government to oil-related issues in Nigeria. Government does so in three main ways, namely: policy, legislation, and punitive (Udoh, 2013). The establishment of the Oil Mineral Producing Development Commission (OMPADEC) in 1992 by Ibrahim Badamasi Banbangida, the Niger Delta Development Commission in 2000 by the government of Olusegun Aremu Obasanjo, and the Niger Delta Ministry in 2008 by Umaru Musa Yar' Adua are a few policy measures made by the federal government in response to threats to national security. For legislations, creation of more states in the Niger Delta; improvement in the derivation policy, from 1.5 to 3% in 1992, and subsequently to 13% in 1999; and a Presidential Amnesty Program for militants remain the legislative responses of the government to deal with threats to national security. Punitive measures are best seen in the use of force by government security agencies and by the support received from MNCs. The hanging of Saro-Wiwa and crackdown on non-violent protests against oil companies by the JTF, Nigerian Police, and Department of State Security (DSS) is a cogent example of this approach to dealing with threats posed to national security. Punitive measures further provoked groups into arming themselves and responding with force. The emergence of the NDVF, Niger Delta Strike Force (NDSF), and those that emerged since 2015 point clearly to the fact that the issues are yet to be resolved.

With swelling global oil consumption reaching 75 million as of the year 2000 and projected to double by 2030, the home countries of MNCs, who are often high consumers or importers, by design or instinct have stakes in what happens to their interests within the industry for which actions might be necessary to check the risk of insecurity. Internal oil-related conflicts in many countries, including Nigeria, Angola, Sudan, and Indonesia, mean struggle for control of oil money. Theories of oil politics suggest that the product shapes politics and policy (Betts, Eagleton-Pierce, & Roemer-Mahler, 2006). Betts and his colleagues captured the relationship between state, business, and politics as follows:

> The oil sector represents a fascinating area for assessing the changing nature of state business relations. Companies such as Exxon Mobil wield enormous political influence within states that require external investments in order to exploit reserves, and the long term nature of contracts give them a significant stake within the provision of public goods and regulation within the state. Haliburton and TotalFina are among the many private oil companies with close links to politics, which has been widely documented (Betts et al., 2006).

Nigeria's current Vice President, Yemi Osibanjo, visited the Niger Delta region in February 2017 to explore ways of addressing issues in the region. He promised opportunities for locals to be more legitimately involved in the oil business through modular refineries. Steps such as this by government, if sustained, will bring peace to the region.

19.5 CONCLUSION

This chapter has discussed possible alliance(s) between MNCs and host governments in the politics of oil. Although this may not be clear at the global front, oil has continued to shape relationships between consuming and exporting countries with respect to politics and economy with attendant security implications. The chapter focused on the Nigerian case by looking at the nature of oil-related alliances in relation to host communities. The venture relationship between the federal government and oil majors means a particular type of relationships defined by interests.

References

Betts, A., Eagleton-Pierce, M., & Roemer-Mahler, A. (2006). Editorial introduction: The international politics of oil. *St. Anthony's International Review*, 2(1), 2−3.

Bucheli, M., & Aguilera, R. V. (2010). Political survival, energy policies and multinational corporations, a historical study for standard oil of New Jersey in Colombia, Mexico and Venezuala in the Twentieth Century. *Management International Review*, 50(3), 347−378.

Coleman, J. (2008). *We fight for oil, a history of U.S petroleum wars*. Available at: https://www.archive.org. Accessed 26.05.17.

Damu, J., & Bacon, D. (1996). Oil rules Nigeria. *The Black Scholar*, 26(1), 51−54.

Ibeanu, O., & Luckham, R. (2006). *Niger Delta, political violence, governance and corporate responsibility in a Petro-State*. Abuja: Centre for Democracy and Development.

Levy, W. J. (1971). Oil power. *Foreign Affairs*, 49(4), 652−668.

Losman, D. L. (2010). The rentier state and national oil companies: An economic and political perspective. *Middle East Journal*, 64(3), 427−445.

Majinge, C. R. (2011). Can multinational corporations help secure human rights and the rule of law, the case of Sudan. *Law and Politics in Africa, Asia and Latin America*, 44(1), 7−31.

Muso, M. (2017). Oil will set us free: The hydrocarbon industry and the Algerian Decolonization Process. In A. W. M. Smith, & C. Jappesen (Eds.), *Britain, France and the Decolonization of Africa*. UK: UCL Press.

Painter, D. S. (2014). Oil and geopolitics: The oil crises of the 1970s and the cold war. *Historical Social Research*, 39(4), 186−208.

Patey, L. A. (2007). State ryles: Oil companies and armed conflict in Sudan. *Third World Quarterly*, 28(5), 997−1016.

Raphael, S., & Stokes, D. (2011). Globalizing West African oil: US' energy security and the global economy. *International Affairs*, 87(4), 903−921.

Sampson, B. P. (2011). The effect of environmental regulations and other government control on oil and gas production. *Energy and Environment*, 22(3), 151−166.

Stopford, J. (1999). Multinational corporations. *Foreign Policy*, 113, 12−21.

Udoh, I. A. (2013). A qualitative review of the militancy, amnesty, and peace-building in Nigeria's Niger Delta. *Peace Review*, 45(2), 63−93.

World Atlas. (2017). Available at: http://www.worldatlas.com/articles/the-world-s-largest-oil-reserves-by-country.html Accessed 29.05.17.

Reactions to Petroleum Exploration From Oil-Bearing Communities: What Have We Learned?

Ebinimi J. Ansa and Ojo A. Akinrotimi

African Regional Aquaculture Center of the Nigerian Institute for Oceanography and Marine Research (ARAC/NIOMR), Port Harcourt, Rivers State, Nigeria

20.1 INTRODUCTION

The oil-bearing communities in the Niger Delta region are located in the southern part of Nigeria in the West African subregion. This region has the largest mangrove distribution in Africa and the third largest in the world (Ashton-Jones, 1998). The inhabitants of these areas derive a wide range of benefits from their natural environment, which include fish, timber, herbal medicine, and mangroves. These ecosystem services support their activities and livelihoods (Akinrotimi, 2012). According to Babatunde (2010), the region consists of important ecosystem components such as stable soil and favorable environment for different flora and fauna. It also serves as a breeding ground for many commercially important fishes.

The commencement of oil explorative activities in the region has led to dilapidation, degradation, and deprivation of the environment and livelihoods of the people, thereby negatively affecting the abundant natural capital and the variety of plants and animals in the environment (Ekire, 2001). The oil-bearing areas have been confronted with a myriad of environmental challenges caused by pollution from oil-related activities (Aworawo, 1999). Although it is obvious that the ecological effects of oil-related activities occur regularly in the region, it is highly imperative to observe that there are some certain environmental issues that are not associated with petroleum-related activities. The United Nations Development Programme report on the Niger Delta (UNDP, 2006), revealed that some environmental issues that are not linked to oil and gas productions nevertheless occur as a result of natural events in the hydrological systems in the Niger Delta. These include flooding, siltation, erosion, and an unusual tidal regime. Ansa,

Uzukwu, Okezie, Aranyo, and Apapa (2011) reported that silt pollution occurred in fish ponds due to daily tidal influence and subsequent settling of suspended silt in calm areas. In another study, it was observed that human endeavors other than oil- and gas-related activities that could change sediment characteristics include extraction of mangrove vegetation, dredging of sand from river bottoms, and expanding rivers and creeks (Ansa & Francis, 2007). However, oil-related environmental effects have been reported as a harbinger of one or more of these occurrences in the region (World Bank, 2007).

The Niger Delta region is a responsive and dynamic ecosystem (Akinrotimi, Edun & Williams-Ibama, 2015). In spite of its abundance of natural resources and great capability to contribute to the financial base, economic advancement, and general development of the nation, the region is subjected to monetary and environmental deprivations as well as societal pressures, which have generated a great deal of reaction from the populace living in the area. Recently, these subtle reactions have graduated to full-blown agitations, like a conflagration that is capable of consuming and crippling the entire economy of the nation (Torulagha, 2007). The thinking of the populace in these areas is that each successive government administration is acting carelessly and negligently in partnership with some oil companies to deny them of their rights, while the valuable ecosystem on which they depend for their survival and livelihoods is constantly being destroyed by crude oil explorative activities. As a result, the condition of things in these communities deteriorates to hostility, and this draws a lopsided response from those in authority, worsening the people's umbrage and sense of isolation. This chapter therefore focuses on the reactions generated by oil exploration and exploitation activities in oil-bearing communities in the Niger Delta region of Nigeria, lessons learned, and the way forward.

20.2 PETROLEUM EXPLORATIVE ACTIVITIES AND ENVIRONMENTAL EFFECTS IN THE OIL-BEARING AREAS OF THE NIGER DELTA REGION

Petroleum explorative activities in the Niger Delta axis and trading of oil and gas resources by the petroleum sector have significantly enhanced the country's financial fortunes for the last 40 years (Ajayi & Ikporukpo, 2005). However, activities associated with crude oil exploration have many damaging effects on the atmosphere, landscape, and aquatic resources in the region (Adeyemi, 2004). Constant release of crude and derived petroleum products have resulted in contamination of fresh and marine environments, unpleasant effects on the well-being of the populace, negative consequences on the regional financial system, socio-economic challenges, and destruction of the sources of livelihood of host communities (Babalola, 1999). The precise environmental consequences of oil explorative activities on the oil-producing areas are examined in this chapter so as to assess the degree of these degradations on the inhabitants' ability to create and maintain their sources of revenue.

20.2.1 Oil Spillage

Oil spillage is the major and most contentious of all the environmental effects of oil-related operations in the Niger Delta (Agahlino, 2000). Constitutional Rights Project (CRP)

(1999) described oil spills as unrestricted flow of every product associated with oil activity, which includes crude oil, chemicals or wastes traced to equipment collapse, operation accident, human mistake, or deliberate breaking of oil facilities. Moreover, oil spillage happens in the course of drilling oil wells or due to leakages from oil pipelines or at the point of loading of oil into the tanks (Adewuyi, 2001). Spills are probably the most destructive oil-related activities on flora, fauna, aesthetics, farmland, and aquatic resources (Aluko, 1999).

UNDP (2006) reported that much of the ecological degradation in oil-producing communities of the Niger Delta may be due to oil spillage, which could be as a result of accidents linked to human error, equipment failure, and deliberate destruction of oil pipes and pipelines. The report also stated that there were 6817 oil spills from 1976 to 2001; with a loss of nearly three million barrels of oil, out of which 70% could not be recovered. Approximately 6% of these spills were on land, 25% on mangrove swamps, and 69% in the offshore environment. The recent happenings (2011–16) in the region where several pipelines are being vandalized almost on a daily basis have exposed the region to untold environmental degradation.

20.2.2 Gas Flaring

Consistent release of gaseous fuel into the environment in the course of oil and gas operations, known as gas flaring, is another important impact of oil exploration on the Niger Delta environment. It creates a ceaseless, excessive, and intense flame that burns continuously (day and night) and further exposes the area to unquantifiable environmental degradation (Eniola, Olusule & Agaye, 1983). One of the by-products of oil extraction is natural gas, which is eliminated from the Earth's crust along with the crude oil. The World Bank (2007) report showed that gas flaring has been known to be the single highest contributor to the problem of global warming in the Niger Delta. Similarly, Orubu (1999) opined that greenhouse gases, including methane and carbon dioxide, emitted during gas flaring make significant contribution to global warming, which may result in sea level rise, accelerate climate change impacts, and if the harsh living conditions on our planet are unchecked, it could also lead to adverse effects on flora, fauna, and the human population. Estimates have shown that the total emission of carbon dioxide (CO_2) from gas flares during oil exploration accounts for the highest quantity of global gas emission (Alakpodia, 1990). The quantity of gas flared during petroleum operations in Nigeria is 3 times the OPEC average and is about 16 times the global average (Ajayi & Ikporukpo, 2005; Egwaikhide & Aregbeyen, 1999).

Gas flaring makes economic sense to the oil manufacturers. Flaring destroys the environment as host communities suffer the adverse effects of gas flaring as well as the accompanying trans-boundary and far-reaching consequences. The negative impact of gas flares is captured in the statement of a youth leader in the Niger Delta who stated that "the roofs of our houses have been severely corroded as a result of acid rain caused by gas flares, which also affects aquatic fauna and flora leading to a reduction in fishery" (Energy Information Administration (EIA, 2006)). Furthermore, Ibeanu (2000) reported that the gas flared in Nigeria in 1991 alone exceeded the global average by 72%; the world average of gas flared as a percentage of total production was 4%, whereas Nigeria flared 76% of her production.

20.2.3 Drilling Activities

Reports have shown that drilling activities produce chemicals capable of contaminating aquatic ecosystems such as streams and rivers through the release of materials into the environment during drill cutting and when drilling mud and fluids are used to stimulate production. When the main components (barite and bentonite clays) of drill fluid are dumped on bare ground, it could prevent or retard plant growth until a new topsoil is developed by natural processes. In aquatic ecosystems, benthic organisms could be suffocated when those materials disperse and sink to the bottom of the water body (ANEEJ, 2004).

20.2.4 Construction of Canals

In oil-bearing communities, many waterways have been created and constructed to bring in heavy drilling equipment. According to the Environmental Rights Action (ERA, 2000), this action affects the hydrology, disturbs the ecosystem, and disrupts the breeding grounds of most fish species, thereby affecting the livelihoods of the people living in these areas. Besides, these artificial canals, according to Onosode (2003), not only aid the intrusion of salt waters from the Atlantic Ocean into freshwater ecosystems, they are also the main cause of the scarcity of drinking water in many areas. In addition, the salt water kills many species of plants, animals, and fishes that are salt-intolerant. The construction of these canals has significantly changed the whole ecosystem, as freshwater is contaminated and silt pollution occurs due to salt water intrusion and agitation of bottom sediments of river courses, respectively.

20.2.5 Dredging Activities

Dredging is the removal of silt at the bottom of rivers, streams, creeks, and seas. It represents yet another environmental impact caused by oil exploitation in the oil-producing regions. Dredging has a dual impact on the ecosystem; it causes destruction of the ecology of the excavated areas as well as the areas where the sands are dumped. Although excavated materials are dumped on land, some get washed back into the aquatic environment through surface runoff, thereby increasing the turbidity of such water bodies and reducing sunlight penetration. Reduction in sunlight penetration into water bodies inhibits photosynthesis; this affects primary productivity by plants and indirectly reduces fish population in natural water bodies such as streams and rivers, whose sustenance is from plant life. Dredged materials in mangrove habitats produce humid acid as the mangrove parts decay. Acid production is facilitated on exposure to atmospheric conditions. Thus, dredged silt from canalization that is dumped on cultivated lands can reduce farm output (Agbo, 2005). In another study, Oribhabor, Opara and Ansa (2011) also reported that canalized water bodies have lower fish diversity than natural rivers; the authors attribute this to a "loss of heterogeneous habitat".

20.2.6 Incidence of Coastal Erosion

A great deal of the oil-bearing areas of the Niger Delta, which borders the Atlantic Ocean, are affected by coastal erosion. UNDP (2006) reported that this erosion may have been caused by oil- and gas-related activities through the construction of canals, jetties, shore crossing pipelines, and moles. Erosion occurs daily as sea waves break on the shore, and soil erodes and washes into the sea. Oil-bearing communities in the Niger Delta are constantly under the threat of coastal erosion (Babalola, 1999).

20.2.7 Discharge of Effluents and Wastes From Oil and Gas Industries

Constant release of wastes into the aquatic environment from oil-related industries is commonplace in many communities in the Niger Delta. Such wastes contain excessive quantities of toxic materials such as heavy metals. These metals are bio-accumulated by aquatic organisms and could be bio-magnified along the food chain, which cuts across aquatic boundaries to terrestrial life. As the contaminants are sequestrated in living systems, there is the tendency to transform into toxic compounds that can easily be consumed by humans at the terrestrial end of the food chain. Some of these toxicants have been marked as carcinogens and teratogens, with potential deleterious effects on human populations. According to Environmental Rights Action (ERA, 2000), a great deal of the underground waters in oil-bearing communities in the Niger Delta are polluted with dangerous metals and chemicals.

There are cases of oil companies that discharge production water that is polluted with crude oil into nearby rivers and creeks without adequate treatment. Toxic chemicals, including sludge that was removed from the bottom of storage tanks during maintenance activities, are also disposed into these water bodies untreated (Okonta & Douglas, 2001). Oil leaks from equipment and storage tanks as well as hydrocarbon vapor that evaporated from tanks also contribute to environmental pollution and have subjected the soil and aquatic ecosystems within the vicinity of the oil terminals to slow but relentless devastation (Orubu, Odusola & Ehwarieme, 2004).

20.3 REACTIONS OF PEOPLE IN OIL-BEARING COMMUNITIES TO EXPLORATIVE ACTIVITIES

Spontaneous and planned reactions to the real and perceived injustice by people in oil-bearing communities in the Niger Delta are numerous, depending on the community, circumstances, and the severity of the case in question. However, Anikpo (1998) identified some reactions common in the oil community, some of which are highlighted below:

1. **Protests and Demonstrations**: These are considered the least severe or relatively harmless type of disturbance by oil-bearing communities. Usually the youths of such communities would cut down logs from trees or use drums or any tangible thing to block strategic roads. In some cases, they may set discarded tires on fire to cause road block and draw attention. They may also raise posters and placards bearing their demands and grievances and restrict human and vehicular movement, especially those associated with oil companies.

2. **Disruption/Stoppage of Operation**: This is a situation in which protesters gain access to the premises of an oil company and interrupt their activities, which may eventually lead the closure of the facility, particularly flow stations.

3. **Closure of Flow Station and Rig/Molestation of Oil Company Staff:** A great deal of risk is involved in this type of crisis, for example, the Ogoni case. This type of crisis can result in the burning down of flow stations as well as the destruction of lives.

4. **Vandalism/Destruction of Facilities:** This type of protest goes beyond molestation of staff and forceful closure of oil installations. The protesters vandalize and destroy oil company facilities in their community (Anikpo, 1998).

5. **Piracy/Temporary Seizure of Vehicles or Boats:** Piracy is the temporary interception of boats and vehicles for financial extortion or forceful dispossession of belongings on waterways as well as the capture and detention of boats and their passengers.

6. **Hostage Taking:** The forceful capture and detention of boats and their passengers, which may lead to the starvation, or murder, of the captured by aggrieved persons. Extreme cases of hostage taking could result in a paramilitary or military intervention to either free the hostage or avenge their killing.

7. **Formation of Millitias:** There has been a transformation of the activities of ethnic militia groups in the Niger Delta region, from struggle against deprivation and marginalization to criminality because of frustration caused by loss of livelihood and environmental degradation (Raji, 2008). The worsening socio-economic conditions of the inhabitants and the lack of political will by the Nigerian State to proffer a permanent solution to the challenges confronting the region has resulted in the proliferation of militia groups, which constantly attacks the Nigerian State (Adejumobi, 2000). The actions of the militia groups on oil installations have affected socio-economic activities in the country and have left the polity in a very precarious state. Sustained exploration and production activities in the region without strategic and holistic efforts aimed at managing the impacts of these activities have resulted in undesirable consequences. As a result, the region has become an unstable area in the Nigerian State where access to revenue and other social amenities has become a source of violence as a result of the constant agitation that is prevalent in the region (Adejumobi, 2002).

8. **The Amnesty Program:** The amnesty program, which started in 2009 by the late ex-President, Umaru Musa Yar'Adua, was seen as a welcome development by members of oil-bearing communities in the Niger Delta region. The outcome of the program does not fully reflect what it was set out to achieve. The militants were asked to drop their arms but were paid for doing so. It is also alleged that the amounts paid in the "cash for arms deal" was actually inflated (Nwaze, 2011). At the end of 2016, militancy is yet to be a thing of the past as kidnaping for ransom, pipeline bombing and vandalism, setting up of illegal refineries, and other criminal vices still thrive in the region.

9. **Poetic Interventions:** In his book *Live 2 Lives* a Niger Delta monarch, His Royal Highness Christian Otobotekere, explains the problems of the effect of oil exploitation and youth restiveness, and the remedies and roles of government, the oil companies, and citizens in resolving issues confronting the very existence of the Niger Delta

(Otobotekere, 2009). Three of the poems in the book have striking information that deals with the topic discussed in this chapter. The titles of the three poems are "Irrepressible Ones", "WAI Universal", and "Closing Fast".

The devastating effects of the exploration and exploitation activities in search of the "liquid gold" on the ecology of the Niger Delta is portrayed in Christian Otobotekere's poem:

"Closing Fast (ii)" (Otobotekere, 2009).
"Closing fast (ii)"
The drama is closing: drama
Of swinging freedom
And of regal chairs;
Drama of unsinkable war boat
In any surf or weather

Of pride of place
And light of excellence
In many a known field,
As much as the ease
Of juicy herbs
With potent flavour

The drama is closing:
The drama of drum and gong
And rhythmic thrill,
Of festive song
With spiralling trill

The drama is closing
Closing fast
Of resources untold,
Even of the whiz and buzz
Of dark skinned beetles
Fan — drying the frothy white
Of foaming raffia palms.

The drama of Wetlands!
Is the drama of treasures
Drama of liquid gold
Siphoned unnoticed,
Closing unnoticed!
The drama is closing
Closing fast.

The "Irrepressible Ones", as the title implies, show the nature of the youths in the Niger Delta. They are uncontrollable and influenced negatively by several vices, frequently agitated and taking laws into their own hands. There is also a call on the leadership, who the author does not absolve from blame, and a call for peace and appreciation of nature.

"Irrepressible Ones" by Christian Otobotekere

Pax vobiscum
Pax vobiscum
Irrepressible ones,
Trigger happy
In and out of season

Peace be unto you
Irrepressible aides
Slotting and slotting
Your dreadful gadgets

Peace be unto you
Warring princes, personae agitate
Hurrying and worrying about
Bout upon bout

Peace… peace… peace
Warriors and palace singers
Victims of strife, slaves to others,
Void stakeholders
Replot your route and ditch
Those finger-staining packets

Opatrons and sponsors
Oppressors and sycophants
Leave them alone,
Let them reform.
Innocent young ones

Dear to society
Your masters are to blame
But step this way
And see mirrored,
On clean waters here,
Designs and mercies
More than earthly riches.

You need to know
Violence makes animals of men
Brutalises decency!
Misled "personae agitate"
Stop it. Stop it.

"Peace... Peace... Peace..."
The balm of blinding rage,
The staff of riper age,
Champion of champions

> *The alarm of breezy conquest*
> *For booty*
> *Is dwarf*
> *To the flag of peace*
> *Peace rules war.*

Brother, sister, you need to
Pull free from the trap
Of sparkling bottles,
From splintering bottles

From sore red eyes
And gashing raw wounds!
From pains and shouts
Into the bliss
Of "kookoo" echo,
A new lullaby sound:

> *A fine echo of peace*
> *Lulling, lulling, lulling*
> *With softest — solfa*
> *Upon this motherly*
> *Silver — toned river*
> *Yes, without strife,*
> *Without pain,*
> *Without stress:*

A soft — fluting tune
Of love — toasting lullaby
For health, beauty, love
Aflow with clear caressing ripples
Upon clean sand reaches
Of love — lapping lullaby:

> *Lapping-lapping-lapping*
> *Lapping-lapping-lapping*
> *Upon the shore of peace.*

III. PETROLEUM INDUSTRY CHALLENGES AND THEIR SOLUTIONS

And in his poem, WAI Universal, the author clearly reacts to the changes that have occurred in the environment previously known to support wildlife, both aquatic and terrestrial, but that has been ruined by all, and he ends this classic poem with the rhetorical question "Should it be?"

"WAI (War Against Indiscipline) Universal" by Christian Otobotekere

The lure of our bush, where
The monkey swings in nimble run,
Where the voiced-parrot, shuttle neck'd;
And the bush squirrel, showing its wavy tail, slowly,
Or the sacred reptile, ambulatory, still exist.
Time was when
These inmates frolicked with our parents,
Over nuts and palm fruits red. Beautiful!
They beckon us still
But, O, unheeded.

The meadowland Eden
Of my mother's farm
And tubers, arm's length, abound
Yet unseen by my daughter young
Who, straining skyward, is
Sweating and oil-lipped with
Fruitless cosmetics.

What of the flowing river?
Your river, my river,
Beauty of beauties now splashy
With concocted chemicals
Of dare-devil youth
Who break the laws of the land
Here, there, and upstream, in trial of
Their puny strength
Also no more of
The land-to-river crocodile
Mystically flouting
And quietly sinking
By day or by moonlight
What have you done
With the village-friendly Iguana
Safely sneaking around village dustbin?

Where and when again can we see
A fat reptile floating, floating,
As easy booty for kids?
Or hippo-kid playing around
At meadow shallows?

Or, as a common day affair
See the great croc,
Five meters long, basking unconcerned
At the cape of a golden sand bank?

Which, as you approach,
Waddle up the sand bank, harmless.
O yet see — at the water's edge
Its senior brother, live!
After a show of its keen eye, glides
Into a long-smooth-wake, into mid-river,
And there, sinking gently, nose last
With all its waves!
Bon voyage' to your
Tiny dug-out.

When and where again this thrill
At the borders of safety?

To add to our losses,
The heartless log man
Bares nature's woodland, suddenly,
Of trees, flower and bird
And of its primal beauty
For a mess of pottage,
A few bottles of toxic wine
Or for new adventures wild

Or is it the oil-man,
The universally branded culprit?

The oil man strapped with swamp
Bulldozer;
From shore to land
From land to shore?
The oil-man, who flanked by
Fat-looking hostage men,
Takes all to the Government,
A Government that knows how to take
But not how to give,
Nay, prevents the oil-man
From serving us decently?
What remedy now?
As man destroys his own heritage,
Daily grows reckless
And with inflexible decrees
Or new-found toys
Rams his own edifice!

O kids, O men of my race,
O men of this pretty little spinning earth,
You are setting fire
To WAI-universal,
Your very nature.
Your very future!
Should it be?

III. PETROLEUM INDUSTRY CHALLENGES AND THEIR SOLUTIONS

10. **Reactions to Educational and Agricultural Support:** In a bid to obtain favor and foster good working relationships with their host communities, several oil companies in the region have taken steps to ameliorate the strained relationships they have with the people hosting them across the Niger Delta region. These actions vary from provision of basic amenities such as electricity and potable water, to educational support through construction of classrooms and the provision of educational books/materials and laboratory equipment in existing community schools. Similarly, some of the companies, including Chevron, Shell Petroleum Development Company and their joint venture partners, provide a limited number of scholarships to promising students from the host communities to enable them to further their education. Other actions taken on the part of the oil companies include provision of agricultural support services and capacity building for individual farmers, and cooperative society members and women's groups.

The Managing Director/Vice Chairman of the Nigerian Agip Oil Company (NAOC) AENR NAE, Mr. Massimo Insulla, during the occasion of the twentieth Farmer's Day celebration that took place on November 19, 2007, stated that the Green River Project (GRP) annually supports 2500 farmers through their plant propagation centers spread across their areas of operation, namely Rivers, Bayelsa, Delta, and Imo States. The centers develop and provide farmers with improved resilient varieties of planting materials, good quality fish fingerlings, and extension services to help boost farmers' production and enhance their socio-economic status. Besides, GRP also supports the farmers by strengthening their abilities to participate in micro credit schemes that provide loans for farm business development. The company also supports value addition to farmed products and assists farmers with needed skills through two schemes, namely the "Agro Skills Acquisition Scheme" and the "Agro Processing Endowment Programme". The pomp and pageantry that are attached to the Farmer's Day celebration by stakeholders and recipients of the farm tools and equipment provided by the NAOC Green River Project indicate that this program is accepted by the people of the Niger Delta. Some of the beneficiaries corroborated this in their reports presented during the twentieth Farmer's Day celebration. The Leader of Ndoni Dynamic Co-operative Society made a statement that the fortunes of her members had changed for the better since coming in contact with the NAOC Green River Project. From receiving technical support from GRP, the co-operative was formally registered with the Ministry of Commerce, Trade and Industry on August 3, 2011. They have made progress in crop farming as well as fish and livestock production. They are also involved in the processing and sale of frozen chicken. Another beneficiary of the NAOC Green River Project scheme is the Nrizuroke Women Farmers Credit and Investment Cooperative, Mgbede, a solely women's group involved in trading and farming of cassava, plantains, cucumber, and the relished indigenous green leafy vegetable, *Telfairia occidentalis*, commonly called *ugu*. The group stated that they were groomed and mentored by the GRP Extension Officer, who also facilitated the prompt registration of their group with the state government. The women's group also received a donation of an improved cassava-processing mill from GRP. In addition, group members received training on "improved farming techniques, seed multiplication, and post-harvest storage" and GRP also gave them improved planting materials to boost their yield. Some of the widows in the group also benefitted from "Green card Micro Credit Facility" to give a fillip to their business.

It is worthy to note that the NAOC is one of the few oil companies that have complied with the 1986 directive of the federal government of Nigeria for oil companies to diversify and carry out projects to support agricultural development. On the other hand, other companies such as SPDC, Chevron, Total, and Mobil have concentrated less on agriculture when compared to the NAOC but more on educational support and capacity building.

20.4 LESSONS AND WAY FORWARD

The sustenance of petroleum exploration and production in the Niger Delta is hinged on a strategic and holistic framework for sustainable management of oil-related impacts on the environment. There is reduction in arable farmlands and pollution of water, which have impacted negatively on aquaculture and fishing activities. This has worsened the prevalence of poverty, unemployment, and general hopelessness in the midst of plenty. The rate of poverty in the region has been on the rise, and regional imbalance in resource allocation further compounds the issue as oil revenue is not equitably distributed, especially among the oil-bearing communities. These have generated a great deal of crisis within the region (Imobighe, 2004). These crises have hindered a great deal of developmental projects in the region. Many viable and flourishing industries have folded up, and many have relocated. Some individuals have moved out of the region due to the fear of being kidnaped. In the midst of these unpalatable situations, there is a need for peace and dialogue as a way out of the crisis (Obi, 2002).

Youth empowerment is one of the major key elements that will bring about lasting peace in the region. In 2009, the government of the Federal Republic of Nigeria declared amnesty from criminal prosecution to militants with a promise to rehabilitate and reintegrate them back into the society through different programs. However, recent events indicate that the amnesty program was not as successful as earlier anticipated. The amnesty program will achieve the desired results if properly managed.

The oil industry should be restructured. This is now necessary so as to give the oil-bearing communities the right over their God-given resources, by amending the outdated 1969 Petroleum Act and replacing it with the Petroleum Industry Bill (PIB) presently before the national assembly. The PIB, when enacted into law, will increase local participation in the petroleum sector as well as empower the oil-producing communities through equity, fairness, and justice. Oil companies must devise strategies and mechanisms to contain and resolve community-induced crises by establishing a community relations department saddled with the task of crises management and conflict resolution. The community relations units must participate in various activities concerning the existence of their companies. They should serve as a link between the company and host communities (Albert, 2001).

20.5 CONCLUSION

The richest part of Nigeria in terms of natural resources endowment is the oil-producing communities of the Niger Delta. The region has large deposits of crude oil and gas, extensive forest, productive agricultural land, and abundant as well as diverse aquatic resources. However, the oil-bearing communities remain marginalized in all ramifications; economic, social, environmental, and political activities in Nigeria. This has generated many reactions, which has had negative impacts on the communities. As the oil-bearing communities can no longer endure the poverty, gross social infrastructural neglect, ecological devastations, and other deprivations they suffer despite their huge contribution to national development, their response to the apparent failure or inability of successive the Nigerian government to protect the people and their land from the hazardous activities of the oil companies such as pollution, incessant gas flaring, oil spills, human right violations, and accompanying economic stagnation and impoverishment has resulted in the clamor for resource control.

In view of the foregoing, there is need for special programs for oil-producing areas that recognize the local communities as key factors in poverty eradication and successful rural development. Emphasis should be on strengthening grassroots organization and resources availability for community development projects. The program that intends to address rural poverty must identify and target the most vulnerable, empowering them to participate effectively in development activities.

Oil companies are obliged to adopt all practicable pre-conditionsm including the provision of up-to-date equipment to prevent pollution and must take "prompt" steps to control and, if possible, mitigate pollution if and when it does occur. All oil installations must be maintained in good condition so as to prevent leakages or avoidable waste of petroleum. For improved workers' performance and peace between oil companies and their host communities, there must be adequate community relations. Oil companies should not spare any effort at fostering good relationships with their host communities, and these can include embarking on community development projects that address the very needs of the people. Community youths should exercise patience with oil companies and the government. Oil companies should also open and maintain channels of communication between them and the oil-bearing communities. Government policies have a major role to play in maintaining peace in the Niger Delta.

The interventions in the Niger Delta region have been described by a crop of Niger Delta authors as mere placebos, since most interventions are superficial and do not tackle the main problems from the roots (Aaron & George, 2010). To maintain peace and stability, there is need for inclusive decision-making processes geared towards the development of the oil-bearing communities in the Niger Delta region, described as the goose that lays the golden egg (Ukiwo, 2010). Between January and February 2017, the Acting President of the Federal Republic of Nigeria, Professor Yemi Osinbajo, was on tour of the region and has had town hall meetings with major stakeholders, including state governments, civil society groups, youth organizations, religious leaders, women's groups, and traditional leaders. It is hoped by the people that these town hall meetings will not be mere talk sprees but will translate into tangible development strategies that will impact on the lives, well-being, and prosperity of the people in the Niger Delta.

Finally, oil revenue distribution in Nigeria requires a major change to reverse the poverty presently ravaging the region, where crude oil has been drilled for the past 50 years. The social and physical infrastructures in the Niger Delta should be improved to meet the growing population of the region. Other interventions include equitable resource distribution, functional local government, effective conflict resolution mechanism, and sustainable resource exploitation. The present Niger Delta situation demands an urgent political, economic, social, cultural, environmental, and spiritual re-engineering, which must take cognizance of the immediate and future needs of the people.

References

Aaron, K., & George, D. (2010). Introduction: Placebo as medicine. In K. Aaron, & D. George (Eds.), *Placebo as medicine: The poverty of development intervention and conflict resolution strategies in the Niger Delta region of Nigeria* (1st ed., pp. 1–18). Port Harcourt: Kemuela Publications.

Adejumobi, S. (2000). The Nigerian crisis and alternative political framework. In S. O. Akhaine (Ed.), *Constitutionalism and national question*. Lagos: Centre for Constitutionalism and Demilitarization.

Adejumobi, S. (2002). The military and the national question. In A. Momoh, & S. Adejumobi (Eds.), *The national question in Nigeria: Comparative perspective*. Aldershot: Ashgate.

Adewuyi, A. O. (2001). The implications of crude oil exploitation and export on the environment and level of economic growth and development in Nigeria. In *A Paper presented at Nigerian Economic Society annual conference* (pp. 119–145).

Adeyemi, O. T. (2004). Oil exploration and environmental degradation: The Nigerian experience. *Environmental Informatics Archives, 2*, 387–393.

African Network for Environment and Economic Justice (ANEEJ). (2004). *Oil of poverty in Niger Delta*. A publication of the African Network for Environment and Economic Justice.

Agahlino, S. O. (2000). Petroleum exploitation and agitation for compensation by oil producing communities in Nigeria. *Geo-Studies Forum: An International Journal of Environmental and Policy Issues, 1*(1 and 2), 11–20.

Agbo, O. (2005). Oil and environmental conflictIn H. A. Saliu (Ed.), *Under democratic rules (1999–2003)* (Vol. 2Ibadan: University of Ibadan Press PLC.

Ajayi, D. D., & Ikporukpo, C. O. (2005). An analysis of Nigeria's environmental vision. *Journal of Environmental Policy and Planning, 7*(4), 341–365.

Akinrotimi, O. A. (2012). Issues limiting the expansion of brackish water aquaculture in the coastal areas of Niger Delta. In *Proceedings of the 26th annual conference of the Fisheries Society of Nigeria* (pp. 169–178), November 28–December 2, 2011, Federal University of Technology, Minna, Niger State, Nigeria.

Akinrotimi, O. A., Edun, O. M., & Williams-Ibama, J. E. (2015). The roles of brackish water aquaculture in fish supply and food security in some coastal communities of Rivers State, Nigeria. *Journal of Agricultural Science Food Technology, 1*(1), 016–019.

Alakpodia, I. J. (1990). Effects of gas flaring on the micro-climate and adjacent vegetation on Isoko Area of Bendel State. Unpunished M.Sc. Thesis, University of Ibadan.

Albert, I. O. (2001). *Introduction to third party intervention in community conflicts*. Ibadan: Petraf/John Archers.

Aluko, M. A. O. (1999). Social dimension and consequences of environmental degradation in the Niger Delta of Nigeria: Suggestions for the next Millennium. In A. Osuntokun (Ed.), *Environmental Problems of the Niger Delta*. Lagos: Friedrich Ebert Foundation.

Anikpo, M. (1998). Communal conflicts in the East Niger Delta: A cultural matrix. *Pan-African Social Science Review, 3*, 1–12.

Ansa, E. J., & Francis, A. (2007). Sediment characteristics of the Andoni flats, Niger Delta, Nigeria. *Journal of Applied Sciences and Environmental Management, 11*(3), 21–25.

Ansa, E. J., Uzukwu, P. U., Okezie, S. O., Aranyo, A. A., & Apapa, E. U. (2011). Silt pollution of tidal fish ponds in the Niger Delta, Nigeria. *Continental Journal of Environmental Sciences, 5*(2), 13–18.

Ashton-Jones, N. (1998). *The human ecosystems of the Niger Delta: An ERA handbook*. Benin City, Nigeria: Environmental Rights Action Publication.

Aworawo, D. (1999). The impact of environmental degradation on the rural economy of the Niger Delta. In A. Osuntokun (Ed.), *Environmental problems of the Niger Delta*. Lagos: Friedrich Ebert Foundation.

Babalola, M. A. (1999). Impact of oil exploration on the environmental protection and the petroleum industry. In Proceedings of environmental protection and the petroleum industry. UNILAG Consultant.

Babatunde, A. (2010). Environmental conflict and the politics of oil in the oil-bearing areas of Nigeria's Niger Delta. *Peace & Conflict Review, 5*(1), 1–13.

Constitutional Rights Project (CRP) (1999). *Land, oil and human rights in Nigeria's Delta*. Lagos: CRP.

Egwaikhide, F. O., & Aregbeyen, O. (1999). Oil production externalities in the Niger Delta: Is a fiscal solution feasible? In *Fiscal federalism and Nigeria's economic development*. Selected papers presented at the annual conference. The Nigeria Economic Society, Department of Economics, University of Ibadan, Ibadan, Nigeria.

Ekire, S. (2001). *Blood and oil*. London: Centre for Democracy and Development.

Energy Information Administration (EIA). (2006). *Nigeria country analysis brief*. Retrieved December 3, 2016 from: <http://www.eia.doc.gov/emeu/cabs/Nigeria/Full.html>.

Eniola, O. A., Olusule, R., & Agaye, G. (1983). Environmental and socio-economic impacts of oil spillage in the petroleum riverine areas of Nigeria. In *Proceedings of the international seminar on the petroleum industry and the Nigeria environment*. Lagos: NNPC.

Environmental Rights Action (ERA). (2000). The Emperor has no cloths. *Report of proceedings of the conferences on the people of the Niger Delta and the 1999 consortium*. Port Harcourt, November 24, 1999.

Ibeanu, O. (2000). Oiling the Friction: Environmental Conflict Management in the Niger Delta, Nigeria. An Environmental Change and Security Profile Report, Issue No 6, Woodrow Wilson Centre.

Imobighe, T. A. (2004). Conflict in the Niger Delta. A unique case or a model for future conflicts in other oil producing countries? In R. Traub-Merz (Ed.), *Oil policy in the Gulf of Guinea: Security and conflict, economic growth, social development*. Bonn, Germany: Fredrich Ebert Stifting.

Nwaze, C. (2011). *Corruption in Nigeria exposed with cases, scams, laws and preventive measures*. Lagos: Control & Surveillance Associates Limited, 419 p.

Obi, C. (2002). Oil and minority question. In A. Momoh, & S. Adejumobi (Eds.), *The national question in Nigeria: Comparative perspective*. Aldershot: Ashgate.

Okonta, I., & Douglas, O. (2001). *Where cultures feasts: Forty years of shell in the Niger Delta*. ERA/FOEN: Benin.

Onosode, G. (2003). *Environmental issues and the challenges of the Niger Delta: Perspectives from the Niger Delta. Environmental Survey process*. Yaba: The CIBN Press Limited.

Oribhabor, B. J., Opara, J. Y., & Ansa, E. J. (2011). The ecological impact of tidal pond channelization on the distribution of tilapia species (Perciformes: Cichlidae) on Buguma Creek, Rivers State, Nigeria. *Aquaculture, Aquarium, Conservation & Legislation International Journal of the Bioflux Society (AACL BIOFLUX), 4*(5), 651–659.

Orubu, C. O. (1999). The exploitation of non-timber forest resources in the Niger Delta: Problems and perspectives. Technical paper, *Niger Delta Environmental Survey*, Port Harcourt, Rivers State, Nigeria.

Orubu, C. O., Odusola, A., & Ehwarieme, W. (2004). The Nigerian oil industry: Environmental diseconomies, management strategies and the need for community involvement. *Humanities and Ecological, 16*(3), 203–214.

Otobotekere, C. (2009). *Live 2 lives* (1st ed., 149 p).). Port Harcourt: Herodotus Publishing Ventures, .

Raji, W. (2008). Oil resources, hegemonic politics and the struggle for re-inventing post-colonial Nigeria. In A.-R. Na'Allah (Ed.), *Ogoni's agonies: Ken Saro-Wiwa and the crisis in Nigeria*. Trenton, NJ: African World Press Inc.

Torulagha, P. S. (2007). *The causes of anger and rebellion in Niger Delta*. Retrieved May 6, 2011 from: <http://nigeriaworld.com/articles/2007/jun/280.html>.

Ukiwo, U. (2010). Conflicts and development in the Niger Delta. In K. Aaron, & D. George (Eds.), *Placebo as medicine: The poverty of development intervention and conflict resolution strategies in the Niger Delta region of Nigeria* (1st ed., pp. 37–61). Port Harcourt: Kemuela Publications.

United Nations Development Programme (UNDP) (2006). *Niger Delta human development report*. Abuja: UN House.

World Bank (2007). *Defining an environmental development strategy for the Niger Delta* (Vol. 2Washington, DC: World Bank.

The Political Economy of the Amnesty Program in the Niger Delta Region of Nigeria and Implications for Durable Peace

Babatunde Abdul-Wasi Moshood[1],
Tarilayefa Ebimo Dadiowei[2] *and Bamidele Folabi Seteolu*[1]

[1]Lagos State University, Ojo, Nigeria [2]College of Education, Sagbama, Nigeria

21.1 INTRODUCTION: UNDERSTANDING AMNESTY

Amnesty is defined as a legislative act by which a state restores those who have been guilty of offenses against it, to a position of innocence (Oluwatoyin, 2011). Olsen, Payne, and Reiter (2010) define amnesty as a process by which a state officially declares that those accused or convicted of human rights violations, whether individual or groups, are excused from prosecution, pardoned for their previous crimes, and subsequently released from prison.

Amnesty as a concept is derived from the Greek word *amnestia*, meaning to cast into oblivion or forgetfulness. This is shared with the medical term *amnesia*, which means a loss of memory (Chigara, 2002). Amnesty is defined as a strategic state policy that takes a form of executive or legislative clemency and in which offenders, or those involved in illegal actions, are formally pardoned. It is presumed that the moment a person or group is granted amnesty all records of the person's accusation, trial, conviction, and imprisonment are summarily closed. In order words, upon amnesty, the antisocial acts of a person are totally wiped from the official record and he/she is considered not only innocent but also as having no legal connection with the crime in the first instance (Schey, 1977).

Furthermore, an amnesty process is one of give-and-take. It requires the recipients to perform certain tasks, such as the willingness to be "amnestized," to provide information, to admit to the truth about their actions, and to show remorse and surrender weapons, as in the case of the militants in the Niger Delta of Nigeria. The conditional amnesty could be individualized, so that the recipients can only benefit from an amnesty program upon successful compliance with its conditions (Ogege, 2011).

In the context of the Niger Delta, amnesty is viewed from diverse perspectives. According to those people interviewed by one of the authors, among whom are environmental activists, academics, and militant leaders, some of them consider amnesty as a fraud, some see it as a stop-gap measure, while others consider it as a state response to militant activity in order for the government to have unhindered access to crude oil exploration. A few of their submissions will suffice here. According to Asari Dokubo:

> The amnesty programme, as far as I am concerned, was for the advantage of the Nigerian state so that the resources of Niger Delta will flow unabated, so that they would bribe some people with sixty five thousand, with going to school, with going to acquire school, and the oil will flow. This is a temporary measure that will fail ultimately, because the fundamental problems and issues of self-determination and resource control have not been addressed. *Personal Communication (Nov. 2014)*

In the same vein, Dr. Uyi Ojo, an environmental activist, opined that:

> The amnesty programme is a government strategy deployed in the phase of the manifestation of a failed state. It is the way of a neo-liberal system to get back on track to continue resource extraction. So the primary objective of amnesty is to continue to drain oil behind military shield. *Personal Communication (Nov. 2014)*

Similarly, Chief Christian Akani, a human rights activist, traditional ruler, and an academic, could not see the rationale for granting amnesty and compensation to those who have inflicted very great pain and havoc on their people. In his words:

> The amnesty programme did not come because of the altruism of the federal government, but because the activities of the militants have affected the economic jugular of the state, hence, something very fundamental has to be done. If not, Nigeria would have collapsed. Unfortunately, this group of people who have wrecked untold hardship on their people were later pardon, to me this is not fair, and amnesty programme itself to me is a fraud. *Personal Communication (Nov. 2014)*

Several literatures have captured the failure of the state to address the fundamental issues of resource mobilization, infrastructural deficit, environmental degradation, and the likes as the raison d'être for the arms struggle in the first instance. Thus, it is arguable that the accusation that the amnesty program has not been able to address the root causes of the conflict may not necessarily be used to assess the success of the program. The amnesty program was initiated to provide immediate cessation to militant activities in the region, so that the atmosphere of peace provided by this will provide room for the development in the region. Speaking in line with this view, Dr. Dion Akhaine, a human rights activist and an academic, said that:

> The amnesty programme is a state response to the insurgency in the Niger Delta. It is a way of persuading the insurgents to disarm, and embrace other peace building measures, and response to some of the demands which they have made over time, such as resource control. It is a halt to further degradation of the environment, and waste of resources, in terms of gas flaring, and all other within the Delta. *Personal Communication (Nov. 2014)*

In the same vein, ex-General Nature, who happens to be one of the ex-militants, conceives the amnesty program as a negotiation brokered between the militants and the state to achieve an immediate truce to enable development to take root. He said:

> Amnesty came as an agreement; it came as the only option for the boys to down their tools so that the federal government can go into negotiation with their leaders to see how they resolve the issue of neglect, and marginalisation of the region. *Personal Communication (Nov. 2014)*

Similarly, ex-General Sunny Clark, a former militant in the Tompolo's camp, sees amnesty as an experiment. Considering the fact that an indigene of the region was the vice president, then the need to give peace a chance through the acceptance of amnesty was worth trying. He said:

> Amnesty started during Yar'Adua, they started using the Ijaw Youth Council (IYC) as the link with the militants. Understanding that Goodluck was going to be the deputy to the President, the militants were persuaded to drop their arms, and we say let us start from there. *Personal Communication (Nov. 2014)*

Corroborating the earlier view, the head of reintegration, Mr. Lawrence Pepple, articulated the governmental perspective during a discussion with one of the researchers. He said that there is no way any development activity could have been embarked upon in the region if there was no cessation of hostility. It would be rather unfair to think that the whole essence of introducing amnesty was to access oil in the region. Although economic explanation could not totally be jettisoned, he nonetheless said that the need to entrench peace, stability, and development in the region was directly behind the introduction of the program. He asked rhetorically if it would have been better to allow the militants, insurgent activities, and the killings to continue when such could be averted by a "win-win" solution that the amnesty program has provided. In his words, amnesty is:

> The Nigeria DDR is otherwise known as amnesty. It is a programme that is internally nurtured, initiated, put in place, sponsored, funded, run exclusively by Nigerians, and no external body. Its uniqueness is that it is only one of the DDRs currently running, initiated by the natives, run by the natives, funded by the natives hundred percent. It was introduced to replace the hitherto weapons that they (the militants) were having with skills and intellect. This home grown DDR is not readily cash for arms as it is alleged, I am not saying it is completely bad, cash for arm has its own benefit of speedily bringing about peace in a community where every business has completely collapsed. *Personal Communication (Nov. 2014)*

As can be deduced from the above submissions, opinion differs on the understanding of the amnesty program, its merits, and its intent—from the perspective of activists, former militants and those in charge of the program. This is understandable given the fact that previous government interventions could not assuage the suffering in the region. Moreover, the quantum of problems that the neglect and laggard approach of successive regimes have bequeathed to the region present the amnesty program with challenges and issues that are not ordinarily within its scope. Notwithstanding the differences of opinion, the amnesty program from the perspective of the Niger Delta conflict is a home-grown peace-building initiative. It was introduced to bring immediate cessation of hostility in the region, to pave the way for eventual development of the environment, which includes both human and infrastructural facilities. Having attempted to conceptualize amnesty from the perspective of the officials and the stakeholders in the region, the next section focuses on the making of amnesty. It discusses the rationale for establishing the program and how it was implemented.

21.2 THE MAKING OF AMNESTY IN NIGERIA

In the early 1990s, the struggles of the Niger Delta ethnic minorities were largely nonviolent and were targeted at both the Nigerian state and oil multinationals. These were driven by demands for self-determination, respect for human and environmental rights, resource control, or a fair share of oil revenues taken from the region. When the initial demands, which were peaceful, were ignored, ethnic minorities such as the Ogoni, through Movement for the Survival of the Ogoni People (MOSOP), embarked upon a successful national and global campaign for self-determination, resource control, ethnic minority, and environmental rights that was eventually met with high-handed repression by the Nigerian military (Amunwa & Minio, 2011; Obi, 2014). By the time Nigeria returned to elected civilian rule in 1999, the ethnic minority struggles of the preceding decade had altered the context of the struggle for resource control. This opened the door for a faction of the Niger Delta elite to gain increased access to power at the federal and regional levels—while also paving the way for some civil society actors, erstwhile resistance activists, and a new generation of militants to become key actors in the region (Agbiboa, 2014; Obi, 2014). More importantly, this gave the Niger Delta elite a moral basis for greater demands and status within the ruling elite and further emboldened them to demand for resource control. With enormous wealth within their control, they have enough leverage to engage in political manipulation. In the struggle that resulted from the introduction of democracy, Obi (2014) noted that there were collaborations between individuals within the Niger Delta elite, between the Niger Delta elite and militant/armed groups in the region, and between the Niger Delta elite and those from other ethnic majority groups in Nigeria. Also, the economy of rebellion and rent-seeking officials equally explains the aggressive dimension the conflict later assumed, because the conflict seemed more lucrative for these collaborators (Idemudia, 2009).

Although the point about the connection between the escalation of violence in the oil-producing region and the creation of an enabling environment for the "complicit union" to thrive is relevant, it is partly reflective of the opportunistic manipulation of deep-seated grievances in the region by a Niger Delta elite and militant youth commanders. These groups of people were keen to strengthen their positions at the local and national levels. In some ways it captures an aspect of the twists and turns in the long-standing conflict in the region, and extends the logic of mutually beneficial opportunism that marks the expedient co-optation of armed groups for political and strategic ends (thugs for intimidating the electorate and rigging elections, attacking political opponents, and waging inter- or intra-community struggles). It also points to the contradictions and complexities that underpin some of the alliances of convenience that are characteristic of the Niger Delta insurgency, in which alliances and tactics are fluid, and actors move across, or straddle, different sides based on exigent calculations of political and personal gain. At different times, and in different places, the elements, motivating factors, and conflict vary in intensity, but remain anchored in the region's history, national politics, and its position in the international political economy of oil.

The legacy of five decades of oil production—alienation, the paradox of plenty, repression, militarization, and high levels of unemployment—and the new opportunities opened up by government co-optation and oil theft have fed into a volatile and complex situation. Of note is the legacy of repression and militarization that had fueled militancy and criminality among Niger Delta youths. The youths now see violence as a means of struggle, negotiation, and survival, and as a way to accumulate political gains. Several commentators have noted the blurred boundaries between resistance and criminality (Ikelegbe, 2011; Obi, 2014). Thus, when a coalition of ethnic minority named the Movement for the Emancipation of the Niger Delta (MEND) took over the struggle in 2006, the conflict tactically assumed a violent dimension. MEND engaged in the kidnapping of expatriate oil workers and publicized the abductions. They carried out armed attacks on government forces and damaged oil installations in the area. This began the era that Obi (2014) refers to as a transition from uncoordinated protests and conflicts into a trans-Delta insurgency.

By 2007, after initially dismissing MEND as criminals, the Nigerian state had recognized the adverse impacts of growing insecurity in the form of huge losses in oil production and revenues, and the strong reverberations of the insurgent attacks on the global oil prices (in the context of a global war on terror). The regime therefore began to consider an alternative to the military approach that had largely failed to halt MEND's attacks on oil infrastructure in the Niger Delta. A few examples may suffice here: Aghalino (2010) reported that as of 1998, there were 92 attacks on the oil industry, which resulted in the fall of crude oil exports down to 1.6 million barrels per day. As a result, the country lost at least $23.7 billion to oil theft, sabotage, and shutdown in production in the first nine months of 1998, and approximately 1000 people were killed within the same period (Report of the Technical Committee on the Niger Delta, Vol. 1, Nov. 2008: 9). Also, the production from the offshore business had dwindled to 300,000 barrels per day as at March 2009 from the initial one million before the crisis in 2004 (Aghalino, 2010). Accordingly, the attack on Bonga, a 43.6 billion floating, production, storage and offloading vessel, and deep water subsea facility, 120 km off the coast, and the attack on Atlas Cove woke the government from their slumber and symbolized the dawn of a new stage in the crisis. In fact, these attacks resulted in the loss of production of 225,000 and 125,000 barrels per day, respectively, which reduced the country's crude oil output by 345,000 barrels per day, with dire consequences for the economy (Aghalino, 2010).

Needless to say that the crisis was multidimensional, as it also increased the spate of criminality in the region. Approximately 40 incidents of piracy were reported in the Delta, including 27 vessels boarded, 5 hijackings, and 39 crew members kidnapped in 2008 (Aghalino, 2010). In 2009, the attack by gunmen on the M.T. Meredith, a tanker carrying 4000 tons of diesel fuel, suggested that the country was becoming the next pirate zone and threatened to seriously damage her reputation (Aghalino, 2010).

Prior to the declaration of amnesty, the military had carried out massive military bombardment in the region. One of the most significant examples was the military escapade carried out in May 2009 by the military Joint Task Force (The Joint Task Force (JTF) is the combination of the army, navy, and other security personnel. It was formed in 2000 to restore peace and order in the region. Its mandate also includes the protection of oil facilities and the prevention of criminality in the region.) in the Gbaramatu area of the western Niger Delta, targeting some militia camps and destroying the camps and communities that were suspected of being sympathetic to their cause (Obi, 2014). The inability of the military to completely neutralize the insurgents made it necessary to consult with the militant leaders when the presidency eventually resolved to address the issue through dialogue. Obi (2014) argued that the consultation, which later involved the militants and their local sponsors, occurred because the militants had demonstrated the capacity to threaten oil interests. Although this is true to a large extent, the military exploit was also strategic, as the continuous bombardment of the militant camps made any alternative other than amnesty seem improbable and unreasonable. This, as Major General Cecil (retired) (General Cesere Cecil was the Chief of Staff to Major General Godwin Abbe (retired), the former Minister of Interior and headed the presidential amnesty and demobilization of the militants.) informed one of the researchers, was to compel the militants to see amnesty as a soft landing. So the Nigerian government used a discreet but subtle military approach to compel obedience. According to Major General Cecil:

> Let me tell you, amnesty offer is an option. That doesn't mean if you refused to key in initially and you now want to get involved in criminal activities that government will allow you to go free. The military onslaught before amnesty was tactically deployed to compel obedience. There are consequences for every action, if you didn't key into amnesty, and you now want to cause problem, of course you get the full wrought of the law. That is why there is still a task force to maintain security in that place. *Personal Communication (Nov. 2014)*

The election in 2007 marked a new dawn because for the first time in the history of the country, an indigene of the Niger Delta, Goodluck Jonathan, emerged as the vice president alongside President UmaruYar'Adua. At this time, there was a great deal of pressure on the presidency regarding the activities of the militants, which had not only derailed normal activities in the region, but also impacted greatly on the oil economy of the country. Obi (2014) reported that one of the first steps embarked upon by President Yar'Adua's administration to curb the militants' activities was a series of consultations involving the state governments of the Niger Delta, the NDDC, the vice president, president, Niger Delta elites, and oil and gas industry operators. This culminated in the inauguration of the Presidential Niger Delta Peace and Conflict Resolution Committee (NDPCRC) in July 2007, with the following terms of reference: recommend to the federal government how to adequately address issues of the Niger Delta; liaise with the groups in the Niger Delta region, the security agencies, and report to the federal government (Nigeriafirst.org 2007 cited in Obi, 2014). A similar body was also inaugurated by the then-governor of Bayelsa state, Timipre Sylvia, to replicate the same peace-building process in the state (Obi, 2014).

Obi (2014) further observed that apart from engaging the source of the threat to continued (legitimate) access to oil and fostering the *status quo ante*, the state also avoided engagement with popular organizations and environmental rights—based groups that had long adopted nonviolent protests (and international campaigns) in pressuring government to address the deep-seated grievances and demands for social justice, equity, and respect for the dignity and human rights of the Niger Delta people. Such groups had earlier successfully opposed the appointment of Ibrahim Gambari, a UN Under-Secretary (and a former federal foreign affairs minister under the military) as mediator of the Niger Delta conflict (Obi, 2014).

Moreover, after the government abandoned the plan for a Niger Delta peace summit in 2008, because it was criticized for being another talk show, given that there have been a series of such summits in the past, their recommendations have never been implemented. Thus, government later set up a 44-member Technical Committee on the Niger Delta (TCND) (The technical committee was constituted in 2008 and its mandate was to collate all previous reports and recommend to the government on how to bring an end to Niger Delta problem.), including several credible stakeholders and activists from the Niger Delta to, among other things, "make recommendations that would assist the federal government to bring about sustainable development, peace, and human and environmental security in the Niger Delta region" (TCND, 2008). It is instructive to note that it was this committee that recommended amnesty for the militants in order to create space for sustainable development.

21.3 AMNESTY: THE NIGERIAN CONTEXT

As stated earlier, one of the recommendations of the 44-member Technical Committee was the introduction of amnesty. Thus, the need to consider the implementation of some of the aspects of the Technical Committee report that would lay the foundation for peaceful coexistence was imperative.

In part, the policy recommendations of the Technical Committee, as contained in the report, state that the federal government should:

1. Establish a credible and authoritative DDR institution and process, including international negotiators to plan, implement, and oversee the DDR programs at regional, state, and local government levels;
2. Provide for open trial and release on bail (with a view to eventually release) of Henry Okah (Henry Okah was the spokesman for the Movement for the Emancipation of the Niger Delta. He was arrested in connection with militant activities in Nigeria and was subsequently released as part of the negotiation for amnesty. He has since been rearrested in connection with the bombing in Abuja on October 1, 2010 and is currently serving a 24-year jail term for acts of terrorism.) and others involved in struggles relating to the region;
3. Grant amnesty to all Niger Delta militants willing and ready to participate in the DDR program;
4. Address short-term issues arising from amnesty to militants, by promoting security for ex-militants and rebuilding of communities destroyed by military invasion;
5. Work out long-term strategies of human capacity development and reintegration for ex-militants; and
6. Reflect on a time line with adequate funds for the DDR program to take place (TCND, 2008).

The findings of the committee were later passed on to another committee for consideration (An unnamed committee within the government was set up to consider the report and issue a white paper. Such a committee is usually constituted within the government ministries.). In June 2009, due to the exigencies of implementing peace in the region, Yar'Adua announced an offer of unconditional amnesty to all militants in the Delta. In addition, the Presidential Committee on Amnesty and Disarmament for Militants under the Minister of the Interior, Major General Godwin Abbe (retired), was established to execute a post-amnesty program (PAP) of socio-economic development in the Niger Delta worth approximately 50 billion Naira (Adeyeri, 2012). The training that former militants received in the Obubra Camp (Obrubra is located in Cross River State in Nigeria. The camp is where the former militants were demobilized, housed, and taught nonviolent methods before they were sent out for further training.) was the first step in transforming the erstwhile militants, and deconstructing their past characterized by militancy. This is expected to give them new orientation about peaceful coexistence and prepare them to become members of civil society. This is to precede the later transformational training that was expected to build the capacity of the militants as useful members of the society. Scholars such as Obi (2014) and Ushie (2014) have argued that the government decided to adopt amnesty from the recommendation because of their continuous interest in oil production, which has been jeopardized by the conflict. Although this is true to a large extent, the need to stabilize the region by ensuring peace cannot be totally ruled out because without peace, there is hardly any developmental project that can be carried out.

In accordance with the amnesty program, the militants were required to give up their arms, and in return they were to receive a Presidential pardon, opportunities for education, training, and general rehabilitation. At the expiration of the October 4, 2009 deadline set by the federal government for the agitators to disarm, virtually all the key militants had yielded to the amnesty deal. Available records indicate that a total of 8299 militants registered with the Presidential Implementation Committee (The Presidential Implementation Committee was established shortly after amnesty was granted in 2009. It was mandated to see to the total implementation of the presidential amnesty program.) from seven of the nine Niger Delta States, with most weapons coming from Bayelsa State, which had 130,877, Rivers 82406, and Delta 52958 (Aghalino, 2010 cited in Agbiboa, 2014). According to the Special Adviser to the President on Niger Delta Affairs, and the Chairman of the Amnesty Committee, Hon. Kuku (Hon. Kuku was the Special Adviser to the President on the Niger Delta and the Chairman of the Amnesty Implementation Committee. He was formerly a lawmaker in the Ondo State House of Assembly in Nigeria, and had worked closely with the former Special Adviser to the President on the Niger Delta, Mr. Timi Alaibe.), during the first phase of the amnesty program, on June 25, 2009, a total of 20,192 former militants accepted amnesty. Another 6166 were added in November 2010, while another 3642 former militants were added in October 2012, to make a total of 30,000. The amnesty program, unlike the previous mechanisms adopted by the government, is aimed at a win-win approach, and incorporated conflict transformation strategies, where the aim was to transform the people and pave the way for the eventual development of the region. The management of the program from the period of disarmament, demobilization, and up to the reintegration stage has been carefully implemented by the officials. However, at the inception, the program suffered some setbacks, especially due to the death of former president Yar'Adua. This initially slowed down the program because the necessary presidential fillip was missing—such as the release of funds and further extension of amnesty to the willing militants. The politics surrounding whether the vice president could assume the presidency after the president's death also contributed to a lull in the program.

In laying down arms, the militants were expected to go to the nearest screening center to turn in their arms and ammunitions, take the oath of renunciation of armed violence, and subsequently receive presidential amnesty, after which the repentant militant would be registered for a rehabilitation and reintegration program (Agbiboa, 2014; Obi, 2014; Oluwatoyin, 2011). Major General Cesere Cecil informed one of the researchers that at this point the ex-militants were given the UN code, which implies that in spite of the fact that the program was home-grown, it has the blessing of the UN. In his words:

> Everybody that went through the amnesty has a UN code, but what I just want you to know is that it is a novel idea; this is one of the few cases of countries that went through the DDR process at its own initiative, at its own expenses, at its own effort. *Personal Communication (Nov. 2014)*

According to Korpamo-Agary, the disarmament and subsequent reintegration of the militants is only a first step towards bringing urgently needed development to the Niger Delta region, since there cannot be development without peace (Agbiboa, 2014). In writing about the amnesty's objectives and deliverables, Agbiboa (2014) reported that the Nigerian government identified the following three phases:

1. A disarmament phase to take place between August 6, 2009 and October 4, 2009 and to include the collection of biometric data of the entire militants;
2. A demobilization and rehabilitation phase to last 6–12 months and to include the provision of, among other things, counseling and career guidance for the participants;
3. A reintegration phase to last up to 5 years and to include the provision of, among other things, occupational training and micro-credit for the participants (Oluwatoyin, 2011).

The biometric data captured were to later serve as a useful security measure to trace the ex-militants in case of recalcitrance. It also serves a dual purpose of knowing the number of people who enrolled in the program. It should be noted that two of the phases have been completed, so far, and that the disarmament and demobilization phases were successful. However, some of the challenges, which include (among others) poor clean-up of arms and poor conditions at the camp, coupled with demobilizing those who are not real militants as a consequence of the dispensing of favoritism, are some of the challenges that the program witnessed. The important thing to note is that the completion of a phase naturally signals the beginning of another phase. During the course of a discussion with one of the researchers, the technical adviser to the Special Adviser to the President on Amnesty and the Head of Reintegration stated that DDR operates a kind of cyclical arrangement, where an end in one phase automatically dovetails into the other. According to Mr. Lawrence Pepple, from the pronouncement on June 25, 2009, a 60-day grace period was provided during which the armed agitators were expected to turn in their weapons; this grace period ended on October 4, 2009. During that period, arms and ammunitions of various descriptions were turned in. On May 22, 2010, all arms gathered were symbolically destroyed at Lapanta near Enugu, in Enugu State, signifying the end of the disarmament component of the DDR program.

This kick-started the demobilization component. The Cross Rivers state government, under the leadership of the former state governor Liyel Imoke, granted the request of the Federal Republic of Nigeria to use the National Youth Service Corps (NYSC) camp in Obrubra, Cross River state as the demobilization camp for the presidential amnesty program. At the end of that program, 21,192,000 agitators had deposited 2000 weapons and ammunition of various description at the camp.

Mr. Lawrence further stated that as the demobilization phase was progressing, by November 2010, there was an upsurge of demands from ex-militants, who said that they turned in their arms but they were afraid to be part of the program, thinking it might be an entrapment exercise by the government. Because of this, another window was opened for 6166 persons who were covered in the second phase of the program, and by 2012, another opening was granted for 3462, bringing the caseload of the program to 30,000 beneficiaries.

During the demobilization phase in Obrubra, there was collaboration with the Martin Luther King Center in the United States (The Martin Luther King Center was established in 1968. It was formed to entrench and teach a nonviolent approach in the resolution of disputes. The center was involved in the teaching of the former militants a nonviolent approach to issues.), as well with as peace-building outfits in South Africa. The ex-militants were taught principles of nonviolence as advocated by Martin Luther King. At the end of that exercise, the reintegration component started. During the demobilization stage, the delegates at that program were brought into camp, processed, and interviewed before the necessary placement was done. According to Mr. Lawrence Pepple:

> You will learn their health concern, their economic concern, their request, and various other things they will need. Some of them as at the time we brought them, you can even be moving them for surgery, because some came with bullet injuries that had stayed with them for years, some of them are tuberculosis infected, some are HIV positive, so, their concern were now taken over by us. *Personal Communication (Nov. 2014)*

The technical head noted that:

> "the militants were being removed from their comfort zone so to speak. They were in their palace, their camps, and their creeks. Removing them, you are removing them from their means of livelihood" *Personal Communication (Nov. 2014)*

In addition, "they have in those camps means of physical security and social security, their weapons were providing them means of livelihood and social security" (Personal Communication, Nov. 2014). To remove their weapons from their hands, you needed to replace them with something in the interim. Because of this, they were placed on "monthly stipends of 65,000 naira, that will cover feeding, some bit of accommodation challenges, while we are settling to move them to the reintegration component of the DDR" (Personal Communication, Nov. 2014).

The reintegration phase was the most challenging of all the phases because, whereas all other phases are tangible and you can count the number of arms collected, the number of people demobilized, and so on, reintegration is more of an attitudinal change that comes as a result of both psychological changes and the opportunity that the program has to initiate those changes. To ensure that the program is well-accepted by the people, state governments were required to support the rebuilding of communities destroyed by military invasion, and establish youth development centers and community demobilization and reintegration committees to enhance reintegration and capacity-building. State governments were also required to provide social amenities, including health centers and schools at the site of former militant camps (TCND, 2008).

Corroborating the above narration by the technical assistant and head of reintegration, Mr. Lawrence Pepple, Agbiboa (2014) reported that in July 2009, a budget of N52 billion (USD145 million) was announced for the amnesty deal. The money was intended to cater to the training and rehabilitation of 20,192 registered militants. According to him, the budget did not state clearly how the money was to be expended. The proportion that was to be allocated to monthly allowances versus the proportion allocated to a broader reintegration and rehabilitation package was not clearly stated. This happened at the initial stage of the program. As the program progressed, Humphrey-Abazie (2014) reported that the way and manner in which the money allotted was to be spent was clearly stated, as the budget goes through some formalization and approval before money is disbursed. According to him:

> The Office of the Special Adviser to the President on Niger Delta (OSAPND) through the Special Adviser to the President on Niger Delta (SAPND) will directly defend the amnesty budget before the National Assembly so as to receive approval as part of the year budget; its disbursement is made directly from the Central Bank of Nigeria to OSAPND. *Humphrey-Abazie (2014).*

He further confirmed a statement credited to Hon. Kingsley Kuku that:

> A total of N234,133,917,560 (USD$14,233,064, 89.72) budgetary allocation has been spent since its actual implementation programme began in March 2010. This budget spending involves overhead cost for staff, delegate's stipends, and DDR project cost. *Humphrey-Abazie (2014).*

Furthermore, former combatants who registered for the 42-month period of training, reintegration, and rehabilitation in government-designated residential training centers received monthly allowances of N65,000 (USD413) (Agbiboa, 2014). Table 21.1 shows the DDR budget description for 2014.

Once the militants have been disarmed and demobilized, the final phase is the reintegration stage. As noted earlier, this is the most critical component of the amnesty program. The aim of the reintegration phase was to enable the ex-militants to acquire real civilian status by providing them with training that would help increase their capabilities to responsibly take control of their lives. These skills are meant to help them gain sustainable employment and income as well as reconcile with local communities. According to a statement released on December 11, 2011, 7556 ex-militants (at home and abroad) have graduated from the program (Agbiboa, 2014). In addition, the Minister of Niger Delta Affairs stated that the Ministry had organized a job fair meant to link the youths in the amnesty program with potential employers (Agbiboa, 2014). In order to prepare the ex-combatants for the reinsertion proper, the amnesty officials organized different vocations and

TABLE 21.1   Amnesty Budget Proposal for 2014

S/N	Project Description	Naira	USD
1	Stipends and allowances of 30,000 ex-agitators	23,625,000,000	144,098,813.25
2	Operational cost	3,699,933,814	22,567,452.77
3	Reintegration of transformed ex-agitators	35,409,859,972	215,979,631.72
4	Reinsertion/transition safety allowances for 3642 ex-agitators (third phase)	546,300,000	3,332,113.51
	Total	63,281,093,786	385,978,011.25

From Humphrey-Abazie (2014) Engaging the Nigeria Niger Delta ex-agitators: The impacts of the presidential amnesty programme to economic development. Workshop organised by Amnesty office on responsible polycentric inequality, citizenship and middle classes *14 EADI General Conference 23–26 June 2014, Bonn, www.gc.2014.org.*

educational programs at home and abroad to build the capacity of the former militants and prepare them for the eventual job market. Some of these are expatiated upon in the next segment.

21.4 EXPLORING THE GAINS OF AMNESTY

The amnesty program creates avenues to transform the former militants into peaceful and civil residents of the society. The reintegration program under amnesty employed fresh idea in dealing with the leadership of the former militants and their foot soldiers. First is the "constructive engagement" with the leadership of the combatants where they were encouraged to have their own private security firms that would be hired to secure oil pipelines and other installations on a multibillion Naira contract agreement (Onapajo & Moshood, 2016). This, according to Onapajo and Moshood (2016), is to find a meaningful avenue for the militant leaders, who have been exposed to great wealth, to find alternative legal sources of income.

The second dimension is the more popular one, which targets the general combatants and noncombatant youths for socio-economic empowerment. The reintegration program also includes nonmilitant youths of the Niger Delta. According to the (then) Head of Reintegration of the amnesty program, Lawrence Pepple, the reintegration process included "people who did not bear arms (noncombatants) that were drawn from what we call 'impacted communities'—communities (in) which militant agitation has dwindled their economy, stopped their family from doing what they are expected to do" (Onapajo & Moshood, 2016).

The reintegration process involves vocational training and formal education programs for its beneficiaries. The vocational training involves a wide range of vocations that would potentially empower ex-combatants upon their reintegration into society. These include piloting, carpentry and furniture making, welding, boat building, marine operations, heavy duty operations, automobile technology, agricultural operations, oil and gas technical operations, electrical and mechanical engineering, and other relevant skills. In formal education, the reintegration process creates the opportunity for ex-combatants willing to pursue formal education to acquire quality education up to tertiary levels (undergraduate and postgraduate) in local and foreign universities. A document from the Amnesty Office indicates that the education and skills training programs cover 157 universities and 22 vocational skills training centers in 30 countries across the world. The local education and vocational training programs involve 9 universities and 19 vocational training centers in 8 states in Nigeria (Onapajo & Moshood, 2016).

As of November 2014, official reports indicated that a total of 18,706 ex-combatants and youths have acquired formal and informal education, out of which 15,392 have graduated in the reintegration program. Another 11,294 individuals are being enlisted into the reintegration program. The gains further include quality training andattitudinal and behavioral transformation of the former militants. This high-quality training provides opportunity for job placement and self-sufficiency of the ex-combatants (Onapajo & Moshood, 2016).

In spite of all these gains, the impression that the amnesty program provides opportunity for the militant leaders to exploit, the amnesty officials to manipulate, and the government to leverage on as political machinery to advance personal interest speaks to the political economic aspect of the program. This has significant impact on the gains of the program, as will be observed in the next section.

21.5 POLITICAL ECONOMY OF AMNESTY

The Marxist political economic theory, which emphasizes the struggle by the oppressed against their oppressors for the control of the means of production, provides an adequate take-off point in the discourse of political economy of amnesty in the Niger Delta. In Nigeria, the state is a key factor in the political economy; it determines the direction of production, distribution, and allocation of resources. The fragile production base and the resultant social forces of production have not been able to support any socio-political transformation that would engineer collective mass action of an active society. The state has been a factor that not only helps in preserving the private bourgeois structures by this act, but perhaps also helps in modifying them (Vajda, 1981). This indicates that the social contract with the Nigerian State has failed because it works and entrenches the interest of the elite class only.

Kuku (2012) argued that in Nigeria, politics defines and determines what happens in the Niger Delta, as the region contributes 80%−95% of foreign exchange earnings. Because oil sustains the national economy, politics is about controlling this economic means whether at the national, state, or local government level. This is because government responsibilities depend on revenue from oil and gas exports. Hence, it is the canvas on which springs the political economy of Nigeria. The above concurs with Daniel Yergin (1991:555−556) who describes the cause of Nigeria-Biafra war thus:

> Civil war broke out in Nigeria. That country's eastern region in which newly developed significant oil production was concentrated, wanted a bigger share of oil revenues. The Nigerian government said no. This clearly puts oil at the core of that war.

Soludo (2005) opined on the rentier nature of Nigeria—the political economy of patronage—in the following words:

> In Nigeria, the excessive dependence on oil is compounded by the concentration of the commanding heights of the economy in the hands of government. Government then becomes the fastest and cheapest means of making quick money, a rentier state emerged, intensifying the politics of sharing rather than production. This created a horde of 'rent –entrepreneurs', that is 'Big Men' without any productive source of livelihood except proximity to state power.

He further affirmed that most Nigerians do nothing for a living other than government patronage, which has led to the distortion of value system. The result is endemic corruption and stalled national development. Therefore, every policy of government becomes an avenue for making quick money by the privileged elite, and the amnesty program easily becomes pliable.

In an article, Obi (2014) explored the PAP, launched in 2009 following the decision of some insurgent militia leaders in the Niger Delta to "drop their weapons in exchange for peace" with Nigeria's federal government. It addressed the following questions: How has the PAP been shaped by the politics of the Nigerian state and elite as well as transnational oil interests? Is the trade-off between peace and justice sustainable when such peace fails to address the roots of the grievances? The article argued that the PAP is an unsustainable state-imposed peace-building project to preserve the conditions for oil extractions by local, national, and global actors (econpapers.repec.org/article/tafrevape/v_3a41_ay2014_3ai_3a140_3a249_263.htm accessed November 6, 2016).

A study by Alapiki, Ekekwe, and Joab-Peterside (2015), which covers Akwa Ibom, Bayelsa, Delta, Edo, Ondo, and Rivers found that "respondents rated the PAP low in terms of its impact on employment generation for ex-militants and members of host communities, community development projects, and access to micro-credit for ex-militants and community youths to start up small-scale businesses and scholarship for ex-militants and youths of impacted (insurgency) communities. The non-inclusion of some ex-militants and community members, hijack of the program by political elites and corruption, and poor implementation of the PAP were also identified as some of the impacts of the program."(Alapiki et al., 2015). This level of dissatisfaction indicates the existence of pent-up frustration that could possibly lead to aggression. Gurr (1973) argued that the sense of dissatisfaction is heightened by the gap between what people expect and what they eventually get. Efforts to build peace should begin with identifying grievances and designing programs that address basic concerns and establish community infrastructure that supports the socio-economic reintegration (Alapiki et al., 2015).

The above succinctly captures the reality with respect to the implementation of the amnesty program in the Niger Delta, as it was characterized by greed and violence. It therefore follows that the conflicted social relations that spring from the prevalent political economy based on the dispensing of favors have engendered a culture of docility among the elites, while a sharp divide between them and the masses is manifest. The desire to have a taste of what the elites enjoy through policy implementation results in blackmail, kidnapping, etc. The implementation of the amnesty program puts the masses, the ex-militants, against their privileged ex-militant generals, politicians, and state officials of the amnesty program. The penchant for diverting government funds for personal use by implementers of government policies was demonstrated at the early stage of the program when it became known that 80% of the funds released for the implementation of the program was spent on consultants and contractors, leaving a paltry 20% to run the program (Nwajiaku-Dahou, 2012). The contracting system and processes were used as the means to create wealth for the designers and implementers at the expense of the masses.

The state of disconnect between the ex-militants and their former bosses, the generals, also underscored the rate of the political economy that enhances conflict and violence. In order to ensure a hitch-free manipulation of the process, militant generals were moved into the elite class of the society through patronage in the forms of contracts and appointments; this is seen as another way of dispensing favors to the ex-generals to consolidate them in power. This marked preferential treatment for the ex-generals endangered the spirit of camaraderie that existed in the hey days of the struggle and marked the beginning of mutual suspicion, distrust, and hatred. Another salient issue that arose from the implementation of the amnesty that borders on the earlier-mentioned political economy is the mode of payment of stipends for the ex-militants. The ex-militant generals submitted names of their former foot soldiers whose monthly stipends were paid through them; this gave room for sharp practices.

It was revealed that most ex-militants are short-paid or are not being paid by their former bosses. The *Vanguard Newspaper* of Thursday September 27, 2012, p. 16, carried a report that seven officials of the amnesty office Abuja were suspended because of fraudulent activities. The report said that the officers colluded with ex-militants to manipulate names of beneficiaries. It is also noted that the demobilization and disarmament processes lacked transparency and were fraught with irregularities that border on corruption. This manifested in the number of missing names of ex-militants who submitted weapons and duly registered with the JTF but are not in the master list. A large number of such ex-militants are not covered by the amnesty program, and so are frustrated, waiting for a third phase all this long. A frustrated ex-militant queried:

> Why did the JTF accept my weapons, if they would not include me in the amnesty programme? I have endured enough. If nothing is done to include me in third phase; I assure you that a novel type of violence is inevitable in the state.

The contracting of the training programs for ex-militants epitomizes the political economic status of amnesty that favored ex-militant generals, program implementers, and notable politicians. The contracting process allowed contractors to scout for institutes and received bulk sums from the program budget to cover the training cost of the number of candidates assigned to such contractors. It smacks of corruption and prebendalism. It is this process that has brought about speculations that beneficiaries are hardly former militants. It was another allocation of slots among politicians and ex-militants. There is a school of thought that argues that the amnesty program has become another arena for factional politics within the Niger Delta elites (Obi & Rustad, 2011). The amnesty officials also see the program as being amenable to achieve their personal gains. There were incidences noticed in the course of the research where some people who were enrolled in the amnesty program were not ex-combatants. Although it was explained to one of the researchers that those who did not bear arms but who were enrolled in the program were people from the impacted communities, that is, those from the communities where militant activities have direct bearing and effects on the people and the area. Although such a gesture is good and laudable, it, however, contradicts the position of Mr. Ledum Mittee, the chairman of the technical committee, who told one of the researchers that many people who were not militants were enrolled into the program because of their connection to one official or former warlord, and those who were supposed to be a part of it were left out, and this calls for questioning. There is no harm in involving people from the region to benefit from the amnesty packages, but there is a problem where those who are supposed to be part of the program are left out.

This is the accusation leveled against some militant leaders who were alleged to have been recruiting and inflating the number of repentant militants for their personal gains (Obi, 2014; Onapajo & Moshood, 2016). All these impacted greatly on the gains of an amnesty program and partly explain the reason militant activities are resurfacing in the region. Moreover, other dissatisfied groups who see the opulence shown by some ex-combatants because of these manipulations are looking for ways to seek attention; thus, militancy is seen as being lucrative and an avenue to quickly obtaining riches.

It is also claimed, e.g., that the political friction between the former Bayelsa state governor, Timipre Sylva, and former presidential adviser on the Niger Delta, Timi Alaibe (allegedly believed to harbor gubernatorial ambitions), was reflected in the ways in which the two men competed in the bid to take sole credit for the surrender of militia leaders. Such views have persisted in the postamnesty phase, with accusations against certain political forces in the Niger Delta for sponsoring ex-militants who marched to Abuja on July 7, 2010 to protest their exclusion from the training program. This followed on the heels of a reported attack on Alaibe by some ex-militants during his visit to the Obubra training camp on July 1, 2010 (Obi & Rustad, 2011).

It is therefore argued that the political economy of amnesty discounted on the expected gains of the amnesty package as a tool for securing durable peace in the state, ditto for the Niger Delta. This is because the implementation phase rather divided the ex-militants and the elites. This has led to a groundswell opposition to the leadership of the militants. There have been occasions where ex-militant generals have been openly accosted by angry ex-militants who perceive them as frustrating their goals. These sharp practices have necessitated ex-militants to report their generals to the State Security Services. A typical example was the one of a camp general from the Southern Ijaw local government area of Bayelsa state who manipulated the payment of his followers. An aggrieved ex-militant complained as follows:

> My brother, our general has changed our names and so we have not been paid our monthly allowance for three months. We have spoken with him several times without results so we decided to report him to the SSS. He is inside with some of our members in the SSS office now (July 16, 2012).

In the face of elite fragmentation, unity of purpose diminished, and various opinions about the amnesty persisted. This led to the relapse into violence and the subsequent return to the creeks for another round of attacks on oil company facilities. Also of paramount importance is the implementation that focused on ex-militants only. The failure of including the fundamental cause of the crisis (addressing the developmental issues of the region) but instead, selectively appeasing the ex-militant generals through the award of multibillion-naira contracts is a high point in the political economy of the amnesty program. This portends great danger for the little successes recorded, and threatens the yearning for durable peace. The result is the emergence of the Avengers and a plethora of other militant groups, attacking oil facilities and raising the conflict credentials of the Niger Delta.

21.6 CONCLUSION

The political economy of amnesty, which was exploited by the ex-militant generals, politicians, and bureaucrats in different areas of the implementation of the program negatively affected the expected gains of the amnesty project as a tool for durable peace in the Niger Delta. What, then, can be done to ensure that the amnesty program does not simply reproduce the conditions it was meant to address, as the continuous spread of agitations is a big cause of concern?

The failure to stop decades of militancy culminated in the introduction of the amnesty program. Amnesty within its 6 years of establishment has achieved so much to rein in the incidences of insurgency. In spite of all the successes recorded, the resurgent militancy currently resurfacing required that a lasting measure be devised to completely stop militancy in the region. Hence, the dialogue option was an opportunity for both sides to correctly diagnose the root causes of the problem and proffer lasting solutions. The golden key, fiscal federalism, was treated perfunctorily. "The region supports the call for true federalism and urges the federal government to treat the matter expeditiously." "There can be no end to this cycle unless Nigeria is run as a proper federation." The Nigerian state no longer has the luxury of taking resources from one part of the country and dispensing crumbs to the indigenes. Other federations, such as Canada, the United States, Australia, India, Malaysia, Switzerland, Belgium, Brazil, and the United Arab Emirates, provide worthy templates of resource control that we should emulate. Every region and state should join the clamor for fiscal federalism to avert bloody conflagration in our increasingly wobbly federal contraption.

An editorial by *New Telegraph* of September 20, 2016 captioned "Umar and the Niger Delta" advised that "There are issues of justice, equity among others at play. Their lands have been devastated and degraded after many years of oil exploration without a corresponding increase in their living conditions. Therefore, it is easy to conclude that the government is only interested in "pacifying" the restive Niger Deltans only to the extent that the action will allow it to continue carrying out its economic activity of oil exploration. So like Umar noted, objective Nigerians and the international community are not likely to support the use of military option in crushing militancy in the Delta. His clincher is that the creeks of the Niger Delta are saturated with oil and that the deployment of heavy explosive materials there will lead to a massive conflagration comparable to Dante's allegorical inferno. Setonji David is a lawmaker representing Badagry Constituency-II in Lagos State where crude oil and gas deposit was discovered in commercial quantity recently. He spoke with Oziegbe Okoeki on the prospects and challenges of oil and gas exploration and said "But I must confess that I received the news with mixed feelings. Mixed feelings in the sense that oil is good, it is positive, but when I look at what has been happening in the Niger Delta, it does not make me happy because I don't want my area to go through that kind of pain the said region has witnessed" (*The Nation*, Sunday, October 2, 2016). This is a confirmation that a substantial number of people reckon with the fact that the state has not been fair to the region.

It is safe to conclude that, the amnesty program, unlike previous conflict resolution and developmental mechanisms that failed, succeeded in bringing the desired result to the region. While the development is incremental, yet, the program paved ways for normal developmental activities to kick start in the region. Unfortunately, the program was badly exploited by the critical stakeholders in the program. This directly reduced the benefits and gains that would have been achieved by the program. The 2-year extension granted the program is laudable, but government should ensure proper probity and auditing of the program to ensure transparency. A sufficient amount of money should be devoted for development of the region and to improve the capacity-building in the region. Whatever happens in the region has direct bearing on the country because of the monolithic nature of the country's economy. The drive towards a diversified national economy should be intensified, and this should not blot the need to entrench development in the immediate, middle, and long-term in the region. Also, the need to properly entrench functional federal practices is at the heart of introducing durable peace initiative in the country.

References

Adeyeri, O. (2012). Nigerian state and the management of oil minority conflict in the Niger Delta, retrospective view. *African Journal of Political Science and International Relations, 6*(5), 97–103.

Agbiboa, E. D. (2014). Transformational strategy or gilded pacification? Four years on the Niger Delta armed conflict and the DDR process of Nigerian Amnesty Programme. *Journal of Asian and African Studies, 50*, 387–411. http://dx.doi.org/101177100219096145300082, 1-25.

Aghalino, S. O. (2010). From NDDC to Amnesty: Change and continuity in the development of the Niger Delta Crisis. In L. E. Otoide (Ed.), *Essays in Honour of Professor Abednego Ekoko* (pp. 43–60).

Alapiki, H., Ekekwe, E. N., & Joab-Peterside, S. (2015). *Post-amnesty conflict management framework in the Niger Delta*. Choba, Port Harcourt, Rivers State, Nigeria: Faculty of Social Sciences, University of Port Harcourt.

Amunwa, B., & Minio, M. (2011). *Counting the cost: Corporations and Human Rights Abuses in the Niger Delta*. London: Platform.

Chigara, B. (2002). *Amnesty in International Law: The legality under International Law of National Amnesty*. Harlow: Longman.

Gurr, T. (1973). Revolution and social change nexus: Some old theories and new hypothesis. *Journal of Comparative Politics, 5*(3), 359–392.

Humphrey-Abazie, I.M. (2014). Engaging the Nigeria Niger Delta ex-agitators: The impacts of the presidential amnesty programme to economic development. *Workshop organised by Amnesty office on responsible polycentric inequality, citizenship and middle classes* 14 EADI General Conference 23–26 June 2014, Bonn, www.gc.2014.org.

Idemudia, U. (2009). The changing phases of the Niger Delta conflict: Implications for conflict escalation and the return to peace. *Journal of Conflict, Security and Development, 9*(3), 307–331.

Ikelegbe, A. (2011). Popular and criminal violence as instrument of struggle in Niger Delta region. In C. Obi, & S. Rustard (Eds.), *Oil and insurgency in the Niger Delta: Managing the complex politics of petro-violence* (pp. 125–135). London: Zed Books Ltd.

Kuku, K. (2012). *Remaking the Niger Delta, challenges and opportunities*. London: Mandigo Publishing.

New Telegraph Tuesday, September 20, 2016. Umar and the Niger Delta.

Nwajiaku-Dahou, K. (2012). The amnesty and the changing face of militancy in special issue: Conflict transformation and the challenge of post-amnesty peace-building in the Niger Delta. *Niger Delta Research Digest, Journal of the centre for Niger Delta Studies Niger Delta University, Wilberforce Island, 6*, No 2. 94—102.

Obi, C. (2014). Oil and the post amnesty programme (PAP): What prospects for sustainable development and peace in the Niger Delta. *Review of African Political Economy, 41*(140), 249—263.

Obi, C., & Rustad, S. (2011). Amnesty and post-amnesty peace, is the window of opportunity closing for the Niger Delta?. In C. Obi, & S. Rustad (Eds.), *Oil and insurgency in the Niger Delta: Managing the complex politics of petro-violence* (pp. 200—210). London and Uppsala: Zed Books and Nordic Africa Institute.

Ogege, S. O. (2011). Amnesty initiative and the dilemma of sustainable development in the Niger Delta Region of Nigeria. *Journal of Sustainable Development, 4*(4), 249—258.

Olsen, D. T., Payne, A. L., & Reiter, G. A. (2010). Transitional justice in the world, 1970-2007: Insight from a new dataset. *Journal of Peace Research, 47*(6), 803—809.

Oluwatoyin, O. O. (2011). Women's protest in the Niger Delta region. In C. Obi, & S. A. Rustard (Eds.), *Oil and insurgency in the Niger Delta: Managing the complex politics of petro-violence* (pp. 150—156). London: Zed Books Ltd.

Onapajo, H., & Moshood, B. A. (2016). The civilianisation of ex-combatants of the Niger Delta: Progress and challenges in reintegration. *African Journal on Conflict Resolution, 16*(1), 35—61.

Schey, P. A. (1977). The pardoning power and the value of amnesty. *Journal of Criminology, 9*, 325—333.

Soludo, C. (2005). The political economy of sustainable democracy in Nigeria. The 5th Nigeria democracy day lecture delivered on May 29, 2005 in Abuja.

TCND. (2008). Report of the technical committee on Niger Delta, November, (pp. 1—147).

The Nation, Sunday October 2, 2016, p. 53.

Ushie, V. (2014). Nigeria's amnesty programme as a peace building infrastructure: A silver bullet? *Journal of peacebuilding & Development, 8*(1), 30—44.

Vajda, M. (1981). *The state and socialism: Political essay*. London: Allison and Busby Ltd.

Vanguard Newspaper, Thursday, September 27, 2012.

Further Reading

Kemedi, D. V., Alagoa, N. C., & Leton, M. (2015). Oil politics and communal conflict in the Niger Delta: A traditional approach to conflict prevention, management and resolution. In O. O. Oshita, A. Ikelegbe, W. Alli, & H. P. Goluwa (Eds.), *Case studies of traditional methods of conflict prevention and resolution in Nigeria*. Abuja, Nigeria: Published by Institute for Peace and Conflict Resolution (IPCR).

Sustainable Exploration of Crude Oil in Nigeria

Adekunle Adedoyin Idowu[1] *and Taiwo Mary Lambo*[2]

[1]Federal University of Agriculture, Abeokuta, Nigeria [2]Lagos State University, Ojo, Nigeria

22.1 INTRODUCTION

Many researchers around the world have at different times carried out studies dwelling on the exploration and exploitation of crude oil resources in different oil-producing areas of the world. Depending on the aim of the study, some of the results showed, in most cases, the abundance and economic benefits of the explored oil in these areas and the nation at large. Its exploitation, however, could also be seen in its negative impacts on the inhabitants and their immediate surroundings, upon which they depend for their means of livelihood.

Crude oil has had critical impacts on the world development as a solitary natural asset because of its unequivocal components in characterizing legislative issues, discussions, and discretion of numerous nations. This point was upheld by Feyide (1986) who observed that the lives of individuals on the planet as well as the condition of countries are reliant on oil explorations. Oil as a key asset sustains industrialized nations' economies by providing revenues required to empower oil exporters to complete plans essential for national and economic growth. In an oil-deprived world, the advancement rate would be hindered and life would be intolerable. For this reason, oil has turned out to be a vital resource to governments, an imperative element of their legislative issues, and it is critically considered in political and conciliatory procedures.

The Delta region (also known as the oil-producing states) is known for its abundant crude oil deposits, and this has been the mainstay of the country's economy since its commercial discovery and exploration. Crude oil was discovered by Shell British Petroleum, now called Royal Dutch Shell, at Oloibiri, a village in Bayelsa State, in 1956 (Anifowose, 2008; Onuoha, 2008).

Kadafa (2012) observed that the Niger Delta ranks amongst the 10 most imperative wetland and marine biological systems on earth. Unquestionably, the oil business situated in

The Political Ecology of Oil and Gas Activities in the Nigerian Aquatic Ecosystem
DOI: http://dx.doi.org/10.1016/B978-0-12-809399-3.00022-7

this district has contributed in no little measure to the development and improvement of the nation. However, careless oil exploration exercises have left this area ranked amongst the five most seriously oil-harmed biological systems on the planet. Furthermore, this region is comprised of various ecosystems, including mangrove, freshwater swamps, and the rain forest, which is Africa's largest wetland, but oil pollution in the region has resulted in the contamination of streams and waterways and the destruction of biodiversity, which has made the region an ecological desert. This has impacted negatively on the livelihood of indigenous people who rely on services provided by these ecosystems for survival. Studies have shown that the amount of oil spills in Nigeria for the past 50 years is approximately 9—13 million barrels, which is equal to 50 Exxon Valdez spills (Federal Ministry of Environment, Nigerian Conservation Foundation, WWF UK, & CEESP-IUCN Commission on Environmental, Economic, and Social Policy, 2006).

Furthermore, pollutants from exploration and exploitation operations and refinery wastes introduced into the environment have constituted hazard potentials to terrestrial and aquatic ecosystems as well as to the atmosphere. Crude oil wastes released into water bodies from oil operations have caused negative impacts on aquatic life stock. In addition, the environmental degradation caused by petroleum exploration activities has made fishing and farming, which are the traditional sources of livelihood of the rural populace in oil-rich Niger Delta (Fig. 22.1), to become nonlucrative. Continuous flaring of gas from oil exploration and production has led to pollution, which has caused health hazards and rendered fishing and farming activities almost impossible. Incidental large oil spills in the region are also a common occurrence, leading to fish kill, destruction of agricultural crops, and pollution of waters, which have made life unbearable for people in the communities (Ekanem & Nwachukwu, 2015) (Fig. 22.1).

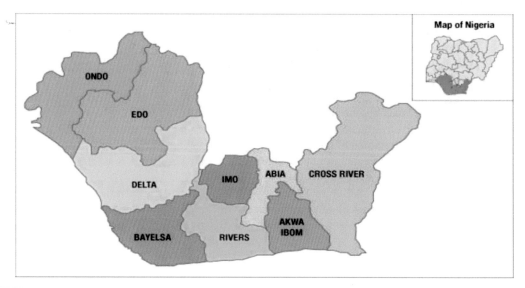

FIGURE 22.1   Map of Nigeria showing crude oil—producing states in the Niger Delta.

22.2 SUSTAINABLE DEVELOPMENT

Sustainability has been used more frequently with respect to human sustenance on earth since the 1980s, and this has brought about the most broadly cited meaning of sustainability as an aspect of the term "sustainable development." The Brundtland Commission of the United Nations defined sustainable development as the development that adequately addresses current needs without affecting the capacity of future generations to address their own issues.

Sustainable development is comprised of striking a balance between global and indigenous efforts to satisfy fundamental human needs without destroying or degrading the natural ecosystem (Kates, Parris, & Leiserowitz, 2005). The issue now is how to create a connection between those necessities and the environment. It involves principles for sustaining limited natural capitals, which are essential in the provision of the needs of the coming generations. It is additionally a procedure that envisions an attractive future for the human race, where their living conditions and resource utilization keep on meeting their needs without undermining the stability, integrity, and magnificence of natural biotic structures.

The United Nations Development Program in 2015 adopted an official agenda for Sustainable Development laying out the 17 Sustainable Development Goals and its related 169 targets. Some of the expected goals include sustainability for food, poverty eradication, health, water availability, habitation, climate, and marine ecosystems. The goals are aimed at conservation and judicious use of aquatic ecosystems, including their resources for sustainable development (UNDP, 2015).

22.3 OIL AND GAS EXPLORATION, PRODUCTION, AND SUSTAINABLE FISHERIES LIVELIHOODS

According to Sophie (2007), oil and gas production has a high-risk potential in the crude oil exploration business. Exploration precedes production yet does not always result in petroleum production. The unpredictable result of exploration partly explains the reason why investors pay little attention to issues that can affect communities' development. Exploration of oil implies a high level of logical, scientific, specialized surveying, industry knowledge, and drilling techniques and technologies. The exploration procedure is driven by demand for the commodity, and when future production conditions turn out to be more defined with time, there are increases in risk changes, the types, number and nature of crude oil, as well as specialized investment outfits.

Sophie (2007) highlighted the significance of exploration stage to fisheries' livelihood as follows:

1. Government should inform and empower its agencies responsible for administrative and technical services, development partners, nongovernmental organizations, and human rights groups on the essence of integrating economic and social objectives, in order to draw lawful policies that link revenue sources, its macro-economics, as well as its poverty mitigation mechanism.

2. There should be a thorough environmental impact assessment at the exploration stage that will provide important information on the opportunity, dangers, weaknesses, and deficiencies that can potentially affect livelihood systems, by identifying the deficiencies of the legislative and policy frameworks that can infringe on customary rights.
3. The exploration licenses should be negotiated and processed towards potential future production licenses. There should be provisions for business intelligence and negotiation preparedness for the mobilization of government and industry to prevent conceivable livelihoods impacts.
4. The importance of social and ecological impacts on fisheries' livelihoods should be considered from the beginning so as to create a conducive political, economic, environmental, and social atmosphere that will boost the country's bargaining power and revenue prospects in taking important foreign investment decisions and priorities.
5. Awareness should be created on the positive and negative impacts of crude oil exploration on the locals and their environment, which may last for several years and continue to affect the abiotic and biotic components of the ecosystem long after the operations started.
6. The exploration phase should be managed in such a way that fishing communities are not unnecessarily displaced.
   If analyses reveal that production is not commercially feasible, then exploration may be the only oil and gas–related activity undertaken.

Oil exploration and production operations clearly affect the livelihoods of fisherfolk. This is commonly seen around production sites near aquatic ecosystems such as fresh and marine waters used for ports, pipeline, and storage facilities that have direct interaction with the fisheries' livelihood; for instance, prevention of direct access to usual fishing grounds, landing sites, as well as processing and marketing areas. Households and local communities are at the epicenter of the sustainable livelihood approach but the local, national, regional, continental, and global scales need to be factored into the process to understand and cause a positive relationship between the fisheries' livelihood and development in the crude oil sector. There has been continuous increase in joint initiatives involving stakeholders such as government, civil society, and the industry instead of independent interventions (Sophie, 2007).

Corporate social responsibility of the oil and gas companies involved in exploration has centerd mainly on peace and security in the oil-producing areas. In its official newsletter, the AAPW (2006) posited that for sustainable livelihood to be effective, the following issues need urgent attention:

1. Low level or no employment, low consideration of award of contracts to local people/businesses
2. Choice of right people to handle the affairs of community projects
3. The adoption of the principle of divide and rule and disrespect for community leadership and hierarchy
4. Paying allowances to chiefs and sectional bias
5. Refusal to recognize, respect, and obey customary rules
6. Refusal to respect and execute Memoranda of Understanding (MoU)
7. Intimidation of locals with security forces

TABLE 22.1   Off-Project Strategies to Enhance Local Economic and Social Performance

Type	General strategies
Employment	Job searchers service
Training	Training on multiple streams of income; Vocational and technical training
Supplier support	Micro-venture business support and funding
Local infrastructure	Unilateral or public—private partnership infrastructure projects, which do not have direct relationship with contract
Institution strengthening	Local institution support aimed at developing capacity and competencies

Source: *Overseas Development Institute and Engineers Against Poverty (ODI & EAP). (2007).* Learning from AMEC's oil and gas asset support operations in the Asia pacific region, with case-study of the Bayu-Undan gas recycle project, Timor leste. Report II Local economic and social performance in low income regions, *13 pp.*

Overseas Development Institute and Engineers Against Poverty (2007) observed that weak local institutions are threats to strong social relationships. However, these can be corrected through mitigation projects that aim to strengthen local institutions by increasing the social content in partnership with civil society groups and government (Table 22.1). There is need for more youth empowerment programs that will give the young people financial freedom in oil-producing regions. Education of the youths, which will make them employable and self-reliant, is equally important.

Negative impacts of oil and gas exploration and production processes can, to a large extent, be avoided or substantially reduced, when the opinions of local fishermen are sought prior to the commencement of the production cycle. This will enable them to intimate the oil producers about the actual time or season of the year that is important for their profession. The following are possible measures for a sustainable mitigation program:

1. "Agree and implement local Notification and Communication Strategies before oil and gas project activities start
2. Ensure consultation at least one-year ahead of time to take seasonal patterns and fishing fleet activities into account
3. Part-payment in-exchange for ecosystem monitoring and reporting tasks
4. Employ fishermen as guard/scout boats
5. Employ fishermen as Fishing Liaison Representatives (FLR)
6. Support at sea and beach monitoring programmes for pollution and biodiversity impacts
7. Vocational training for safety, including sponsored awareness and information radio programmes
8. Preferential employment of ex-fishermen for seafaring jobs
9. Compensation funds towards fishermen/processors and traders professional organization strengthening
10. Community schemes, including educational and health programmes directed at the most vulnerable" (Sophie, 2007).

Ibama and Eyenghe (2015) affirmed that for the inhabitants of Ogoni Land (a major oil-producing area in the Niger Delta region of Nigeria) to be pacified to a large extent, the following mitigation measures need to be put in place by the government to enhance sustainability and effective livelihood of the rural dwellers:

1. Monitoring of the oil companies and their activities
2. Mediation between the host communities and petroleum companies
3. Formulation of good policies and their implementation
4. Facilitate the development of oil-producing region by the oil companies
5. Ensure that oil companies honor and implement the Memoranda of Understanding (MoU) they sign with their host communities
6. Repeal all laws that are repugnant to human welfare and environmental serenity (e.g., Land Use Act of 1978).

22.4 CONCLUSION

It is generally believed that oil exploration comes with its negative impacts on the environment; prevention and mitigation measures of these impacts as highlighted above come in handy in maintaining a high degree of sustainability and in strengthening the means of livelihood of the community inhabitants.

Kadafa (2012) pointed out that monitoring oil company activities in the Niger Delta is important, but this is lacking. This situation has persisted because of a number of reasons: inadequate qualified personnel in monitoring agencies, difficulty in accessing the production areas, poor revenue allocation to monitoring agencies, among others. Omorede (2014) believes that it is not enough for the government to formulate policies for oil exploration; it is equally important for them to put in place effective monitoring and enforcement mechanisms that will check the activities of the oil companies, punish offenders appropriately, and compensate those who comply with the guidelines and regulations.

Government should channel the natural gas produced from crude oil exploration into power generation, as a stable power supply in Nigeria holds the key to effective and efficient industrial activities. In addition, government should diversify the economy by developing the agricultural and other nonoil sectors to reduce the country's over-dependence on crude oil (Inomiesa, 2015).

References

Academic Associates Peace Works (AAPW). (May 2006). Peace works news. *Official Newsletter of Academic Associates Peace Works 6*(1), 12 pp. Available at <http://www.aapw.org/aapw_may_2006_newsletter.pdf>.

Anifowose, B. (2008). Assessing the impact of oil and gas transport on Nigeria's environment. In *U21 postgraduate research conference proceedings 1*, University of Birmingham, UK.

Ekanem, J., & Nwachukwu, I. (2015). Sustainable agricultural production in degraded oil-producing and conflict-prone communities of Niger Delta, Nigeria. *Journal of Agriculture and Sustainability, 8*(1), 14—28.

Federal Ministry of Environment, Nigerian Conservation Foundation, WWF UK, & CEESP-IUCN Commission on Environmental, Economic, and Social Policy (31 May 2006). *Niger delta resource damage assessment and restoration project.*

Feyide, M.O. (1986). *Oil in world politics: Proceedings of 1986 public lecture at University of Lagos.*

Ibama, E., & Eyenghe, T. (2015). An evaluation of the effects of petroleum exploration and production activities on the social environment in Ogoni Land, Nigeria. *International Journal of Scientific & Technology Research, 4*, 04.

Inomiesa, O. (2015). *Sustainable exploration of oil and gas in the United Kingdom and Nigeria.* (Ph.D. thesis). Liverpool John Moores University. 226 pp.

Kadafa, A. Y. (2012). *Environmental impacts of oil exploration and exploitation* in the Niger Delta of Nigeria. *Global Journal of Science Frontier Research Environment & Earth Sciences, 12*, 3.

Kates, R., Parris, T., & Leiserowitz, A. H. (2005). What is sustainable development? Goals, Indicators, Values, and Practice. *Environment, 47*(3), 8–21.

Overseas Development Institute and Engineers Against Poverty (ODI & EAP). (2007). *Learning from AMEC's oil and gas asset support operations in the Asia pacific region, with case-study of the Bayu-Undan gas recycle project, Timor leste. Report II local economic and social performance in low income regions*, 13 pp.

Omorede, C. K. (2014). Assessment of the impact of oil and gas resource exploration on the environment of selected communities in Delta State, Nigeria. *International Journal of Management Economics and Social Sciences, 3* (2), 79–99.

Onuoha, F. C. (2008). Oil pipeline sabotage in Nigeria: dimensions, actors and implications for national security L/C. *African Security Review, 17*(3).

Sophie, C. (2007). *Mitigating the impact of oil exploration and production on coastal and wetland livelihoods in Central and West Africa. Sustainable fisheries livelihoods programme.* <ftp://ftp.fao.org/fi/document/sflp/sflp_publications/english/oil_fish.pdf>.

UNDP. (2015). *Life below water.* <http://www.ng.undp.org/content/nigeria/en/home/post-2015/sdg-overview/goal-14.html>. Retrieved 28.09.15.

Further Reading

UNDP. (2006). *United Nations development programme: Niger delta human development report.*

Dealing with Oil Spill Scenarios in the Niger Delta: Lessons from the Past

*Chibuike Allison[1], Godwin Oriabure[1],
Prince Emeka Ndimele[2] and Jamiu Adebayo Shittu[2]*

[1]Appleseed Consulting Limited, Lagos, Nigeria [2]Lagos State University, Lagos State, Nigeria

23.1 INTRODUCTION

The Niger Delta region located in Nigeria has a size of approximately 70,000 km² with a total wetland area of approximately 20,000 km², and it is mainly made of deposits of sediment. Twenty million people, comprised of 40 diverse ethnic groups, inhabit this region. Its floodplain is the largest in Africa, representing 7.5% of the total land mass of Nigeria. Its drainage basin is the third largest in Africa. The Niger Delta region comprises four ecological zones, namely: coastal inland water, mangrove swamp zone, freshwater zone, and low rainforest zone.

This blessed and unique ecosystem supports plants and wildlife, and it is one of the most diverse ecosystems in the world. Its land is arable, supporting growth of crops for food and trees for timber. It also has the highest concentration of freshwater species when compared to other ecosystems in West Africa. However, activity such as extensive dam construction in this region has limited its productivity, and it is predicted that in about 30 years' time, it could have lost 40% of its habitable land. This problem can also be accelerated by laxity in the oil industries, as oil spills are not treated with the urgency they deserve. This is shown in the 1983 report by the Nigerian National Petroleum Corporation (NNPC) even before the recent turmoil experienced in this region.

351

23.2 OIL SPILLS

23.2.1 Extent of Oil Spills

There are variations in the reports on oil spills in the Niger Delta. It was estimated by the Department of Petroleum Resources (DPR) that the quantity of crude oil spilled in the Niger Delta was 1.89 million barrels in a total of 4835 incidents between 1976 and 1996 (Vidal, 2010). Another report by United Nations Development Programme (UNDP) stated that there were approximately 6817 incidents between 1976 and 2001, which resulted in the loss of approximately 3 million barrels of crude oil, with less than 30% recovered (UNDP, 2006). Most of these spills occurred offshore (69%), 25% were in swamps, and the rest on land.

The volume of oil spilled annually into the Niger Delta according to NNPC was given at 2300 m^3, with 300 incidents on the average. However, the World Bank argued that the figure may be much higher as the official figures do not account for "minor" spills (Moffat & Linden, 1995). The explosion of a Texaco offshore station in 1980 and the tank failure of the Royal Dutch Shell's forcados terminal were the two largest individual spills in Nigeria. The Texaco event had an estimate of 40,000 barrels (64,000 m^3) of crude oil dumped into the Gulf of Guinea while the Royal Dutch produced a spillage of approximately 580,000 barrels (92,000 m^3) (Nwilo & Badejo, 2001). It was also reported by Baird (2010) that since 1958, the total quantity that has been spilled into the Niger Delta environment is between 9 million and 13 million barrels. Another report claims that 100 million barrels was spilled between 1960 and 1967.

23.2.2 Causes of Oil Spills

Oil spills are very prevalent in Nigeria (Fig. 23.1). Approximately 50% of the spills are attributable to pipeline vandalism and tanker accidents. Sabotage is said to be responsible for approximately 28% of spills while oil production operations accounts for 21%. The remaining 1% of the spills is attributed to deficient equipment used in production. Corrosion results from the rupturing of pipelines and tankers or leakages from old production facilities that have been neglected and left uninspected. It also accounts for a significant percentage of oil spills due to small oil field size in the region. Extensive form of pipeline network and massive small flow line networks exist between oil fields. The latter is narrow diameters of pipes conveying oils from wellheads to flow lines, which allows leakages and spills. Several pipelines and flow lines in onshore areas are constructed above the ground surface, making them susceptible to leakages. Reports have shown that the life span of most facilities constructed between 1960 and 1980 was approximately 15 years. Many of the pipelines are used beyond 20–25 years and as such are susceptible to corrosion.

Further degradation is caused by oil sabotage and theft. Sabotage is done by bunkering where the saboteur aims to tap the pipeline, which can lead to pipeline damage during the extraction process. The damage may be undiscovered for several days and their reconstruction takes a longer time. Stolen oil readily has its own market (the black market), which makes oil siphoning a business that is lucrative.

FIGURE 23.1   Selected spillages in the Niger Delta. *Source: Pictures courtesy of UNEP (2006).*

In recent times, spills also happen as a result of deliberate blowing up of pipeline and oil platforms by militant groups with a hope to sabotage the oil production operations so as to prevent the federal government from generating increased revenue. This has become prevalent since armed struggle became a tool in the quest to make the federal government aware of the high level of under-development in this region. Several groups have been engaged in different militant actions against the government and oil exploring companies in the Niger Delta, whom they see as collaborators and partners in the continued degradation of the environment of the region.

23.2.3 Categories of Oil Spillage

Oil spills can be grouped into four categories according to the quantity of oil spilled:

1. *Minor spills*: This is when the quantity of spilled oil is less than 25 barrels in inland waters such as lakes or less than 250 barrels on land or coastal water. This spill does not usually threaten public health or human welfare.
2. *Medium spills*: This occurs when the spilled volume is not more than 250 barrels in inland waters or in a range of 250–2500 barrels in coastal waters.
3. *Major spills*: This takes place when the spilled volume is above 250 barrels in inland waters and 2500 barrels in coastal waters.
4. *Catastrophic spills*: In this case, there is a serious threat to human welfare and health due to unchecked well blowout, rupturing of pipelines, or failure of storage tanks.

23.2.4 Consequences of Oil Spills in the Niger Delta

The ecosystem is strongly impacted by oil spillage. The release of crude oil into an ecosystem may cause destruction of the ecosystem. Oil spill is partly responsible for the wiping out of about 5%–10% of the Nigerian mangrove ecosystem. It has also led to the destruction of the rainforest habitat occupying a land portion of 7400 km^2.

The results of oil spills are contamination of underground water, pollution of soil, and most especially losses experienced in crop production and aquaculture. Bacteria that are responsible for decomposition of oil use the available oxygen during this process, which leads to drastic reduction in dissolved oxygen required for fish growth, and this could cause fish kill. Oil spill can lead to the destruction of the equivalent of a year's food supply in agricultural communities in one incident (Table 23.1). Unavoidable oil spills in this region are increasingly reducing the sustainability factor of this region.

Health problems such as difficulty in breathing and skin lacerations are common in affected areas. Basic human rights such as health, access to food, clean water, and an ability to work have been lost by so many people in this area. Shell Petroleum Development Company (SPDC) was accused of polluting the Niger Delta by a Dutch court on January 30, 2013.

- Loss of mangrove forests

   The Niger Delta is characterized by mangrove forests, swamp forests, and rainforests. Five thousand square kilometers of the total 8580 km^2 of land is covered by extensive mangrove forest. Mangrove is important for the survival of indigenous people because they harvest them for fuel, and it is also essential to organisms inhabiting the aquatic ecosystems in the region because they provide breeding grounds for these organisms. Pollution activities have led to the disappearance of approximately 5%–10% of these mangrove ecosystems.

   Oils that are highly volatile, penetrating, and viscous in nature have wiped out large areas of vegetation. Spilled oil can extend to areas with vegetation if spills occur close to or within the drainage basin. This is caused by the hydrologic force of the river as well as the tides.

   Mangrove forests have a complex trophic structure such that death of a particular organism in a level will pose problem to the balance of the trophic structure and indirectly affect a host of other organisms. The plant community depends on factors such as nutrient recycling, clean water, light, and suitable substrate. The mangrove ecosystem, under optimum condition, provides conducive habitat structure and energy input through photosynthesis for organisms in that environment. Oil spills on mangroves result in acidification of soils, inhibiting vital oxygen in roots and hindering cellular respiration and activity.

   The loss of mangrove will not only have side effects on the survival of plants and animals but will also affect humans who depend on them. Indigenous people residing in this region place high value on these ecosystems. These ecosystems have been a major source of wood. They are also essential for the conservation of species endemic to them and to people who unfortunately do not benefit from petroleum.

TABLE 23.1   Some Severely Oil-Polluted Sites in the Niger Delta

Location	Environment	Impacted area (ha)	Nature of incident
BAYELSA STATE			
Biseni	Freshwater swamp forest	20	Oil spillage
Etiama/Nembe	Freshwater swamp forest	20	Oil spillage and fire outbreak
Etelebu	Freshwater swamp forest	30	Oil spill
Peremabiri	Freshwater swamp forest	30	Oil spill
Adebawa	Freshwater swamp forest	10	Oil spill
Diebu	Freshwater swamp forest	20	Oil spill incidence
Tebidaba	Freshwater swamp forest mangrove	30	Oil spill
Nembe creek	Mangrove forest	10	Oil spill
Azuzuama	Mangrove forest	50	Oil spill
Nine sites			
DELTA STATE			
Opuekebe	Barrier forest island	50	Salt water intrusion
Jones Creek	Mangrove forest	35	Spillage and burning
Ugbeji	Mangrove	2	Refinery waste
Ughelli	Freshwater swamp forest	10	Oil spillage-well head leak
Jesse	Freshwater swamp forest	8	Product leak/burning
Ajato	Mangrove		Oil spillage
Ajala	Freshwater swamp forest		Oil spillage
Uzere	Freshwater swamp forest		Oil spillage
Afiesere	Freshwater swamp forest		Oil spillage
Kwale	Freshwater swamp forest		Oil spillage
Olomoro	Freshwater swamp forest		QC
Ughelli	Freshwater swamp forest		Oil spillage
Ekakpare	Freshwater swamp forest		Oil spillage
Ughuvwughe	Freshwater swamp forest		Oil spillage
Ekerejegbe	Freshwater swamp forest		Oil spillage
Ozoro	Freshwater swamp forest		Oil spillage
Odimodi	Mangrove forest		Oil spillage

(Continued)

TABLE 23.1   (Continued)

Location	Environment	Impacted area (ha)	Nature of incident
Ogulagha	Mangrove forest		Oil spillage
Otorogu	Mangrove forest		Oil spillage
Macraba	Mangrove forest		Oil spillage
20 sites			
RIVERS STATE			
Rumuokwurusi	Freshwater swamp	20	Oil spillage
Rukpoku	Freshwater swamp	10	Oil spillage
Ogoni	Mangrove forest		Oil spillage

Source: Compiled from data courtesy of FME, NCF, WWF UK, CEEP-IUCN 2006 Niger Delta Resource Damage Assessment and Restoration Project.

These problems have received a great deal of attention worldwide because the value of this ecosystem in monetary terms could be as much as trillions of Naira. International and local groups have made both funds and labor available for the remediation and restoration of these damaged mangrove ecosystems. In the year 2000, the federal government of Nigeria established the Niger Delta Development Commission (NDDC), which intends to reduce the environmental and ecological impacts of petroleum on these areas. Technology has also been developed by governmental and non-governmental organizations to identify sources and motion of petroleum during spills, which will help in their containment.

- Depletion of fish populations

For essential proteins and other nutrients that fish provides to the human population, the fishing industry is invariably important to Nigeria's sustainability. However, the high demand of fish has resulted in the decline in fish stock from the wild because the rate of fishing exceeds the rate of natural fish stock recruitment. The shortfall in fish supply is further compounded by other factors, such as climate change, loss of habitat, and pollution. There is a need to support the growth of aquaculture to reduce the gap between fish demand and supply. Aquaculture involves artificial rearing of fish. Oil pollution reduces fish population, which will lead to a reduction in income and overall well-being of people who rely on them for survival.

The Niger Delta ecosystem has approximately 250 fish species of which 20 species are endemic to the region. This rich biodiversity is one of the reasons why this ecosystem must be protected and conserved. The death of these endemic species will mean a complete wipeout of these species from planet Earth. With climate change and habitat loss, it is necessary to prevent temperature increase so as to maintain the stability of the aquatic ecosystem. Efforts put into restoring habitats should be complemented with even greater efforts aimed at reducing pollution. Agricultural fields should be sited far away from the local waterways, and natural pesticides can be used in order to reduce pollution arising from the use of chemical pesticides on agricultural fields.

- Invasion of water hyacinth

  Water hyacinth is an invasive aquatic macrophyte well known for its prolific growth rate in freshwater ecosystems. It was initially introduced to Africa as an ornamental plant. Water hyacinth possesses a fibrous root system that can clog the waterways where it thrives. Water hyacinth now grows in the Niger Delta, inhibiting the navigation of fishing boats and also depriving the aquatic organisms that live there both sunlight and oxygen.

  Water hyacinth competes for sunlight with native plants in any ecosystem it permeates, thereby reducing energy supply within the ecosystem. This has threatened the survival of many native plants. Apart from this, the growth of water hyacinth also reduces the available oxygen in this water body, as it is used up during photosynthesis.

23.3 NATURAL GAS FLARING

It is believed that Nigeria flares a great deal of the natural gas that accompanies the oil extraction process. A total of 2.5 billion cubic feet (70,000,000 m^3), which is approximately 70% of a total of 3.5 billion cubic feet (100,000,000 m^3), gas produced yearly is lost to flaring. This quantity is approximately 25% of the total natural gas consumed in the United Kingdom and equivalent to approximately 40% of the total consumption of gas in Africa in 2001 (Tawari & Abowei, 2012). Although these data are unreliable, Nigeria is still believed to be losing approximately USD2 billion annually through flaring of gas. Due to weak regulation and enforcement, operators prefer to flare associated gas instead of separating it from oil. This is to avoid the cost of harnessing the associated gas.

Gas flaring has a substantial effect on climate change due to the release of poisonous gases such as methane. In the western part of Europe, almost all associated gases are used or re-introduced into the ground. It can be said that gas flaring started at the same time as oil exploration by Shell-Bp in 1960s. Gas flaring can be reduced by practicing re-injection or storage.

Large amounts of methane, carbon dioxide, and other major greenhouse gases that have high global warming potential are released in the process of gas flaring. It was estimated in 2002 that Nigeria has released more than 34.38 million metric tonnes of methane. This statistics depict 50% of the total emission from industries and 30% of the total emission of CO_2. Although all countries in the western world are trying to limit the problem caused by gas flaring, this problem has grown at the same rate as oil production in Nigeria.

This may soon change, as major stakeholders such as the international community and especially the Nigerian government and the oil community recently decided to take actions that will curb this problem. Gas flaring was declared illegal in 1984 under section 3 of the "Associated Gas Re-injection Act" in Nigeria. In 2004, the World Bank reported the quantity of gas flared by Nigeria to be 75% of the total gas produced. However, OPEC and Shell, the key players in gas flaring, argued that it was 50% of the total gas that was flared. However, there are systematic efforts, currently, to stem the tide of flaring by the establishment of the Nigerian Liquified Natural Gas (NLNG) company.

Gas flares emit toxic chemical substances such as nitrogen dioxides, sulfur dioxide, and highly volatile compounds such as benzene, toluene, xylene, and hydrogen sulfide. Carcinogenic compounds such as benzapyrene and dioxins are also emitted. These substances have the potential of negatively affecting human health and livelihood. Chronic bronchitis and respiratory problems are the common problems caused by exposure to these poisonous chemicals. Benzene emitted during gas flaring causes leukemia. Climate Justice reported that eight new cases of cancer should be expected as a result of benzene exposure.

Gas flaring in the Niger Delta region has resulted in acid rain, which corrodes roofing made especially of zinc. This has paved way for the use of asbestos for roofing. Asbestos has a better repelling power to acid rain. However, the usage of this asbestos has increased the risk of contracting diseases such as cancer of the lung, pleural and peritoneal mesothelioma, as well as asbestosis (Burdorf, Järvholm, & Siesling, 2007), thereby contributing to the declining health of the people and their environment.

The relationship between gas flare and acid rain has been subject to argument. Some studies established that sulfur dioxide and nitrous dioxide emitted from most flares are not sufficient enough to cause acid rain. Some other studies from the Energy Information Administration (EIA) in the United States concluded that gas flaring has resulted in air pollution and falling of acid rain (Bronwen, 1999). Older flares that occurred near villages have led to the coating of viable lands with soot, destroying vegetation. Currently, growth of vegetation is impossible in areas surrounding the flares as a result of heat emitted from these sources.

The federal high court of Nigeria gave an order in November 2005 that gas flaring activities must be stopped, as it defies basic rights to life in the constitution. Justice C.V Nwokorie gave the ruling in a case against Shell Petroleum Company of Nigeria. The judgment notwithstanding, the flaring activity still continued until May 2011.

23.4 ENVIRONMENTAL ASSESSMENT OF OGONILAND BY UNITED NATIONS ENVIRONMENT PROGRAMME AND THE LAUNCH OF THE $1 BILLION CLEAN UP AND RESTORATION PROGRAMME BY THE FEDERAL GOVERNMENT OF NIGERIA

There was a credible effort made by the federal government of Nigeria in June 2016 on the restoration and remediation of Ogoniland—a polluted area in the Niger Delta. This clean-up process has a total worth of a billion dollars. The financial and legal frameworks of the project were also disclosed. This region is well known as a site used by oil industries for their operations since the late 1950s. The area covers $1000 \, km^2$ in River State, southern Nigeria. The extent of contamination of this site due to long-term oil spills and oil fires has not been adequately documented.

The environmental assessment of this site cut across the affected ecological zones such as groundwater, surface water, air, sediments, and also human welfares such as health, industry practice, and institutional issues. The assessment by United Nations Environment

Programme (UNEP) was for a period of 2 years, which involved desk review, field work, and laboratory analysis. This study by UNEP on Ogoniland is one of the most complex assessments carried out by the body. The hierarchy of work involves an upper body comprising senior UNEP managers to create the necessary guidance and supervision. The next comprises a group of international experts recruited from different fields (land, water, and health). Then another group of experts from the local area, which comprises academics, a support team, logistics, and security personnel.

Over a period of a year and 2 months, the UNEP team surveyed more than 200 affected locations comprising 122 km of pipelines, oil spill sites, oil wells, and abandoned or out-of-use facilities. The information about the surveyed sites was supplied by government agencies, the SPDC, and members of the community in Ogoniland.

The investigation showed 69 affected sites of those that had their soils and groundwater investigated (Fig. 23.2). Samples from drinking water, sediments, air, and different water sources such as rainwater and surface water as well as organisms such as fish were collected and analyzed. The total samples amounted to 4000 and consisted of 142 groundwater-monitoring wells constructed for the study purpose and soil samples collected from 780 boreholes. Medical reports of approximately 5000 people were also analyzed, and several meetings were held with the community members with a total attendance of 2300 in 264 meetings (Fig. 23.3).

FIGURE 23.2 Contaminated River at Sugi Bodo, Gokana LGA. The report provides baseline information on the scale of the challenge for Ogoniland and priorities for action in terms of clean-up and remediation. *Source: Pictures courtesy: UNEP (2006) Report.*

FIGURE 23.3   One of the meetings UNEP held with community members during the project aimed at ensuring participation of the community. *Source: Pictures courtesy: UNEP (2006) Report.*

23.4.1 Summary of United Nations Development Programme Findings

It was concluded from various analyses and observations that there has been severe contamination of the environment by crude oil, which has negative impacts on the ecosystem and man. The summary of the result is presented in the following sections.

23.4.1.1 *Contaminated Soil and Groundwater*

A high degree of oil contamination in land, sediments, and swamps was observed. Most spills were from oil pollution, although three locations recorded contamination by refined petroleum products. It was also concluded that land in Ogoni lacks a continuous clay layer, which can serve as a shield protecting lower layers from oil spilled on the upper layers. This means that the groundwater is exposed and in turn polluted. It was observed in 49 cases that the least depth at which oil was observed was 5 m, and this determines the remediation process. More than 66% of the analyzed soils close to oil industries had an oil concentration that was above the limit set by the Environmental Guidelines and Standard for Petroleum Industries in Nigeria (EGASPIN, 2002).

The most severe damage of oil pollution was experienced in a place close to the NNPC at Nisisioken Ogale, in Elene LGA. It was observed that there was an 8-cm layer of oil floating on the groundwater, which supplies the well used in that community.

23.4.1.2 *Vegetation*

Hydrocarbon contamination in numerous tidal streams leaves bare mangrove parts such as leaves and stem. The root region is observed to be covered with a substance having a thickness of at least 1 cm and that is similar to bitumen. Spawning activities and a nursery for young fish take place in mangroves. Pollution destroys this habitat as well as these organisms. Vegetation in affected areas is damaged, and root crops such as cassava become unstable.

Subsequent farming in affected areas results in a decline in yield as plants show different stress signs. Recovery of affected vegetation is almost impossible because of explosion from oil spills leading to the death of vegetation and producing a crust of earth. Satellite images have shown widened channels as a result of dredging activities. If adequate restoration measures are not taken, colonization by Nipa palm, an invasive species having better pollution resistance than native species, will take place.

Artisanal refining that took place at Bodo West (Bonny LGA) between 2007 and 2011 led to a 10% loss of productive mangrove of a total size of 307,381 m^2 in area. If not curbed, the result will be irreversible mangrove loss in the area.

23.4.1.3 Aquatic Resources

The UNEP survey showed that hydrocarbons are present in water bodies located in the creeks. The oil floating on the water bodies ranged from thick black in nature to a thin luster layer. The highest reading of 7.420 g/L concentration was recorded at Ataba Otokroma (a border to the Gokoma and Andoni LGAs).

The movement of fish away from the polluted area will have an effect on the fishermen's catch. The fishermen will now have to search upstream or downstream for better fishing grounds. It was also deduced that the concentration of hydrocarbon in fish is not as high as people expected in Ogoniland. So it is not really causing negative health impacts yet. Pollution of creeks, which makes them inappropriate for fishing activities, and destruction of the fish habitat in the mangroves is adversely affecting the fishing industry. Fish farms set up in or near the creeks have been covered by a layer of floating oil that is always present, thereby ruining the businesses of the fish farmers. The report concludes that unless adequate laws and policies are in place, restoration of these polluted and degraded wetlands will not be effective.

23.4.1.4 Public Health

Exposure to these poisonous substances is mostly through air and drinking water, although exposure can also be through skin contacts with polluted soils, surface water, and sediments. The life expectancy of people in Nigeria is given at less than 50 years; we can, therefore, conclude that the Ogoni people have lived their entire lives with incessant oil pollution. A case study is that of the Nisisioken Ogala community where the drinking water has a benzene level (a carcinogen) 900 times greater than the WHO set limit. Urgent actions must be taken on health issues of these people before other remediation efforts.

Waters gotten for 10 communities (containing 28 wells) were found to be polluted with oil. Seven wells had hydrocarbon levels 1000 times greater than the Nigerian standard limit of 3 µg/L. Despite the awareness of this water pollution problem, the people still use water for their activities (drinking, washing, cooking) because no alternative is available. Air samples had a benzene level that ranges from 0.155 to 48.2 µg/m^3. Ten percent of samples had benzene concentration higher than the WHO limit; and the USEPA reported a

1:10,000 cancer risk from using such water (USEPA, 2010). The benzene concentrations detected in Ogoniland are similar to those observed in other places in the world where there is constant crude oil exploration, especially in the third world countries. However, benzene concentration in the developed world where there is strict regulation and enforcement of environmental laws is low.

23.4.1.5 Institutional Issues

Regulation and guidelines to prevent adverse environmental impacts of oil-producing industries on the environment were first introduced in 1992 by the Environmental Guidelines and Standard for Petroleum Industries in Nigeria (EGASPIN). This legislation has shortcomings when it comes to the most important measure—"intervention" and "target" values. These two aspects form the basis for oil spill and its management. They are the two criteria that prompt remediation action and its termination. The study observed that there was a wrong interpretation given to these guidelines by the DPR and the National Oil Spill Detection and Response Agency (NOSDRA). It was observed that the remediation process was not fully completed to the point where contamination is completely removed and soil quality is restored to its former unpolluted state for proper human survival.

Insufficient resources and lack of qualified technical experts have limited the efforts of the Nigerian government and agencies. For instance, since the establishment of NOSDRA, allocation of funds from the government has not been sufficient for its oil spill detection operations. So, it has to depend on industry operators for funds. Also, human settlement is interwoven with the oil field in Ogoniland. This shows weakness in government regulation when it comes to land allocation. This creates huge vulnerability and hinders control of installations by the pipeline operators.

The UNEP also exposed flaws in the area of waste generation and disposal. It revealed $1000-1500 \, m^3$ of waste bags containing cuttings from crude oil drilling activities, which were dumped in an open environment.

23.4.1.6 Oil Industry Practices

This study also showed the inadequate maintenance and dismantling of oil field facilities. Public health safety has been of concern since good industrial practices and the SPDC's procedures have been neglected. Remediation by Enhanced Natural Attenuation (RENA), which is the sole remediation practice observed, has not been effective. The initial assumption that oil will not penetrate deeper than 5 m due to its nature, given by SPDC, was found to be untrue. The report showed that the missing clay layer has made oil penetrate deeper and get to underground water in most locations. It was also observed that values for remediation closure were exceeded in 10 out of 15 sites investigated. These sites were initially recorded to have completed remediation. Contamination to the underground water level was observed in eight of the sites.

This led to the formation of a new remediation management system in January 2010. This system was adopted by all Shell companies (exploration and production) in Nigeria. However, they are still not up to international standards, as they are short of requirements for local regulation.

23.5 MANAGEMENT OF OIL SPILLS IN NIGERIA

23.5.1 Lessons From the Past

A great deal of dispute arising from oil spillage between the oil majors and the local communities where they operate is documented in the literature. The most celebrated case between the Ogoni people and Shell continues to linger. Oil exploration and production by Shell in Nigeria dates back to the early 1950s. Ninety six oil wells were drilled, but operations stopped in Ogoniland as a result of conflict in the region, although there are still pipelines that pass through this region. Oil installation in this region has been abandoned, existence of oil spill is common, illegal oil practices and improper maintenance of oil infrastructure have defined this region over time, and the remediation program has not been effective (UNEP, 2006).

There are several laws and policies formulated to manage incidents of oil spills at local and international levels. These laws are as follows:

- Oil Pollution Act (OPA) of 1990

  This law deals with the prevention of oil spills and the development of adequate responses in the case of any event. It provides management procedures for the oil industries and government on mitigation, prevention, and clean-up actions for oil spill. This regulation was made to ensure a drastic reduction in oil spill cases. It also provides guidelines on allocation of funds for clean-up actions and compensation to those affected by the spill. In addition, it establishes the readiness of an oil spill response system put in place by the government. The response system also extends to the oil tankers and industries. This law was targeted to enhance the national response system to oil spill scenarios and the invention of contingency plans.

- NOSDRA

  The NOSDRA was established by the Federal Executive Council of Nigeria to implement the National Oil Spill Contingency Plan (NOSCP). The introduction of this contingency plan was in conformity with international convention signed by Nigeria on Oil Pollution Preparedness, Response and Cooperation (OPRC 90). This agency also ensures that the drilling fluid and mud systems used by oil companies are environmentally friendly as well as ensuring compliance to gas flare-out regulation and standards.

- The NDDC

  The NDDC was created by the federal government of Nigeria to implement development programs in the Niger Delta region. This agency came into existence by an act of the National Assembly in the year 2000 because of violence in the Niger Delta, which caused increased vandalization of oil installations in the region.

This act of the National Assembly gave the following mandate to the commission:

1. Prompt survey of the area in order to determine criteria needed for the development of its physical and socio-economic segments.
2. Establishments of schemes for the physical development of the area.

3. Identification of factors restraining the progress of the region, and aid in the establishment of policies that ensure the sustainable management of Niger Delta resources.
4. Assess and provide adequate report on any funded project carried out by oil and gas industries as well as non-governmental organizations to ensure proper fund utilization
5. Solve problems pertaining to the ecosystem and the environment as a result of oil exploration.
6. Establish a working relationship with mineral- and gas-producing countries on issues pertaining to prevention and control of pollution.

23.6 PETROLEUM-RELATED LAWS AND REGULATIONS

The following laws and international agreement are in use in Nigeria and are relevant to the management of oil production and its environment.

- Endangered Species Decree Cap 108 LFN 1990
- Federal Environmental Protection Agency Act Cap 131 LFN 1990
- Harmful Waste (Special Criminal Provisions, etc.) Cap 165 LFN 1990
- Petroleum (Drilling and Production) Regulations, L.N. 69 of 1969
- Mineral Oil (Safety) Regulations, L.N. 45 of 1963
- International Convention on the Establishment of an International Fund for
- Compensation for Oil Pollution Damage, 1971
- Convention on the Prevention of Marine Pollution Damage, 1972
- African Convention on the Conservation of Nature and Natural Resources, 1968

23.7 THE ENVIRONMENTAL IMPACT ASSESSMENT DECREE NO 86 OF 1992

This decree was established for the protection and sustenance of our natural ecosystems. This law stipulates that EIA should be carried out on projects that may have negative impacts on the environment (Ntukekpo, 1996; Olagoke, 1996). The EIA process identifies the possible environmental effects of a particular project from all aspects over the short and long term and provides measures to curb these adverse impacts (Ozekhome, 2001). The EIA is carried out by the environmental protection agency at both federal and state levels.

23.8 FEDERAL AND STATE AGENCIES

Problems of pollution are dealt with in Nigeria by several agencies, including the DPR, the Federal Ministry of Petroleum Resources, Ministries of Environment at both federal and state levels, and the Nigerian Maritime Administration and Safety Agency (NIMASA).

23.9 EFFORTS OF THE OIL COMPANIES AND NON-GOVERNMENTAL ORGANIZATIONS

The Clean Nigeria Associates (CNA) was formed in 1981. The CNA comprises 11 oil companies and the NNPC. The persisting oil pollution problems led to this development. The CNA is meant to build strategies to curb oil pollution and other forms of pollution (Nwilo & Badejo, 2005). The Niger Delta Environmental Survey (NDES) was established by Shell in conjunction with members of Oil Producers Trade Section (OPTS) of the Lagos Chambers of Commerce. This is to increase focus on Shell's operations in Nigeria. Funding for the operations of the NDES is provided by Shell, the OPTS, Rivers, and Delta State governments.

The followings are the mandate of the NDES:

- An adequate description of ecological zones, boundaries, and all forms of natural resources (renewable or non-renewable) in the area
- An in-depth review on the status of the environment and how the local people are affected
- An investigation of the linkages between usage of land, settlement plan, companies, and also the environment in order to provide a foundation for subsequent development
- A workable plan for Niger Delta Management and progress (NDES, 1996)

23.10 OIL TRAJECTORY AND FATE MODELS FOR OIL SPILL DISASTER MONITORING

Models have been used to determine the fate of oil spill incidents and to assess the impacts they have on the environment. This model is known as an oil simulation model, and it is one of the tools used in oil response and contingency planning (Rossouw, 1998). When oil spill occurs, slick can be predicted if adequate meteorological information is obtainable (Rossouw, 1998). Other processes such as mechanical system use, burning, sinking, dispersion, and absorbent mechanisms can be employed in the removal of oil spills. Natural processes can also aid oil removal. This includes evaporation, photochemical oxidation, and dispersions (Wardley-Smith, 1977). Bioremediation can also be employed in oil spill management (Atlas, 1995; Hoff, 1993; Ndimele, Jenyo-Oni, Chukwuka, Ndimele, & Ayodele, 2015; Ndimele & Ndimele, 2013).

23.11 NIGERIA SAT 1

A universal early-warning satellite network producing current information about natural disasters such as earthquakes and drought, is known as the Disaster-Monitoring Constellation. Nigeria Sat 1 has now aligned with this network. This implies that aside from the geographical mapping function of the satellite, information on oil spill incidents will also be provided by the satellite. This will help in curbing the vandalization problem and aid in oil spill management. Information on the spill sites, which serves as the input data and also on the degree of pollution in coastal water will be important for quick and effective remediation operations carried out on the affected areas.

23.12 INTERNATIONAL CO-OPERATION

There is increased international co-operation in the fight against oil piracy. A typical example is the donation of three 56-m (180-ft.) refitted World War II-era patrol boats by the United States of America to the Nigerian Navy. The United Nations also informed the country that the United States still has an additional four vessels to donate. The refurbishments of the boats are funded by the Pentagon to the tune of $3.5 million each. These and other international co-operation have begun yielding results, as there have been interceptions of several illegal oil tankers by the Nigerian Navy.

23.13 GEOGRAPHIC INFORMATION SYSTEM FOR MANAGING OIL SPILL INCIDENTS

The success of a management and remediation process relies on a rapid response system, which is in existence from when the event is initiated to the termination point. GIS creates opportunities for incorporating oil drift forecast (wind and current influence prediction) in a computer program framework (Milaika, 1995). GIS can help store all important information in a systemized way onto a map that can help solve some of the proposed problems. GIS can also be used to determine how adequate a particular oil spill contingency plan is (Parthiphan, 1994).

23.14 ENVIRONMENTAL SENSITIVE INDEX MAPPING

This produces a map showing how susceptible a particular area is to a particular stress factor (oil exposure) on a scale of 1—10; the numbers are directly proportional to the degree of sensitivity. The map may also contain biological, socio-economic, and geographical features. Some specific information needed by the response team on proposed boom or skimmers location can also be provided. The color assigned to the symbol representing a given feature determines the sensitivity to a stress factor. The Environmental System Research Institute (ESRI) has recommended standards for the development of maps on the environmental sensitivity index for the Nigerian coast. These standards have been adopted by oil companies to develop Environmental Sensitive Index (ESI) maps for their operational regions in the country.

23.15 AWARENESS CREATION

The people must be aware of the adverse effects caused by oil spill. This should be an important component of any oil spill management plan. This is done by Government (federal and state) and also by agencies such as the NDDC. Awareness is also being carried out by several civil society organizations.

23.16 CONCLUSION

The country has been agonizing since the discovery of crude oil in 1956 due to the negative environmental impacts of its exploration and production. Oil sabotage is a problem that has led to several oil spill incidents in the country. Land, vegetation, and water have been destroyed by these persistent incidents. Crisis and turmoil in affected areas, especially in the Niger Delta, have grown, basically because of the degradation of the environment by petroleum-related activities and their impacts on human health. However, there are several efforts, laws, and policies in place to mitigate the adverse effects of oil spill, although these measures have not been totally effective. Some of the policies and efforts are:

- The NOSDRA, formed by the Federal Executive Council of Nigeria
- Relevant acts and regulations that are in existence to curb oil spillage
- The passing into law of the NDDC
- The formation of the Niger Delta Environment Survey
- Developing models to determine the movement of oil, which can aid in management policy
- The setting of standards for the development of maps of environmental sensitivity for the Nigerian coast

These efforts have made progress in detecting and managing oil spill incidents along the Nigerian coast. However, lack of technical expertise, weak regulatory framework, and inadequate/inconsistent enforcement have ensured that Nigeria remains negatively impacted by oil spill.

References

Atlas, R. M. (1995). Petroleum biodegradation and oil spill bioremediation. *Marine Pollution Bulletin, 31*, 178−182.

Baird, J. (2010). *Oil's Shame in Africa*. Newsweek: 27, 26 July 2010.

Bronwen, M. (1999). The price of oil human rights watch. *Perception and reality: Assessing priorities for sustainable development in the Niger River Delta*. Retrieved 3 October 2011.

Burdorf, A., Järvholm, B., & Siesling, S. (2007). Asbestos exposure and differences in occurrence of peritoneal mesothelioma between men and women across countries. *Occupational and Environmental Medicine, 64*(12), 839−842.

Environmental Guidelines and Standards for the Petroleum Industry in Nigeria (EGASPIN). (2002). Revised Edition. Environmental Studies Unit, Department of Petroleum Resources (DPR), Lagos.

Hoff, R. Z. (1993). Bioremediation: an overview of its development and use for oil spill cleanup. *Mar. Pollut. Bull, 29*, 476−481.

Milaka, K. (1995). *Use of GIS as a tool for operational decision making, implementation of a national marine oil spill contingency plan for estonia*. Glostrup, Denmark: Carl Bro International a/s.

Moffat, D., & Linden, O. (1995). Perception and reality: Assessing priorities for sustainable development in the Niger River Delta. *Ambio, 24*(7-8), 527−538.

Ndimele, P. E., & Ndimele, C. C. (2013). Comparative effects of biostimulation and phytoremediation on crude oil degradation and absorption by water hyacinth (*Eichhornia crassipes* [Mart.] Solms). *International Journal of Environmental Studies, 70*(2), 241−258.

Ndimele, P. E., Jenyo-Oni, A., Chukwuka, K. S., Ndimele, C. C., & Ayodele, I. A. (2015). Does fertilizer ($N_{15}P_{15}K_{15}$) amendment enhance phytoremediation of petroleum-polluted aquatic ecosystem in the presence of water hyacinth (*Eichhornia crassipes* [Mart.] Solms)?. *Environmental Technology, 36*, 2502–2514.

Niger Delta Environmental Survey (NDES). (1996). *The Niger Delta environmental survey: Terms of reference*, April 3, 1996.

Ntukekpo, D. S. (1996). *Spillage: Bane of petroleum, ultimate water technology & environment.*

Nwilo, P. C., & Badejo, O. T. (2001). *Impacts of oil spills along the Nigerian coast.* The Association for Environmental Health and Sciences.

Nwilo, P. C., & Badejo, O. T. (2005). Oil spill problems and management in the Niger Delta. In *International oil spill conference*. Miami, FL, USA.

Olagoke, W. (1996). *Niger Delta environmental survey: Which way forward water technology & environment.*

Ozekhome, M. (2001). *Legislation for growth in the Niger Delta, midweek pioneer.*

Parthiphan, K. (1994). *Oil spill sensitivity mapping using a geographical information system.* Department of Geography, University of Aberdeen. EGIS Foundation.

Rossouw, M. (1998). *Oil spill simulation: Reducing the impact.* START/IOC/LOICZ.

Tawari, C. C., & Abowei, J. F. N. (2012). Air pollution in the Niger Delta area of Nigeria. *International Journal of Fisheries and Aquatic Sciences, 1*(2), 94–117.

United Nations Development Programme (UNDP). (2006). *Niger Delta human development report.* Abuja: United Nations Development Programme, p. 229. <http://hdr.undp.org/sites/default/files/nigeria_hdr_report.pdf>.

United Nations Environment Programme Report (UNEP). (2006). *Environmental assessment of ogoniland*, pp. 8–17.

United States Environmental Protection Agency (USEPA). (2010). *Summary and analysis of the 2010 gasoline benzene pre-compliance reports [EPA Report].* (EPA-420-R-10-029). Washington, DC.

Vidal, J. (2010). *Nigeria's agony dwarfs the gulf oil spill. The US and Europe ignore it.* The Observer. Government's national oil spill detection and response agency (Nosdra). Retrieved 27 July 2010.

Wardley-Smith, J. (1977). *The control of oil pollution on the sea and inland waters.* United Kingdom: Graham and Trotman, Ltd.

Remediation of Crude Oil Spillage

Prince Emeka Ndimele[1], Abdulwakil O. Saba[1], Deborah O. Ojo[2], Chinatu C. Ndimele[2], Martins A. Anetekhai[1] and Ebere S. Erondu[3]

[1]Lagos State University, Ojo, Lagos State, Nigeria [2]University of Ibadan, Oyo State, Nigeria [3]University of Port Harcourt, Rivers State, Nigeria

24.1 INTRODUCTION

Oil spill is a regular occurrence in the Niger Delta region of Nigeria. Nigeria's oil-rich Niger Delta is characterized by a prevalence of both old and new oil spills totaling about 9343 incidents in the last 10 years, according to official records by National Oil Spill Detection and Response Agency (NOSDRA). This translates to an average of nearly a thousand spills yearly, the highest rate of spills globally. NOSDRA reported that, within the period from 2006 to 2015, there were over 5000 spillage sites from the over 9000 spills. The average volume of oil spill annually is about 115,000 barrels, which is worth about $4.58 million (current oil price is: $39.84). The most vulnerable segment of the population is the women, which make up about 75% of the workforce, and their dependent children. The situation is worsened by the reluctance of oil companies to clean up the environment after spillage. Even when they attempt a cleanup exercise, they use conventional methods that do more harm to the aquatic ecosystem than the oil spill itself. Oftentimes, the spill is allowed to attenuate naturally; this can take years and the damage to the environment is incalculable.

Recently, biological-based approaches to crude oil remediation have been used and the results have been encouraging, though there is still room for improvement. This chapter discusses the conventional methods of crude oil spill cleanup as well as explores bioremediation as a viable alternative to these conventional methods.

24.2 WHAT IS OIL SPILLAGE?

Oil spillage can be defined as the release of crude oil hydrocarbons into the environment. It is an important environmental disaster of global concern that usually occurs

369

accidentally or intentionally, mostly resulting from everyday human activities that release crude oil into coastal waters and land. Oil spillage is a controversial aspect of oil exploration and it occurs during crude oil exploration, oil pipeline leakage, vandalization, illegal tampering of oil-well heads, transfer of oil into vessels, and during the transport of oil in tankers. Oil spills are taken seriously by the oil industry at both upstream and downstream sites. The oil spreads out rapidly on water surfaces to form a thin layer called an oil slick. As the oil continues spreading, the oil slick layer becomes thinner and thinner and finally becomes a very thin layer called a sheen, which often looks like a rainbow. This is a frequent occurrence, particularly because of the daily use of petroleum products by man and this poses a multifaceted problem presently ravaging oil-producing communities all over the globe.

In developed countries, many oil spill events are reported and prompt actions are taken to remedy the affected ecosystem. However, many of the oil spills are not reported in developing countries and many times concise efforts are not made to restore the ecosystem to its previous state even when the oil spills are accounted for (Anna, 2013). Spill statistics are collected by registered agencies in any country. The International Tanker Owners Pollution Federation Limited (ITOPF) has maintained a worldwide database of spills from tankers since 1974. In the United States, the US Coast Guard maintains a database of spills, while the Bureau of Safety and Environment (BSEE) maintains records of spills from offshore exploration and production activities. Official records from the NOSDRA, the agency saddled with the responsibility of preparedness, detection and response to all cases of oil spillage in Nigeria, indicates that there were over 5,000 spillage sites from over 9000 spills in Nigeria over the period from 2006 to 2015 (Dave & Ghaly, 2011). Some of the biggest oil spills in the world were reported in Italy, Canada, South Africa, Angola, France, Persian Gulf, Uzbekistan, West Indies, Mexico, and Kuwait. The largest oil spill in history happened in Kuwait during the first Gulf War in 1991. Some 240 million gallons of oil flowed into the Persian Gulf, and the resulting slick spanned an area just larger than the size of the island of Hawaii (Michel & Fingas, 2015).

Crude oil spills have damaged vulnerable ecosystems across the world. An offshore oil spill is a major cause of concern because of its hazardous impact on marine life (Obi, Kamgba, & Obi, 2014). The effects are generally more catastrophic at sea than on land, since the spills can spread faster for hundreds of nautical miles and form a thin coating of oil on water bodies. Hence, marine oil spills, particularly large-scale spill accidents, have received greater attention. For example, the spill of 11 million gallons (37,000 metric tons) of North Slope crude oil into Prince William Sound, Alaska, from the Exxon Valdez in 1989 led to the mortality of many intertidal and subtidal organisms and long-term environmental impact (Tewari & Sirvaiya, 2015).

24.3 CAUSES OF CRUDE OIL SPILLS

Oil spills into the environment through natural means or as a result of anthropogenic (man-made) activities can be deliberate or accidental. Natural disasters like earthquakes, adverse weather conditions, and hurricanes are responsible for the dumping of crude oil and gas into the oceans. Anthropogenic activities are the activities of man that cause oil to be released into the environment. These could be the results of pipeline vandalism, acts of terrorism, intentional discharge of oil during wars, and washing of tankers and vessels into the sea. The discharges from offshore drilling, mechanical faults of machines used in drilling, tank cleanups and leakages from oil tankers are other causes of oil spills.

24.4 WHY CLEAN UP THE ENVIRONMENT AFTER OIL SPILL?

Oil spillage has enormous adverse effects on the environment. It impacts the environment by causing physical changes, ecological changes, and chemical toxicity. The severity of these impacts depend on the extent of the pollution, the physical and chemical properties of the oil spilled, the ambient condition of the environment, and the sensitivity of the living organisms and their habitats to extraneous factors. The toxic effects may be acute or chronic on the living organisms depending on the period of exposure and the concentration of the oil spill. Acute means toxic effects occurring within a short period of exposure in relation to the life span of the organism, and generally the result of exposure to high contaminant concentrations. For example, acute toxicity to fish could be an effect observed within four days of exposure after which the contaminant is withdrawn from the organism's environment. Chronic toxicity refers to long-term effects, and is generally the result of exposure to low contaminant concentrations that are usually related to changes in such things as metabolism, growth, reproduction, behavior, or ability to survive.

24.5 EFFECTS OF CRUDE OIL CONTAMINATION ON SOIL

Oil contamination affects the quality of soil (texture) and its fertility. Soil porosity is affected due to the fact that oil tends to force soil particles together, thereby decreasing porosity. Furthermore, the availability of carbon dioxide for a living organism's respiration is depleted due to the coat formed on the soil surface by an oil spill (Ezeji, Anyadoh, & Ibekwe, 2007). Crude oil contamination on soil distorts germination rates, degrades farmland, kills plants and small trees, permeates into groundwater, and harms aquatic life. Large areas of mangrove have been destroyed by oil spills.

24.6 EFFECTS OF CRUDE OIL CONTAMINATION ON WATER

Water is a gift of nature and essential for life. Its contamination has adverse effects on life in general. Oil can affect aquatic animals in many ways, including changing their reproductive and feeding behavior, and causing tainting and loss of habitat. Tainting occurs when an organism ingests enough hydrocarbons to cause an "off" flavor in the seafood, making it unsuitable for human consumption until this taint disappears. Oil has significant effects on marine larvae, causes damage to the structure and value of wetlands, destroys the insulating ability of fur-bearing mammals such as sea otters, and the water repelling ability of bird feathers, thus exposing these creatures to harsh elements. Oil spill also affects aquatic food chains, ecosystem biodiversity, aquatic nursery grounds, and leads to poisoning of aquatic life.

24.7 EFFECTS OF CRUDE OIL CONTAMINATION ON THE SOCIETY

The environmental degradation caused by oil spillage has socioeconomic impacts on the oil-producing communities. This pollution destroys the esthetic values of water bodies. It also affects other qualities of water such as drinking, recreation, swimming, fishing, and domestic use. Eventually, there is loss of aquatic lives, which has a devastating impact on the livelihood of fisherfolk households who solely depend on fishing. Webler and Lord (2010) noted that humans can be affected by oil spills in three major ways: oil can affect ecological processes that cause direct harm, e.g., health impacts from eating seafood with bioaccumulated oil toxins; oil spill stressors can change intermediary processes, e.g., economic impacts on fishers caused by oil spill damage to fisheries; and stressors can directly harm humans, e.g., health impacts from breathing oil vapors. It is therefore important to restore the ecosystem, ensure conservation, and protect human health.

24.8 CONVENTIONAL METHODS OF ENVIRONMENTAL REMEDIATION

Since oil spill accidents are not uncommon, it is important to have a good knowledge of their adverse environmental impacts, in order to find effective ways of remediating the environment after oil spillage. Remediation could be done ex situ or in situ through various technologies. The ex situ method involves the removal of the contaminated soil and/or water to clean up on another surface, while the in situ method refers to decontaminating the soil and/or water at the site of pollution.

24.9 REMEDIATING TECHNOLOGIES FOR MARINE OIL SPILL

Remediating techniques have been developed to clean up the marine environment after an oil spill incident. These techniques have been adopted over the years as a means to reduce the toxic impacts of crude oil spillage on the environment. Conventional countermeasures for oil spill in the aquatic environment include various physical, chemical, thermal, and biological processes.

24.9.1 Physical Methods

Also referred to as mechanical methods, these methods do not require the use of chemicals and are preferred in many countries. The physical methods are in form of barriers mainly used to control the spread of an oil spill.

24.9.1.1 Booms

Booms are containment techniques for oil spills. They act as physical barriers that enclose floating oil and prevent it from spreading. Booms are also used to divert oil from biologically sensitive areas, or to concentrate oil and maintain an adequate thickness so that skimmers can be used, or other cleanup techniques. These processes facilitate oil recovery from the spill site. The majority of boom designs fall into two categories: curtain booms and fence booms. Fire booms are specifically constructed with fireproof metal that can withstand very high temperatures generated by burning oil. The fire booms can be of either fence or curtain design. Other types of special purpose booms are the shoreline seal booms, ice booms for light ice conditions, sorbent booms, and tidal seal booms.

Fence booms are floating fence-like structures having a flat cross-section, which has a vertical position in the water by means of integral or external buoyancy. They are lightweight, easy to handle, clean, and store in minimal storage space and resistant to abrasion. However, some of the limitations of the fence booms include low stability in strong winds and currents, low flexibility for towing, and low efficiency in high waves.

Curtain booms consist of a subsurface skirt that remains under water to contain the oil and is supported by a large air- or foam-filled flotation chamber usually of circular cross-section. The diameter of the circular chamber ranges from 100 to 500 mm, the skirt length ranges from 150 to 800 mm; the floating chamber prevents oil from being washed over the boom. Curtain booms are made up of polyurethane, polystyrene, bubble wrap, or cork. They have high flexibility in towing and perform better than fence booms, but are more difficult to clean and store (Fig. 24.1).

FIGURE 24.1 Boom used in the Gulf of Mexico oil spill. *Source: https://commons.wikimedia.org/wiki/File: Shrimp_boats_tow_fire-resistant_oil-containment_boom_in_Gulf_of_Mexico_2010-05-03_1.jpg*

Booms are often arranged in a U-configuration for the purpose of containment. The U-shape is created by the current pushing against the center of the boom. The critical requirement is that the current flowing perpendicular to the boom does not exceed 0.35 m per second (m/s) or 0.7 knots, which is referred to as the critical velocity (Michel & Fingas, 2015). Above this velocity, turbulence causes viscous oil to be lost from the boom by entrainment when the droplets of the oil are carried down under the boom into the water.

24.9.1.2 Skimmers

Skimmers are usually used together with booms to remove floating oil from the water surface without causing changes in its properties, and transfer it to the storage tanks onboard the vessel. This process does not change the physical and chemical properties of oil so it can be reprocessed and reused. The type and thickness of the oil spill and weather conditions generally determine the success of skimming. Skimmers are generally effective in calm waters, and subject to clogging by floating debris. They are categorized into oleophilic, suction, and weir skimmers. Skimmers generally rely on specific gravity, surface tension, and a moving medium to remove floating oil. They may be self-propelled, used from shore, or operated from vessels.

Oleophilic skimmers employ materials that have the ability to adhere to oil. They trap the oil from the surface, and the oil is scraped or squeezed out into the collecting recovery tank. The oleophilic materials can be of different shapes, such as disc, drum, belt, brush or rope-mop. Suction skimmers have the simplest design and are most widely used for the recovery of oil from beaches, confined areas, or removal of oil from land surface. Made in the form of vacuum pumps, suction skimmers suck up oil through wide floating heads and transfer it into storage tanks. Weir skimmers use gravity action to collect floating oil from water. They act like dams to trap the oil, then the oil is pumped out of the weir central sink through a pipe or hose to a storage tank for recycling purposes.

24.9.1.3 Adsorbent Materials

Adsorbent materials facilitate conversion of a liquid to a semisolid phase and are used after the skimming operation to clean the remaining oil. These adsorbents can be natural organic, natural inorganic, or synthetic materials. Natural organic sorbents include peat, hay, vegetable fibers, feathers, kapok, sawdust, milkweed, straw, etc. These readily available and less expensive materials have been used by researchers to achieve maximum absorption. They can soak up from 3 to 15 times their weight in oil. The major disadvantage for their use is the difficulty in collecting adsorbents after spreading on the oil spill water for disposal. Natural inorganic sorbent includes vermiculite, clay, wool, glass, sand, vermiculate and volcanic ash. They are also readily available and can absorb up to 4 to 20 times their weight. Synthetic absorbents are the most widely used sorbent materials. They include materials similar to plastic like polyethylene, polypropylene, polyurethane foam, polyester foam, polystyrene, and nylon fibers. They can absorb up to 70–100 times their weight in oil due to their hydrophobic and oleophilic nature, but are nonbiodegradable.

24.9.2 Thermal Method

This method involves the burning of oil with minimal specialized equipment like fire-resistant boom or igniters. In situ burning is most successful immediately after the oil spill has occurred. It is an effective oil spill response in the open water with calm winds, and is best applicable for refined oil products that will burn quickly without causing danger to marine life. It is usually ascertained first that there are no floating vessels nearby such as ships, speedboats, and oil tankers before the thermal method is used. The major constraints in the use of this method are: fear of secondary fires, destruction of vegetation and aquatic lives adjacent to the site, which can be affected by burning, long-term alteration in aquatic plants and animals, and risk to human health due to the gases emitted from thermal combustion. Thermal remediation methods are usually affected by water temperature, speed, wave amplitude, wind direction, slick thickness, oil type, and amount of weathering and emulsification that have occurred (Tewari & Sirvaiya, 2015).

24.9.3 Chemical Methods

Chemical means of remediating the marine environment do not only block the spreading of the oil spill, but also protect the shoreline and sensitive marine habitats. These methods are, therefore, among the best remediation techniques available for both onshore and offshore (Tewari & Sirvaiya, 2015). Dispersants and solidifiers are used together with physical methods to treat oil spill by changing the physical and chemical properties of the oil.

24.9.3.1 Dispersants

Dispersants have surface-active agents known as surfactants. The main aim of dispersants application is to weather the oil slicks into small droplets, which submerge into the depth of the water column and become rapidly diluted and easily degraded. Examples of concentrated types of dispersants include: SlickgoneNS, Neos AB3000, Corexit 9500, Corexit 8667, Corexit 9600, SPC 1000™, Finasol OSR 52, Nokomis3-AA, Nokomis 3-F4, Saf-Ron Gold, ZI-400, and Finasol OSR 528 (USEPA, 2011). Dispersants are applied by spraying the water with the chemical from vessels or aircrafts; they can be used on rough seas where there are high winds that will enhance the proper mixing of the chemical with the water. Each dispersant molecule contains both oleophilic (attracted by oil) and hydrophilic (attracted by water) parts. When applied on the oil spill site, the solvent transports the dispersants to the oil/water interface where the molecules rearrange so that the oleophilic part is in the oil and the hydrophilic part is in the water. This process reduces the surface tension of the oil/water interface, which together with wave energy, results in droplets separating from the oil slick. Dispersants also allow for rapid treatment of polluted water, slow down the formation of oil–water emulsions, make the oil less likely to stick to surfaces (including animals such as sea birds), and accelerate the rate of natural biodegradation. However, the inflammable nature of most dispersants is hazardous to human health and marine life.

24.9.3.2 Solidifiers

Solidifiers are those hydrophobic polymers enhanced by Van der Waals forces, which on reaction with oil, convert it into a solid rubber state that does not sink and can be easily removed by physical means. They are hydrophobic polymers (oleophilic) that can be applied as dry particulate or semisolid materials. There are three types of solidifiers, each having unique characteristics and properties: polymer sorbents, cross-linking agents, and polymers with cross-linking agents. Examples of solidifiers are Rawflex, Norsorex, Oil Bond, Molten wax, Elastol, Gelco 200, CI agent, Rubberizer, Jet Gell, SmartBond HO, etc. (Fingas & Fieldhouse, 2011).

24.9.4 Biological Method

The biological method refers to bioremediation, in which biological processes (microorganisms) are used to degrade and metabolize chemical substances and restore environment quality. It is a cheap, sustainable remediation method, and yet benign to the environment. Biostimulation and bioaugmentation are two types of bioremediation methods. In biostimulation, nutrients are added to stimulate the growth of the microorganism, while in the case of bioaugmentation, microorganisms are added to existing native oil-degrading population. There are only a few microorganisms capable of digesting the petroleum hydrocarbons. The microorganisms assimilate organic molecules into their cell biomass and release carbon dioxide, water, and heat as byproducts. In the case of a marine oil spill, microorganisms with the ability to degrade hydrocarbons are ubiquitous in the indigenous oil spill site. Both paraffinic and aromatic hydrocarbons can be degraded by a variety of microorganisms but have different degradation rates. The biodegradation of an oil spill in the marine environment is mainly affected by the bioavailability of nutrients, the concentration of oil, time, and the extent to which the natural biodegradation had already taken place.

24.10 REMEDIATION TECHNOLOGIES FOR CONTAMINATED SOIL AND GROUNDWATER

Contamination of soil by crude oil has remained an emerging issue mainly because of the costly damage it has on soil and vegetation. The need to clean up crude oil-contaminated soil becomes crucial. Different methods have been developed over the years, and attempts have been made to find the most efficient and eco-friendly method for the cleanup of crude oil-contaminated soil. There are four preliminary steps that should be taken before environmental remediation is carried out. The steps are: site assessment, identification of possible remediation measures, its feasibility and effectiveness, and the selection of the most appropriate remediation technique. In the first step, the site of crude oil contamination is assessed, the source of contamination is defined and the threats that the pollution poses to the environment and living organisms are itemized. The second stage involves the identification of possible remediation methods that will address the threats recognized in the first stage. A feasibility study on the remediation techniques selected is carried out in the third stage in order to analyze their effectiveness. The fourth stage is the selection of the most appropriate remediation technique peculiar to the demand of that contaminated environment.

24.10.1 Soil Excavation or Dredging

This method refers to the removal of the contaminated soil to an off-site location for burying or burning. This method requires additional space; it is labor-intensive and expensive. Moreover, there is high risk of exposure to hazardous compounds during excavation, and the excavated land is prone to environmental disaster such as erosion.

24.10.2 Incineration

Incineration refers to burning contaminated soil to destroy the crude oil in the soil and reduce toxic elements to nontoxic substances. This method is usually done ex situ, in which the contaminated soils are normally first excavated and carried to off-site facilities before incineration. This method requires high operational cost and large space, and it can cause environmental pollution.

24.10.3 Soil Washing

Soil washing is mostly used as a pretreatment method for cleaning up soils. It is an ex situ treatment process that involves the use of liquid/water sometimes combined with chemical additives and a mechanical instrument to scrub soils (Ezeji et al., 2007). This process mechanically separates contaminated sand from uncontaminated soil. Soil washing also separates small contaminated soil particles from larger soils and gravel. The disadvantage to this method is that it requires a lot of water, and then the water has to be treated to remove contamination. The efficiency of this process increases when hot water is used.

24.10.4 Thermal Desorption

This method is also called thermal stripping, low temperature thermal volatilization, or soil roasting. It involves heating the contaminated soil to very low temperatures of about $200-1000°F$ to enhance the vaporization and physical separation of contaminants with low boiling points from the soil (Ezeji et al., 2007). Depending on the organics present and the temperature of the thermal desorption, this process can cause complete or partial decomposition of some organic contaminants. In case of partial decomposition, the byproducts are reignited or treated in a secondary treatment unit.

24.10.5 Soil Vapor Extraction

Soil vapor extraction is the technique in which contaminants in the soil are allowed to vaporize at the soil temperature by applying vacuum in soil (Tewari & Sirvaiya, 2015). This method can be done by putting perforated pipes into the contaminated soil and pulling air through the soil and into the pipes. This creates pore spaces and increases the air flow in soil. Contaminants are extracted in gaseous form through a vapor extraction well, which may then be treated before being released. That is why this process is also called vacuum extraction, or enhanced in situ volatilization (Tewari & Sirvaiya, 2015). The soil is periodically sampled to see if it meets the cleanup level. This technique works well if the contaminant is a volatile compound like gasoline, which easily turns into a vapor.

24.10.6 Pump and Treat

Pump and treat system is a commonly adopted remediation technique. In this method, the contaminated water is pumped out of the ground to the surface and treated in surface water facilities. It could be passed through a filter or other treatment system to remove the contamination and the treated water can be reused. The treatment depends on the kind of contaminants present in the groundwater. If nonaqueous phased liquid is present, then phase separation is done by adding another pumping unit. This technique can also be used with vapor extraction technique.

24.10.7 Advantages and Disadvantages of Conventional Methods

The mechanical method, although cheap and readily available, cannot be applied under the conditions of rough sea waves, high wind velocity and high water waves. The major disadvantages of using natural adsorbent materials are that they are not advisable for water surface, they are labor-intensive, and they adsorb water along with oil, which leads to their sinking. Many natural inorganic adsorbents such as clay and vermiculite are loose materials and very difficult to apply in windy conditions, and they are associated with potential health risks if inhaled. Chemical dispersants are very effective but have a toxic nature and long-term environmental effects. Biological techniques are encouraged because of the advantage of soil sustainability and the possibility of the soil to be restored to its original form.

24.11 THE USE OF BIOLOGICAL TECHNIQUES IN CRUDE OIL SPILL CLEANUP

Biological oil spill cleanup involves the use of plant and animals to carry out remediation of polluted sites and particularly, the cleanup of crude oil spills. As a result, biological methods such as bioremediation and phytoremediation have been applied as biological techniques to achieve the cleanup of oil-polluted sites. According to Frick, Farrell, and Germida (1999), bioremediation can be described as the detoxification of polluted sites with the aid of biological processes. Bioremediation uses microorganisms and their yields to eliminate contaminants from the environment (Leung, 2004; USEPA, 2000), whereas phytoremediation enhances this process in the presence of plants. Consequently, remediation of crude oil pollution has recorded varying levels of success across the globe.

24.11.1 Bioremediation

Modern bioremediation has been defined as the use of microbes in the detoxification of polluted environments and is credited, partly, to George Robinson. George Robinson successfully utilized microbes to contain an oil spill along the coast of Santa Barbara, California during the late 1960s. Thereafter, especially since the 1980s, oil spills and hazardous waste bioremediation has received much attention (Shannon & Unterman, 1993). According to Mohammed (2004), the bioremediation method has been accepted globally as an in situ therapy for polluted sites. In bioremediation, microorganisms and their products are employed in the removal of contaminants from the environment (Leung, 2004).

These microorganisms eliminate, transform, contain, or reduce the potency of contaminants present in the environment. According to Shannon and Unterman (1993), the process targets the chemicals via transformation, mineralization, or alteration. Yakubu (2007) noted that compared to the use of bioremediation for wastewater treatment, its purposeful use in the reduction of harmful waste materials is still a recent development. During bioremediation, energy is produced in a redox reaction consisting of respiration and other biological functions required for reproduction and maintenance of microbial cells within which the reactions take place.

Various techniques have been used to enhance the efficacy of bioremediation processes. Of important note is bioaugumentation/microbial seeding and biostimulation.

- *Bioaugmentation/microbial seeding*: beneficial microbes with an affinity towards specific pollutants are utilized. These microbes are usually suspended by a stabilizing agent and remain inactive until stimulated in solution and operated jointly with stimulants and micronutrients (Yakubu, 2007). One characteristic of this method is that the nature of the biomass can be controlled by ascertaining that the appropriate selection of microbes is available in the polluted environment in adequate quantity, type, and compatibility in order to effectively confront the constituents and disintegrate them into their most elementary constituents (Venosa *et al.*, 2002).

  Because indigenous microbes may not be efficient in the degradation of the multitude of potential substrates resident in complex mixtures such as petroleum (Leahy & Colwell, 1990) or may be stressed as a result of a current exposure to the oil spill (Yakubu, 2007), the introduction of oil-degrading microorganisms to supplement the indigenous populations (bioaugmentation) has been proposed as a complementary technique for the bioremediation of oil-contaminated sites.

  According to Forsyth, Tsao, and Bleam (1995), bioaugmentation may also be deemed necessary when the indigenous population responsible for hydrocarbon degradation is minimal and when the rate of degradation, which is a primary factor, is slow. The effectiveness of bioaugmentation is measured by the ability of the seed microbes to degrade most petroleum constituents, preserve genetic viability and steadiness during storage, endure in unfamiliar and harsh environments, successfully contend with native microorganisms, and efficiently target the contaminants (Goldstein, Mallory, & Alexander, 1985).

  Yakubu (2007) stated that microbial species are varied in their enzymatic potentialities and predilections for oil compounds degradation. Some target linear, branched, or cyclic alkanes, while others select mono- or polynuclear aromatics. Others mutually destroy both aromatics and alkanes. Successful selection of microorganisms with the ability to degrade and produce compounds with biotechnological applications in the oil and petrochemical industry is enhanced by investigation of microbes in bioremediation systems.

- *Biostimulation*: the environment is modified to incite existing bacteria that can efficiently carry out bioremediation (Adams, Tawari-Fufeyin, & Igelenyah, 2014). In this case, limiting electron acceptors and nutrients including oxygen, nitrogen, phosphorus, and carbon are introduced (Ndimele, Jenyo-Oni, Chukwuka, Ndimele, & Ayodele, 2015;

Rhykerd, Crews, McInnes, & Weaver, 1999). Also, Perfumo, Banat, Marchant, and Vezzulli (2007) described it as a situation whereby the site is inundated with oxygen, nutrients or other electron acceptors and donors so as to step up the number and action of indigenous microorganisms available for bioremediation. In the opinion of Margesin and Schinner (2001), biostimulation is a form of biological remediation capable of improving conditions such as aeration, addition of nutrients, controlling temperature and pH to improve contaminant remediation. The main benefit of this method is that the process of bioremediation will be carried out by indigenous microorganisms, which are not only well distributed within the subsurface but also highly suitable to the surrounding. The added nutrients may serve to enhance the development of microorganisms which are heterotrophic but not innate degraders to hydrocarbons, and as such establish competition between the inhabitant microflora (Adams *et al.*, 2014; Ndimele *et al.*, 2015).

• *Phytoremediation:* this could be classified first as a separate method of biological remediation or as a form of bioremediation. Frick *et al.* (1999) defined phytoremediation as a form of biological remediation technique which involves the use of plants to detoxify polluted sites. According to Raskin (1996), phytoremediation is essentially the utilization of plants in the elimination of pollutants from the environment or attenuation of their potential to cause harm. Certain species of plants have been demonstrated to possess the ability not only to grow in contaminated soils, but extract pollutants from the medium on which they grow. Some plants accumulate toxic heavy metals in their tissues (Ndimele, 2003), while others transform pollutants into less harmful substances and volatilize them (Brooks, 1998; Terry & Zayed, 1994).
Some other aquatic plant roots are capable of filtering pollutants from water (Brooks & Robinson, 1998).

To heighten the efficacy of phytoremediation, rate-limiting factors like nutrient and oxygen can be integrated into the process. These factors will facilitate the multiplication and efficiency of the microorganisms in charge of petroleum degradation (Ndimele & Ndimele, 2013). Therefore, phytoremediation, which is basically a form of bioremediation, can be achieved not only by plants alone based on some characteristics, but also and more effectively, in the presence of microorganisms. Ndimele and Ndimele (2013) noted that certain factors determine the suitability of plants for phytoremediation. These include the growth rate, root system, as well as the ability to grow in polluted environment and extract the pollutants.

Some of the plants that have been considered for their capability to absorb petroleum and remediate oil-polluted site include ryegrass (*Lolium perenne* L.); alfalfa (*Medicago sativa* L.); arctared red fescue (*Festuca rubra* var. Arctared); water hyacinth (*Eichhornia crassipes* [Mart.] Solms) (Ndimele, Jenyo-Oni, Ayodele, & Jimoh, 2010; Reynolds & Wolf, 1999). For example, Ndimele *et al.* (2010) carried out an experiment to examine the ability of water hyacinth (*Eichhornia crassipes*) to detoxify crude oil-polluted aquatic environments and concluded that water hyacinth can absorb crude oil and can be a remedial option in the mitigation of crude oil pollution in aquatic ecosystems.

24.11.2 Advantages and Disadvantages of Biological Remediation

24.11.2.1 Advantages

According to Yakubu (2007), the following are the pros of biological remediation:

1. Being a natural process, the disintegration of hydrocarbon pollutants is carried out essentially by microbes at times in the presence of plants in the case of phytoremediation, thus transforming contaminants and not switching them from medium to medium.
2. An extensive array of contaminants is disintegrated through concurrent and multifarious action with some level of resistance to toxins.
3. The process is independent of exotic and nonnative materials, being basically a local process.
4. Another very interesting factor is the low cost involved.
5. Finally, in comparison with other techniques involved in contaminant remediation, labor, and the risk involved are reduced with an elevated level of safety (Wagenet & Bouma, 1996).

24.11.2.2 Limitations

The procedure is confronted with a number of challenges.

24.11.2.3 For Phytoremediation:

1. According to Ndimele et al. (2010), phytoremediation is still a relatively new and not totally comprehended technology.
2. In some cases, plants require chelators to mobilize metal ions for uptake by plant roots, and the outcome of this process can be unpredictable.
3. Finally, small-sized plants which do not produce high biomass are the best accumulators of pollutants (Banuelos et al., 1997).

24.11.2.4 For Bioremediation

1. Bioremediation is cumbersome to determine the particular level of the microbes contribution to the process of degradation, bearing in mind that there is concurrence in certain processes (chemical transformation and volatilization) inside the scheme.
2. When genetically modified microbes are utilized, there is a challenge of differentiating remediation involving the GMM because resident microbial community is not absent.
3. It is impossible to statistically conclude on the efficacy of bioremediation owing to the great heterogeneity in the spread of pollutants.
4. When oil spills are massive or sunken, bioremediation is usually unproductive.
5. Abiotic factors in the environment including deficient oxygen, minimal temperatures, as well as nonelevated nutrient quantity greatly limits bioremediation.
6. Local subsurface geology determines the availability of additives to subsurface microorganisms in biostimulation. If the subsurface lithology is tight and impermeable, it could be challenging to spread additives throughout the entire area affected by the contamination.

24.12 ECONOMICS OF AQUATIC REMEDIATION

Due to its importance, remediation of the aquatic environment should always be done, but in the most economical sense so as to ensure sustainability of the process. While considering remediation, analysis of the level of impact or damage imposed by the contamination should be done to identify the most apt method with regard to effectiveness and cost implication. Therefore, biological methods which have been identified as cost-effective should be prioritized except if they are suspected not to be more effective than other methods highlighted. The level and rate of a particular pollutant may also inform how much resource would be deployed in combating it. Aquatic oil pollution poses a great danger since the organisms residing in the environment utilize water for most of their metabolic processes and when contaminated, it easily leads to deleterious effects on these organisms. It is therefore pertinent for nations to prioritize the cleanup of oil-polluted water bodies and their surrounding soils to avoid a greater cost of losing our precious and invaluable aquatic organisms. In addition, aquatic oil pollution should be quickly attended to, in order to nip it in the bud so as to save the cost of remediating a larger area occasioned by a spread resulting from late response.

24.13 RECENT STRATEGIES FOR BIOREMEDIATION

24.13.1 Use of Fungi in Remediation

Fungi species have proved to possess the ability to degrade petroleum. For example, Adedokun and Ataga (2014) assessed the possibility of fungi to degrade crude oil in liquid medium. This was investigated using three indigenous mushroom species: *Pleurotus tuberregium*, *Pleurotus pulmonarius*, and *Lentinus squarrosulus*. A combination of mineral salts at different concentrations of crude oil (0.5%, 1.0%, 1.5%, and 2.0%) was injected into the mushrooms' mycelia in triplicates. It was demonstrated that three native mushrooms disintegrated crude oil at varying rates. The injection with mycelia mixture of the three fungi was the most active in oil disintegration and was significantly different ($P < .05$) from other treatments.

24.13.2 Use of Certain Genetically Engineered Microorganisms

The use of some microorganisms, which are genetically engineered to stimulate their ability to exploit precise contaminants such as hydrocarbons and pesticides, is attaining wide acceptance. This technique had been referenced in the late 1980s and early 1990s. This feat is achievable due to the likelihood of exploring microorganisms' genetic diversity and metabolic adaptability (Fulekar, Singh, & Bhaduri, 2009). The required scheme for encoding the gene can be found in extra chromosomal and chromosomal microbial DNA.

24.13.3 Aquatic Zooremediation

Gifford, Dunstan, O'Connor, Koller, and MacFarlane (2007) introduced some concepts of zoological correspondence with phytoremediation and opined that animals, although scarcely considered for the remediation of polluted sites due to some moral or human concern, have the potential to remediate contaminated sites. Some of the instances cited include the use of oyster pearls. Therefore if further research is carried out on this, many animals with potential to remediate polluted sites may be discovered.

24.14 CONCLUSION

Remediation of oil spillage should be given greater attention to reduce the resultant disaster occasioned by oil spill. Methodologies for restoration should be carefully chosen to ensure efficiency and effectiveness. Although, various physical, chemical, thermal, and biological processes have proved valuable in achieving this important goal, biological remediation is highly beneficial owing to the fact that it has proved not only to be more sustainable but also economical.

References

Adams, G. O., Tawari-Fufeyin, P., & Igelenyah, E. (2014). Bioremediation of spent oil contaminated soils using poultry litter. *Research Journal in Engineering and Applied Sciences, 3*(2), 124–130.
Anna, M. A. (2013). *Evaluation of the methods for the oil spill response in the off-shore arctic region* (Bachelor's thesis). Bachelor of Engineering Degree Programme in Environmental Engineering, Helsinki Metropolia University of Applied Sciences, 52 pp.
Banuelos, G. S., Ajwa, H. A., Mackey, B., Wu, L., Cook, C., Akohoue, S., et al. (1997). Evaluation of different plant species used for phytoremediation of high soil selenium. *Journal of Environmental Quality, 26,* 639–646.
Brooks, R. R. (1998). Phytoremediation by volatization. In R. R. Brooks (Ed.), *Plants that hyperaccumulate heavy metals: their roles in phytoremediation, microbiology, archeology, mineral exploration and phytomining* (pp. 289–312). Oxon, UK: CAB International.
Brooks, R. R., & Robinson, B. H. (1998). Aquatic phytoremediation by accumulator plants. In R. R. Brooks (Ed.), *Plants that hyperaccumulate heavy metals: Their roles in phytoremediation, microbiology, archaeology, mineral exploration and phytomining* (pp. 203–226). Oxon: CAB International.
Dave, D., & Ghaly, A. E. (2011). Remediation technologies for marine oil spills: A critical review and comparative analysis. *American Journal of Environmental Sciences, 7*(5), 423–440.
Ezeji, U. E., Anyadoh, S. O., & Ibekwe, V. I. (2007). Clean up of crude oil contaminated soil. *Terrestrial and Aquatic Environmental Toxicology, 1*(2), 54–59.
Fingas, M., & Fieldhouse, B. (2011). Review of solidifiers. *Oil Spill Science and Technology,* 713–733.
Forsyth, J. V., Tsao, Y. M., & Bleam, R. D. (1995). When is augmentation needed? In R. E. Hinchee, J. Fredrickso, & B. C. Alleman (Eds.), *Bioaugmentation for site remediation.* Columbus, OH: Battelle Press.
Frick, C. M., Farrell, R. E., & Germida, J. J. (1999). *Assessment of phytoremediation as an in situ technique for cleaning oil-contaminated sites.* PTAC Petroleum Technology Alliance Canada Calgary.
Fulekar, M. H., Singh, A., & Bhaduri, A. M. (2009). Genetic engineering strategies for enhancing phytoremediation of heavy metals. *African Journal of Biotechnology, 8,* 529–535.
Gifford, S., Dunstan, R. H., O'Connor, W., Koller, C. E., & MacFarlane, G. R. (2007). Aquatic zooremediation: deploying animals to remediate contaminated aquatic environments. *Trends in Biotechnology, 25*(2), 60–65.
Goldstein, R. M., Mallory, L. M., & Alexander, M. (1985). Reasons for possible failure of inoculation to enhance biodegradation. *Applied and Environmental Microbiology, 50,* 977–983.
Leahy, J. G., & Colwell, R. R. (1990). Microbial degradation of hydrocarbons in the environment. *Microbial Reviews, 53*(3), 305–315.
Leung, M. (2004). Bioremediation: Techniques for cleaning up a mess. *Journal of Biotechnology, 2,* 18–22.
Margesin, R., & Schinner, F. (2001). Bioremediation (natural attenuation and biostimulation) of diesel oil-contaminated soil in an alpine glacier skiing area. *Applied and Environmental Microbiology, 67,* 3127–3133.

Michel, J., & Fingas, M. (2015). *Oil spills: Causes, consequences, prevention, and countermeasures. Fossil Fuels* (pp. 159–201). Columbia: Research Planning, Inc.

Mohammed, M. A. (2004). Treatment techniques of oil-contaminated soil and water aquifers. International conference on water resources & arid environment, King Saud University, Riyadh.

Ndimele, P.E. (2003). *The prospect of phytoremediation of polluted natural wetlands by inhabiting aquatic macrophytes (Water hyacinth).* (M.Sc. thesis). Nigeria: University of Ibadan.

Ndimele, P. E., & Ndimele, C. C. (2013). Comparative effects of biostimulation and phytoremediation on crude oil degradation and absorption by water hyacinth (*Eichhornia crassipes* [Mart.] Solms). *International Journal of Environmental Studies, 70*(2), 241–258.

Ndimele, P. E., Jenyo-Oni, A., Ayodele, A. I., & Jimoh, A. A. (2010). The phytoremediation of crude oil-polluted aquatic environment by water hyacinth (*E. crassipes* [Mart.] Solms). *African Journal of Livestock Extension, 8,* 62–65.

Ndimele, P. E., Jenyo-Oni, A., Chukwuka, K. S., Ndimele, C. C., & Ayodele, I. A. (2015). Does fertilizer ($N_{15}P_{15}K_{15}$) amendment enhance phytoremediation of petroleum-polluted aquatic ecosystem in the presence of water hyacinth (*Eichhornia crassipes* [Mart.] Solms)? *Environmental Technology, 36,* 2502–2514.

Obi, E. O., Kamgba, F. A., & Obi, D. A. (2014). Techniques of oil spill response in the sea. *IOSR Journal of Applied Physics, 6*(1), 36–41.

Perfumo, A., Banat, I. M., Marchant, R., & Vezzulli, L. (2007). Thermally enhanced approaches for bioremediation of hydrocarbon-contaminated soils. *Chemosphere, 66*(1), 179–184.

Potter, S., & Morrison, J. (2008). *World catalogue of oil spill response products* ((9th ed.)). Ottawa, Canada: S.L Ross Environmental Research Limited.

Rasking, I. (1996). Plant genetic engineering may help with environmental clean-up. *Proceedings of the National Academy of Sciences, 93,* 3164–3166.

Reynolds, C.M., & Wolf, D.C. (1999). Microbial based strategies for assessing rhizosphere-enhanced phytoremediation. Presented at the phytoremediation technical seminar, Calgary, Ottawa, 31 May–1 June 1999.

Rhykerd, R. L., Crews, B., McInnes, K. J., & Weaver, R. W. (1999). Impact of bulking agents, forced aeration and tillage on remediation of oil-contaminated soil. *Bioresource Technology, 67,* 279–285.

Shannon, M. J., & Unterman, R. (1993). Evaluating bioremediation: Distinguishing fact from fiction. *Annual Review of Microbiology, 47,* (Annual 1993): pp715(24).

Terry, N., & Zayed, A. M. (1994). Selenium volatilization by plants. In J. R. Frankenberger, & S. Benson (Eds.), *Selenium in the Environment* (pp. 343–367). New York, NY: Marcel Dekker.

Tewari, S., & Sirvaiya, S. (2015). Oil spill remediation and its regulation. *International Journal of Research in Science & Engineering, 1*(6), 1–7.

United States Environmental Protection Agency (USEPA) (2000). The quality of our nation's waters: A summary of the national water quality inventory: 1998 report to congress. EPA 841-S-001, Office of Water, U.S. Environmental Protection Agency.

United States Environmental Protection Agency (USEPA) (2011). National contingency plan product schedule. Retrieved from http://www.epa.gov/emergencies/content/ncp/prod uct_schedule.htm.

Venosa, A. D., Lee, K., Suidan, M. T., Garcia-Blanco, S., Cobanli, S., Moteleb, M., ... Hazelwood, M. (2002). Bioremediation and biorestoration of a crude oil contaminated freshwater wetland on St. Lawrence River. *Bioremediation Journal, 6,* 261–281.

Wagenet, R.J. & Bouma, J. (1996). The role of soil science in interdisciplinary research, SSSA. Special Publication Number 45.

Webler, T., & Lord, F. (2010). Planning for the human dimensions of oil spills and spill response. *Environmental Management, 45,* 723–738. Available from http://dx.doi.org/10.1007/s00267-010-9447-9.

Yakubu, M. B. (2007). Biological approach to oil spills remediation in the soil. *African Journal of Biotechnology, 6* (24), 2735–2739.

Further Reading

Adedokun, O. M., & Ataga, A. E. (2014). Oil spills remediation using native mushroom—A viable option. *Research Journal of Environmental Sciences, 8,* 57–61.

CASE STUDIES

The Values of Mangrove Ecosystem Services in the Niger Delta Region of Nigeria

Adeniran Akanni[1], John Onwuteaka[2], Michael Uwagbae[3], Richard Mulwa[4] and Isa Olalekan Elegbede[5]

[1]Lagos State Ministry of Environment, Lagos, Nigeria [2]Rivers State University of Science and Technology, Port Harcourt, Nigeria [3]Wetlands International-Nigeria, Port Harcourt, Nigeria [4]University of Nairobi, Nairobi, Kenya [5]Brandenburg University of Technology, Cottbus-Senftenberg, Germany

25.1 INTRODUCTION

Mangroves are trees and shrubs that grow in saline coastal habitats usually confined to tropical and sub-tropical regions of the world, and they offer a vast array of economic and ecological benefits. Saenger, Hegerl, and Davie (1983) described mangroves to include a wide variety of trees and shrubs (approximately 80 species), that share characteristics of being adapted to conditions of high salinity, low oxygen, and changing water levels.

Mangrove forests occupy approximately 15.2 million hectares of tropical coast world-wide (Spalding, Kainuma, & Collins, 2010), and Africa represents approximately 19% of this mangrove cover, totaling some 3.2 million Ha. According to Wetlands International—Africa (WIA), there are 17 species of mangroves in Africa, which are equally divided between the east and west. Furthermore, mangroves also function as nurseries for many fish species; many commercially caught fish have spent part of their lives in mangroves. Mangroves are also home to terrestrial fauna, including mammals, reptiles, and avian species, especially water birds.

Globally, half of all mangrove forests have been lost since the mid-20th century, with one-fifth since 1980 (Spalding et al., 2010). The loss of mangroves has not been unconnected to poor perception and inadequate knowledge by the general public about the economic and ecological values of mangroves (Feka & Ajonina, 2011), which include provision of ecosystem services, such as fish, flood prevention, erosion prevention, water regulation, and timber products.

According to Donato et al. (2011), a major reason for mangrove ecosystem valuation is to offer strong reason for decision-makers for sustainable management and preservation of this pristine ecosystem and not to determine its market worth for anticipated sale, thereby factoring this value in decision-making regarding land use and land use change. This could help halt the degradation of this ecosystem, which is put currently at 30%−50% globally (Donato et al., 2011).

The Niger Delta region of Nigeria, which is the focus of this chapter, is characterized by diverse ecosystems, abundant natural resources, including vast area of mangrove ecosystems. This region is known for its large deposits of crude oil and gas. The Niger Delta accounts for more than 80% of Nigeria's total annual earnings. However, oil and gas (O&G) developments have been highly damaging to the Niger Delta's environment, destroying habitats, threatening plant and animal species, as well as destroying the livelihoods for many resource users, whose source of living is based on farming and fishing (UNEP, 2011). Lack of data on economic valuation of mangroves of the Niger Delta region may impair effective decision-making by policymakers regarding trade-off between conservation and development.

25.2 THE MANGROVES OF THE NIGER DELTA

The Niger Delta region of Nigeria is located on the Atlantic coast of Southern Nigeria, which extends between latitudes 4°2′ and 6°2′ north of the equator and is 5°2′ east of the Greenwich meridian (Adekola & Mitchell, 2011; Davies, Davies, & Abowei, 2009). The region is said to have a yearly rainfall ranging between 3000 and 4500 mm and an average temperature of 27°C. The wet season is relatively long, occurring from March to October, and the dry season from November to February (NDRMP, 2006; Okonkwo, Kumar, & Taylor, 2015; World Bank, 1995).

The Niger Delta mangrove ecosystem is composed of salt-tolerant and low-oxygen soil−adapted multifarious plant representatives occurring in sheltered coastline areas along estuaries, lagoons, and creeks in the region (Omogoriola et al., 2012). The mangrove ecosystem of the Niger Delta situated in the lower tidal floodplain includes more than 20 large estuaries and 7700 km^2 of mangrove forests, comprising more than 70% of Nigeria's estimated 10,000 km^2 of mangrove forests (NDBP, 2012) and making up nearly 35% of the total cover for West Africa (Omogoriola et al., 2012; UNEP-WCWC, 2006).

The Niger Delta mangrove is adjudged to be Africa's largest mangrove ecosystem and the world's third largest after India and Indonesia, the third largest drainage basin in Africa, and Africa's largest river delta (Adekola & Mitchell, 2011; Ajibola & Awodiran, 2015; Ajonina, Diamé, & Kairo, 2008; Dupont, Jahns, Marret, & Ning, 2000; Macintosh & Ashton, 2003; NDBP, 2012; Omogoriola et al., 2012; Okonkwo et al., 2015; Uluocha & Okeke, 2004; Umoh, 2008).

The most emblematic mangrove species in Nigeria comprised of six species in three families. These are: Rhizophoraceae (*Rhizophora racemosa, R. harrisonii and R. mangle*), Avicenniaceae (*Avicennia africana*), and Combretaceae (*Laguncularia racemosa* and

Conocarpus erectus) (NDBP, 2012). In terms of biodiversity, researchers have shown that the Niger Delta mangrove ecosystem is considered to be one of the richest wetlands globally. The mangrove zone of the Niger Delta is said to traverse parallel to the coast and reaches between 15 and 45 km inland. This deep belt of mangrove forest protects the freshwater wetlands in the Inner Delta (NDBP, 2012).

25.3 DESCRIPTION OF THE ECOSYSTEM OF THE STUDY AREA

In describing the mangrove ecosystem in the Niger Delta, three states of the Niger Delta were chosen as a study area, namely: Rivers, Bayelsa, and Delta States. The sites are in the Asarama community in Andoni LGA of Rivers State; Opume and Abobiri communities, situated in Ogbia LGA of Bayelsa State, and Obi-Ayagha Community in Ughelli South LGA in Delta State, respectively (Fig. 25.1).

The Asarama community (Fig. 25.2) had four wetland types, which were Mangrove/mudflats, Nypa, Freshwater swamp forest, and Estuary. The Mangrove/mudflats and Nypa habitats occupied a total area of 6.48 km^2, while the Freshwater swamp forest occupied a total area of 8.89 km^2.

The Abobiri community (Fig. 25.3) had three wetland types which were Mangrove, Freshwater swamp forest, and Estuary. The Mangrove occupied a total area of 2.63 km^2; the freshwater swamp forest, 16.21 km^2; and the Estuary, a total of 1.19 km^2.

The Opume community (Fig. 25.4) had four wetland types, which were Mangrove, Nypa, Freshwater swamp forest, and Estuary. The Mangrove/Nypa occur together and

FIGURE 25.1   Study area of the Niger.

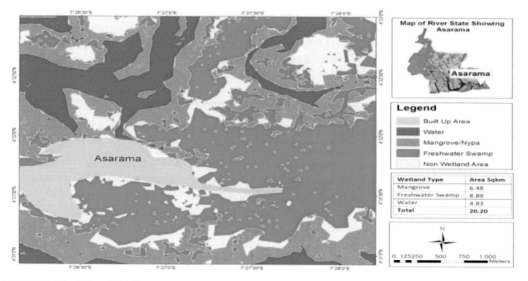

FIGURE 25.2 Map of the Asarama community in Rivers State.

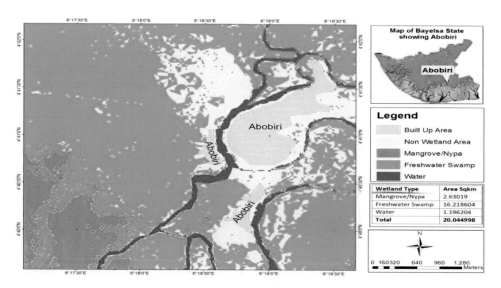

FIGURE 25.3 Map of the Abobiri community in Bayelsa State.

cover a total area of 0.88 km², the freshwater swamp forest, 18.52 km², and the Estuary, 0.07 km².

The Obi-Ayagha community (Fig. 25.5), on the other hand, had only two wetland types, namely, Freshwater swamp forest and Estuary. The Freshwater swamp forest occupied a total area of 11.54 km², while the estuary occupied 0.21 km².

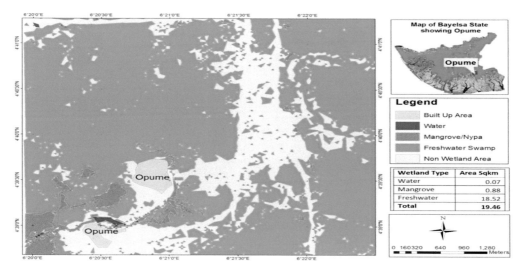

FIGURE 25.4   Map of the Opume community in Bayelsa State.

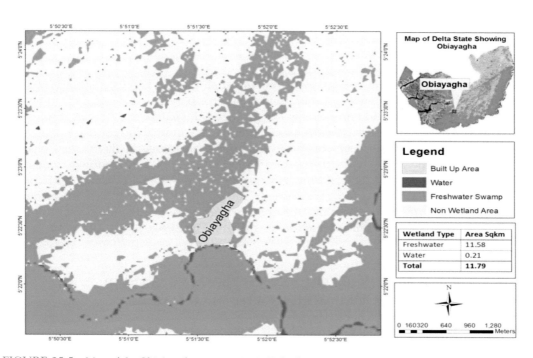

FIGURE 25.5   Map of the Obi-Ayagha community in Delta State.

FIGURE 25.6   Prop roots providing attachment sites for oyster (*Crassostrea gasar*).

25.4 MANGROVE ECOSYSTEM SERVICES IN THE NIGER DELTA

The Millennium Ecosystem Services report (MA, 2005) had identified four categories of ecosystem services, which are:

- Regulating services
- Provisioning services
- Cultural services
- Supporting services

The mangroves in the Niger Delta perform almost all the services listed above, ranging from atmospheric and climate regulation, flood and erosion control, wood and timber for cooking fuel and construction, to benefits such as aesthetic value, sacred sites, traditional medicine, and supporting services such as nutrient cycling and habitat for fish nurseries.

Specifically, the ecosystem services provided by the mangrove in the Niger Delta arc numerous, including niche types that support the existence of a rich fauna, from large individuals to juvenile and larval forms. Numerous burrows contain diversified fauna of polychaetes, crabs, shrimps, and fishes. These are supported by the rich food sources peculiar to the mangrove. For humans, stilt roots of mangrove plants provide substrate for the attachment of edible organisms such as the oysters (Fig. 25.6).

25.5 IMPACTS OF OIL POLLUTION ON MANGROVE MICROCLIMATE IN THE NIGER DELTA

There has been oil exploration in the Niger Delta since the 1950s. The consequences of oil exploration in the area include oil spill, fire caused by oil wells resulting in vegetation loss, as well as threat to human health from contaminated drinking water while also affecting the viability and productivity of the ecosystem with attendant consequence on

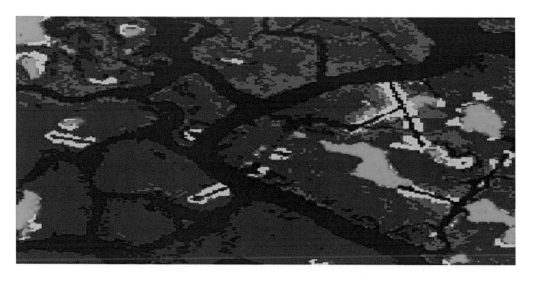

Dredge Spoil

FIGURE 25.7 Dredge spoils microhabitats within the mangrove zone. Slots and pipeline areas are easily visible from satellite imagery.

the livelihood of the people. In the environmental assessment report conducted by UNEP (UNEP, 2011), oil contamination was found to be widespread in the area impacting many components of the environment. The ecosystem services offered by these mangroves such as spawning areas for fish and nurseries for juvenile fish is impaired, thus affecting fish life cycle (UNEP, 2011). Furthermore, occasional fire outbreaks were also reported as a result of oil spill, and this kills vegetation. Suffice it to say that oil pollution in the Niger Delta also has an effect on the mangrove ecosystem.

One of the impacts of oil pollution is dredge spoils within the mangrove zone, which arise out of oil and gas activities. Canal construction for pipeline laying or channel widening for large craft movement results in the deposition of sediments from the river bed on the shoreline (Fig. 25.7). Over the years, a number of vegetation associations have developed on these spoils; the type depends on the soil types, which vary from peat mud, sandy clay, or sand (Fig. 25.8). Dredge spoils have become a utility microhabitat providing shelter for wildlife such as antelopes, otters, iguanas, and crocodiles. In some deserted spoils around the Niger Delta area, large stocks of water birds, especially ducks (*Dendrocygna viduata*), use them as roosting sites (Fig. 25.9).

25.6 VALUATION OF THE MANGROVE ECOSYSTEM

Economic valuation has been defined as a process of attaching monetary value or price to nonmarketed environmental goods and services (Rao, 2000). Economic valuation plays important role in decision-making, although it has limitations. For instance, the cost of a mangrove ecosystem in Nairobi, Kenya cannot be compared cost-wise with the same size

FIGURE 25.8   Vegetation types on a dredge spoil.

FIGURE 25.9   Mangrove vegetation on a dredge spoil.

of mangrove ecosystem in Lagos, Nigeria. This is because, although they are the same mangroves, they both have different values based on the ecosystem services they offer.

Valuation of mangrove ecosystem services is the method of examining the impacts associated with the interaction of ecosystem services and human activities, which gives additional information for environmental regulations and sustainable management of the resources. Valuation of ecosystem services is one of the concluding levels for critical evaluating and analysis of ecosystem services impacts (Turner, Morse-Jones, & Fisher, 2010).

Fig. 25.10 shows the steps involved in the valuation of ecosystem services. Requisite to ecosystem calculation, the detailed understanding of the ecosystem services needs to be considered, which will prompt the consideration of stakeholders that are peculiar to the type of respective services to be evaluated. Evaluation of ecosystem services are engaged based on

FIGURE 25.10 Schematic presentation of steps in mangroves ecosystem valuation approach. *Adapted from Sennye, M. (2014). Guidelines for the economic valuation of the mangrove ecosystem in the Limpopo river estuary. For the USAID Southern Africa Resilience in the Limpopo River Basin (RESILIM) Programme.*

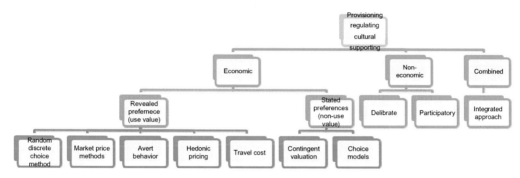

FIGURE 25.11 Ecosystem services valuation methods for the Niger Delta (Price, 2007).

the categories of these ecosystem services, provisioning (resources and product derived from the ecosystem), regulating (opportunities from monitoring and regulating ecosystem components), cultural (human benefits as a result of way of life), and supporting services (benefit that enhance production of ecosystem services). The Niger Delta region is endowed with several ecosystem services, ranging from mangroves, water, and other natural and artificial resources. There are various ecosystem services evaluation techniques that can be adopted for this region. However, this will depend on the type of ecosystem services approach from the MA framework, also considered in line with the qualitative and quantitative values of the data available. Ecosystem services valuation techniques can be done in economic and non-economic terms (Elegbede, Abimbola, Saheed, & Damilola, 2015).

The economic valuation method is mounted on the auspice of the total economic value (TEV), which considers benefit transfer as an important platform for appreciating and analyzing sustainability values in the decision-making process. The TEV is valued by revealed preference or stated preference (Fig. 25.11).

The TEV tries to adopt entirely marginal values change for ecosystem services, according to the additional values derived from the total estimation of willingness to pay and willingness to accept of some environmental commodities. TEV is the combination of direct, indirect, option, existence and bequest value, altruistic value, quashi option value (which are based on use value), and nonuse ecosystem values of services (Price, 2007). The TEV of world mangrove ecosystem services is around USD200 billion (Vo, Kuenzer, Vo, Moder, & Oppelt, 2012).

Direct use includes tangible and intangible ecosystem services use purposely for consumption or non-consumption, which can be market based or non-market-based (Fig. 25.12). These activities of indirect use value are services that enhance production and consumption of resources. Option values analyzes the cost of willingness to pay for resources yet to be used, in order to avoid consequences of not been able to utilize it

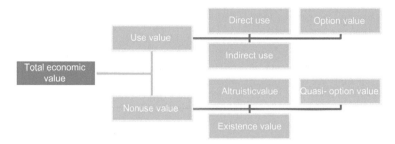

FIGURE 25.12   Categories of total economic values for ecosystem services (Albani & Romano, 1998; GIZ, 2012; Price, 2007).

subsequently. Nonuse value is a non-active use that is based on the understanding of the situation of natural environments and entails bequest value, altruistic value, and existence value. Bequest value is when ecosystem resources beneficiaries give values by extending ecosystem resource to next regenerations. It is the value of satisfaction derived from preserving a natural environment or a historic environment. Altruistic value is a passive value where humans consider the abundance of resources to manage for the next generation. Existence value considers the existence of ecosystem services and does not intend to use it or utilize its value. Quasi-option value is considered as the value of information from ecosystem services that hold decisions when there are various uncertainties around the data available for the ecosystem services. This ensures delaying final conclusions on it (Albani & Romano, 1998; GIZ, 2012; Price, 2007).

25.7 MAPPING OF MANGROVE ECOSYSTEM IN THE NIGER DELTA

Various ecosystem approaches for quantification are available, and quantification of the various ecosystem services requires data as fuel to enable the analysis of the ecosystem services. Secondary data and primary data are obtainable (Table 25.1). Analyzing ecosystem services in the Niger Delta requires data that have attributes on the peculiar ecosystem. Each method has its peculiarities and its weakness. Hedonic pricing, cost-based, and benefit transfer fit well for regulating ecosystem services, and can be well used for an indirect class of valuation techniques.

The provisioning services type of ecosystem can be used in direct and indirect usage pattern, which, as well, fit for market prices, effect on production method, and benefit transfer. Recreation and the aesthetic ecosystem services adapt well for market prices, travel costs, stated preferences, and benefit transfer. The ecosystem mapping provides useful input in estimating values directly computed in the Nigerian currency, the Naira.

25.8 STAKEHOLDER ANALYSIS METHOD

The assessment of stakeholders' interests and characteristics was carried out through a variety of data collection methods. Various tools were then used to identify stakeholders: informal, semistructured interviews of key informants, focus group discussions, and

TABLE 25.1 Adaptation of Valuation Techniques for Ecosystem Services (Spurgeon & Cooper, 2011)

Ecosystem services	Selected TEV	Market prices	Effect on production	Travel costs	Hedonic pricing	Cost-based	Stated preference	Benefit transfer
Regulating	Indirect use		x		x	x		x
Provisioning	Direct use, nonuse	x	x					X
Recreation, aesthetic		x		x			x	x

secondary data. In general, the following methodology was undertaken in conducting the stakeholders' mapping exercise:

1. Review of relevant documents, including studies conducted on the sites
2. Develop a socioeconomic mapping questionnaire
3. Identification of various stakeholder groups
4. Conduct a public survey within the communities
5. Conduct structured interviews with representatives of different stakeholders' groups working in the community, in order to complete the questionnaire. Information was also gathered through the following means:
   a. Transect walks
   b. Semi-structured discussions
   c. Direct observations
   d. Focus group discussions with the Mixed Group of Community Development Committee, the Project Implementation Committee, Adult Male group, Adult female group, Youth Male group, and Youth Female group
   e. Key informant interviews
   f. Mobility and community map

The stakeholder analysis from gathered information was undertaken using the following methodology:

1. A comprehensive listing of stakeholders organized into different institutions of decision makers and individuals influencing policy on all administrative levels within the community
2. Identification of stakeholders defined in relation to their involvement in livelihoods connected with ecosystems within the community
3. Identification, categorization, and ranking of Ecosystem Services provided by Mangrove, Nypa Palm, River Estuary, and Estuarine Tidal flats according to a five-level Likert scaling of: Strongly Disagree, Disagree, Neither Agree nor Disagree, Agree, and Strongly Agree.

25.9 IDENTIFICATION OF MANGROVE USE THROUGH SELECTION OF ECOSYSTEM SERVICES ATTRIBUTES

The choice experiment (CE) attributes used in the valuation comprised of ecosystem services listed from the socioeconomic valuation exercise in the selected target communities and modified following discussions with experts to better reflect the local context. Key criteria for the selection of the ecosystem service attributes for inclusion in the CE included:

1. Those services that provide nonmarket benefits

    The focus of the CE was to explore the nonmarket ecosystem service benefits delivered by the habitats in the selected target communities. Services that are delivered through economic markets (e.g., commercial food and water purification) were not included in the CE since the value of these can be measured using alternative market-based approaches.

2. Those services that were considered to be important to the livelihood systems in the Niger Delta

    A series of focus groups were organized with members of the target communities to identify which of the ecosystems services they considered to be important within the local context. Services considered not important were not included in the CE.

3. Only those services relevant in a regional and national context

    There is an acknowledgment that, with a developing country context in mind, some of the services identified would be directly relevant on a regional context within the foreseeable future. The public focus groups during the stakeholder and socioeconomic valuation exercise were used to identify which services were relevant in a regional context and therefore included in the CE.

To develop the attributes to be used in the CE, four (4) to five (5) key ecosystem services were identified from each community during the stakeholder and socioeconomic survey from each wetland ecosystem (Table 25.2). These were selected to include provisioning, regulating, cultural services, and supporting services. For each, the *status quo* value was used to define the levels of water quality (NTU value using a turbidity meter), fish nursery and breeding (number of fish caught per year), water hyacinth presence (tonnes/person/year), shoreline erosion (meters/year), flooding extent (meters/year), trees felled (trees logged/year), and carbon sequestration (kg/ha/year) from the following literatures: (Bravo, LeMay, Jandl, & Gadow, 2008; Carlos, 2014; Pandey & Pandey, 2013; UNEP, 2014). Periwinkle catch (tonnes/person/year), shrimp catch (tonnes/person/year), mudskipper catch (tonnes/person/year), and soil salinity (pH using a pH meter).

25.10 RESULTS AND DISCUSSION OF A FIELD STUDY ON ECOSYSTEM VALUATION

25.10.1 Ecosystem Services and Livelihood Linkages

An ecosystem is a community of organisms interacting with one another and with the chemical and physical factors that make up their environment. The chemical and physical factors include sunlight, rainfall, soil nutrients, climate, salinity, etc. Ecosystems are functional units that result from the interactions of abiotic, biotic, and cultural (anthropogenic) components (CRA, 2006). Ecosystems exist wherever plants, animals, and people have an interdependent relationship within the context of their physical environment, and these relationships result in the services made available by the ecosystem.

TABLE 25.2   Ecosystem Attributes in the Four Communities Studied

Community/attributes	TEEB/MA categories	*Status quo* value
ASARAMA COMMUNITY		
Carbon sequestration		(kg/ha/year)
Fish catch	Provisioning	(tonnes/person/year)
Periwinkle catch	Provisioning	(tonnes/person/year)
Mudskipper catch	Provisioning	(tonnes/person/year)
Soil salinity	Supporting/regulating	(pH)
ABOBIRI COMMUNITY		
Water quality (turbidity)	Regulating	NTU
Water hyacinth presence	Provisioning	(tonnes/person/year)
Shoreline erosion	Supporting/regulating	(meters/year)
Flooding extent	Regulating	(meters/year)
OPUME COMMUNITY		
Water quality (turbidity)	Regulating	NTU
Fish catch	Provisioning	(tonnes/person/year)
Shrimp catch	Provisioning	(tonnes/person/year)
Shore erosion	Supporting/regulating	(meters/year)
Flooding extent	Supporting/regulating	(meters/year)
OBI-AYAGHA COMMUNITY		
Water quality (turbidity)	Regulating	NTU
Soil erosion	Supporting/regulating	(tonnes/person/year)
Trees felled	Regulating	(Trees logged/year)

Ecosystem services, which refer to the collection of benefits provided by nature, have been described by UNDP (2004) as the life-blood of human societies, economies, and identities around the world.

A summary table of key ecosystem services for the three States of the Niger Delta, based on primary data network and analysis of secondary sources, is presented below in Tables 25.3 and 25.4.

As outlined in the tables, this is drawn from a range of published secondary sources and input from field work. The data focus on providing a generic overview of dominant livelihood systems within the three states in the survey. These livelihood systems, which are connected with various wetlands, are known to provide transportation, fisheries, ecotourism, a large diversity both in flora (many indigenous medicinal plants) and fauna with endemic fishes, rich wildlife biodiversity with endemic species of mammals and

TABLE 25.3   Ecosystems Goods and Services (Provisioning) in the Selected Niger Delta States

Ecosystem services	Wetland type							Rivers State	Delta State	Bayelsa State
	Estuary	Freshwater swamp	Nypa	Mangrove	Tidal mudflat estuarine/ coastal	Tidal freshwater	Freshwater river/ stream			
PROVISIONING SERVICES										
Sugar cane		*				*				*
Marine snails					*			*	*	*
Freshwater snail		*						*	*	*
Bivalve (mollusk)					*			*	*	*
Fruits/Vegetables		*		*				*	*	*
Bush mango		*						*	*	*
Anadara spp.					*			*		
Honey		*						*	*	*
Periwinkle				*				*	*	*
Oysters				*				*	*	*
Banana		*								*
Plantain		*								*
Rice						*				*
Fishes	*	*	*	*	*	*	*	*	*	*
Palm wine		*						*	*	*
Palm nuts		*						*	*	*
Cane		*						*		
Groundwater supply										
Surface water	*	*			*	*	*	*	*	*
Timber		*						*	*	*
Fuel		*						*	*	*
Wood		*	*	*				*	*	*
Fodder		*						*	*	*
Tannin		*	*	*				*	*	*
Aggregates		*						*	*	*

birds that qualify as Ramsar sites (IBAs: Bird life International, 2002), and other important species (Olowoye & Onwuteaka, 2012; Powell, 1995).

Specifically, an expert opinion–driven approach was adopted and augmented with the scientific literature to populate the Ecosystem Services Matrix. In summary, Tables 25.3

TABLE 25.4   Ecosystems Goods and Services (Regulating, Cultural, and Supporting) in the Selected States

| Ecosystem services | Wetland type | | | | | | | Rivers State | Delta State | Bayelsa State |
	River Estuarine	Freshwater swamp	Nypa	Mangrove	Intertidal estuarine/ coastal	Tidal freshwater	Freshwater River/ stream			
REGULATING SERVICES										
Coastal/Shoreline protection				*				*	*	*
Biological control	*	*		*	*	*	*	*	*	*
Maintenance of soil fertility	*	*		*				*	*	*
Regulation of water flows		*	*	*			*	*	*	*
Climate regulation	*	*	*	*	*	*	*	*	*	*
Erosion prevention		*	*	*			*	*	*	*
Waste processing/ treatment	*	*	*	*	*	*	*	*	*	*
Pollination		*	*	*				*	*	*
CULTURAL SERVICES										
Aesthetic enjoyment		*		*			*	*	*	*
Recreation and tourism	*			*			*	*	*	*
Educational opportunities	*	*	*	*	*	*	*	*	*	*
SUPPORTING SERVICES										
Gene pool protection	*	*	*	*	*	*	*	*	*	*
Habitat services	*	*		*	*	*	*	*	*	*
Biodiversity	*	*		*	*	*	*	*	*	*

and 25.4 illustrate the range and importance of ecosystem services derived from the three Niger Delta States. At an aggregate level, key provisioning services common across all the states comprise aquatic resources, especially fisheries, marine snails, freshwater snail, bivalve (mollusk), shrimps, periwinkle, oysters, and crabs; forest resources such as fruits/ vegetables, bush mango, honey, palm wine, palm nuts, genetic resources/wildlife, pharmaceuticals, timber, fuel, wood, fodder, and aggregates; plus a range of regulatory and

supporting services. In general, the study observed that despite the abundance of livelihood schemes, a marginal level of exploitation due to many challenges has kept the dependent population of over 80% in poverty.

25.10.2 Density Mapping of Ecosystem Goods and Services

In the Asarama community, the mangrove/nypa habitats for fish (Fig. 25.13) are over and more abundant than periwinkle and mudskipper. The benthic bivalve (*Anadara*) is also a good, which is marginally harvested for both commercial and domestic uses.

In Abobiri, within the areas of estuary and tidal fresh water, fisheries goods dominate over shrimps and other freshwater clams such as the *Etheria* species. In addition, the cutting of wood for carpentry and domestic uses dominate other uses such as traditional herbal uses (Fig. 25.14).

In the Opume community, most of the goods are more of fishes and few of wood and herbs (Fig. 25.15), which makes the dominant livelihood to be fishing, and this is mostly practiced by men folk.

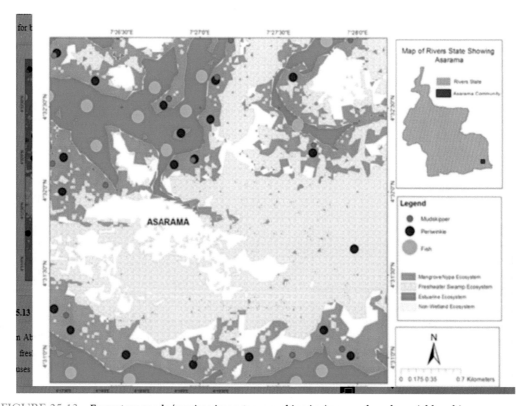

FIGURE 25.13   Ecosystem goods/services importance ranking in Asarama based on yield and income.

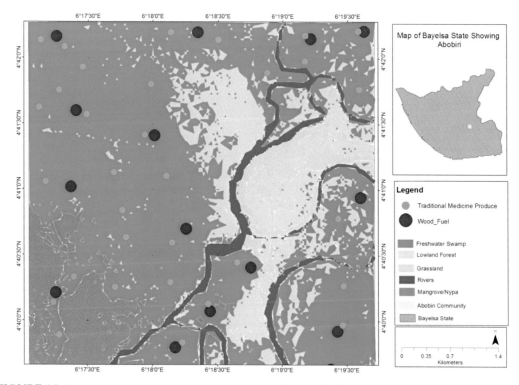

FIGURE 25.14   Ecosystem goods/services importance ranking in Abobiri based on yield and income.

In the Obi-ayagha community, the dot density map indicates that goods from the forest wetland are linked to livelihoods dominated by cassava followed in decreasing order by palm oil, palm fruit, plantain, banana/oranges/yam, rubber, and cocoyam (Fig. 25.16). In all, the female folks dominate the farming and harvesting of these goods while the males are dominant in oil palm—related livelihoods. Notable within the freshwater ecosystem is the prevalence of illegal logging dominated by the male folk. In the freshwater wetland type, most of the goods are fishes and consequently, the dominant livelihood is fishing dominated by men folk.

25.10.3 Modeling of Ecosystem Services With Direct Use Value

The modeling of ecosystem services is not within the scope of this chapter. However, for Rivers State, the ecosystem services expressions for each criterion were 63% for wide distribution, 36% for health benefits, 6.7% with high income yield, 33% of seasonal occurrence, 80% of low environmental cost of extraction, and 87% with very high potential for sustainability. Ninety percent (90%) of the ecosystem goods/services considered had a very high importance index score. For Bayelsa State, the ecosystem services expressions for each criteria were 50% for wide distribution, 60% for health benefits, 55% with high

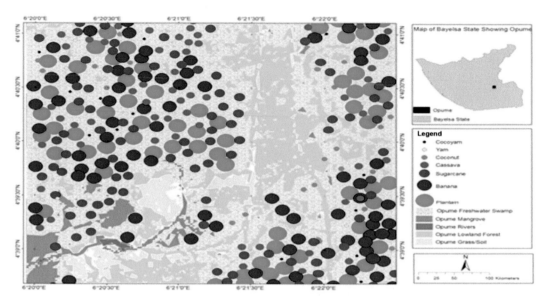

FIGURE 25.15   Ecosystem goods/services importance ranking in Opume based on yield and income.

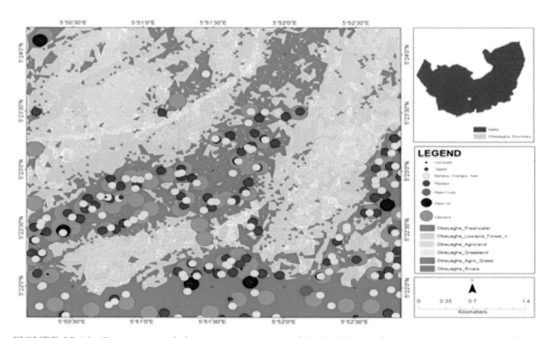

FIGURE 25.16   Ecosystem goods/services importance ranking in Obi-Ayagha community based on yield and income.

income yield, 73% of seasonal occurrence, 67% of low environmental cost of extraction, and 67% with high potential for sustainability. Fifty three percent (53%) of the ecosystem goods/services considered had a very high importance index score. For Delta State, the ecosystem services expression for each criteria were 14% for wide distribution, 81% for health benefits, 27% with high income yield, 58% of seasonal occurrence, 95.5% of low environmental cost of extraction, and 91% with high potential for sustainability. Eighty six percent (86%) of the ecosystem goods/services considered had a very high importance index score. In general, for all the three states considered, model output shows that the ecosystem services with very low importance value are: rubber and sand mining in Rivers State; rubber, logging, fuel wood, and sand mining in Bayelsa; and sand mining and cane in Delta State.

25.10.4 Ecosystem Services With Indirect Use Value

GIS mapping was used to demonstrate the capacities of Niger Delta wetlands to supply ecosystem services. Details of the modeling are beyond the scope of this chapter. However, in the three examples of indirect use values (IUV) such as breeding grounds, forest use, and ecotourism use, the results reveal generic patterns of the wetlands to provide ecosystem services under changing use as determined through the matrix method. An indicator of a breeding ground such as periwinkle shows in a step-wise manner relevance in the tidal freshwater portion of the marsh deltaic forest, very high relevance in the mangrove/nypa zone, with no relevance in the Niger flood plain, marsh deltaic forest, and the coastal freshwater forest.

Furthermore, an indicator of breeding ground such as *Anadara* shows in a step-wise manner high relevance in the mangrove/nypa zones, with no relevance in the marsh deltaic, the Niger Delta flood plain, and the coastal freshwater forest. An indicator of a breeding ground such as fish shows relevance in the coastal freshwater forest, high relevance in the Niger flood plain, marsh deltaic forest, and very high relevance in the mangrove/nypa zone. An indicator of a breeding ground such as snail indicates very high relevant capacity in the marsh deltaic forest, Niger flood plain, and coastal freshwater forest, with no relevance in the mangrove/nypa zones.

25.10.5 Stakeholder Mapping Analyses

25.10.5.1 *The Asarama Community*

Table 25.5 shows the composition and percentage of stakeholders in livelihood systems in different ecosystems found in Asarama (freshwater swamp forest, river estuary, mangrove, nypa, and intertidal mud).

The stakeholders of the freshwater forest include:

- farmers
- hunters
- loggers
- palm wine tappers
- fuel wood cutters

TABLE 25.5   Composition and Percentage of Stakeholders in Livelihood Systems (Asarama)

Stakeholders	Composition	Percentage involvement (%)
Crop farmers	Adult male	5
	Adult female	60
	Male youth	2
	Female youth	3
Hunters	Adult male	5
	Male youth	5
Loggers	Adult male	3
	Adult female	5
	Male youth	10
Palm wine tappers	Adult male	2
Firewood cutters	Adult male	30
	Adult female	60
	Male youth	35
	Female youth	80

The stakeholders associated with the river estuary include:

- the fisher folks
- marine transporters

The stakeholders associated with nypa palm are children who are seldom involved in the collection of mudskippers. The estuarine intertidal wetland area has periwinkle pickers as the only stakeholders. The mangrove wetland area has no stakeholders that derive any livelihood, as it is protected by customary law and prohibits any livelihood activity by any member of Asarama community.

The details of the composition of stakeholders in relation to livelihoods in Table 25.5 shows that the women (youth and adults) dominate the periwinkle picking, crop farming, and fuel wood cutting, whwereas the males dominate the hunting, logging, palm wine tapping, and marine transportation aspects of the identified wetland types. It is also evident that all the wetland types support the fishing livelihood system.

Table 25.6 shows the quantified indicators of ecosystem goods and services in the Asarama community. The goods and services provided by the ecosystems are mainly Provisioning, Cultural, and Supporting services extracted from mangrove, nypa, seasonal freshwater swamp, river estuary, and intertidal mud. There were no identified indicators of regulatory services from interactions with the stakeholders.

TABLE 25.6 Quantified Indicators of Ecosystem Goods and Services (Asarama)

Ecosystem service	Mangrove	Nypa palm	Seasonal freshwater swamp	River Estuary	Tidal Estuarine habitat
PROVISIONING					
Food: production of fish, algae and invertebrates, crops		₦200 worth of mudskipper/ person/week	• 150,000 kg of cassava/ annum, 40 kg of bush mango/hectare/annum • 120 kg of cocoyam/ hectare/annum • 1000 kg of banana/ hectare/annum • 1000 kg of plantain/ hectare/annum, 1000 kg of fresh fruit bunches of palm fruit/ hectare/annum • 2500 kg of yam/ hectare/annum	₦6000 fish catch/ person/ week	₦300 worth of periwinkle/ person/day
Fiber fuel Raw materials: timber, fuel wood, peat, fodder, aggregates	No information available	No information available	60,000 kg of fuel wood/ annum	No information available	No information available
Fresh water: storage and retention of water; provision of water for irrigation and for drinking	No information available	No information available	4 L of drinking water/ person/day	148 L for other domestic uses/ household/ day	No information available
CULTURAL					
Cultural heritage and identity: sense of place and belonging	Cultural heritage and identity for the Asarama people	No information available	No information available	No information available	No information available
Spiritual and artistic inspiration: nature as a source of inspiration for art and religion	Reverence for sacred shrine	No information available	No information available	No information available	No information available
Recreational: opportunities for tourism and recreational activities	No information available	No information available	No information available	No information available	No information available
Educational: opportunities for formal and informal education and training	3 scientific studies/ annum	No information available	No information available		3 scientific studies/ annum

(Continued)

IV. CASE STUDIES

TABLE 25.6   (Continued)

Ecosystem service	Mangrove	Nypa palm	Seasonal freshwater swamp	River Estuary	Tidal Estuarine habitat
SUPPORTING					
Biodiversity and nursery: Habitats for resident or transient species	1.63 ha of mangrove forest reserve	No information available	No information available	No information available	No information available

The bulk of the goods fall under provisioning service, and this includes food, fiber fuel, raw materials, and fresh water in various quantities. The seasonal freshwater swamp accounts for the agricultural crops, fuel wood, and drinking water in quantities as much as 150,000 kg, 60,000 kg and 4 L, respectively. The river estuarine provides up to #6000 worth of fish per person weekly and 148 L of water daily per person for domestic usage. The intertidal mud provides up to #300 worth of periwinkle daily.

The cultural services are provided by the mangrove, which covers approximately 1.63 ha. The age-old custom that has preserved this area of mangrove from any exploitation gives the Asarama community a sense of identity and spiritual inspiration. The supporting services are also associated with the mangrove wetland which has, through preservation, conserved the gene pool of many species of plants and animals, thereby providing educational and scientific value of immense importance.

25.10.6 Wetland Values

The wetland types identified within the Asarama community include the nypa palm, which was observed by respondents to provide very limited services. The responses showed that the palm has become invasive for more than 40 years, out-competing the mangrove, which was the native species. The community's perception of the nypa ecosystem is that of a "curse" since it does not provide them any instrumental, intrinsic, or constitutive value. Only children are known to extract service from nypa occasionally by collecting mudskippers, which they sell for a paltry sum of ₦300. The spread of nypa into the Asarama area is seriously lowering the biodiversity as well as affecting people's livelihoods through a reduced fish catch and the total absence of shellfish, tilapia, and oyster collections and the picking of periwinkles.

25.10.6.1 The Abobiri Community

The stakeholders in livelihood systems in different ecosystems (mangrove, river estuary, and freshwater swamp forest) found in Abobiri are as shown below.

The stakeholders of the freshwater forest include:

- hunters
- fishermen/women
- crop farmers

TABLE 25.7   Composition and Percentage of Stakeholders in Livelihoods (Abobiri)

Stakeholders	Category	Percentage involvement (%)
WETLAND TYPE: MANGROVE		
Firewood cutters	Adult male	4
	Male youth	35
	Female youth	50
	Adult female	60
Fishermen/women	Adult male	50
	Male youth	30
	Female youth	20
	Adult female	70
Traditional doctors	Adult male	5
	Male youth	3
	Adult female	2
Peat/Chikoko miners	Adult male	1
	Male youth	3
Tannin and resin collectors	Adult male	10
	Adult female	7
	Male youth	10
Lumberers	Male youth	10
	Adult male	5
	Adult female	3
WETLAND TYPE: FRESHWATER SWAMP FOREST		
Hunters	Adult male	2
	Male youth	5
Fisher folk	Adult male	50
	Male youth	30
	Female youth	20
	Adult female	70
Crop farmers	Adult male	30
	Adult female	50
	Male youth	5
	Female youth	40
	Adult male	5

(Continued)

TABLE 25.7   (Continued)

Stakeholders	Category	Percentage involvement (%)
Traditional doctors	Male youth	3
	Adult female	2
	Adult male	4
Firewood cutters	Adult female	60
	Male youth	35
	Female youth	50
	Adult male	5
Lumberers	Adult female	3
	Male youth	10
Rubber tappers	Adult male	10
Palm wine tappers	Male youth	5
	Adult male	50
WETLAND TYPE: ESTUARY		
Fisher folk	Male youth	30
	Female youth	20
	Adult female	70
	Adult male	5
Sand miners	Adult female	1
	Male youth	10
	Female youth	1
	Adult male	5
Marine transporters	Male youth	3

- traditional doctors
- firewood cutters
- lumberers
- rubber tappers
- palm wine tappers

The stakeholders associated with the mangrove include:

- fuel wood cutters
- fishermen/-women

- traditional doctors
- peat/chikoko miners
- lumberers
- tannin
- resin collectors

The stakeholders associated with the River Estuary include:

- marine transporters
- sand miners
- fisher folk

The details of the composition of stakeholders' in relation to livelihoods in Table 25.7 shows that the women (youth and adults) dominate crop farming and fuel wood cutting, whereas the males (youth and adults) dominate the hunting, logging, rubber tapping, peat/chikoko/sand mining, lumbering, palm wine tapping, and marine transportation aspects in the identified wetland types. It is also evident that all the wetland types support the fishing livelihood system.

The wetlands of Abobiri—mangrove, seasonal freshwater swamp/forest, and river estuary provide different goods and services. These include provisioning, cultural, and supporting services. The bulk of the goods falls under provisioning services, which include food, timber, raw materials, and fresh water in various quantities, while the others are associated with cultural and supporting services.

The seasonal freshwater swamp, which covers an area of 1621 ha, provides agricultural crops, fuel wood, and drinking water in quantities as much as 1,095,000 kg, 190,000 kg, and 4 L/person/day, respectively. It also provides spiritual inspirational service and serves as a cultural heritage for the people of Abobiri.

The mangrove, which covers an area of approximately 263 ha, provides 500 kg of herbs annually; ₦3000 worth of fish per person per week, and it has the potential for various scientific studies and biodiversity conservation.

The River Estuary provides up to ₦3000 worth of fish per person per week and 52 L of water for domestic use per person per day. Table 25.8 shows quantified indicators of ecosystem goods and services in the Abobiri community.

25.10.6.2 *The Opume Community*

Different stakeholders derive livelihoods from different ecosystems found in Opume (mangrove, nypa, river estuary, and freshwater swamp forest).

The stakeholders connected to the freshwater forest include:

- crop farmers
- hunters
- palm wine tappers
- palm fruit cutters
- traditional doctors
- timber merchants
- rubber tappers
- fisher folk

TABLE 25.8   Quantified Indicators of Ecosystem Goods and Services (Abobiri)

Ecosystem service	Mangrove	Seasonal Freshwater swamp	River Estuary
PROVISIONING			
Food: production of fish, algae and invertebrates, crops	₦3000 fish catch/person/week	1. 1000 kg of banana/hectare/annum 2. 1000 kg of plantain/hectare/annum 3. 90 kg of bush mango/hectare/annum 4. 1000 kg of pepper/hectare/annum 5. 15,000 kg of cucumber/hectare/annum 6. 24,000 kg of okra/hectare/annum 7. 1400 kg of coconut/hectare/annum 8. 1,095,000 kg of sugarcane/hectare/annum	₦3000 fish catch/person/week
Fiber fuel and other raw materials: production of timber, fuel wood, peat, fodder, aggregates	No available information	190,000 kg of fuel wood/annum	No available information
Biomass fuel	No available information	No available information	No available information
Fresh water: storage and retention of water; provision of water for irrigation and for drinking.	No information	4 L of drinking water/person/day	52 L of water for other domestic use/person/day
Traditional medicine products/Biochemical products and medicinal resources	500 kg of herbs/hectare/annum	500 kg of herbs/hectare/annum	No available information
REGULATING			
Air quality regulation: for example, capturing dust particles	No available information	No available information	No available information
Erosion protection: retention of soils	Chikoko/Peat dredging for shoreline protection	No available information	No available information

(Continued)

TABLE 25.8   (Continued)

Ecosystem service	Mangrove	Seasonal Freshwater swamp	River Estuary
CULTURAL			
Cultural heritage and identity: sense of place and belonging	No available information	32.8 m^2 Sacred Pond	No available information
Spiritual and artistic inspiration:nature as a source of inspiration for art and religion	No available information	3 ha of cemetery as a place for religious practice	No available information
Recreational: opportunities for tourism and recreational activities	No available information	No available information	No available information
Aesthetic: appreciation of natural scenery (other than through deliberate recreational activities)	No available information	No available information	No available information
Educational: opportunities for formal and informal education and training	3 scientific studies/ annum	3 scientific studies/ annum	3 scientific studies/ annum
SUPPORTING			
Biodiversity and nursery: habitats for resident or transient species	263 ha of mangrove	16.22 ha of seasonal freshwater forest	No available information
Soil formation: sediment retention and accumulation of organic matter	No available information	No available information	No available information

- raffia collectors
- fuel wood cutters

The stakeholders associated with the river estuary include:

- the fisher folks
- loggers
- sand miners
- marine transporters

The stakeholders of the mangrove include:

- fuel wood cutters
- fisher folk
- crab collectors
- traditional doctors
- chikoko/peat miners
- tannin collectors
- lumberers

The wetland containing nypa is linked with livelihoods that crop the palm fronds for roof construction and maintenance.

Overall, the composition of stakeholders in relation to livelihoods shows that the males dominate the lumbering, hunting, palm fruit cutting, palm wine tapping, rubber tapping, raffia collection, sand mining, log raft transportation, and water taxi transportation. In other livelihood types, women play approximately an equal role (Table 25.9).

TABLE 25.9   Composition and Percentage of Stakeholders (Opume)

Stakeholders	Category	Percentage involvement (%)
A. WETLAND TYPE: MANGROVE		
Firewood cutters	Male youth	35
	Adult male	4
	Female youth	50
	Adult female	60
Fisher folk	Adult male	80
	Adult female	70
	Male youth	50
	Female youth	30
Crab collectors	Male youth	60
	Adult male	25
	Female youth	20
	Adult female	40
Traditional doctors	Adult male	5
	Adult female	6
Chikoko/peat mining	Adult male	10
	Adult female	7
	Male youth	10
Tannin & resin collectors	Adult male	9
	Adult female	8
	Male youth	3
Lumberers	Adult male	3
	Male youth	10
B. WETLAND TYPE: FRESHWATER SWAMP FOREST		
Hunters	Adult male	2
	Male youth	5

(Continued)

TABLE 25.9 (Continued)

Stakeholders	Category	Percentage involvement (%)
Fisher folk	Adult male	80
	Adult female	70
	Male youth	50
	Female youth	30
Crop farmers	Adult male	10
	Adult female	60
	Male youth	5
	Female youth	30
Palm fruit cutters	Adult male	1
	Male youth	5
Traditional doctors	Adult male	4
	Adult female	12
Firewood cutters	Male youth	35
	Adult male	4
	Female youth	50
	Adult female	60
Timber merchants	Adult male	3
	Adult female	15
Rubber tappers	Male youth	10
	Adult male	10
	Female youth	5
Palm wine tappers	Adult male	10
Raffia collectors	Adult male	2
	Male youth	7
C. WETLAND TYPE: ESTUARY		
Fisher folk	Adult male	80
	Adult female	70
	Male youth	50
	Female youth	30
Log transportation	Male youth	10
	Adult male	3

(*Continued*)

TABLE 25.9   (Continued)

Stakeholders	Category	Percentage involvement (%)
Sand miners	Male youth	23
Water transportation	Male youth	30
	Adult male	5
D. WETLAND TYPE: NYPA		
Nypa frond collectors	Adult male	11
	Male youth	8
	Adult female	15

Table 25.10 shows the quantified indicators of ecosystem goods and services in the Opume community. The goods and services provided by the ecosystem include Provisioning, Regulatory, Cultural, and Supporting services, which comprise the following wetland types such as Mangrove, Nypa, Seasonal Freshwater Swamp, and River Estuary. The majority of the goods are identified under Provisioning services and these include food, fiber fuel, raw materials, and fresh water in various quantities, whereas the others are related to the cultural and supporting services. For Provisioning services, the seasonal freshwater swamp accounts for the agricultural crops such as plantain, cocoyam, banana, yam, coconut, and sugar cane worth approximately 1.9 million naira per hectare annually with wood, medicines, and drinking water in quantities as much as 113,000 kg, 500 g and 4 L, respectively.

The mangrove wetland provides fish worth ₦2000 per person daily, 225,000 kg of fuel wood, and 500 kg of traditional medicine annually. Likewise, the estuary accounts for approximately ₦2000 of fish per person daily and provides up to 60 L of water per person daily. The Nypa wetland accounts for up to 360 kg of palm frond annually, which is used as roofing sheets. Together, the mangrove, estuary, and fresh water forest support a certain amount of research annually and make up the biodiversity that is present in Opume community.

25.10.7 Indirect Values of the Niger Delta Ecosystem Service

The economic values associated with wetland goods and services can be categorized into distinct components of the TEV according to the type of use. Direct use values (DUV) are derived from the uses made of a wetland's resources and services, for example, wood for energy or building, water for irrigation, and the natural environment for recreation. IUV are associated with the indirect services provided by a wetland's natural functions, such as storm protection, nutrient retention, carbon sequestration, etc. Nonuse values of wetlands are unrelated to any direct, indirect, or future use, but rather reflect the economic value that can be attached to the mere existence of a wetland (Pearce & Turner, 1990).

TABLE 25.10 Quantified Indicators of Ecosystem Goods and Services (Opume)

Ecosystem service	Mangrove	Nypa	Seasonal Freshwater swamp	River Estuary
PROVISIONING				
Food: production of fish, algae and invertebrates, crops	₦2000 worth of fish catch/person/day	No information available	1. ₦1.9 million worth of plantain/hectare/annum 2. 500 kg of cocoyam/hectare/annum 3. ₦1million worth of banana/hectare/annum 4. 4350 kg of yam/hectare/annum 5. 150,000 kg of cassava/annum 6. 18,500 kg of coconut/hectare/annum 7. 1,095,000 kg of sugarcane/hectare/annum	₦2000 worth of shrimps and fish/person/day
Fiber fuel and other raw materials: production of timber, fuel wood, peat, fodder, aggregates	225,000 kg of fuel wood/annum	360 kg of nypa palm/annum used in roofing of houses	112,000 kg of fuel wood/annum and 113,000 kg of timber/annum	No information available
Biomass fuel	No information available	No information available	No information available	No information available
Fresh water: storage and retention of water; provision of water for irrigation and for drinking.		No information available	4 L of drinking water/person/day	60 L of water for other domestic use/person/day
Traditional medicine products/Biochemical products and medicinal resources	500 kg of herbs/hectare/annum	No information available	500 kg of herbs/hectare/annum	No information available
REGULATING				
Air quality regulation: e.g., capturing dust particles	No information available	No information available	No information available	No information available

(Continued)

TABLE 25.10 (Continued)

Ecosystem service	Mangrove	Nypa	Seasonal Freshwater swamp	River Estuary
Erosion protection: retention of soils	5–10 tonnes of chikoko used for shoreline protection	No information available	No information available	No information available
CULTURAL				
Cultural heritage and identity: sense of place and belonging	No information available	No information available	No information available	No information available
Spiritual and artistic inspiration: nature as a source of inspiration for art and religion	No information available	No information available	No information available	No information available
Recreational: opportunities for tourism and recreational activities	No information available	No information available	No information available	No information available
Aesthetic: appreciation of natural scenery (other than through deliberate recreational activities)	No information available	No information available	No information available	No information available
Educational: opportunities for formal and informal education and training	2 scientific studies/annum	No information available	2 scientific studies/annum	2 scientific studies/annum
SUPPORTING				
Biodiversity and nursery: habitats for resident or transient species	88 ha of mangrove	No information available	1852 ha of seasonal fresh water forest	7 ha of river estuary
Soil formation: sediment retention and accumulation of organic matter	No information available	No information available	No information available	No information available

Use values can further be divided into DUV, which refer to actual uses such as fishing, timber extraction, etc.; IUV, which refer to the benefits derived from ecosystem functions such as a forest's function in protecting the watershed or water quality regulation; and option values (OV), which is a value approximating an individual's willingness to pay to safeguard an asset for the option of using it at a future date (Pearce & Moran, 1994). Nonuse values (NUV) can be divided into bequest value (BV), which is the benefit accruing to any individual from the knowledge that others might benefit from a resource in the future; altruistic value (AV), which is the satisfaction of knowing other people have access to nature's benefits; and existence or passive use value (XV), which relates to values derived simply from the existence of any particular asset (Brander, Gómez-Baggethun, Martin-López, & Verma, 2010; Pearce & Moran, 1994). This section of the report will focus on estimation of IUV. Fig. 25.17 is another illustration of the structure of TEV of an ecosystem.

Given the different services offered by wetlands and their values, there is a large number of studies attempting to value the partial or TEV of numerous wetland sites (Brander et al., 2010; Schuijt, 2002). A diverse range of valuation approaches have been applied to value wetland services. These approaches can be broadly categorized into: (1) direct market valuation approaches, (2) revealed preference approaches, (3) stated preferences approaches (Chee, 2004). Under each of these valuation approaches are valuation methods that have been used in valuing different wetland goods and services. Heimlich, Weibe, Claassen, Gadsy, and House (1998), Kazmierczak (2001) and Boyer and Polasky (2004) provide extensive overviews of wetland valuation studies that also capture a broad variety of valuation approaches and methods. These include the contingent valuation method

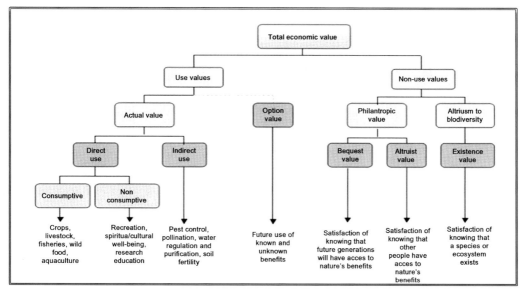

FIGURE 25.17   Types of value in TEV approach. *From Brander, L., Gómez-Baggethun E., Martin-López, B., Verma, M. (2010). The economics of valuing ecosystem services and biodiversity. In R. David Simpson (Ed.),* The ecological and economic foundations *(p. 133).*

(e.g., Bateman & Langford, 1997), hedonic pricing (e.g., Doss & Taff, 1996), travel cost method (e.g., Cooper & Loomis, 1993), production function method (e.g., Acharya & Barbier, 2000), net factor income method (e.g., Schuyt, 2004), opportunity cost method (e.g., Sathirathai & Barbier, 2001), and replacement cost (Emerton & Kekulandala, 2002). It should, however, be noted that the applicability of each of these methods depends largely on the wetland goods or services being valued and the type of value associated with it (Freeman, 1993). For the purposes of this study, we used the CE method, which is a stated preference approach.

25.10.8 Choice Experiment Results

Although the analysis of the CE is not within the scope of this book, it was approached from a utility perspective. The argument was that different ecosystem services in the Niger Delta confer utility to the communities around the different wetlands, and the conservation and/or enhancement of the services will improve utility and hence, social welfare of the communities, which is an aggregation of individual utilities. We therefore used CEs to determine the values of these services. CEs are an example of the stated preference approach to environmental valuation, since they involve eliciting responses from individuals in constructed, hypothetical markets, rather than the study of actual behavior (Hanley, Robert, & Alvarez-Farizo, 2006).

The first step in CE design is to define the goods or services to be valued in terms of its attributes and their levels. In this study, the services to be valued were the wetland services of flood protection, water quality regulation, carbon sequestration, fishing et cetera, depending on the region in the delta. The choice of attributes is a very important task for a number of reasons. Firstly, the attributes should in one way or another be relevant for the policy-making process. Secondly, the respondents must perceive the attributes as relevant. Perception of the attributes and/or levels as nonrelevant might influence the responses negatively, and the number of valid responses would decline (Bennett, Rolfe, & Morrison, 2001).

Based on different studies and a reconnaissance survey in the region, attributes used in the CE were selected. In addition to the identified attributes, a monetary attribute—volunteer time in hours per day that one would be willing to give to conservation efforts—was included to help in estimating welfare changes (Birol, Karousakis, & Koundouri, 2008). Attributes and levels were then assigned into choice sets using a fractional factorial design. Each respondent answered 5 choice questions, and each question consisted of a three-way choice: Option A and Option B, which gave an improvement in at least one attribute for a positive cost; and the zero-cost-zero-improvement *status quo*.

25.10.9 Willingness to Pay Estimates

In this study, the parameter estimates for the attributes and characteristics can be obtained by appropriate statistical analysis of respondents' choices. In our case, we have chosen the conditional logit model. It is important to note that other statistical approaches such as mixed logit, multinomial logit, and alternative specific conditional logit (ASCL) can also be used. From the statistical findings, the analyst can obtain estimates of WTP. For example, in Asarama, one can assume that the utility or well-being (of the ith respondent for the j different alternatives in the choice set) depends simply (and linearly) on the attributes of the choices presented to respondents and unobserved factors as follows:

$$U_{ij} = \beta_0 + \beta_1(\text{Carbon})_{ij} + \beta_2(\text{Fish})_{ij} + \beta_3(\text{Periwinkle})_{ij} + \beta_4(\text{Mudskipper})_{ij} + \beta_5(\text{Salinity})_{ij} + \mu_{ij}$$

(25.1)

25.10.9.1 Asarama

The analyses for the different regions were carried out using the conditional logit model. Table 25.11 presents the results of an attribute-only choice model with volunteer time. All coefficients, except fish catch, are statistically significant at the 1% level. As expected, the utility of the environmental protection alternatives is positively related to carbon sequestration, fish catch, periwinkle catch, and mudskipper catch, whereas this utility is negatively related to salinity. In the conditional logit model, the model coefficients and marginal effects are the same, and hence model coefficients can be interpreted as marginal effects, which represent the absolute change in the choice probability for one of the environmental protection options that results from a one unit or 1% increase in the explanatory variable.

Table 25.12 shows individual maximum willingness-to-pay (WTP) per attribute in terms of volunteering time per month or monthly opportunity costs. The results show that individuals are on average willing to spend approximately 87 h/month for an additional ton of sequestered carbon, 18.42 h/month for an additional ton of fish catch, 19.58 h/month for an additional ton of periwinkle, 3.8 h for an additional ton of mudskipper, and 17.63 h/month to prevent an increase in a unit of salinity.

TABLE 25.11   Estimation Results of an Attribute-Only Model With Volunteer Time (Asarama)

Variable	Coefficient
Volunteer time	−0.12187***
Carbon sequestration	0.00035***
Fish catch	0.00007
Periwinkle catch	0.00796***
Mudskipper catch	0.00154***
Salinity	−0.07163***
Psuedo-R^2	31.83
Log-likelihood	699.39***
Observations	3000

Notes: *** stands for significant at the 1% level.

TABLE 25.12   Maximum Willingness-to-Pay Per Attribute in Terms of Volunteering Time per Month or Monthly Opportunity Costs (Asarama)

	WTP in time (h)	WTP in opportunity costs (Naira/month)	WTP in opportunity costs (USD/month)
Carbon sequestration	87.37	9283	46.65
Fish catch	18.42	1957	9.83
Periwinkle catch	19.58	2080	10.45
Mudskipper catch	3.79	403	2.02
Salinity	17.63	1873	9.41

TABLE 25.13   Maximum Willingness-to-Pay per Attribute in Terms of Volunteering Time per Year or Annual Opportunity Costs (Asarama)

	WTP in time (days/year)	WTP in opportunity costs (Naira/year)	WTP in opportunity costs (USD/year)
Carbon sequestration	131	111,397	559.78
Fish catch	28	23,486	118.02
Periwinkle catch	29	24,965	125.45
Mudskipper catch	6	4832	24.28
Salinity	26	22,478	112.96

Source: Authors' computation.

The figures about willingness to pay in volunteer time are translated to opportunity costs amounts by multiplying hourly time with the local average hourly wage. For this analysis, we assumed a mean rural wage of Naira850 per day, which translates to 106.25 Naira/hour. This results in monetary WTP values of Naira7645 per ton of sequestered carbon, Naira1612 per ton of fish, Naira1714 per ton of periwinkle, Naira331 per ton of mudskipper, and Naira1543 prevented increase in a unit of salinity.

From Table 25.12, on average, each household is willing to contribute an equivalent of between USD46.65 to ensure carbon sequestration; USD9.83 (An exchange rate of 1USD = Naira 199 was used) per month to conserve fish catch; USD10.45 per month to ensure continued periwinkle catch; USD.02 per month to conserve mudskipper catch; and USD9.41 per month to reduce salinity. These figures can also be expressed on an annual basis. This was done by getting the product of the monthly willingness to pay in Table 25.12 and 12 (months). The values—both in Naira and USD—are as shown in Table 25.13. Therefore, every single year—assuming constant returns to scale of the wetland values—each household will be willing to contribute between USD24.3 and USD560 for the different ecosystem services.

These unit values can be aggregated to evaluate the total benefits of a given option to conserve Asarama wetland. This can be done if the physical increases or reduction in the attribute levels (from the baseline) are known. This can then be multiplied by the implicit price for that attribute. For example, for an option that aims at capturing the total benefits from the wetland, the total benefits per household of this policy is calculated as the willingness to pay per year multiplied by change in attribute levels (highest—baseline). For example, the change in attribute level for carbon sequestration is $(1.8 - 1.39 = 0.41$ tons) (Table 25.14).

Therefore, to capture the full benefits of the wetland, each individual household would be willing to pay between USD80.12 for mudskipper catch, and USD789.29 for carbon sequestration.

For policy- and decision-making, these annual household values need to be expressed from the perspective of the selected sample and the total number of people benefiting from the wetland. From the Asarama, we sampled 200 households, and calculations show that the sampled households have a total annual willingness to pay that ranges between USD16,024 and USD157,858 for the different ecosystem services (Table 25.15).

We also considered the society—strictly the number of households directly benefiting from the different ecosystem values. This figure for Asarama is estimated at (Population

TABLE 25.14   Maximum Willingness-to-Pay per Attribute for Attaining the Highest Attribute Level (Asarama)

	Change in attribute levels	WTP in opportunity costs (Naira/year)	WTP in opportunity costs (USD/year)
Carbon sequestration	0.41	157,070	789.29
Fish catch	1.8	42,275	212.44
Periwinkle catch	2.3	57,420	288.54
Mudskipper catch	3.3	15,946	80.12
Salinity	1.3	29,221	146.85

Source: Authors' computation.

TABLE 25.15   Sampled Households' Maximum Willingness-to-Pay per Attribute Annual Opportunity Costs (Asarama)

	WTP in opportunity costs (Naira/year)	WTP in opportunity costs (USD/year)
Carbon sequestration	31,414,000	157,858
Fish catch	8,455,000	42,488
Periwinkle catch	11,484,000	57,708
Mudskipper catch	3,189,200	16,024
Salinity	5,844,200	29,370

Source: Authors' computation.

of 11,224 people divided by average household size of 4.5 persons) 2500 households and was multiplied by the households' **mean** willingness to pay per year (households' **mean** willingness to pay per year is obtained by dividing households' willingness to pay per year (Table 25.15) by the number of households sampled (200 in this case)). The results are shown in Table 25.16, and they indicate that the annual total willingness to pay for the society ranges from USD200,300 for mudskipper to USD1.97 million for carbon sequestration. Therefore, the total indirect use value for Asarama wetland is USD3.79 million.

These values were converted to monetary values per hectare by dividing the respective values for each ecosystem service with the total acreage of 2020 hectares of the wetland in Asarama. These values also known as average value per ha and are shown in Table 25.17. Therefore, assuming each hectare contributes equally to the various IUV, one hectare of the wetland in Asarama contributes an equivalent of USD977 towards carbon sequestration, USD263 for preserving fish catch, USD357 towards preserving periwinkle catch,

TABLE 25.16   Societies' Maximum Willingness-to-Pay per Attribute in Annual Opportunity Costs (Asarama)

	WTP in opportunity costs (Naira/year)	WTP in opportunity costs (USD/year)
Carbon sequestration	392,675,000	1,973,225
Fish catch	105,687,500	531,100
Periwinkle catch	143,550,000	721,350
Mudskipper catch	39,865,000	200,300
Salinity	73,052,500	367,125
Total Indirect Use Value	754,830,000	3,793,100

Source: Authors' computation.

TABLE 25.17   Average Indirect Use Values Across Different Ecosystem Services (Naira and USD per Year) (Asarama)

	WTP in opportunity costs (Naira/year)	WTP in opportunity costs (USD/year)
Carbon sequestration	194,394	976.84
Fish catch	52,321	262.92
Periwinkle catch	71,064	357.10
Mudskipper catch	19,735	99.16
Salinity	36,165	181.75
Average Indirect Use Value	**373,678**	**1878**

Source: Authors' computation.

USD99 towards improved mudskipper catch, and USD182 towards reduction in salinity. The total indirect use value per hectare is therefore USD1878 (Table 25.17).

25.10.9.2 Abobiri

The most important identified wetland attributes in Abobiri were water quality improvement, hyacinth elimination, shore line erosion protection, and flood protection. Table 25.18 presents the results of an attribute-only choice model with volunteer time in Abobiri. All coefficients, except volunteer time, were statistically significant at the 1% level. As expected, the utility of the environmental protection alternatives was negatively related to water quality regulation, hyacinth extension, shoreline erosion, and flood protection.

Table 25.19 shows individual maximum willingness-to-pay (WTP) per attribute in terms of volunteering time per month or monthly opportunity costs. The results show that individuals are on average willing to spend approximately 46 h/month to reduce water quality deterioration, 31.5 h/month to prevent a 0.1 increase in hyacinth growth (Note that

TABLE 25.18 Estimation Results of an Attribute Only Model with Volunteer Time (Abobiri)

Variable	Coefficient
Water quality	−0.13598***
Hyacinth extension	−0.93961***
Shore erosion	−0.07418***
Flood protection	−0.06562***
Volunteer time	−0.08955**
Psuedo-R^2	37.30
Log-likelihood	818.67***
Observations	2997

Notes: *** stands for significant at the 1% level; ** significant at 5%.

TABLE 25.19 Maximum Willingness-to-Pay per Attribute in Terms of Volunteering Time per Month or Monthly Opportunity Costs (Abobiri)

	WTP in time (h)	WTP in opportunity costs (Naira/month)	WTP in opportunity costs (USD/month)
Water quality	45.55	4840	24.32
Hyacinth extension	31.48	3345	16.81
Shore erosion	24.85	2640	13.27
Flooding	21.98	2,335	11.73

hyacinth extension is bounded between 0 and 1), 24.9 h/month to prevent a meter of shoreline erosion, and 22 h to prevent a meter of flooding.

The figures about willingness to pay in volunteer time are translated to opportunity costs amounts by multiplying hourly time with the local average hourly wage using a mean wage of Naira, 850 per day, which translates to 106.25 Naira/h. This translates in monetary WTP values of Naira 4840 for water quality enhancement, Naira 3345 for prevention of 0.1 increase in hyacinth growth, Naira 2640 for prevention of a meter of shore erosion, and Naira 2335 for preventing a meter of flooding (Table 25.19).

From Table 25.19, on average, each household is willing to contribute an equivalent of USD24.32 per month to ensure quality water; USD16.81 per month to prevent hyacinth extension; USD13.27 per month to prevent shoreline erosion; and USD11.73 per month to reduce flooding. These figures can also be expressed on annual basis. This was done by multiplying the monthly willingness to pay in Table 25.19 by 12 (months). The values—both in Naira and USD—are as shown in Table 25.20. Therefore, every single year, assuming constant returns to scale on the wetland values, each household will be willing to contribute between USD141 and USD292 for the different ecosystem services.

We aggregated these values to evaluate the total benefits of a given option to conserve the Abobiri wetland. This was done by comparing physical increases or reductions in the attribute levels with the baseline. These were multiplied by the implicit price for that attribute. For example, for an option that aims at capturing the total benefits from the Abobiri wetland, the total benefits per household of this policy is calculated as the willingness to pay per year multiplied by change in attribute levels (highest − baseline). For example, the change in attribute level for carbon sequestration is (1.5 − 0.5 = 1 m) (Table 25.21).

For policy- and decision-making, these annual household values need to be expressed from the perspective of the selected sample and the total number of people benefiting from the wetland. For Abobiri, 200 households were sampled, and calculations show that the sampled households have a total annual willingness to pay that ranges between USD40,344 and USD84,456 for the different ecosystem services (Table 25.22).

We also considered the society—strictly the number of households directly benefiting from the different ecosystem services. This figure for Abobiri was estimated at (Population of 5115 people divided by average household size of 4.5 persons)1140 households and was multiplied by the households' mean willingness to pay per year {households' **mean** willingness

TABLE 25.20   Maximum Willingness-to-Pay per Attribute in Terms of Volunteering Time per Year or Annual Opportunity Costs (Abobiri)

	WTP in time (days/year)	WTP in opportunity costs (Naira/year)	WTP in opportunity costs (USD/year)
Water quality	68	58,080	291.84
Hyacinth extension	47	40,140	201.72
Shore erosion	37	31,680	159.24
Flooding	33	28,020	140.76

Source: Authors' computation.

TABLE 25.21   Maximum Willingness-to-Pay per Attribute for Attaining the Highest Attribute Level (Abobiri)

	Change in attribute levels	WTP in opportunity costs (Naira/year)	WTP in opportunity costs (USD/year)
Water quality	1.0	58,080	291.84
Hyacinth extension	1.0	40,180	201.72
Shore erosion	2.5	79,200	398.10
Flooding extend	3.0	84,060	422.28

Source: Authors' computation.

TABLE 25.22   Sampled Households' Maximum Willingness-to-Pay per Attribute Annual Opportunity Costs (Abobiri)

	WTP in opportunity costs (Naira/year)	WTP in opportunity costs (USD/year)
Water quality	11,616,000	58,368
Hyacinth extend	8,036,000	40,344
Shore erosion	15,840,000	79,620
Flooding	16,812,000	84,456

Source: Authors' computation.

TABLE 25.23   Society's Maximum Willingness-to-Pay per Attribute in Annual Opportunity Costs (Abobiri)

	WTP in opportunity costs (Naira/year)	WTP in opportunity costs (USD/year)
Water quality	66,211,200	332,697.60
Hyacinth extension	45,805,200	229,960.80
Shore erosion	90,288,000	453,834.00
Flooding extending	95,828,400	481,399.20
Total Indirect Use Value	**298,132,800**	**1,497,891.60**

Source: Authors' computation.

to pay per year is obtained by dividing households' willingness to pay per year (Table 25.22) by the number of households sampled (200 in this case)}. The results are shown in Table 25.23, and they reveal that the annual total willingness to pay for the society ranges from USD229,960 for preventing hyacinth extension to USD481,399 for prevention of flooding. Therefore, the total indirect use value for Abobiri wetland is USD1,497,892.

TABLE 25.24   Average Indirect Use Values Across Different Ecosystem Services in Abobiri (Naira and USD per Year) (Abobiri)

	WTP in opportunity costs (Naira/year)	WTP in opportunity costs (USD/year)
Water quality	33,040	166.02
Hyacinth extend	22,857	114.75
Shore erosion	45,054	226.46
Flooding	47,819	240.22
Average indirect use value	**148,770**	**747.45**

Source: Authors' computation.

These values were converted to monetary values per hectare by dividing the respective values for each ecosystem service with the total acreage of 2004 hectares of the wetland in Abobiri. The values obtained are shown in Table 25.24. Therefore, assuming each hectare contributes equally to the various IUV, one hectare of the wetland in Abobiri contributes to an equivalent of USD166 towards water quality improvement, USD115 towards preventing water Hyacinth extension; USD226 towards preventing shoreline erosion, and USD240 towards reduction in flooding. The total indirect use value per hectare is therefore USD747 (Table 25.24).

25.10.9.3 Opume

The key attributes identified in Opume are water quality improvement, improved fish catch, shrimp catch, reduced shore erosion, and flood prevention. Table 25.25 presents the results of an attribute-only choice model with volunteer time in Opume. All coefficients, except volunteer time, are statistically significant at the 1% level. As expected, the utility of the environmental protection alternatives is negatively related to water quality regulation, flooding, and shoreline erosion.

Table 25.26 shows individual maximum willingness-to-pay (WTP) per attribute in terms of volunteering time per month or monthly opportunity costs. The results show that individuals are on average willing to spend approximately 58 h/month to reduce water quality deterioration, 93 h/month to increase the fish catch by an extra 100 kg, 84 h/month to increase shrimp catch by 100 kg, 56 h/month to prevent shore line erosion by 1 m, and 61 h/month to prevent a meter of flooding.

The figures on willingness to pay in volunteer time are translated to opportunity costs amounts by multiplying hourly time with the local average hourly wage using a mean wage of 850 Naira per day, which translates to 106.25 Naira/h. This translates in monetary WTP values of 5067 Naira for water quality improvement, Naira 8121 for fish catch, Naira 7352 for shrimp catch, Naira 4873 for shore erosion protection, and Naira 5330 for flood prevention (Table 25.26).

From Table 25.26, on average, each household is willing to contribute an equivalent of between USD30.92 per month to ensure water quality, USD49.55 per month to conserve fish catch, USD44.87 per month to conserve shrimp catch, USD29.73 per month to prevent

TABLE 25.25   Estimation Results of an Attribute Only Model With Volunteer Time (Opume)

Variable	Coefficient
Water quality	−0.1289***
Fish catch	0.0021***
Shrimp catch	0.0019***
Shore erosion	−0.1240***
Flooding	−0.1356***
Volunteer time	−0.0668**
Psuedo-R^2	29.99
Log-likelihood	655.71***
Observations	2985

Notes: *** stands for significant at the 1% level; ** significant at 5%.

TABLE 25.26   Maximum Willingness-to-Pay per Attribute in Terms of Volunteering Time per Month or Monthly Opportunity Costs (Opume)

	WTP in time (h)	WTP in opportunity costs (Naira/month)	WTP in opportunity costs (USD/month)
Water quality	57.91	6,153	30.92
Fish catch	92.81	9,861	49.55
Shrimp catch	84.03	8,928	44.87
Shore erosion	55.69	5,917	29.73
Flooding	60.92	6,473	32.53

Source: Authors' computation.

shoreline erosion, and USD32.53 per month to reduce flooding. These figures can also be expressed on an annual basis. This was done by multiplying the monthly willingness to pay in Table 25.26 by 12 (months). The values—both in Naira and USD—are as shown in Table 25.27. Therefore, every single year, assuming constant returns to scale on the wetland values, each household will be willing to contribute between USD357 and USD595 for the different ecosystem services.

We aggregated these values to evaluate the total benefits of a given option to conserve Opume wetland. This was done by comparing physical increases or reductions in the attribute levels with the baseline. These were multiplied by the implicit price for that attribute. For example, for an option that aims at capturing the total benefits from the Opume wetland, the total benefits per household of this policy is calculated as the willingness to pay per year multiplied by change in attribute levels (highest—baseline) (Table 25.28). For instance, the change in attribute level for fish catch improvement is $(2.5 - 0.73 = 0.77)$.

TABLE 25.27 Maximum Willingness-to-Pay per Attribute in Terms of Volunteering Time per Year or Annual Opportunity Costs (Opume)

	WTP in time (days/year)	WTP in opportunity costs (Naira/year)	WTP in opportunity costs (USD/year)
Water quality	87	73,836	371.04
Fish catch	139	118,332	594.6
Shrimp catch	126	107,136	538.44
Shore erosion	84	71,004	356.76
Flooding	91	77,676	390.36

Source: Authors' computation.

TABLE 25.28 Maximum Willingness-to-Pay per Attribute for Attaining the Highest Attribute Level (Opume)

	Change in attribute levels	WTP in opportunity costs (Naira/year)	WTP in opportunity costs (USD/year)
Water quality	1.8	132,905	667.87
Fish catch	1.77	209,448	1052.44
Shrimp catch	1.67	178,917	899.19
Shore erosion	1.5	106,506	535.14
Flooding	1.5	116,514	585.54

Source: Authors' computation.

For policy- and decision-making, these annual household values need to be expressed from the perspective of the selected sample and the total number of people benefiting from the wetland. In Opume, 200 households were sampled, and calculations show that the sampled households have a total annual willingness to pay that ranges from USD107,028 to USD210,488 for the different ecosystem services (Table 25.29).

We also considered the society—strictly the number of households directly benefiting from the different ecosystem services. This figure for Opume is estimated at (Population of 7913 people divided by average household size of 4.5 persons) 1760 households and was multiplied by the households' mean willingness to pay per year (households' **mean** willingness to pay per year is obtained by dividing households' willingness to pay per year (Table 25.29) by the number of households sampled (200 in this case)). The results are shown in Table 25.30 and they show that the annual total willingness to pay for the society ranges from USD941,846 for shoreline erosion protection to USD1,852,298 for preservation of fish catch. Therefore, the total indirect use value for the Opume wetland is USD6.58 Million.

These values were converted to monetary values per hectare by dividing respective values for each ecosystem service with the total acreage of 1946 hectares of the wetland in

TABLE 25.29 Sampled Households' Maximum Willingness-to-Pay per Attribute Annual Opportunity Costs (Opume)

	WTP in opportunity costs (Naira/year)	WTP in opportunity costs (USD/year)
Water quality	26,580,960	133,574
Fish catch	41,889,528	210,488
Shrimp catch	35,783,424	179,839
Shore erosion	21,301,200	107,028
Flooding	23,302,800	117,108

Source: Authors' computation.

TABLE 25.30 Society's Maximum Willingness-to-Pay per Attribute in Annual Opportunity Costs (Opume)

	WTP in opportunity costs (Naira/year)	WTP in opportunity costs (USD/year)
Water quality	233,912,448	1,175,455
Fish catch	368,627,846	1,852,298
Shrimp catch	314,894,131	1,582,583
Shoreline erosion	187,450,560	941,846
Flooding	205,064,640	1,030,550
Total indirect use value	**1,309,949,626**	**6,582,732**

Source: Authors' computation.

Opume. The outcomes, also known as average value per ha, are shown in Table 25.31. Therefore, assuming each hectare contributes equally to the various IUV, one hectare of the wetland in Opume contributes an equivalent of USD604 towards water quality enhancement, USD951 in preserving fish catch, USD813 towards preserving shrimp catch, USD484 towards reduction in shoreline erosion, and USD530 towards reduction in flooding. The total indirect use value per hectare is therefore USD3383 (Table 25.31).

25.11 SUMMARY OF FINDINGS

The analysis focused on the values that people in different wetlands in the Niger Delta place on different ecosystem services. When the responses of all the communities were analyzed, it was found that individuals were in support of conserving the wetlands. The varying questions imposed on the respondents elicited various conclusions, which at the

TABLE 25.31   Average Indirect Use Values Across Different Ecosystem Services (Naira and USD per Year) (Opume)

	Average WTP in opportunity costs (Naira/Year)	Average WTP in opportunity costs (USD/Year)
Water quality	120,202	604.04
Fish catch	189,428	951.85
Shrimp catch	161,816	813.25
Shore erosion	96,326	483.99
Flooding	105,378	529.57
Average indirect use value	**673,150**	**3,383**

Source: Authors' computation.

end of the day agree with the preceding statement. Although preferences differ with respect to community, other variables such as age, sex, and years one has lived in the community did not have much disparity in their choices. Results from the willingness to pay estimates also show that different communities attach different values on different ecosystem services and would be willing to donate varying time periods for their preservation. For instance, enhancement of water quality was valued differently in the different communities. The different values attached to the different wetlands constitute the marginal and average values of the wetland. From the results, it was shown that the average wetland values per hectare vary from USD747 per ha/year in Abobiri to USD3383 in Opume. The values are in agreement with other wetland valuation studies done elsewhere. The estimates from the different wetlands in the Niger Delta are comparable with other estimates obtained elsewhere in the world. In Kenya for instance, Kabubo-Mariara, Mulwa, and Di Falco (2015) estimated an average value of USD2662 per ha/year for the Yala Wetland in Kenya. Morris and Camino (2011) carried out a study on the ecosystem valuation of freshwater, wetland, and floodplain ecosystem services in the whole of UK. Using the benefit/value transfer method, their overall estimate for all inland wetlands was USD473 per ha/year with USD421 per ha/year and USD520 per ha/year for lowlands and upland wetlands, respectively. The same study recorded values for England wetlands at USD1306 per ha/year, with USD3252 per ha/year for lowland wetlands and USD824 per ha/year for upland wetlands.

Seyam, Hoekstra, and Ngabirano (2001) estimated the freshwater wetland values of the Zambezi basin using market-price method. Their study covered 10 wetlands in the basin with a total area of 2,982,000 ha. From their analysis, they obtained a TEV of USD123 million, which translates to approximately USD41.25 per ha/year, a figure that is considerably low compared to other studies. Schuyt and Brander (2004) provide estimates of TEVs of wetlands in different parts of the world. The authors are not categorical on whether the values were primary studies or from value transfer methods. However, these values are useful for comparative analysis between this and other studies in the world. These values range from as low as USD88 per ha/year in Malawi to as high as USD8628 per ha/year in the Netherlands. Brander, Florax, and Vermaat (2006) reported average wetland values from different wetland values estimated in all the continents. From their data set, these values are highest in Europe, followed by North America, Australasia, Africa, Asia, and finally South America. Their analysis shows that the average annual wetland value is slightly over USD2800 per ha (in 1995). The figures are, however, different for the different wetland types, with un-vegetated sediment wetlands having the highest average value of over USD9000 per ha/year, whereas mangroves have the lowest average value of just over USD400 per ha/year.

25.12 CONCLUSION

From this study, we can therefore conclude that there is the need for varied conservation policy options across communities and involvement of community members in wetland preservation, including the mangrove. The results imply that the model was well specified, and important attributes in the wetlands of the communities were captured in the survey. In addition, results showed that respondents are aware to a significant degree of the importance of their wetland characteristics and its impacts on livelihood preservation. The result also revealed that wetland (mangrove) protection by host communities will lead to improvement in its positive characteristics such as fish and periwinkle catch, et cetera, and reduction in its negative characteristics such as flooding, erosion, et cetera, and vice versa. This underscores the urgent need for wetland (mangrove) protection based on government—community partnership. The WTP in opportunity costs across communities portray their threshold of commitment towards protecting their wetlands. Hourly sacrifices and implications in monetary terms beyond those levels might be unacceptable by the communities. Hence, policymakers need to decide on the level of protection desired in the wetland per community before environmental protection policies are made. Another important implication is the unit or marginal effect of each hourly sacrifice on wetland protection, which will guide policymakers in matching desired protection levels and expected results per community. For example, 87 h/month is perceived to achieve an additional ton of sequestered carbon; 18.42 h/month, an additional ton of fish catch; 19.58 h/month is perceived to achieve an additional ton of periwinkle; 3.8 h, an additional ton of mudskipper; and 17.63 h/month is perceived to prevent an increase in a unit of salinity. Governmental contribution should start beyond community threshold levels for each attribute in each community. Closely related to this is that respondents perceived that government involvement in wetland protection is expected to be much higher than community participation. For example, the highest value of time-based sacrifice is 87.5 h, which is 12% of the total hours in the month. The ranking of wetland characteristic in order of importance as perceived by communities to guide policy can also be obtained by the total hour of sacrifice/WTP for the attribute. This is portrayed by their willingness to offer a higher number of hours to achieve improvements in that particular characteristic, which is an indication of their peculiar need in order of importance per community. It is also important to note that the degree of sacrifice or opportunity cost differ for each wetland characteristic and across communities. This implies that policies to target environmental protection must differ among communities and specific wetland characteristic in order to be effective.

Also, the work undertaken in this study illustrates the big potential for using limited data in combination with limited field studies to develop a coherent and integrated ecosystem services valuation assessment. This report presents a first step in developing awareness of a list of indicators, which can be used together with a map of wetland ecosystems to evaluate ecosystem conditions and services in relation to wetlands in the Niger Delta. The pilot studies have provided a five step working structure that can be adopted to guide future work on ecosystem services pilot cases. The operational structure of the five step approach are (1) Mapping of the concerned wetland ecosystems; (2) Assessment of the condition of the wetland ecosystems; (3) Quantification of the services provided by the wetland ecosystems in pilot sites; (4) Stakeholder mapping and their metrics of the wetland ecosystems and (5) Compilation of these into an ecosystem valuation using an empirical CE administered specifically for the study.

The report has provided a basic understanding of the mangrove habitat types with characteristics. It has also produced an awareness of the need to complete their inventory, documentation and a classification that would ensure a coherent approach to support their integration into a wetland accounting system across member states in the Niger Delta. The study has also shown evidence of the use of pluralism and grass root participation from the community stakeholders in the pilot sites to value and quantify indirect use ecosystem services for integration into traditional cost-benefit analysis.

Finally, the study highlighted several issues that remain to be resolved in the future. Arguably, this report presents the best possible mangrove ecosystem valuation work providing guidance to deliver a new initiative on ecosystem valuation in the Niger Delta. Clearly, the activities around this pilot initiative will require continuous improvement and need the guidance on the current limitations to a delta-wide policy initiative for sustainable protection and conservation of mangrove ecosystems and the services they provide.

Acknowledgment

We authors gratefully acknowledge Wetlands International for financing the "Sustainable Livelihoods and Biodiversity Project in Nigeria's Niger Delta." Part of the data from this chapter was from the project.

References

Acharya, G., & Barbier, E. B. (2000). Valuing groundwater recharge through agricultural production in the Hadejia-Nguru wetlands in northern Nigeria. *Agricultural Economics, 22*, 247–259.

Adekola, O., & Mitchell, G. (2011). The Niger Delta wetlands: threats to ecosystem services, their importance to dependent communities and possible management measures. *International Journal of Biodiversity Science, Ecosystem Services and Management, 7*(1), 50–68.

Albani, M., & Romano, D., (1998). Total economic value and evaluation techniques. In R. C. Bishop (Ed.), *Environmental resource valuation: Applications of the contingent valuation method in Italy* (pp. 47–71).

Ajibola, M. O., & Awodiran, O. O. (2015). Assessing wetland services in the Niger Delta, Nigeria. *International Journal of Humanities and Social Science, 5*(1), 268–277.

Ajonina, G., Diamé, A., & Kairo, J., (2008). Current status and conservation of mangroves in Africa: an overview. World Rainforest Mov Bull [Internet]. [Cited 2016, Aug. 29]; 133. Available from: http://wrmbulletin.wordpress.com/2008/08/25/current-status-and-conservation-of-mangroves-in-africa-anoverview.

Bateman, I. J., & Langford, I. H. (1997). Non-users' willingness to pay for a national park: An application and critique of the contingent valuation method. *Regional Studies, 31*(6), 571−582.

Bennett, J., Rolfe, J., & Morrison, M. (2001). Remnant vegetation and wetland protection: Non-market valuation. In J. Bennett, & R. Blamey (Eds.), *The choice modelling approach to environmental valuation* (pp. 93−114). Cheltenham: Edward Edgar.

Bird Life International (2002). *Important bird areas and potential Ramsarsites in Africa.* Cambridge, UK: BirdLife International.

Birol, E., Karousakis, K., & Koundouri, P. (2008). Using a choice experiment to account for preference heterogeneity in wetland attributes: The case of Cheimaditida wetland in Greece. In E. Birol, & P. Koundouri (Eds.), *Choice experiments informing environmental policy.* UK: Edward Elgar Publishing Limited.

Boyer, T., & Polasky, S. (2004). Valuing urban wetlands: A review of non-market valuation studies. *Wetlands, 24* (4), 744−755.

Bravo, F., LeMay, V., Jandl, R., & Gadow, K. (2008). Managing forest ecosystems. *The Challenge of Climate Change, ,* 297.

Brander, L. M., Florax, R. J. G. M., & Vermaat, J. E. (2006). The empirics of wetland valuation: A comprehensive summary and a meta-analysis of the literature. *Environmental and Resource Economics, 33,* 223−250.

Brander, L., Gómez-Baggethun E., Martin-López, B., & Verma, M., (2010). The economics of valuing ecosystem services and biodiversity. In R. David Simpson (Ed.), *The ecological and economic foundations* (p. 133).

Carlos, E. Q. A. (2014). *Carbon sequestration in tidal salt marshes and mangrove ecosystems.* Master of Science degree in Environmental Management at the University of San Francisco, 204 pp.

Centre for Resource Analysis Limited (CRA) (2006). *Ecosystems, ecosystem services and their linkages to poverty reduction in Uganda* (p. 56). National Environmental Management Authority (NEMA).

Chee, Y. E. (2004). An ecological perspective on the valuation of ecosystem services. *Biological Conservation, 120*(4), 549−565.

Cooper, J., & Loomis, J. (1993). Testing whether waterfowl hunting benefits increase with greater water deliveries to wetlands. *Environmental & Resource Economics, 3*(6), 545−561.

Davies, R. M., Davies, O. A., & Abowei, J. F. N. (2009). The status of fish storage technologies in Niger Delta Nigeria. *American Journal of Scientific Research, 1,* 55−63.

Donato, D. C., Kauffman, J. B., Murdiyarso, D., Kurnianto, S., Stidham, M., & Kanninen, M. (2011). Mangroves among the most carbon-rich forests in the tropics. *Nature Geoscience, 4,* 293−297.

Doss, C. R., & Taff, S. J. (1996). The influence of wetland type and wetland proximity on residential property values. *Journal of Agricultural and Resource Economics, 21*(1), 120−129.

Dupont, L. M., Jahns, S., Marret, F., & Ning, S. (2000). Vegetation change in equatorial West Africa: Time-slices for the last 150 ka. *Palaeogeography, Palaeoclimatology, Palaeoecology, 155,* 95−122.

Elegbede, I. O., Abimbola, L. H. M., Saheed, M., & Damilola, O. A. (2015). Wetland resources of Nigeria: Case study of the Hadejia-Nguru Wetlands. *Poultry, Fisheries & Wildlife Sciences, 2,* 123. Available from http://dx.doi.org/10.4172/2375-446X.1000123.

Emerton, L., & Kekulanda, L. D. C. B., (2002). Assessment of the economic value of Muthurajawela wetlands. Occasional Paper No. 4 Colombo. IUCN Sri Lanka Country Office.

Feka, N. Z., & Ajonina, G. N. (2011). Drivers causing decline of Mangroves in West Africa: A review. *International Journal of Biodiversity Science, 7*(3), 217−230.

GIZ Deutsche Gesellschaft für Internationale Zusammenarbeit, (2012). Economic valuation of ecosystem services. Eschborn, Germany. https://www.giz.de/expertise/downloads/giz2013-en-biodiv-economic-valuation-ecosystem-services.pdf.

Hanley, N. W., Robert, E., & Alvarez-Farizo, B. (2006). Estimating the economic value of improvements in river ecology using choice experiments: an application to the water framework directive. *Journal of Environmental Management, 78*(2), 183−193.

Heimlich, R. E., Weibe, K. D., Claassen, R., Gadsy, R., & House, R. M. (1998). *Wetlands and agriculture: Private interests and public benefits.* Washington, DC: U.S.D.A. Economic Research Service.

Kazmierczak, R., Jr. (2001). *Economic linkages between coastal wetlands and water quality: a review of the value estimates reported in the published literature.* Baton Rouge, LA: Department of Agricultural Economics and Agribusiness, Louisiana State University.

Kabubo-Mariara, J., Mulwa, R., & Di Falco, S. (2015). *Adaptation to climate change and its implications on food security among smallholder farmers in Kenya.* Mimeo. Environment for Development.

IV. CASE STUDIES

Macintosh, D. J., & Ashton, E. C., (2003). Report on the Africa regional workshop on the sustainable management of mangrove forest ecosystems. ISME/cenTER/CAW (eds.).

Millennium Ecosystem Assessment (MA) (2005). *Ecosystems and human well-being: Synthesis.* Washington, D.C: Island Press.

Morris, J., & Camino, M., (2011). Economic assessment of freshwater, Wetland and Floodplain (FWF) Ecosystem Services. UK National Ecosystem Assessment Working Paper.UK NEA Economics Analysis Report. 81.

Niger Delta Biodiversity Project (NDBP), (2012). A United Nations Development Programme (UNDP) and Global Environment Facility (GEF) funded project. 172.

Niger Delta Regional Master Plan (NDRMP) (2006). *A report of the development strategies for the Niger Delta region* (p. 258). Niger Delta Development Commission (NDDC).

Okonkwo, C. N. P., Kumar, L., & Taylor, S. (2015). The Niger Delta wetland ecosystem: What threatens it and why should we protect it? *African Journal of Environmental Science and Technology, 9*(5), 451−463.

Olowoye, A., & Onwuteaka, J., (2012). Biodiversity: A tool for environmental management in oil and gas operations: the Brass LNG case study. The 15th International Biennial HSE Conference on the Oil and Gas Industry in Nigeria.

Omogoriola, H. O., Williams, A. B., Ukaonu, S. C., Adegbile, O. M., Olakolu, F. C., Mbawuike, B. C., & Ajulo, A. A. (2012). Survey, biodiversity and impacts of economic activities on mangroves ecosystem in eastern part of Lagos Lagoon, Nigeria. *Nature and Science, 10*(10), 30−34.

Pandey, C. N., & Pandey, R. (2013). Carbon sequestration by mangroves of Gujarat, India. *International Journal of Botany and Research (IJBR), 3*(2), 57−70.

Pearce, D. W., & Turner, R. K. (1990). *Economics of natural resources and the environment* (p. 374). Baltimore, MD: Johns Hopkins University Press.

Pearce, D. W., & Moran, D. (1994). *The economic value of biodiversity. IUCN Biodiversity Programme* (p. 172). London: Earth Scan Publications Ltd.

Powell, C. B. (1995). *Wildlife study 1. Final report submitted to Environmental Affairs Department* (p. 154). Port Harcourt: Shell Petroleum Development Company of Nigeria, Ltd. East Div.

Price, P. (2007). *An introductory guide to valuing ecosystem services.* Department for Environment, Food and Rural Affairs. Nobel House.

Rao, K. P. (2000). *Sustainable development−Economics and policy.* Oxford: Blackwell Publishers.

Saenger, P., Hegerl, E. J., & Davie, J. D. S., (1983). Global status of mangrove ecosystems by the Working Group on Mangrove Ecosystems of the IUCN Commission on Ecology in cooperation with the United Nations Environment Programme and the World Wildlife Fund, Vol. 3. The Environmentalist, 1−88.

Sathirathai, S., & Barbier, E. B. (2001). Valuing mangrove conservation in Southern Thailand. *Contemporary Economic Policy, 19*(2), 109−122.

Schuijt, K. (2002). *Land and water use of wetlands in Africa: Economic values of African wetlands.* Austria: International Institute for Applied Systems Analysis, (Interim Report IR-02-063).

Schuyt, K., & Brander, L. (2004). *The economic values of the world's wetlands.* Gland, Switzerland: WWF-International.

Sennye, M. (2014). *Guidelines for the economic valuation of the mangrove ecosystem in the Limpopo river estuary.* For the USAID Southern Africa Resilience in the Limpopo River Basin (RESILIM) Programme.

Seyam, I. M., Hoekstra, A. Y., & Ngabirano, H. H. G. (2001). *The value of freshwater wetlands in the Zambezi basin* (p. 22). Delft, the Netherlands: UNESCO-IHE, Institute for Water Education.

Spalding, M., Kainuma, M., & Collins, L. (2010). *World atlas of mangroves* (p. 319). London: A collaborative project of ITTO, ISME, FAO, UNEP-WCMC, UNESCO-MAB, UNU-INWEH and TNC, Earth scan.

Spurgeon, J., & Cooper, E. (2011). *Corporate ecosystem valuation additional notes B selection & application of ecosystem valuation techniques for CEV.* London: WBCSD.

Turner, R. K., Morse-Jones, S., & Fisher, B. (2010). Ecosystem valuation. *Annals of the New York Academy of Sciences, 1185*(1), 79−101.

Uluocha, N., & Okeke, I. (2004). Implications of wetlands degradation for water resources management: Lessons from Nigeria. *Geo Journal, 61*, 151−154.

United Nations Environment Programme (UNEP) (2011). *Environmental assessment of Ogoni Land.* Nairobi, Kenya: United Nations Environment Programme.

United Nation Environmental Programme (UNEP), (2014). Carbon Pools and Multiple Benefits of Mangroves in Central Africa Assessment for REDD, 72.

UNEP-WCMC (2006). *Spatial data layer of mangrove distribution derived through Landsat image.* Cambridge, UK: Classification, UNEPWCMC, Data analysis.

Umoh, S. G. (2008). The promise of wetland farming; evidence from Nigeria. *Agricultural Journal, 3,* 107−112.

Vo, Q. T., Kuenzer, C., Vo, Q. M., Moder, F., & Oppelt, N. (2012). Review of valuation methods for mangrove eco-system services. *Ecological Indicators, 23,* 431−446.

World Bank (1995). *Defining an environmental strategy for the Niger Delta, Nigeria: World Bank Industry and Energy Operations Division* (p. 76). West Central Africa Department.

Further Reading

Freeman, A. A. (1993). *The measurement of environmental and resource values.* Baltimore: Resources for the Future Press.

Schuyt, K. (2004). Economic values of global wetlands. *Ecological Economics, 33,* 1−6.

United Nations Development Programme, (2004). Human development report: Uganda http://hdr.undp.org/statistics/data/cty/cty_f_UGA.html.

Bunkering Activities in Nigerian Waters and Their Eco-Economic Consequences

Bolaji Benard Babatunde[1], Nenibarini Zabbey[1], Ijeoma Favour Vincent-Akpu[1] and Gabriel Olarinde Mekuleyi[2]

[1]University of Port Harcourt, Port Harcourt, Rivers State, Nigeria [2]Lagos State University, Lagos, Nigeria

26.1 GENESIS OF CRUDE OIL DISCOVERY IN NIGERIA

The first oil in Nigeria was extracted in Oloibiri village of Bayelsa State, by Shell British Petroleum. However, the commercial production became mundane in 1958. Since then, many companies have been engaged in the exploitation of Nigerian crude oil. Agitation for complete recovery of crude oil reserves in Nigeria has been a crucial agenda for government. A few years ago, more than nine hundred million barrels of crude oil reserves were identified. In terms of gas reserves, Nigeria is recognized as one of the top 10 natural gas enrichments in the world, due to her outstanding natural gas reserves. However, as a result of inadequate usage of infrastructure in Nigeria, a large percentage of the natural gas produced is usually flared, while approximately 12% are re-injects to enhance oil recovery.

The Niger Delta is blessed with wetland and coastal marine ecosystems enriched with massive oil deposits. More than $600 billion has been generated from oil exploited in this region since the 1960s, but this affluence does not reflect in the standard of living of the majority of the Niger Deltans. The region has suffered from several deprivations such as lack of drinkable water, poor health care, and unemployment. Their penury status in contrast to the affluence generated from oil in this region is a good illustration of what is called "resource curse." A sequel to Nigerian laws, rights to access oil and gas reserves are not applicable to local indigenes; rather, licenses are given to prospecting oil companies, which are empowered by law to explore the oil.

The Political Ecology of Oil and Gas Activities in the Nigerian Aquatic Ecosystem
DOI: http://dx.doi.org/10.1016/B978-0-12-809399-3.00026-4

Oil spills, waste dumping, and gas flaring are notorious phenomena in the Niger Delta. Hundreds of thousands of people are affected by this pollution, especially the poorest and those who rely on fishing and agriculture as livelihoods. Records have shown that exploration and exploitation of crude oil in Nigeria have brought several environmental menaces, which include, among others, pollution of water and land, destruction of habitats, and loss of terrestrial and aquatic biodiversity. These problems must be stopped now to avoid irreversible changes in the environment that will largely put humans on the list of endangered species by severing the opportunity of tomorrow's people to have access to the same natural resources that have sustained us.

26.2 FUNDAMENTALS OF OIL AND GAS OPERATIONS IN NIGERIA

Two basic groups of operation have been identified in the oil and gas industry. These are upstream (in which crude oil exploration, well servicing, and production are done) and downstream (which involves refining, production, and distribution of petroleum products to consumers).

26.2.1 Exploration and Production Companies

For years, the government of Nigeria has established policy that permits participation of both indigenous and foreign companies in the oil business. Notable among these companies are Shell Petroleum Development Company (SPDC), Mobil, NAOC, Total, Chevron, Addax, Pan Ocean, Conoco Phillips, Conoil, Oando, NPDC, and Sterling. They are fully involved in activities that take place in different terrains where the oil reservoirs are located both at the onshore (land) and offshore (sea or shallow water).

26.2.2 Services Company

Exploration and production companies operate in collaboration with several drilling companies. They engage in various activities such as casing and cementing, logging of wells, and general maintenance of facilities. The revenue of many public companies (such as Schlumberger, Halliburton, Transocean, Baker Hughes, Oil data, AOS/Orwell, Petrodynamics, and Geoplex) that are involved in well-service activities depend largely on the oil and gas industry.

26.3 INTRODUCTION TO TRENDS IN OIL AND GAS OPERATIONS IN NIGERIA AND ASSOCIATED ENVIRONMENTAL ISSUES

Production of oil from the oil field began in 1958, and approximately 5100 bpd of crude oil was produced. Other exploration and production companies that joined Shell after 1960 led to more discoveries of oil reservoirs and new entrants. Circa 1965, foreign companies with the right to extract and explore oil in onshore and offshore areas of the Niger

Delta increased. Also, the Nigerian government—owned company, the Nigeria National Petroleum Corporation (NNPC), was established in 1977.

26.3.1 Oil Spills

Occurrences of oil spill in Nigeria became inevitable as a result of equipment failure during drilling, natural seepage, sabotage, and accidents during production and transportation. The adverse effects of oil spills in oil-producing communities and the environment are terrible. Many farmlands, rivers, streams, etc, have been affected by oil spills in Nigeria. Based on records, the Niger Delta is among the 10 most important global wetland and marine ecosystems. However, careless exploration activities have rendered the Niger Delta region one of the first five most severely petroleum-damaged ecosystems in the world. The reports of Manby (1999) and Nwilo and Badejo (2001) have revealed the leading companies causing oil spill in Nigeria. Two major operational spills that occurred at Bodo Creek in 2008—09 had severe livelihood consequences (AI and CEHRD, 2011; Pegg & Zabbey, 2013).

> The spill of oil in the Niger Delta has caused untold hardship already to the inhabitants of the region through degradation of farm lands causing soil fertility loss and poor yield, water pollution, fish kill, biodiversity loss and species extinction, and unprecedented diseases that are still killing people.
>
> Many scientists have declared the Niger Delta "An Ecologically Waste Land." Additional pollution and environmental impact from activities of artisanal refining of crude oil can only mean suicide for people of this region. A question arise!, *Do they know this??*

26.4 SOURCES OF OIL SPILL

Accidents have contributed in no small measure to oil spill incidents. Accidents occur in all crude oil operations in the oil and gas industry; blow out may occur during drilling, an explosion may occur at a storage facility, pipeline break, and leaking underground storage tanks, and tankers may either collide or run aground during transportation. All of these will definitely lead to oil spill. Deliberate operational discharge from oil platforms, ships, and pipelines are other sources of oil spill.

- *Natural sources*
    Oil spill can occur through natural means; there may be natural seeping of hydrocarbon to the surface.
- *Sabotage and vandalism*
    Oil spill could occur as a result of criminal activities such as sabotage and pipeline vandalism and illegal bunkering activities as well as activities of local refining points.

26.4.1 Oil Spill in Marine Environment

According to an Natural Resources Charter (NRC) report, four main media of oil spill such as natural seeps, extraction, transportation by vessels, and spill during utilization of oil at sea and on land pollute marine habitat. Other ways are destruction of pipelines and operational discharges during dry docking and scrapping.

26.4.2 Oil Spill on Land

Land is polluted with oil via discharges of untreated municipal sewage and urban run-off, untreated waste water from coastal industries and refineries, filling stations, and vehicles.

26.4.3 Oil Bunkering and Crude Oil Theft in Nigeria

Theft of crude oil in Nigeria is at different levels, ranging from individual (such as militants and traders) to industrial scale as well as government officers. Other transnationally organized crime is also involved. The crude oil stolen in Nigeria is either refined locally or exported almost in the same manner as the official routes.

A layman cannot locate or break open crude oil pipelines, siphon crude, truck it to vessels, and export to countries such as the United States, China, Singapore, and Brazil, among others. To do this successfully, all the players mentioned above must be involved. Therefore, this action is best described as official theft or "White-Collar Oil Theft" whereby the proceeds are shared among the players instead of remitting them to the federal government accounts. Elsewhere, this action should be considered a treasonable felony and the people involved tried and hung but instead, they would hang the genuine fighters against environmental degradation and human right violations.

Poor governance characterized with corruption has strengthened this organized crime. A major indicator of poor governance has been a lack of corporate social responsibility by the joint venture, which gave rise to the incessant social vices such as militancy in the Niger Delta, with its attendant consequences. Over the years, especially beginning from the military era, these community boys watch the oil in their backyard being stolen officially by a few highly placed individuals who engage their services for peanuts. This activity later fueled the increase in crude oil theft and gave rise to the illegal refineries that dot the Niger Delta region of Nigeria.

26.5 EXPORT OF STOLEN CRUDE OIL

In Nigeria, the oil industries, government officials, and grouped criminals are culpable of exporting stolen crude oil. This sinister activity was unique because there are large-scale theft networks with well-organized security (Chatham House, 2013). The volume of crude oil stolen and exported offshore in Nigeria is not clearly reported. However, some reports have put the value at between 30 and 300,000 barrels per day, which is equivalent to about #1.78 trillion Naira per annum at the $50 per barrel and present exchange rate.

This volume of trade cannot be possible if there is no coordinated business platform. The irony of the matter is that nations, including the United States, China, Singapore, and many West African countries that are patronizing the legal sales from the country also partake in purchasing the stolen crude oil. This is a mystery that needs to be solved. Money gotten from stolen oil is often used to augments political and legal hazards. Indeed, the oil theft has destabilized the Niger Delta and could do so again if proper measures are not taken.

26.6 ARTISANAL CRUDE OIL REFINING

Artisanal refiners skillfully use rudimentary equipment and improvised techniques that suggest some basic knowledge of health and safety procedures in the oil industry. The research suggested that a standard artisanal refinery produces approximately 40–60 drums of diesel a day. The refining process leads to a significant quantity of wastage being dumped in rivers and streams or on land. Sites vary in size and reflect different levels of investment. Small-scale sites tend to attract women and people with very low levels of capital to invest, whereas large-scale sites involve entrepreneurs who own relatively big production sites. Research has identified artisanal refiners as young entrepreneurs who did not necessarily rely on elite patrons (elders and community leaders) to set up refineries. These refineries are set up with zero considerations for human and environmental health, and beginning from the operators who are directly exposed to toxic components of crude oil on a daily basis, the chain of irreversible destruction extends to the ecosystems, damaging the already-depleted biodiversity. A typical activity at artisanal refinery is shown in Fig. 26.1A–D.

26.6.1 Environmental Impacts of Artisanal Crude Oil Refining

Artisanal refining of crude oil has become widespread in the Niger Delta region due to several reasons, including livelihood sustenance borne out of neglect of the people by oil companies and the government, scarcity of refined products, among others. The artisanal refiners rely on the crude knowledge used to refine locally made gin called "ogogoro," and thus basic safety and environmental considerations are completely ignored and as such, beginning from collection of crude oil from the pipelines, some oil is spilled into the environment, creating the same effects as the general spills. During the refining process, operators are directly exposed to toxic components of crude oil on a daily basis through skin contact, inhalation, and possibly ingestion. According to expert judgment, two drums of crude oil translate into one drum of product once refined, thereby indicating that half of the original is a waste expunged into the environment. If this is multiplied by the number of drums of crude oil refined per day in one refinery and multiplied by all artisanal refineries in the Niger Delta, then the ecological damage that may arise from such discharges may lead to irreversible changes in the environment.

Environmental assessment of Ogoni land carried out by the United Nations Development Programme (UNDP) and United Nations Environment Programme (UNEP)

FIGURE 26.1   (A) Burning the crude oil, (B) personnel cladded in crude oil and waste discharged into river, (C) fuel the burning process, (D) an area view of environmental destruction caused by artisanal refining of crude oil.

identified artisanal oil refining as one of the principal causes of environmental degradation in the Niger Delta. Beyond polluting the groundwater and making fishing and farming grounds inaccessible, several types of vegetation and wild animals are reportedly disappearing or dying mysteriously. The cake-like remains of bitumen and tar are absorbed by the ground or used to add fuel to the fire that is heating up the refining process. This adds to air pollution and further impacts on the health of humans.

26.7 ECONOMIC LOSSES AS A RESULT OF CRUDE OIL BUNKERING IN NIGERIA

The socio-economic impacts of oil bunkering in Nigeria cannot be over-emphasized. Studies have shown that the government of Nigeria loses approximately $1.7 billion monthly from crude oil theft. In the past, this amount is less than the entire annual national budget for the country. So it is safe to say that the individuals involved in the illicit business are as rich as the country itself. If just 20% of such funds have gone into community development, the Niger Delta would have become one of the world's greatest cities by now, stemming all vices and paving the way for monumental development. The negative impacts of oil theft result in a shortfall in budget implementation and efficiency (Ogbeifun, 2014). The activities of oil thieves in the Niger Delta have also led to several shutdowns of pipelines and crude oil production by international oil companies, resulting in loss of revenues (Alawode & Ogunleye, 2013). Studies have shown over the years that, in order to curtail these illegal activities, especially in Ogoni land, oil companies consistently declared force majeure on its operations.

26.7.1 Losses of Natural and Environmental Resources

Pipeline destruction and illegal oil refining in the Niger Delta has eroded several natural biota on land and in water. Farms lands, forests, mangroves, and game reserves are also affected. Many households and families of the Niger Delta remained impoverished as a result of this scale of environmental devastation.

26.7.2 Loss of Livelihood and Consequences

There are many dimensions to loss of livelihood among communities in the oil-rich region due to oil theft. Some youth are usually engaged by the perpetrators of this heinous crime on a temporary basis. When they are relieved, they usually cannot go back to their former activities of fishing and farming due to enhanced economic status during the engagement. The breaking of pipelines usually led to massive oil spills on water and land, devastating farmlands, fishing grounds, and routes as well as rendering many helpless. The loss of livelihood has led to the incessant militancy, community clashes, and cultism in the region.

26.8 CONCLUSION

The crude oil reserves of Nigeria are enormous. However, the past events have portrayed this natural resource as more of a curse than a blessing. This is due mainly to poor governance, corruption, and complicated international relations founded on deceit. No doubt, crude oil theft impacted the massive deficit on the country's foreign exchange earnings, and deprived the government from embarking on laudable developmental projects with such funds. The illegal crude oil trade in Nigeria has also rendered the environment and livelihoods of many poor communities in an incorrigible dilemma. This hopelessness instigated violent protests, tribal wars, community clashes, militancy, killings, rape, and other vices. In order for these complex problems to be nipped in the bud, the government of Nigeria and its anticorruption agencies must collaborate with well-meaning international governments to stall the activities of criminals engaged in oil theft.

26.9 RECOMMENDATIONS

There is a necessity to stop oil theft in Nigeria if there will be any meaningful improvement in the environment, livelihoods, and socio-economic growth of the nation. Therefore, the following steps and measures are recommended.

- There should be frequent orientation and an awareness program to sensitize the people that their environment is their life and future.
- A jail time punishment with no option of fine should be given as punishment to any crude oil thief.
- Government should provide basic social amenities for the Niger Delta people.
- The percentage of federal monthly allocation given to the Niger Delta should be increased.
- Government should establish a multi-stakeholder scheme that would compel both refiners and shippers to vet the oil that they buy.

It is hoped that all the aforementioned steps will not only reduce oil theft drastically in Nigeria but would also reduce other forms of transnationally organized crime.

References

Alawode, J. A., & Ogunleye, O. I. (2013). Maintenance, security, and environmental implications of pipeline damage and ruptures in the Niger Delta Region. *The Pacific Journal of Science and Technology, 12*(1), 565–573.

Amnesty International and CEHRD. (2011). *The true tragedy, delays and failures in tackling oil spills in the Niger Delta*. Amnesty International Index No. AFR/44/018/2011.

Chatham House. (2013). *Nigeria's criminal crude: International options to combat the export of stolen oil*. The Royal Institute of International Affairs. <https://www.chathamhouse.org/publications/papers/view/194254>.

Manby, B. (1999). *The price of oil: Corporate responsibility and human rights violations in Nigeria's oil producing communities* (p. 202). New York: Human Rights Watch.

Nwilo, P. C., & Badejo, O. T. (2001). *Impacts of oil spills along the Nigerian coast*. Retrieved on 17 November 2016 from <http://web.archive.org/web/20080430164524/http://www.aehsmag.com/issues/2001/october/impacts.htm>.

Ogbeifun, B. (2014). *Why Nigeria should not treat oil theft with kid gloves*. Vanguard in Sweet Crude, January 7. Retrieved on 15 July 2014 from <http://www.vanguardngr.com/2014/01/Nigeria-treat-oil-theft-kid-gloves-ogbeifun>.

Pegg, S., & Zabbey, N. (2013). Oil and water: The Bodo spills and the destruction of traditional livelihood structures in the Niger Delta. *Community Development Journal, 48*, 391–405.

Index

Note: Page numbers followed by "*f*" and "*t*" refer to figures and tables, respectively.

Printed in the United States
By Bookmasters